NOBEL LECTURES

PHYSIOLOGY OR MEDICINE

2016–2020

NOBEL LECTURES

INCLUDING PRESENTATION SPEECHES
AND LAUREATES' BIOGRAPHIES

PHYSICS

CHEMISTRY

PHYSIOLOGY OR MEDICINE

ECONOMIC SCIENCES

NOBEL LECTURES

INCLUDING PRESENTATION SPEECHES
AND LAUREATES' BIOGRAPHIES

PHYSIOLOGY OR MEDICINE

2016–2020

EDITOR

Bo Angelin

Karolinska Institutet, Sweden

World Scientific

NEW JERSEY • LONDON • SINGAPORE • BEIJING • SHANGHAI • HONG KONG • TAIPEI • CHENNAI • TOKYO

Published by

World Scientific Publishing Co. Pte. Ltd.
5 Toh Tuck Link, Singapore 596224
USA office: 27 Warren Street, Suite 401-402, Hackensack, NJ 07601
UK office: 57 Shelton Street, Covent Garden, London WC2H 9HE

NOBEL LECTURES IN PHYSIOLOGY OR MEDICINE (2016–2020)

ISBN 978-981-12-6063-6 (hardcover)
ISBN 978-981-12-6064-3 (ebook for institutions)
ISBN 978-981-12-6065-0 (ebook for individuals)

For any available supplementary material, please visit
https://www.worldscientific.com/worldscibooks/10.1142/12973#t=suppl

Printed in Singapore

PREFACE

In this volume, exciting discoveries in Physiology or Medicine are presented based on the Nobel Prizes awarded in this field for the years 2016–2020. As always, they represent the wide area of science related to the understanding of human biology in health and disease, and to the development of novel therapies. In addition to their lectures, the Laureates have also provided biographical sketches describing their life histories and their development as scientists. The introductory speeches given at the Prize ceremonies by representatives of the Prize-awarding Nobel Assembly at Karolinska Institutet summarize the ground-breaking discoveries which were:

- "Mechanisms of autophagy" to Yoshinori Ohsumi, who identified the cellular machinery that controls removal of degraded molecules from cells — in 2016.

  In a series of elegant experiments, he carefully identified the genes responsible for the regulation of controlled breakdown of molecules within individual cells commonly referred to as autophagy ("self-eating"). This important mechanism is of vital importance for how our metabolism is adjusted in response to fasting and feeding, and disturbances in its function have been shown to be associated with many human disease entities including cancer, diabetes, and various infections and neurological conditions.

- "Molecular mechanisms controlling the circadian rhythm" to Jeffrey C. Hall, Michael Rosbash and Michael W. Young — in 2017.

  Their work identified how the circadian rhythms present in living organisms are regulated by an intricate interaction between different genes and proteins, resulting in the generation of a robust approximately 24-hour "body-clock" function. This timekeeper is obviously fundamental to our adaptive capacity and health, and is also important in metabolic and sleep disorders, cancer, and several neurological diseases.

- “Cancer therapy by inhibition of negative immune regulation” to James P. Allison and Tasuku Honjo — in 2018.
  Their bold experiments showed that by producing a more powerful endogenous immune reaction against cancer cells through treatments releasing the inhibition of immune active cells normally present, dramatic effects could be achieved in the treatment of several cancer forms. Their ideas were rapidly developed into novel clinical therapies which have revolutionized the possibility to successfully treat previously very resistant and deadly malignancies.
- “How cells sense and adapt to oxygen availability” to William G. Kaelin, Jr., Sir Peter J. Ratcliffe and Gregg L. Semenza — in 2019.
  Their work clarified in detail the sensitive machinery that ensures how cells in the body can receive the exact amount of oxygen needed for optimal function. In addition to the obvious relevance of understanding physiological responses important for adaptation to varying environments, interaction with these regulatory pathways holds promise for important developments of therapy in conditions such as anemia and cancer.
- “Hepatitis C virus” to Harvey J. Alter, Michael Houghton and Charles M. Rice — in 2020.
  Their studies identified a new form of contagious liver inflammation mainly spread through blood transfusions. Compared to previously known forms of virus-induced hepatitis, it was more insidious and had a high frequency of late complications such as cirrhosis and cancer. Utilizing novel techniques, they could isolate the causative factor, hepatitis C virus, leading to the development of specific diagnostics, antiviral therapy, and eventually vaccines. Their work has led to a dramatic improvement in global health.

The announcement and delivery of the Prizes generally follow formal procedures which have been established since the very first prize ceremonies in 1901. Due to the Covid-19 pandemic, the year 2020 became very different. The award ceremony in the Blue Hall at the Stockholm City Hall on 10 December 2020 was streamed, with only a limited audience. The laureates received their medals and diplomas in their home countries, and their lectures were generally given later.

The Prize in Physiology or Medicine should be given to “the person who made the most important discovery within the domain”. The wide range of scientific achievements in Physiology or Medicine is obvious from the breadth of awarded discoveries during these years. A distinct observation is also that development of clinical treatments frequently is the result of more basic science-oriented research. While more and more medical science becomes dependent on large-scale collaborations between experts in different fields, Nobel's vision to identify and award individuals making "paradigm-shifting" research remains a unique aspect of the Prizes. While two or three Laureates often share an award, this is most commonly the result of work performed in a sequential way, with critical questions being explored by separate groups.

As always, the personal histories of the Laureates are very interesting to take part in. While each one is obviously unique, some themes seem to recur irrespective of background: an early interest in natural science, great curiosity, and tenacity, together with sheer joy in making discoveries. The importance of being stimulated by teachers/mentors, both at a very young age and when choosing a more definite path, also seems to be of high relevance for future success.

Bo Angelin
Senior Professor of Clinical Metabolic Research
Former Chairman of the Nobel Assembly
and the Nobel Committee

CONTENTS

Preface v

2016 **Yoshinori Ohsumi**

Presentation by Nils-Göran Larsson 3

Biography of Yoshinori Ohsumi 7

Molecular Mechanisms of Autophagy in Yeast 22

2017 **Jeffrey C. Hall, Michael Rosbash and Michael W. Young**

Presentation by Carlos Ibáñez 61

Biography of Jeffrey C. Hall 65

The Little Flies and their Genes: How Research Exploiting Drosophila has led to Prize-wining Advances in Genetic and Biological Knowledge, all the way to uncovering Mysteries of the Circadian Clock 74

Biography of Michael Rosbash 107

Circadian Rhythms and the Transcriptional Feedback Loop 119

Biography of Michael W. Young 157

2018 **James P. Allison and Tasuku Honjo**

Presentation by Klas Kärre 171

Biography of James P. Allison 175

Immune Checkpoint Blockade in Cancer Therapy 188

Biography of Tasuku Honjo 205

Serendipities of Acquired Immunity 221

2019 **William G. Kaelin, Jr., Sir Peter J. Ratcliffe and Gregg L. Semenza**

Presentation by Anna Wedell 253

Biography of William G. Kaelin, Jr. 257

The von Hippel-Lindau Tumor Suppressor Gene: Insights into Oxygen Sensing and Cancer 272

Biography of Sir Peter J. Ratcliffe 315

Elucidation of Oxygen Sensing Mechanisms in Human and Animal Cells 328

Biography of Gregg L. Semenza 357

Hypoxia-Inducible Factors in Physiology and Medicine 368

2020 **Harvey J. Alter, Michael Houghton and Charles M. Rice**

Presentation by Gunilla Karlsson Hedestam 409

Biography of Harvey J. Alter 413

Hepatitis C Virus: From Hippocrates to Cure 426

Biography of Michael Houghton 445

The Identification of the Hepatitis C Virus 469

Biography of Charles M. Rice 499

Hepatitis C: From Discovery to Cure 515

Physiology or Medicine 2016

Yoshinori Ohsumi

"for his discoveries of mechanisms for autophagy"

The Nobel Prize in Physiology or Medicine, 2016

Presentation speech by Professor Nils-Göran Larsson, Member of the Nobel Assembly at Karolinska Institutet, Member of the Nobel Committee for Physiology or Medicine, 10 December 2016.

Your Majesties, Your Royal Highnesses, Esteemed Nobel Laureates, Ladies and Gentlemen,

Today we have gathered for the awarding of this year's Nobel Prizes in the Stockholm Concert Hall, one of Sweden's architectural masterpieces, which was inaugurated in 1926. The concert hall is in good shape despite intensive use over many decades, thanks to repairs and maintenance. Today it is also newly cleaned and spotless. The consumables that previous guests have left behind have been collected and hopefully recycled. We view this daily maintenance as something self-evident, but if the concert hall had not been repaired or cleaned in 90 years this would of course have been obvious to us. Similarly, our body's cells are constantly replacing defective components and removing waste materials. This year's Nobel Prize in Physiology or Medicine is about how the cell maintains its interior and recycles its own components in order to function optimally and adapt to new circumstances.

The cell's degradation station, the lysosome, was identified by biochemical methods as early as 1955 by the Belgian scientist Christian de Duve. For this discovery, he was awarded the Nobel Prize in Physiology or Medicine in 1974. Shortly after the discovery of the lysosome, several researchers – using electron microscopy – showed that lysosomes seemed to exist in all organisms having nucleated cells. They also observed that lysosomes could sometimes be full of normal cellular components in various stages of decomposition. Remarkably, it appeared as if the cell could break down some of its own contents. In 1963 Christian de Duve named this process autophagy, which means self-eating. In the following decades, some progress was made in our understanding of autophagy, but fundamental questions remained unanswered: Is autophagy crucial to normal cell function? If so, how is autophagy regulated? Is autophagy important in human disease?

When Yoshinori Ohsumi was in his early 40s, he was eager to make new major scientific advances. In the late 1980s he therefore decided to take on a whole new area of research: cell degradation processes. Ohsumi predicted that degradation might very well be a process just as important as the production of new components in a cell. For his studies, he chose to use ordinary baker's yeast. Yeast cells are unicellular eukaryotes, nucleated organisms that can be used as model systems for understanding fundamental processes also found in human cells. By employing a brilliant experimental strategy, Ohsumi was able to show that autophagy actually does occur in yeast. He then carried out a series of very elegant studies that enabled him to identify a number of genes that are vital in regulating autophagy.

Ohsumi went on to show new, completely unexpected biochemical mechanisms that are controlled by these genes. It also soon turned out that the cellular machinery that controls autophagy is highly conserved and is found in all organisms containing nucleated cells.

The discovery of the molecular machinery of autophagy revolutionised Ohsumi's entire field of research, and we know today that autophagy is a very important maintenance process that ensures the normal functioning of our cells. For example, autophagy plays a key role when we adjust our metabolism to fasting and starvation. Impairments in the regulation of autophagy are linked to a variety of human diseases, such as diabetes, cancer, infections and severe disorders of the nervous system. Thanks to Ohsumi's fundamental discoveries, we now know that autophagy plays a fundamental role in many different biological processes.

Professor Ohsumi:

Your groundbreaking research has solved a long-standing enigma in biology and revealed mechanisms for how our cells recycle components and combat disease. With a series of brilliant experiments you have opened up a new field in biology. On behalf of the Nobel Assembly at the Karolinska Institute, it is my great privilege to convey to you our warmest congratulations. I now ask you to step forward to receive your Nobel Prize from the hands of His Majesty the King.

Yoshinori Ohsumi. © Nobel Prize Outreach AB. Photo: A. Mahmoud

Yoshinori Ohsumi

Biography

1. EARLY LIFE AND INFLUENCES

I was born in 1945 in Fukuoka, half a year before the end of the World War II, to my father Yoshio and mother Shina. My father was employed at Kyushu University as a professor of mining engineering. At that time, everyone in Japan was equally poor, and even food was not readily available. I was a weak child who continually fell ill, a condition that was perhaps exaggerated by the malnourishment of the era. My mother, after giving birth to me, contracted tuberculosis and spent a long time battling the disease in bed. I can still remember the sight of her in plaster after developing spine caries. Thankfully, a friend of my parents in Hawaii sent just-developed antibiotics which allowed my mother to recover. Words such as streptomycin and aureomycin stuck in my mind as a child, although I didn't fully understand their meaning. When I was about 8 years old, my mother was finally able to resume a normal life. My brother, Kazuo, who is 12 years my senior, entered the Faculty of Literature at the University of Tokyo as I began elementary school. I was brought up through the efforts of my father, my two sisters Reiko and Junko and many other people who assisted through a difficult period for our family. If I had met a wonderful doctor at such a time, I might have chosen a medical education. However, I had a dream to become a scientist that was fuelled by the expectations of my parents.

I was brought up in a house surrounded by nature, with paddies, streams, hills and the sea all nearby. My youngest years were spent outdoors, catching fish and picking plants. During elementary school, I became interested in insect collecting and spent much time searching for unusual specimens. I also loved to look up at the arching night sky and remembered the names of stars and constellations. But without anyone around to foster these interests, they were left behind as childhood hobbies rather than lifelong passions. Every time my brother returned home from Tokyo during university holidays, he would bring a

Figure 1. Family photo at Fukuoka taken in 1951 (parents, elder brother and two elder sisters).

book, including thoughtfully chosen scientific books that broadened my awareness of the world beyond what we were taught at school. Being a weak child, I had no talent for sports and was neither artistic nor gifted in literature. But I managed to achieve good results at school.

During my high school years, I was a member of the chemistry club, where we would mix chemicals and enjoy watching the reactions between them. I distinctly remember the beauty of the Liesegang phenomenon in agar, but having no idea how to analyse it. I hoped to become a chemist and entered the University of Tokyo. However, the chemistry classes did not rouse my interest and I went through an uncertain period during which I agonised over my future. Thankfully, at that time the exciting field of molecular biology was just being established, and I decided to major in it. As a graduate student, I joined the laboratory of Kazutomo Imahori, who was one of the few scientists running a molecular biology laboratory in Japan at the time. Imahori's special field was physicochemistry to assess interactions between proteins and nucleic acids. Under the direction of Akio Maeda, I began to study the role of ribosome subunits in protein synthesis. Although I was unable to produce particularly interesting results, my time in the Imahori lab was exciting as we participated in the work of a rapidly developing

Figure 2. Chemistry Club at High School in 1967.

field and I got my first taste of the joy of experimental work. The experience also provided me with a strong sense of the continuous nature of protein synthesis within cells.

The period when I was a graduate student was characterised by large student protests, and Tokyo University was right in the middle of the social debate and change of the period. I wasn't able to completely focus on my lab work as I participated in the discussions and demonstrations. However, I believe that I gained a sense that we must be aware of social developments and the direction of society during this time.

From the second year of my doctoral studies, I decided to move to Kyoto University, where Maeda had joined the just-founded Department of Biophysics as an associate professor. Kyoto University provided a very free environment, and I was stimulated by talented new students and friends in biochemistry lab in the Department of Chemistry. In my research, I became interested in colicin E3, which is able to pass through the membrane of bacterial cells and instantly inhibits protein synthesis. I think I was also influenced by my growing interest in biological membranes during this period.

One year later, in 1971, I married Mariko Nakazawa, who joined Imahori's laboratory two years after me. The year we spent in Kyoto after we got married was very memorable for me. However, Mariko unexpectedly became pregnant,

(a)

(b)

Figure 3. (a) Kazutomo Imahori and (b) Yasuhiro Anraku

and she therefore returned to Tokyo after less than a year. In order to support our new family, she successfully applied to the newly formed Mitsubishi Life Science Institute. I also returned to Tokyo to follow Imahori, who had moved to the Faculty of Agriculture at the University of Tokyo, and I barely managed to finish a thesis to graduate with my PhD.

At the time, it was not easy to get a good academic position. Imahori advised me that the field of cell biology was just taking off and suggested that I work in the laboratory of G. M. Edelman at The Rockefeller University. At the time, Ichiro Yahara, a senior student of Imahori's, was playing an active part there, so I was immediately offered a position. It was my first time leaving Japan and was also a very different field to my previous work, so I felt very uncertain. Mariko decided to leave the Institute and joined me in New York. Fortunately, Mariko soon was able to find a research position at the same university, in the laboratory of Norton Zinder.

Initially I was to study mitogenesis in lymphocytes, but Edelman decided to shift the focus of the lab to mouse development, so I became involved in setting up an in vitro mouse fertilisation system. I had only ever handled *Escherichia coli*, so I would agonise daily over how to capture the process of early development with only a small number of eggs. Before long, Mike Jazwinski transferred to our laboratory from Arthur Kornberg's lab. Taking inspiration from the elegant uncovering of the *cdc* mutants by Hartwell, a project began to uncover the

Figure 4. Yoshinori and Mariko Ohsumi in 1971 and 2015.

mechanism of DNA replication initiation in yeast, which I joined. We had first planned to use isolated nuclei in this project using density-gradient centrifugation, but it proved difficult to obtain undamaged nuclei. I noticed a white layer at the top of the tube, and when I looked at this layer under the microscope out of mere curiosity I realised that we were also isolating highly purified vacuoles. This left a strong impression on me and was one of those serendipitous steps that led me to where I am today.

In the autumn of 1977 Yasuhiro Anraku, a professor at Tokyo University's Faculty of Science, asked me to return to Japan as an assistant professor in his laboratory, and while I'd just started our work on yeast DNA replication initiation, I decided to hurry back to Japan. In December 1977, I brought my soon to be 6-year-old son back to Japan, while Mariko stayed for several months longer with our recently born second son in New York to finish her work in Zinder's laboratory.

Anraku's laboratory was renowned for its work on transporters and the respiratory mechanism in *E. coli*, but Anraku allowed me to start a project on yeast. Looking back, I really appreciate Anraku's decisiveness in affording me this opportunity. I eventually decided to study not the plasma membrane but rather the membrane of the vacuole, a cellular organelle. At the time, the vacuole was thought of as little more than a garbage dump of the cell, and there were few scientists interested in it, but my position in the Department of Botany as well as location next to Masashi Tazawa's laboratory went some way in convincing me that the vacuole was an underappreciated, fascinating organelle. I'm sure that many yeast researchers thought it a strange choice for a project.

Soon I had developed a procedure for the isolation of vacuole membrane vesicles from highly purified vacuoles. Although the vacuole makes up approximately one quarter of the volume of the cell, I could only obtain a surprisingly small amount of vacuolar membrane, but was able to demonstrate the active transport of amino acids and calcium over the vacuolar membrane. In addition, we were the first to show the novel proton pump V-type ATPase, which gives the driving force for vacuolar transport, works that were the fruit of collaboration with the late Yoshimi Kakinuma and Etsuko Uchida. Following this, structural and functional studies of V-type ATPase became an active area of research that continues to this day.

2. MOVING TO KOMABA AND DISCOVERING AUTOPHAGY

Following this decade of work on the vacuolar membrane, I was offered a place at the University of Tokyo's College of Arts and Sciences (Komaba Campus) as an associate professor and started my own laboratory—consisting of only myself. My lab was very modestly equipped, with just a shaker, incubator, spectrophotometer, basic light microscope and a few other instruments. I realised that this was a really rare opportunity to start an entirely new subject, and at our first

Figure 5. First laboratory members at the University of Tokyo.

departmental meeting announced that I would examine the lytic function of the vacuole. However, I didn't have any meticulously crafted strategy for how I could study this question. Since the vacuole is an acidic compartment that contains a range of lytic enzymes, I hypothesised that this organelle would play a similar role to the mammalian lysosome. The segregation of the potentially hazardous degradation process within a membrane compartment makes biological sense, but I knew at the time that if this were true material destined for degradation would have to be delivered across the vacuolar membrane, suggesting that it would require a cell biological approach and not be easy to tackle. My work on protein degradation originates from this research on the vacuole; I was 43 years old.

The lysosome was discovered by Christian de Duve in 1955, and soon after researchers at The Rockefeller University observed the delivery of both extracellular materials and cytoplasmic materials to the lysosome by electron microscopy. De Duve coined the term autophagy to describe the latter process, from the Greek for 'self-eating'. Although some, most notably Glenn Mortimore and Per Seglen, had conducted intricate analyses of the regulation and complex membrane dynamics of autophagy, there were many more questions than answers when I joined the field. The complexity of lysosome and membrane dynamics in mammalian cells prevented the uncovering of the autophagy mechanism. De Duve's lab was actually in the same building as Edelman's, and while I had friends from his lab, I never had the chance to meet with this great scientist at The Rockefeller University.

Light microscopes today, including the development of fluorescence microscopy, are advanced instruments, but at that time the power of light microscopy was limited and little information about the intracellular events could be obtained, especially for the tiny yeast cell. However, the yeast vacuole could be clearly observed under phase contrast optics. The inside of the vacuole is somewhat like a salt solution with a very low protein concentration, so I knew that it would be easy to detect structures within the vacuole due to the difference in refractive index. As I would observe the vacuole every day at the microscope, I think I probably spent longer watching yeast cells than anyone else during these years!

First of all, I searched for a condition where yeast would break down their own proteins and settled on sporulation as the best candidate. This dramatic change, from vegetatively growing cell to four spores via meiosis, is induced by the environmental depletion of nitrogen. As nitrogen is key ingredient of protein, I reasoned that the protein synthesis required for such drastic cellular remodelling must depend on the turnover of the cell's own proteins. I therefore closely observed the morphology of the vacuole during sporulation, but wasn't able to observe any obvious changes.

One day, I conceived that if I were to use a cell lacking vacuolar proteases, structures delivered to the vacuole for degradation would remain in this organelle and might be visible. Fortunately, Elizabeth Jones had already constructed strains lacking vacuolar proteases and, in an example of the sharing ethos of the yeast community, had deposited them into the Yeast Genetic Stock Center at UC Berkeley, so I immediately requested that they be sent to my laboratory. When they arrived, I subjected them to nitrogen starvation and examined them under the microscope. I observed small, spherical structures moving vigorously within the vacuole. These structures appeared within 30 minutes of the onset of nitrogen starvation and had completely filled the vacuole within a few hours. In the otherwise tranquil yeast cell, this dramatic phenotype really struck me, and I distinctly remember staring down the microscope for hours, unable to tear my gaze away. These structures are immediately degraded in wild-type cells, which is why they had never been properly documented. I was fortunate for two reasons: first, the structures inside the vacuole, which we now call autophagic bodies, were big enough to see under the microscope; and second, the intense movement of autophagic bodies by Brownian motion made their accumulation in the vacuole easy to notice, even with a basic light microscope. The observation of this phenomenon determined the course of the rest of my career.

I decided to immediately determine what causes this phenomenon by electron microscopy and enlisted the help of Misuzu Baba and Masako Osumi. The images they captured were overwhelmingly beautiful and provided unequivocal evidence of the membranous phenomena characteristic of autophagy in yeast. These images showed that autophagic bodies contain the same density of ribosomes as the cytoplasm as well as various cytoplasmic structures and organelles, suggesting that autophagy is a non-selective degradative process. These experiments provided us with our current understanding of the membrane dynamics associated with autophagy. Although the vacuole is vastly larger than the mammalian lysosome, the entire process that we observed was topologically the same as macroautophagy in mammalian cells, confirming that yeast could be used as a model organism in the study of autophagy.

The most attractive feature of yeast as a model organism is its genetic tractability, lending it great power in the unpicking of complex biological phenomena. Two important examples that strongly influenced me were Lee Hartwell's identification of the *cdc* mutants and their function in the cell cycle, and Randy Schekman's isolation of the *sec* mutants and their role in membrane trafficking. As we knew nothing of the genes and proteins that are involved in the autophagy, the first step was therefore to isolate autophagy-defective mutants. But the only means we had of monitoring autophagy was to observe autophagic body accumulation in the vacuole. My first Master's degree student, Miki Tsukada, was central at this stage of our work. She took thousands of mutagenised cells, individually subjected them to starvation and then checked for mutants lacking autophagic bodies in the vacuole. This approach led us to identify the first autophagy mutant strain, which we called *apg1*. A failure to induce protein degradation in this strain convinced us that autophagic bodies are indeed responsible for degradation during starvation. As expected, we also found that the homozygous diploid mutant was unable to sporulate. However, under nutrient rich conditions no difference was observed between the *apg1* strain and wild-type cells, and there appeared to be no obvious defect in vacuolar function or secretion.

Soon we found that in comparison to their wild-type counterparts, *apg1* cells rapidly lose viability during nitrogen starvation. Assuming that this phenotype was caused by a defect in autophagy, we next conducted a preliminary screen assessing viability. By isolating strains in which viability was reduced and then examining these strains for morphological defects in autophagy, we were able to identify roughly 100 autophagy-defective strains in a single screen. Further genetic analysis showed that 14 complementation groups were among these strains. Looking from our current understanding of autophagy, it's clear that this screen was extremely effective. We initially predicted that these mutants would be characterised by defects at various stages of the autophagy process, but surprisingly all were essential for the formation of the autophagosome, the most important step in autophagy.

Next, we set out to clone and identify the genes implicated in the mutant phenotypes observed in the screen. Tsukada first succeeded in the cloning of the *APG1* gene, through which we learned that this gene encodes a protein kinase. However, we also found that the remaining genes encoded uncharacterised proteins essential for autophagy, but not required for growth under rich conditions. Therefore, we were unable to deduce anything about their function and went through a difficult period where we were unable to publish our exciting findings. Although we made rapid progress, autophagy genes had somehow escaped the attention of other researchers in spite of the comprehensive genetic analyses of yeast all around the world. I suspect that this is because researchers were interested in essential genes, conventionally defined by inviability in nutrient rich medium. At this time, Takeshi Noda also developed a quantitative assay of autophagic activity, the still widely-used Pho8Δ60 assay. These developments together laid the foundations for our subsequent studies on the molecular mechanism of autophagy.

This is how my career in autophagy research began in that small laboratory at the University of Tokyo. The whole process, from uncovering autophagy to beginning the characterisation of the Atg genes, was achieved within eight years in Komaba.

3. NIBB

The next phase began in 1996, when I was chosen as a professor at the National Institute for Basic Biology (NIBB) in Okazaki. This facility provided an environment very supportive of basic biology, and I selected three researchers to join our team: Tamotsu Yoshimori, originally of Kansai Medical School, who would work to incorporate our findings in yeast into mammalian cells; Takeshi Noda, who was the first doctoral student to graduate from our lab at Komaba; and Yoshiaki Kamada, who had studied at Johns Hopkins University in the US. The following year, Noboru Mizushima, previously a clinician at Tokyo Medical and Dental University, also joined the laboratory as a JSPS postdoctoral researcher and later as an assistant professor. From this point, graduate students and postdocs from all over Japan joined our lab and we entered the most fruitful age of our research. At the NIBB, we were able to produce a truly unique research environment where we extended our work to include mammalian cells and plant cells, which was important in the development of the field. Most members of our laboratory lived

Figure 6. The Ohsumi Laboratory at the NIBB in 2003.

near the NIBB, and as both Yoshimori and I were working away from home during this period we would often work late into the night, discussing science over a beer.

I was expecting that it would take a long time to isolate all the genes required for autophagy, but thanks to our collaboration with my wife Mariko's laboratory, which helped with cloning at the Teikyo University of Science, as well as the sequencing of the entire yeast genome in 1996, we were able to elucidate the genetic context of nearly all the *APG* genes in a relatively short period.

After that, we undertook to identify interacting proteins of the characterised Apg proteins, and were able to confidently identify all 18 of the genes essential for autophagy under nitrogen starvation conditions. A group led by Michael Thumm was studying the reduction of protein levels using antibodies under starvation conditions and was close behind our group, naming genes they identified using the '*aut*' alias. Meanwhile, Dan Klionsky's group was investigating the delivery of aminopeptidase Ape1 by cytosol to vacuole targeting, now known as the Cvt pathway, identifying genes using the '*cvt*' alias. This pathway is in fact a biosynthetic pathway responsible for the maturation of Ape1. Through the collaboration of our groups, we determined that this was occurring through similar membranous phenomena that depend on the core autophagic machinery and very selectively isolate Ape1 for delivery to the vacuole. Klionsky's group thus uncovered the Cvt pathway, which served as an important model of what was to become the major field of selective autophagy.

With the continued isolation of autophagy-defective mutant strains in yeast by many groups, we collectively decided to unify gene names under the *ATG* alias to simplify nomenclature in the field. As the majority of the key autophagy-related genes were identified by our group, the gene numbers used by our group when assigning *APG* gene names was retained in this system. With the identification of further autophagy-related genes, most of which are involved in selective autophagy, the number of *ATG* genes has now reached 41.

During our time at the NIBB, important findings came one after the next as we began to look at the *ATG* gene products to gain insights into their function. Mizushima made the first important breakthrough in his investigation of the only very recently cloned Atg12 protein. He discovered an intricate conjugation system that functions in an ubiquitin-like manner to yield a dimer of Atg12-Atg5-Atg16 complexes essential for autophagy. Two graduate students, Takayoshi Kirisako and Yoshinobu Ichimura, also uncovered another ubiquitin-like conjugation system that, unusually for such a system, results in the conjugation of Atg8 with a membrane phospholipid, phosphatidylethanolamine. Atg8, homologues of which were soon identified in mammals by Yoshimori and in plants by our group, remains associated with the autophagosome throughout all stages of autophagy and is therefore widely used as a marker of autophagy progression. We therefore learned that nearly half of the essential Atg proteins were involved in just two conjugation reactions. Important structural details of the complexes involved in these reactions were solved thanks to our close collaboration with Fuyuhiko Inagaki and Nobuo Noda's groups.

Further details came with the identification of the autophagy-specific PI3 kinase complex, which we termed PI3 kinase complex I. Akio Kihara's work showed that Atg6 had in fact only just been identified as Vps30, a protein of the vacuole protein sorting pathway, by Scott Emr's group. Vps30 is a component of the PI3 kinase complex, and we subsequently found that Atg14 affords this complex its specific role in autophagy. Yet another specific component of PI3 kinase complex I, Atg38, was uncovered more recently by Yasuhiro Araki. Meanwhile, we also discovered the important role of Atg13 in the activation of Atg1 kinase function, and identified Atg17, Atg29 and Atg31 as forming a complex required for autophagy induction under starvation conditions. With our investigations of Atg9, Atg18 and Atg2, we were before long able to classify all 18 genes essential for autophagy under starvation conditions into six functional groups. This was further explored by Kuninori Suzuki, who used fluorescence microscopy to discover the pre-autophagosomal structure (PAS), a dynamic structure to which all Atg proteins at least partially assemble and intricately co-ordinates autophagosome formation. Suzuki's work also demonstrated that the Atg proteins associate with the PAS in a hierarchical manner to perform their essential role in autophagy, shedding further light on the mechanisms of autophagy.

Identification of the *ATG* genes has completely revolutionised autophagy research. One important example of this was Mizushima's demonstration of the progression of autophagy in mammals by using a GFP-tagged LC3 transgenic mouse, as well as his successful generation of an autophagy knockout mouse, the first for Atg5. Along with the knockout of Atg7 in mice by Masaaki Komatsu, these studies changed our understanding of autophagy in mammalian cells and spurred the widespread genetic manipulation of *ATG* genes in various cells, tissues, organs and whole organisms all over the world. This movement helped to uncover a variety of physiological functions of autophagy. Of these, the recycling function of autophagy is the most fundamental in the cell, helping to survive the commonly-encountered challenge of nutrient limitation. But autophagy is not only important for nutrient recycling: it is also implicated in the often selective purging of unnecessary or harmful intracellular components, which is important in the maintenance of the intracellular environment. Recently, the selective elimination of proteins, supramolecular structures, and organelles have come to the fore as physiologically important applications of the core autophagy machinery. A particularly salient example of this has been the intensely-studied field of selective mitochondrial autophagy, or mitophagy. The removal of old and potentially damaged mitochondria by autophagy ensures efficient energy production and a healthy intracellular environment, underscoring the nuanced integration of autophagy into the cell.

4. STARTING MY FINAL LAB AT THE TOKYO INSTITUTE OF TECHNOLOGY

Even in an organism as apparently simple as the yeast cell, many mysteries remain unresolved. My lab has therefore been working exclusively in yeast for 13 years since Yoshimori and Mizushima formed their own groups.

In 2009, another important change occurred in my career. The Tokyo Institute of Technology offered me a position as a specially appointed professor in a laboratory with excellent facilities, so I decided to return to the Tokyo area. With Hitoshi Nakatogawa and Suzuki as assistant professors, and later Hayashi Yamomoto, we started a new laboratory. I brought along many colleagues from Okazaki to this new facility. We have been focussed one subject, how the autophagosome is formed, through the functional analysis of Atg proteins. Recently we found how the early steps of PAS formation are organised. Another area of interest to us was the mechanism of selective autophagy in yeast. In this field, we were also able to identify various receptors for target selection. Hitoshi Nakatogawa identified novel receptors for ER and nucleus degradation.

Figure 7. Ohsumi research group members at the University of Tokyo in 2015.

Now, my lab has shifted its focus to the physiological meanings of autophagy. In order to return to the original questions that intrigued me at the start of my career—the what, when, how and why of autophagic degradation—we need to systematically consider cells as they exist in an array of environmental contexts. This year we have set up a research unit at the Tokyo Institute of Technology, and with Tomoko Kawamata as an assistant professor we are now running a laboratory consisting mainly of postdoctoral researchers that will be my last laboratory. Here we are tackling those original questions that I believe are important to deepen our understanding of autophagy. Yeast is still an extremely powerful model organism, and with the continued development of new biochemical techniques I believe that we still have many ground-breaking contributions to make, especially in identifying the induction conditions and various modes of autophagy. These contributions will necessarily assess the products that arise from autophagic degradation, the mechanism of their export from the vacuole and their implications for cellular metabolism. As a fundamental cellular process, autophagy is linked directly and indirectly to a range of cellular pathways, and the development of quantitative methods that will allow us to grasp the subtle characteristics of this mechanism are essential. A better understanding of autophagy promises to help characterise the ambiguous relationship between genetics, phenotype and disease.

Autophagy research is still at an early stage, and our understanding of the physiological role of autophagy in particular is only in its infancy. Of course, I'm very pleased that we have managed to unveil such intricate detail about autophagy and its associated phenotypes, and the attention the field has attracted from around the world is very exciting. But I am most appreciative for the many wonderful colleagues I have worked with over the years. I was always driven by curiosity rather than immediate practical outcomes in my work, and have been lucky to work with postdoctoral researchers and graduate students who appreciate the importance of basic research. Many of them have shared my passion for studying difficult projects that do not always give immediate outcomes, and fortunately many of them continue to advance the field at a range of universities as researchers. I have also been very fortunate to have worked with many collaborators over the years who have helped us along the way. In particular I want to note the contribution of two researchers who have collaborated closely with us for over a decade in the systematic structural analysis of the Atg proteins: Nobuo Noda, who works now at the Institute of Microbial Chemistry, and Fuyuhiko Inagaki, who regretfully passed away in 2016. I deeply appreciate their collaboration over the years.

Figure 8. The Ohsumi family in 2016.

I am deeply moved by the interest that has been shown in autophagy research, which for me has represented an incredible journey over the last 28 years. This was only possible thanks to grant support of our fundamental research, which was indispensable in this undertaking. It is my greatest pleasure and honour as a basic scientist if our work was able to trigger a development in our understanding of life. I await the continued development of the field over the years to come with great anticipation.

Molecular Mechanisms of Autophagy in Yeast

Nobel Lecture, December 7, 2016
by Yoshinori Ohsumi
Tokyo Institute of Technology, Tokyo, Japan.

SCIENCE IS A SYSTEM OF KNOWLEDGE that is gradually accumulated over many years by society, but it is also an inherently human activity. I believe that every scientist is a product of the era in which they live. So I'd like to start from a brief introduction of my life, and then give a historical overview of my scientific work.

Early life

I was born in Fukuoka, southern Japan, in 1945, half a year before the end of World War II. This was a very challenging time in Japan, and everyone had difficulty getting basic daily necessities, including food. I myself suffered from severe malnutrition and was a very sickly child. Around this time my mother got tuberculosis and spent a long period bed-bound, but she was miraculously able to recover thanks to a gift of just-developed antibiotics from a family friend in Hawaii. When I was about 8 years old, my mother was finally able to resume a normal life. My brother, Kazuo, who is 12 and a half years my senior, left home to enter the Faculty of Literature at the University of Tokyo as I began elementary school. I was brought up through the efforts of my father and two sisters, Reiko and Junko, and many other people throughout what was a difficult time for our family.

My childhood home was surrounded by nature, with rice paddies, streams, hills and the sea all nearby. I spent a lot of time outdoors catching fish and picking plants. As an elementary school student, I was engrossed in collecting insects and watching the night sky. I would say that this intimacy with nature had a strong influence on me. My brother, every time he returned home from Tokyo during university holidays, would bring a book, such as carefully chosen works of George Gamow and Michael Faraday that broadened my awareness of the world beyond what we learned at school (Faraday, 1861; Gamow, 1940). Being a weak child, I had no talent for sports and was neither artistic nor gifted in literature. But I managed to achieve good results at school. At the time, I had a vague longing to be a scientist that was influenced by the expectations of my parents.

University & first experience of research

In high school, I joined the chemistry club, where I first experienced the wonder of chemical reactions. Partly thanks to this experience, I entered the University of Tokyo, where I initially hoped to become a chemist, but I had difficulty finding a field to specialise in. Thankfully, this was the time when the central dogma of molecular biology was just being established. Molecular biology immediately fascinated me, so I decided to join the lab of Kazutomo Imahori, who was one of the few scientists running a molecular biology laboratory in Japan at the time. Imahori's laboratory had adopted a physicochemical approach to assess interactions between proteins and nucleic acids. Under the direction of Akio Maeda, I began to study the role of ribosome subunits in protein synthesis. Although I was unable to produce particularly interesting results, my time in the Imahori lab was exciting as we participated in the work of a rapidly developing field and I got my first taste of the joy of experimental work. The experience also provided me with a strong sense of the continuous process of protein synthesis within cells.

From the second year of my doctoral studies, I decided to move to Kyoto University, where Maeda had joined the just-founded Department of Biophysics as an associate professor. Kyoto University provided a very free environment, and I was stimulated by talented new students and friends I met in the biochemistry lab of the Department of Chemistry. Throughout my graduate studies, the Summer Seminar for Young Biochemists also allowed me to make a large number of friends from all over Japan who would have a great impact on my life. In my research, I became interested in colicin E3, a toxic protein that is able to pass through the membrane of bacterial cells and instantly inhibits protein synthesis. I think I was also influenced by my growing interest in biological membranes during this period.

Post-doc life in New York

After I finished my doctoral studies, I enrolled in the lab of Gerald Edelman at The Rockefeller University, New York. I was asked was to establish an in vitro fertilisation system for mouse egg cells, which I reluctantly accepted. Experimentally it was not too difficult, but I had no idea how to examine such fascinating yet few early embryos. After some time, Mike Jazwinski joined our group from Arthur Kornberg's lab, where he had been studying phage DNA duplication. Taking inspiration from the elegant study of the cell cycle through the *cdc* mutants by Hartwell (Hartwell, 1974), a project to uncover the mechanism of DNA replication initiation in yeast began in Edelman's lab, which I joined in my final year. This was to be my first encounter with yeast as an experimental organism.

We had first planned to use isolated nuclei in this project, but it was difficult to obtain these organelles in an undamaged state. During our use of density-gradient centrifugation to purify nuclei, I noticed a white layer at the top of the centrifuge tube. Out of mere curiosity I looked at this white layer under the microscope, and realised that it was a highly-enriched vacuole fraction. The ease with which this organelle could be purified left a strong impression on me, and I think helped to decide the direction of my research.

Return to Japan and work on the vacuole

At the end of 1977 I returned to Japan as an assistant professor in Yasuhiro Anraku's lab at the University of Tokyo. Anraku's entire lab was tasked with studying the transporters and the respiratory chain in *E. coli*. In spite of this, Anraku allowed me to start a project on yeast, and in retrospect I really appreciate his decisiveness in affording me this opportunity, which resulted in the beginning of my nearly 40 years of work in yeast (Figure 1). In those days, the plasma membrane and its transport mechanisms were the focus of much research. But I have never been a competitive person, and prefer to work on a subject that few people are interested in. After a period of reflection, I decided instead to work on transport across the membrane of an intracellular organelle, the vacuole. While we now know that elaborate transport systems exist on the vacuolar membrane, this was by no means a fashionable topic.

At that time the vacuole was thought to be no more than a garbage dump in the cell, but I thought it must play yet unknown roles in cell physiology. First I established procedures for purifying vacuoles and making vacuolar membrane vesicles, and was able to show active transport systems of amino acids and calcium ions over the vacuolar membrane, providing evidence that the vacuole is

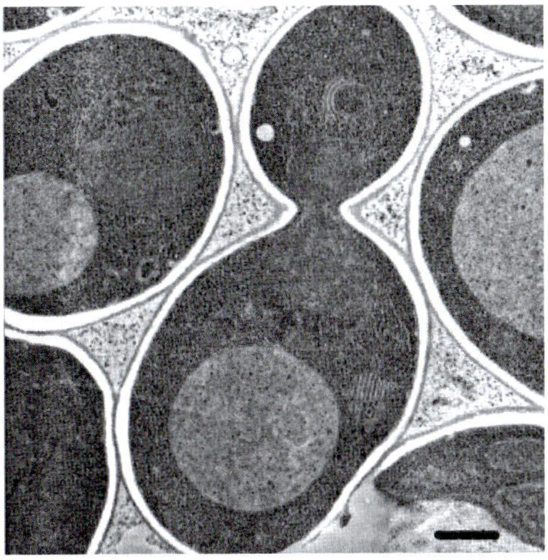

Figure 1. Thin-section electron micrograph of a yeast cell. The vacuole is visible as the large, pale region near the centre of each cell.

important for homeostasis of metabolites and ions (Ohsumi and Anraku, 1981; Ohsumi and Anraku, 1983). We also described a novel proton-pump on the vacuolar membrane, the V-type ATPase, which generates a proton gradient across the vacuolar membrane as a driving force of these transport systems (Kakinuma et al., 1981; Uchida et al., 1985) (Figure 2). These experiences played an important role in the development of my scientific interests, leading into my continued study of the vacuole and foreshadowing my research on autophagy. Since then,

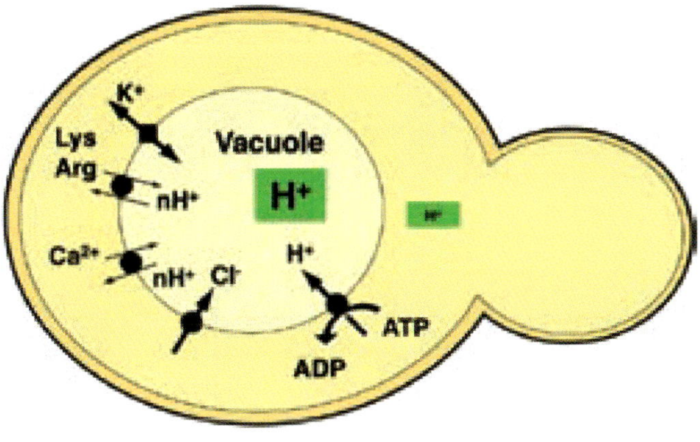

Figure 2. Vacuolar membrane transport systems in yeast. The V-type ATPase maintains the acidity of the vacuole by pumping H+ into the vacuole, while various active transporters modulate the flow of calcium, amino acids and ions between the cytoplasm and this organelle.

the structure and function of the V-type ATPase has become a major field of cell biology research in its own right.

STARTING MY OWN LAB AT TOKYO UNIVERSITY, KOMABA CAMPUS

In 1988 I was able to start my own lab—consisting of just myself—at the College of Arts and Sciences at the Komaba campus of the University of Tokyo. My lab here was modestly equipped, without any fancy instruments. I decided to devote myself to a related but completely novel task: the lytic function of the yeast vacuole. Since the vacuole is an acidic compartment, due to the function of V-type ATPase, and contains various hydrolytic enzymes including proteases, I thought it might be homologous to the lysosome of mammalian cells. However, I had no well-articulated idea how to test this hypothesis.

Thinking about the life of proteins

Before going into experimental details, I would like to briefly outline protein metabolism in cells. Proteins, which are a polymer of amino acids, are the key players in all biological processes. The central dogma of molecular biology states that protein is synthesised according to the genetic information encoded in DNA through RNA. As a consequence of this, many researchers throughout the latter half of the 20th century worked diligently to understand gene expression and its regulation, and the mechanisms of protein synthesis. Another important consideration that subsequently came to scrutiny was the compartmentalisation of the cell, as proteins are able only able to properly function after being delivered to the appropriate part of the cell. This spurred an effort to understand the localisation of each protein within the cell. Thanks to the efforts of researchers, we now know that each protein is intricately and precisely trafficked to the correct destination within the cell according to an intrinsic localisation signal (Figure 3).

However, one aspect that was missing from this picture when I started my own lab was the fate of proteins, or in other words the question of how intracellular protein degradation occurs. It was known that each protein has a distinct lifetime, ranging from a few minutes to several months. But for many years, the degradation of protein was thought to be a passive process that is not particularly important or interesting.

The signicance of protein turnover

I used to begin biology class for first-year undergraduate students with this question: how many red blood cells are made within just one second in your body?

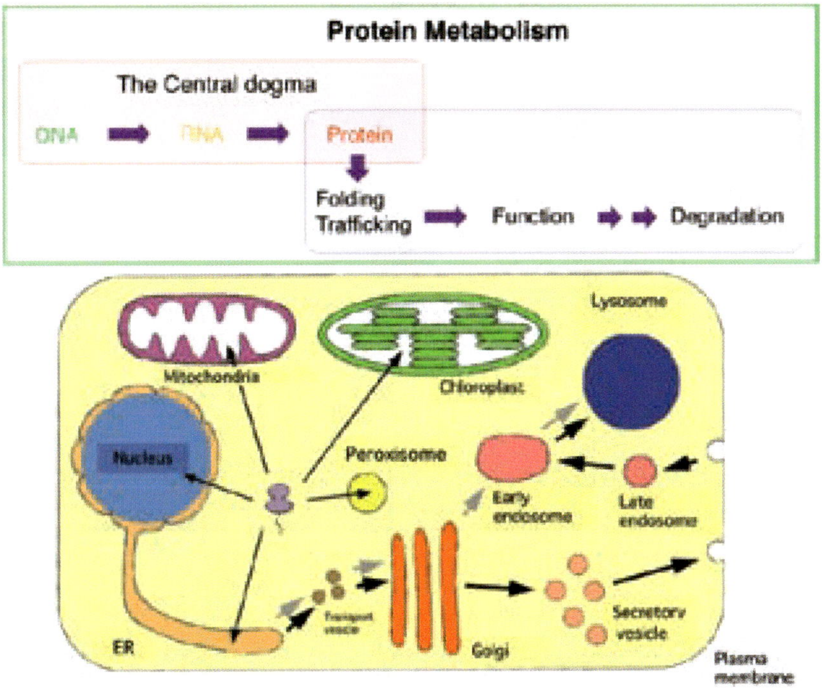

Figure 3. Overview of protein metabolism in cells. In addition to the emphasis on synthesis provided by the central dogma, the delivery, function and eventual degradation of proteins are important aspects of their life in the cell. The delivery of proteins to various cellular compartments following translation by the ribosome (purple) is represented at the bottom of the figure.

The calculation is quite simple and gives an answer of about three million cells per second. How about hemoglobin in the red blood cell? This number can also be easily derived, giving about $1x10^{15}$—one million billion—hemoglobin molecules per second. It follows that exactly the same amount of cells and proteins are degraded. The key message is that living things are extremely dynamically maintained.

The Japanese climate is characterised by four distinct seasons that influence our culture, and students at school learn that all things are in a cycle of continuous change. Plant leaves offer an excellent example of this concept. In spring, leaves develop and then actively synthesise starch by sunlight, but in autumn leaves become red or yellow and fall off (Figure 4). What we see as beautiful autumn colours is actually caused by the degradation of the leaf's photosynthetic machinery: green chloroplasts are completely degraded, and the resulting amino acids are transported to the trunk, leaving only the red or yellow-coloured

Figure 4. Ginkgo trees throughout the four seasons at the University of Tokyo. The removal of green chloroplasts in autumn, leaving golden accessory pigments in the leaves, is one striking example of the importance of degradation in the dynamic cycle of life.

accessory photosynthetic pigments in leaves. Similarly, rice leaves turn yellow at harvest time. All the protein in the leaves is degraded and transported to make proteins in rice grains for the next generation. These examples demonstrate that degradation is not an adverse process, but rather is essential for new construction or regeneration. Let's think about the human body. Our bodies make about 2–300 grams of protein every day, but we only intake about 70–80 grams of

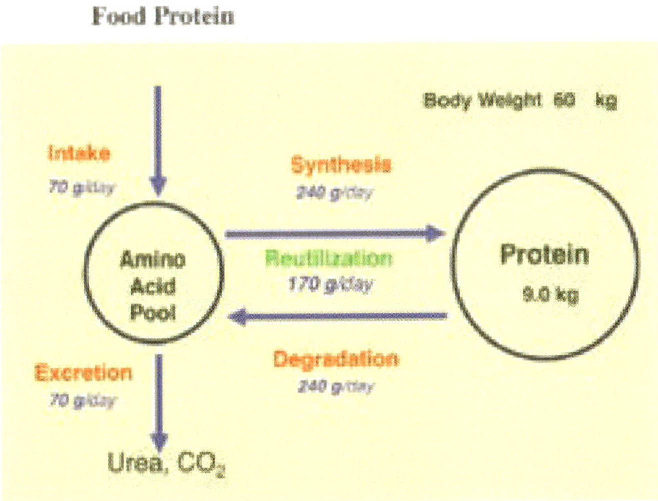

Figure 5. Protein dynamics in the human body. The bulk of amino acids required for synthesis are met not through dietary intake, but through the turnover of existing proteins in the cell.

protein from our diet (Figure 5). The amino acids required for protein synthesis mostly come from the degradation of the body's own proteins. Life is maintained by a tightly controlled balance between synthesis and degradation. Just two years ago I experienced this personally when I learned that my wife Mariko's mother had lived without any food and even water for 13 days. This was another shocking reminder that the recycling of cellular constituents is a potent force in the robust nature of life.

Historical perspectives on protein turnover

The establishment of the central role of proteins in biology was a very important development in modern biological science. In contrast to the process of synthesis emphasised by the central dogma, when I began my research the degradation of proteins was seen as a passive process that was not particularly important by many of my contemporaries. In this regard Rudolph Schoenheimer was a pioneer in the field, as he had proposed the concept of "protein turnover" in the 1930s following his development of isotopes to follow metabolic processes (Schoenheimer et al., 1939; Schoenheimer, 1942) (Figure 6A). Unfortunately, however, his contention was controversial and not widely accepted. After this, Christian de Duve, recipient of the Nobel Prize in 1974, used cell fractionation to discover the lysosome, an organelle containing a range of degradative enzymes (De Duve

(a) (b)

Figure 6. (a) Rudolph Schoenheimer and (b) Christian de Duve.

et al., 1955) (Figure 6B). Researchers at The Rockefeller University subsequently observed the delivery of extracellular material to the lysosome using electron microscopy (Novikoff et al., 1956). The delivery of intracellular material including organelles such as mitochondria to the lysosome was also observed. De Duve referred to these processes as 'heterophagy' (now known as endocytosis) and 'autophagy', respectively (de Duve, 1963). Autophagy, from the Greek for 'self-eating,' was described by The Rockefeller University researchers as having several hallmarks. Firstly, a membrane sac appears and enwraps a portion of the cytoplasm, forming a double-membrane structure, the autophagosome. Next, the outer membrane of the autophagosome fuses with the lysosome, and finally the inner membrane and its contents are degraded within the lysosome.

Autophagy research was continued by two key groups: one, led by Glenn Mortimore (Figure 7A), conducted detailed physiological and biochemical analyses of the phenomenon using liver perfusion in rats (Mortimore and Ward, 1976; Mortimore et al., 1983), while a second group, led by Per Seglen, examined the effect of nutrient starvation on autophagy and the regulation of autophagy, mainly using cultured cells (Seglen and Gordon, 1982; Seglen et al., 1980) (Figure 7B). However, as electron microscopy was the only means of assessing autophagy at the time and the lysosome is a dynamic organelle, most works focused on the analysis of membrane dynamics in autophagy, leaving the genes and proteins implicated in the autophagy mechanism unidentified for many years. A simpler model system was required to make further progress in the field.

(a) (b)

Figure 7. (a) Glenn Mortimore and (b) Per Seglen.

In the meantime, another intracellular degradation system, the ubiquitin/proteasome system, was discovered. As it became clear that this system plays important roles in the regulation of a variety of cellular processes, protein degradation suddenly attracted a huge amount of attention in biological research. This was ultimately recognized by the Nobel Prize in Chemistry in 2004, which was awarded to Aaron Ciechanover, Avram Hershko and Irwin Rose, but the field of lysosomal degradation and autophagy remained quiet. Now it is thought that these two systems function together as the major means of cytoplasmic protein degradation.

Discovering autophagy in yeast by light microscopy

Assuming that the vacuole functions as a lytic compartment, the questions I had were when, what and how cytoplasmic constituents go across the vacuolar membrane and become accessible to vacuolar enzymes. First, I searched for a condition where massive protein degradation might occur in the yeast life cycle. Spore formation is a meiotic process that generates four spores and is induced by the depletion of nitrogen, the essential element of amino acids (Figure 8). I reasoned that this dramatic cellular remodelling should require the degradation of pre-existing proteins to make the proteins necessary for this cell differentiation.

I have to admit that I've always loved to observe yeast cells by light microscopy. The vacuole is relatively large, is the only clearly visible organelle in the cell and contains a salt-like solution that has a very low concentration of protein, so I knew that it would be easy to see structures if they were inside. I probably spent more time sitting at the light microscope than anyone else during my studies of

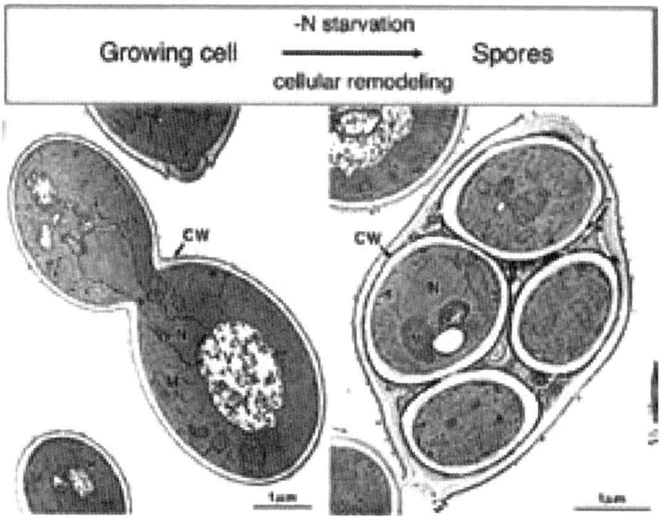

Figure 8. Morphological changes associated with sporulation in yeast. The onset of nitrogen starvation causes diploid cells to undergo meiosis and cell division, forming four spores in a process that requires vast remodelling of the protein component of the cell. N = nucleus, V = vacuole, CW = cell wall, M = mitochondrion.

the vacuole, looking for morphological changes that might provide hints about the function of this dynamic organelle.

I carefully observed the early stages of the sporulation process, but couldn't see any obvious changes in the vacuole. I thought that this may be due to the degradation of structures in the vacuole. I conceived that if I were to use a cell lacking vacuolar proteases, structures delivered to the vacuole for degradation would remain in this organelle and might be visible. Fortunately, the well-known yeast geneticist Elizabeth Jones had donated many proteinase mutants to the Yeast Genetic Stock Center at UC Berkeley, so I obtained these strains and shifted to nitrogen depleted medium.

Looking down the microscope, I saw many spherical structures vigorously moving around inside the vacuole. These structures appeared within 30 minutes and almost completely filled the vacuole after 3–4 hours (Figure 9). For me, this observation marked the exact starting point of more than 27 years of autophagy research. The reason that I was able to discover this phenomenon with only a common low magnification light microscope owes to the vigorous movement of these structures by Brownian motion. I was also fortunate that the structures, which we now call autophagic bodies, were large enough to be visible under a light microscope. In the otherwise tranquil yeast cell, this dramatic

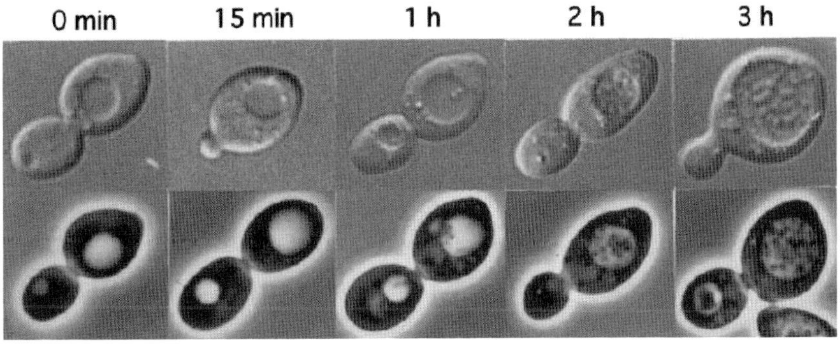

Figure 9. Accumulation of autophagic bodies within the vacuole under nitrogen starvation, as observed by bright-field (top) and phase contrast (bottom) light microscopy.

phenotypic change really struck me, and I remember staring down the microscope for some hours, unable to tear my gaze away. I knew that I had encountered an unknown and fascinating phenomenon; this was the starting point of my autophagy research.

Electron microscopy and the morphology of autophagy

I next called on the excellent electron microscopists Misuzu Baba and Masako Osumi to capture the morphology of this phenomenon by electron microscopy. Their efforts produced beautiful images that provided irrefutable evidence of the topological features of the entire autophagy process (Baba et al., 1994), as shown in Figure 10.

A thin section image of a nitrogen starved cell is shown in Figure 10A. Spherical structures can be seen within vacuole, and the high-magnification image presented in Figure 10B shows that these structures are bound by a single membrane. We can see that their contents are exactly the same as the cytoplasm, containing the same density of ribosomes and various cytoplasmic structures, such as the rough ER observed in Figure 10C, suggesting nonselective sequestration of cytoplasmic components. A membrane sac next to the vacuole, the isolation membrane, is also visible in Figure 10C enwrapping a portion of the cytoplasm. After completion of membrane expansion and sealing, a double membrane structure, the autophagosome, is formed (Figure 10D). Freeze-fracture imaging (Figure 10E) shows the fusion event of an autophagosome with the vacuolar membrane, where the outer membrane of the autophagosome is continuous with the vacuolar membrane (Baba et al., 1995), and the inner membrane structures that we named autophagic bodies are also present. Autophagic bodies

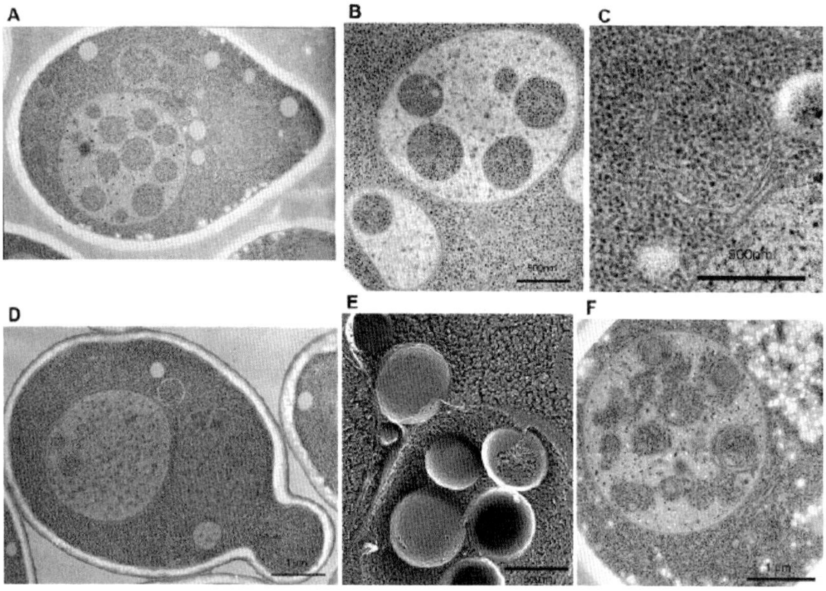

Figure 10. Electron micrographs showing topological phenomena associated with nitrogen starvation-induced autophagy. (A) Autophagic bodies within the vacuole, (B) high-magnification image of autophagic bodies, (C) cup-shaped isolation membrane adjacent to the vacuole, (D) an autophagosome adjacent to the vacuole, (E) freeze-fracture image of an autophagosome fusing with the vacuole, (F) mitochondria within autophagic bodies delivered to the vacuole.

occasionally contain mitochondria, as can be seen in Figure 10F, indicating that autophagy can degrade not only cytosolic protein but also supramolecular structures, such as ribosomes and even entire organelles, which is a characteristic feature of autophagy. This is in contrast to the specific recognition of smaller, individual targets by the ubiquitin/protease system. Autophagosomes are transient, ephemeral structures, and autophagic bodies are degraded immediately in the vacuole in wild-type cells, features that prevented their identification for many years.

A scheme of the autophagic process in yeast is presented in Figure 11. When cells are faced with starvation conditions, a small membrane sac appears next to the vacuole and expands to enwrap a portion of cytoplasm. Then, a double membrane structure, the autophagosome, targets the vacuole and fuses with the vacuolar membrane, releasing the inner membrane structure into the vacuole. In wild type cells these structures are immediately disrupted by vacuolar enzymes, and their contents are degraded for reuse. The proteinase-deficient mutant allowed us to follow the progression of autophagy as the accumulation of autophagic bodies.

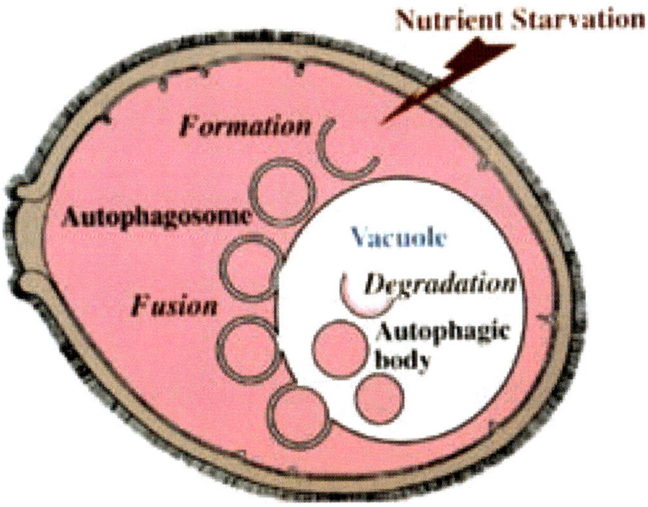

Figure 11. Schematic diagram of nitrogen starvation-induced autophagy in yeast.

Importantly, while the vacuole is a vastly larger organelle than the mammalian lysosome, the entire process that we observed was topologically the same as macroautophagy in mammalian cells, confirming that yeast could be used as a model organism in the study of autophagy (Baba et al., 1994). Kazuhiko Takeshige was also able to show early on that this phenomenon is not restricted to sporulation, but is a general cellular response to a range of nutrient starvation conditions (Takeshige et al., 1992). These factors gave the first indication that autophagy is a general process that is induced in diverse organisms in response to a range of stimuli, further convincing me that we had discovered an important cellular process.

Identifying autophagy-related genes in yeast

Yeast is a very good organism for the dissection of complicated biological phenomena through genetic analyses. In this regard I was inspired by the beautiful works of Lee Hartwell, whose group uncovered details of the cell cycle through their analysis of the *cdc* mutants (Johnston et al., 1977), and also by the intricate analysis of the secretory pathway using *sec* mutants by Randy Schekman's group (Novick et al., 1980). These contributions were recognized with the Nobel Prize in Physiology or Medicine in 2001 and 2013, respectively.

As we knew nothing of the genes and proteins that are involved in the autophagy mechanism, I thought that the isolation of mutants defective in autophagy was the first step in opening up the field of autophagy. However, we had no idea what an autophagy-defective phenotype would look like. At this point, we decided to use autophagic body accumulation in the vacuole as a morphological indicator of autophagy progression. My first Master's course student, Miki Tsukada, was central at this stage of our work and did an excellent job isolating autophagy defective mutants. We took mutagenised cells, individually subjected them to starvation and checked for the accumulation of autophagic bodies, aiming to isolate cells in which autophagic bodies were not observed. Using this approach, we identified the first autophagy mutant strain, which we called *apg1*, now referred to as *atg1* (Tsukada and Ohsumi, 1993). We then confirmed, by the failure to induce protein degradation in this strain, that the autophagic body is indeed an intermediate structure of the degradation process. We also found that, as expected, the homozygous diploid mutant was not able to sporulate. However, under standard growth conditions in the nutrient rich medium YEPD, no difference was observed between the *atg1* strain and wild-type cells, and there appeared to be no obvious defect in vacuolar function or secretion.

It seemed impossible that such an intricate pathway of delivery to the vacuole could be controlled by a single gene, and I was convinced that we should be able to identify many more. We found that *atg1* cells that were subjected to prolonged nitrogen starvation lose viability faster than their wild-type counterparts. Assuming this loss of viability phenotype was caused by the defect in autophagy, we next conducted a primary screen looking at the viability of mutant strains. By isolating strains in which viability was reduced and then performing a secondary screen examining these strains for vacuolar morphology defects, we were able to identify roughly 100 autophagy-defective strains at once (Tsukada and Ohsumi, 1993) (Figure 12). Further genetic analysis showed that 14 complementation groups were among these mutants. As we now know that in nitrogen starved cells 18 genes are essential for autophagy, I can say that this refined approach was extremely effective. We initially predicted that these mutant strains would be characterised by defects at various stages of the autophagy mechanism, but as it turned out all of these genes were essential for the formation of the autophagosome, the most important process in autophagy.

This was a morphological screen, based on the inability to accumulate autophagic bodies in the vacuole. We therefore eliminated mutants characterised by a severe growth defect, cells lacking essential genes and also strains with abnormal vacuolar morphology. In addition, as we used the decline in viability as a primary screen, which only appears with a complete block of autophagy function, we only identified autophagy null mutants, to the exclusion of partially defective mutants.

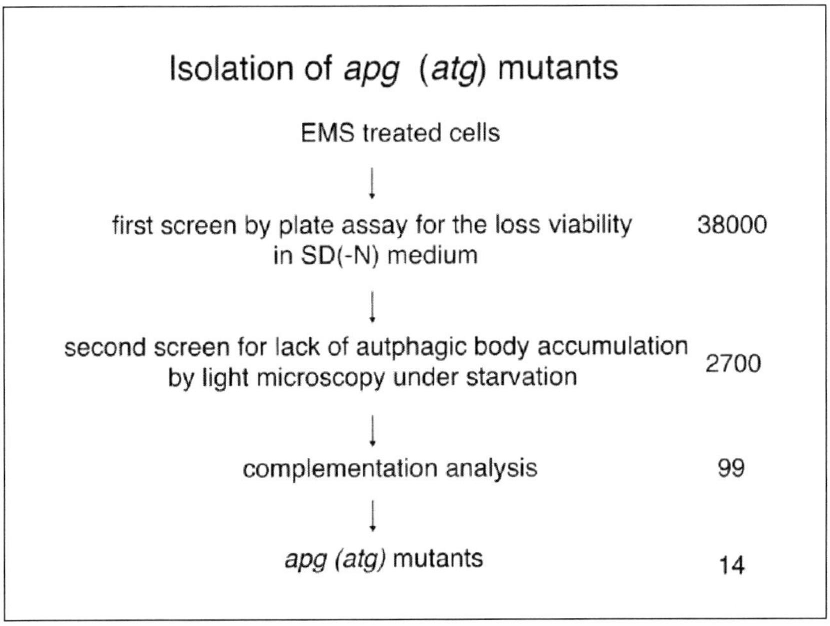

Figure 12. Overview of the experimental approach used to isolate autophagy-defective mutants. This screen allowed us to identify most of the *ATG* genes in a short period of time.

Next, we set out to clone and identify the genes implicated in the mutant phenotypes observed in the screen. Tsukada was the first to succeed, cloning the *ATG1* gene. Through the sequencing of *ATG1*, we learned that this gene encodes a protein kinase (Matsuura et al., 1997). However, as we identified the remaining genes, starting with Kametaka's work on *ATG5* (Kametaka et al., 1996) and *ATG6*, we found that all the other *ATG* genes encoded unidentified proteins that were otherwise not essential for growth in nutrient replete conditions and had no phenotypic implications for cells other than a defect in autophagy. Therefore, although we had identified nearly all the genes responsible for autophagy, we were unable to deduce anything about their function from their amino acid sequences and went through a difficult period where we were unable to officially announce our findings.

Yeast had been the subject of comprehensive genetic analysis throughout the world for many years at this point, and it was very surprising that autophagy genes had somehow escaped the attention of other researchers. I suspect that this is because researchers were interested in essential genes, and the conventional definition of an essential gene was the cell's inability to grow in nutrient rich medium when the gene in question was disrupted.

One important product of our time at the College of Arts and Sciences of Tokyo University that deserves mention was Takeshi Noda's establishment of an assay of autophagic activity (Noda and Ohsumi, 1998; Noda and Klionsky, 2008). Noda constructed a strain in which the precursor of the vacuolar alkaline phosphatase Pho8 lacks its N-terminal vacuolar targeting sequence, and expressed this precursor in the cytosol. Upon the induction of autophagy, a portion of the modified version of Pho8 would be taken up by the non-selective activity of autophagy and delivered to the vacuole. Here, the precursor is processed to yield the mature, enzymatically active form of Pho8. Therefore, we were able to quantitatively measure the extent of autophagy by assessing Pho8 activity. This assay, in which degradation can be quantitatively detected through the appearance of enzymatic activity, has been an excellent and widely-used tool in the analysis of autophagic activity that is still used today. I hope that this type of enzymatic assay can also be developed for use in other organisms, especially mammalian systems.

In this way, my career in autophagy research began in that small laboratory at the Komaba campus. The discovery of autophagy by light microscopy, its morphological analysis by electron microscopy, the isolation of mutant strains implicated in autophagy and the beginning of the identification of genes responsible for the process were all achieved within eight years.

A GOLDEN AGE OF AUTOPHAGY RESEARCH AT NIBB

In 1996, I became a professor at the National Institute for Basic Biology (NIBB) in Okazaki, which provided us with a very conducive research environment. I asked Tamotsu Yoshimori to join our lab as an associate professor to start work on autophagy in mammalian cells, and also Takeshi Noda and Yoshiaki Kamada to join us as assistant professors. The following year, Noboru Mizushima came to my lab as a postdoctoral researcher, and later as an assistant professor. Year by year, talented postdoctoral researchers and graduate students gathered to my lab. In 1997, just after we started at the NIBB, we also organised the first International Symposium on Autophagy (ISA) meeting, which we held at the institute. This forum for the discussion of autophagy research has proven to be very successful, and we will hold the 8th ISA in 2017, the largest yet.

Thanks to the NIBB, we were able to produce a truly unique research environment where our work in yeast was accompanied by studies in mammalian and plant cells. I was expecting that it would take a long time to identify all the genes required for autophagy, but thanks to our collaboration with Mariko's laboratory at the Teikyo University of Science, which helped with cloning, as well as the sequencing of the entire yeast genome in 1996, we were able to elucidate the

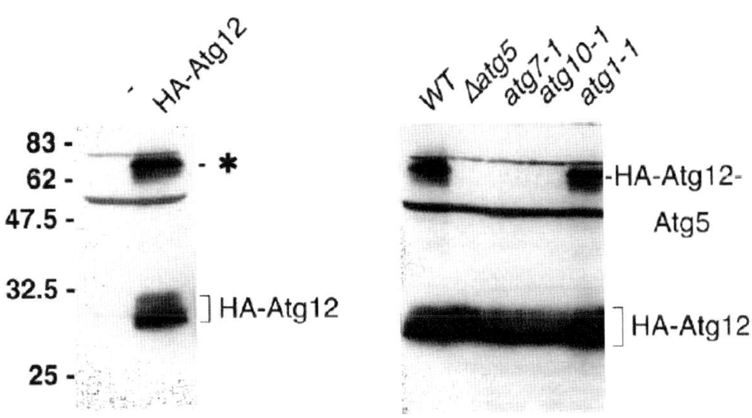

Figure 13. Immunoblot analysis of the Atg12 protein. (Left) The appearance of a band of much higher molecular weight than expected for Atg12 is indicated by an asterisk. (Right) The high molecular weight band disappears in cells lacking *ATG5*, *ATG7* or *ATG10*, but not *ATG1*.

genetic context of nearly all the *ATG* genes in a relatively short period. With our subsequent effort to identify interacting proteins of the characterised *ATG* gene products, we were able to confidently identify all 18 of the genes essential for autophagy under nitrogen starvation conditions. From this point, we focused on the characterisation of the Atg proteins and the mechanistic details of autophagy.

The Atg12 conjugation system

The first breakthrough was made by Mizushima when he examined the only just cloned Atg12 protein (Mizushima et al., 1998a). Using immunoblotting, he noticed two bands corresponding to Atg12, one of which indicated a much greater apparent molecular weight than that predicted from the protein sequence (Figure 13). In the absence of Atg5, Atg7 or Atg10, however, the high molecular weight band disappeared. This was the first indication we found of a relationship between Atg proteins, and within a short period we were able to demonstrate that these proteins together constituted a unique ubiquitin-like conjugation pathway. Although Atg12 is a vastly larger protein than ubiquitin, it ends with a glycine residue at its C-terminus and is expressed as a mature protein without any additional residues. Atg12 is then activated by Atg7, an E1 enzyme, to yield a thioester intermediate that is then transferred to the E2 enzyme Atg10, before it ultimately forms an isopeptide bond at a lysine residue at the centre of the Atg5 molecule. This Atg12-Atg5 conjugate then binds to a dimeric form of Atg16, yielding a functional dimer of Atg12-Atg5—Atg16 molecules (Figure 14A). A

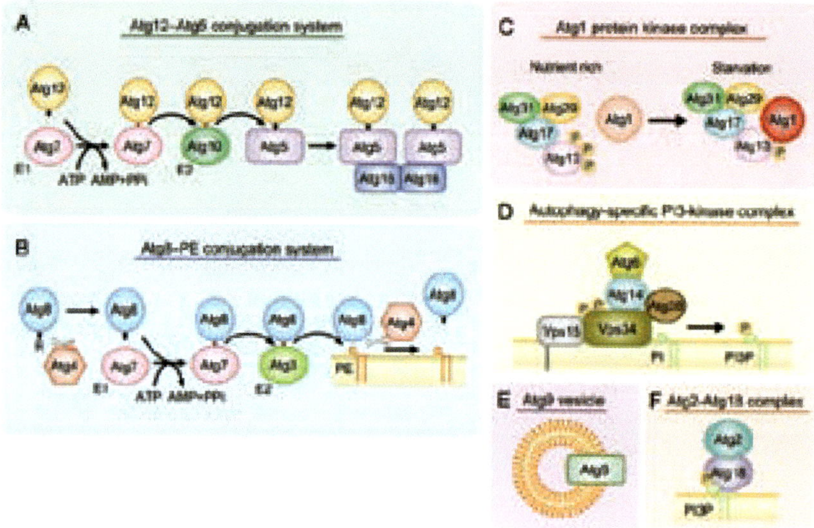

Figure 14. The six functional groups of Atg proteins: (A) the Atg12-Atg5 conjugation system, (B) the Atg8-PE conjugation system, (C) the Atg1 protein kinase complex, (D) the autophagy-specific PI3 kinase complex I, (E) Atg9 vesicles and (F) the Atg2-Atg18 complex. The concerted action of all groups is necessary for autophagosome formation.

unique feature of this conjugation reaction is that Atg5 is the sole target of the system and the reaction is not regulated by environmental conditions, such as starvation. With the elucidation of this conjugation system, we were able to characterise the functions of five Atg proteins in one series of experiments. However, why such an elaborate conjugation system is required to form this functional complex remains a mystery.

The Atg8 conjugation system

We were also interested in the small protein Atg8. We knew from immunoelectron microscopy that Atg8 localises to the isolation membrane and autophagosome, and were therefore hopeful that this protein could be used as a marker of membrane formation in autophagy (Figure 15). During these studies, the work of two graduate students, Takayoshi Kirisako and Yoshinobu Ichimura, unveiled a second ubiquitin-like conjugation system (Ichimura et al., 2000). Atg8 is another ubiquitin-like protein that is involved in autophagy that contains an arginine residue at its C-terminus. This arginine is cleaved by the cysteine protease Atg4 to expose a glycine residue, giving the mature form of the protein (Kirisako et al., 2000). The mature Atg8 molecule is then activated and transferred to Atg3

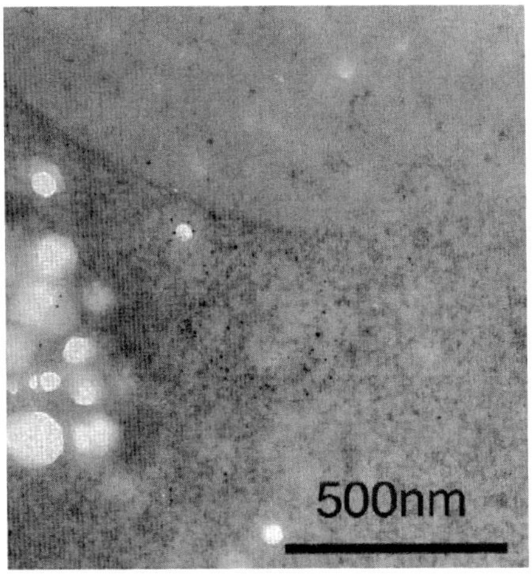

Figure 15. Immunoelectron micrograph showing Atg8 (dark spots) residing on the membrane of an autophagosome.

by Atg7, which is the same E1 enzyme as found in the Atg12 conjugation system. However, we weren't able to detect the final product of this reaction by conventional SDS-PAGE analysis. We soon found that most Atg8 resides as a tightly membrane-bound form, and when we purified this bound form we determined that the apparent molecular weight had increased by approximately 635 daltons. Through our collaboration with Toshifumi Takao of Osaka University, we were also able to attribute the binding of Atg8 to a membrane phospholipid, phosphatidylethanolamine, which we confirmed by isolating Atg8-PE using urea SDS-PAGE (Figure 14B). This attracted a lot of attention because the target of conjugation was not a protein but a lipid, which is very unusual for this type of reaction. We were also able to reconstitute the Atg8 conjugation system *in vitro* (Ichimura et al., 2004), which eventually allowed us to show that Atg8 is involved in membrane tethering and hemifusion that supports the expansion of the isolation membrane (Nakatogawa et al., 2007).

Thus, to our surprise, we found that about half of the *ATG* genes are involved in these two conjugation systems. Both Atg12 and Atg8 are activated by the same E1 enzyme, Atg7, before the protein is transferred to distinct E2 enzymes, Atg10 or Atg3, respectively. In addition, we also found that the Atg12 system is required for the conjugation of Atg8 to PE, further hinting at the close relationship between these two systems (Hanada et al., 2007). Recently, it has been demonstrated that the Atg12-Atg5 conjugate induces a conformational change in Atg3, promoting Atg8-PE formation (Sakoh-Nakatogawa et al., 2013). Structural details of the complexes involved in these reactions were solved thanks to our

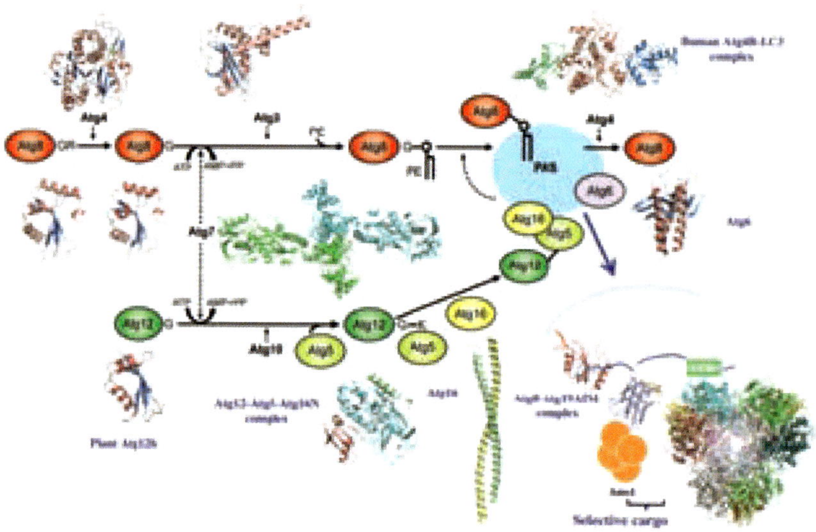

Figure 16. Schematic overview of the two ubiquitin-like conjugation systems involved in autophagosome formation. Structures of the associated proteins are included in the figure.

close collaboration with Fuyuhiko Inagaki and Nobuo Noda's groups, but details of their functions in the cell remain to be uncovered (Figure 16).

Events during autophagy induction and their regulation

In 1998, Takeshi Noda found that rapamycin induces autophagy even when cells are grown in rich medium (Noda and Ohsumi, 1998). As rapamycin is an inhibitor of TOR kinase, which is an important nutrient sensing complex, we therefore proposed that TOR kinase is the most upstream regulator of autophagy. Meanwhile, work by our group led by Tomoko Funakoshi and Yoshiaki Kamada also discovered the important role of Atg13 in the activation of Atg1 kinase function (Kamada et al., 2000). Atg13 is one of the targets of TOR kinase and functions as the key regulator of autophagy induction (Figure 14C). When nutrients are replete in the environment, Tor kinase heavily phosphorylates Atg13, blocking its interaction with Atg1. Upon starvation, Atg13 is quickly dephosphorylated, however, and associates with Atg1, subsequently facilitating its interaction with other Atg proteins. This crucial upstream event is essential for the induction of autophagy. Kamada also led the discovery of Atg17 (Kamada et al., 2000), which interacts with Atg13. Subsequently two other proteins, Atg29 and Atg31, were found by Tomoko Kawamata and Yukiko Kabeya (Kawamata et al., 2005; Kabeya et al., 2007). These latter two proteins form a stable complex with Atg17, and together with Atg1 and Atg13 this complex, known as the Atg1 complex, is essential for the induction of autophagy under conditions of starvation, as described below.

The autophagy specic PI3 kinase complex I

Although every other *ATG* gene encoded novel proteins that had not been described elsewhere, *ATG6* turned out to be the same gene as *VPS30*, which had only just been discovered by Scott Emr's group as a protein of the vacuolar protein sorting (Vps) pathway (Seaman et al., 1997). This attracted our attention, and through biochemical analyses led by Akio Kihara we discovered that this protein is actually a component of a PI3 kinase complex essential for autophagy, comprising Vps34, Vps15, Atg14 and Vps30, which we termed PI3 kinase complex I (Kihara et al., 2001) (Figure 14D). Interestingly, PI3 kinase complex II, which is involved in the Vps pathway, contains Vps38 in the place of Atg14, and these subunits are responsible for the appropriate localisation of the complexes to play their specific function. Following this, Yasuhiro Araki uncovered yet another specific component of PI3 kinase complex I, Atg38 (Araki et al., 2013). This protein localises to the isolation membrane and is essential for the incorporation of the product of PI3 kinase, PI3P, into the isolation membrane, with which it is ultimately delivered to the vacuole.

Atg9 and the supply of membranes to the isolation membrane

An unsolved mystery in the field is the origin of membranes that are supplied to the autophagosome during its formation. A protein that attracted much attention for its role in membrane expansion is Atg9, the only multiple-pass transmembrane protein among the Atg proteins (Figure 14E). There has been much speculation about the localisation of Atg9 within the cell, but we observed that Atg9 resides mostly on small membrane vesicles arising from the Golgi apparatus that freely move throughout the cytoplasm (Yamamoto et al., 2012). Our analyses indicated that a small number of Atg9 vesicles are required for the early stages of the formation of a single autophagosome, suggesting that these vesicles form seed membranes that incorporate membranes from other sources for isolation membrane expansion.

Atg2 and Atg18

In addition, Atg18, which was first identified by Michael Thumm's group, contains many WD repeats and binds with PI3P to form a complex with Atg2, thereby exerting its essential function in both the Cvt pathway and autophagy (Barth et al., 2001) (Figure 14F). Atg2, on the other hand, is very large protein that while essential for autophagy does not contain any known functional domains, and its role in the mechanism of autophagy remains poorly understood. Considering all these findings, we therefore concluded that a total of 18 Atg proteins are essential for autophagy, and that these proteins can be classified into the six functional groups presented in Figure 14: the Atg1 kinase and its regulators, the specific PI3 kinase complex, two conjugation systems, the sole membrane spanning membrane protein Atg9, and the Atg2-Atg18 complex.

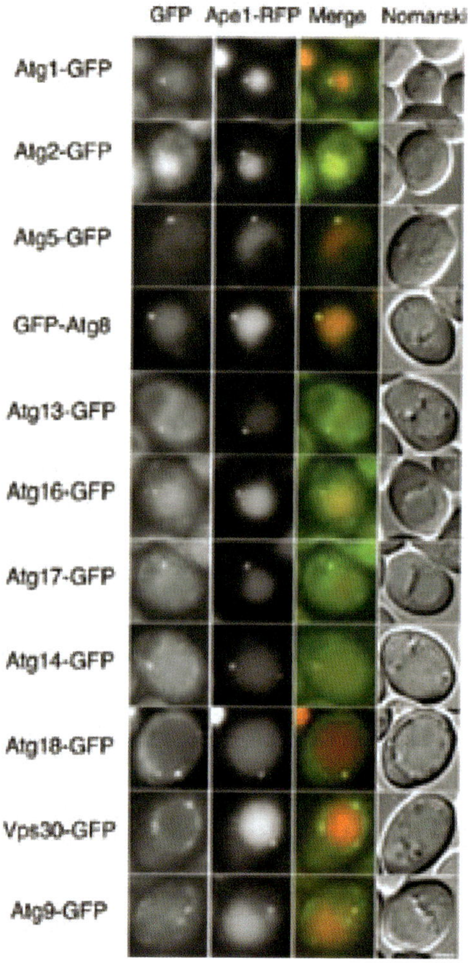

Figure 17. Atg proteins, tagged with GFP, at least partially localise to the PAS, indicated by RFP-tagged Ape1. In response to an induction signal, Atg proteins, usually moving freely throughout the cytosol, associate with the PAS near the surface of the vacuole, facilitating the progression of autophagy.

Identication of the PAS

Although we had identified the core machinery of autophagosome formation, the next questions were where and how these proteins function in the cell. Kuninori Suzuki, who was observing Atg8 localisation under starvation conditions, noticed that these proteins form a dot near the vacuole from which the autophagosome subsequently generated (Suzuki et al., 2001). We called this dot the Pre-Autophagosomal Structure (PAS). By fusing each Atg protein with GFP and confirming their continued function, Suzuki found that all Atg proteins at least partially localise to the PAS under starvation conditions (Suzuki et al., 2007) (Figure 17). We next devised a systematic approach to understand the relationship between each Atg protein and the PAS by assessing the localization of Atg proteins in strains lacking individual *ATG* genes. Through this work, we learned that the six functional groups of Atg proteins associate with the PAS in a hierarchical manner (Suzuki et al., 2007) (Figure 18). Atg17 exists at the top of

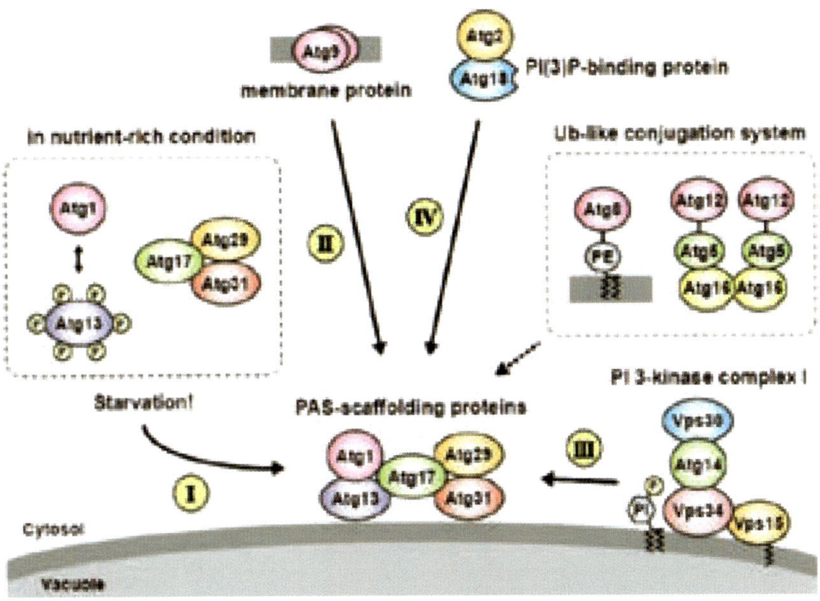

Figure 18. Overview of the sequential nature of Atg protein recruitment to the PAS. Upon starvation, a scaffold complex is formed through the interaction of dephosphorylated Atg13 with Atg1 and the Atg17-Atg29-Atg31 complex at the vacuolar membrane (I). Membrane recruitment then begins as Atg9 vesicles associate with the PAS (II) before PI3 kinase complex I joins the PAS to contribute PI3P to the expanding isolation membrane (III). Atg2 and Atg18 then interact with the PAS through an unknown mechanism that is essential for autophagy progression (IV). Finally, Atg8-PE, a product of the Atg12 and Atg8 conjugation reactions, plays a role in membrane expansion.

this hierarchy, and the association of other Atg proteins follows a strict order of precedence. This study was at the forefront of the many molecular studies of the PAS that followed, becoming a field in its own right.

Thanks to these fluorescence microscopy studies of the spatiotemporal dynamics of Atg proteins, we now know that most Atg proteins exist in a diffuse localisation during nutrient replete conditions, free to move randomly through the cytosol. The onset of starvation causes a fraction of these proteins to be recruited to the PAS. The hierarchical relationship amongst Atg proteins describes the order and timing of each Atg protein's recruitment to the PAS.

Mammalian ATG genes and the expansion of autophagy research

Also during our time at NIBB, Mizushima and Yoshimori began to study mammalian autophagy, and successfully showed that the Atg12 conjugation system is well conserved in mammalian cells (Mizushima et al., 1998b). Yoshimori definitively proved that LC3, the yeast Atg8 homologue, can be used as a very effective marker of autophagy progression (Kabeya et al., 2000). At around this time, two of my colleagues, Hideki Hanaoka and Kohki Yoshimoto, succeeded in identifying *ATG* genes in plant cells (Yoshimoto et al., 2004). Although the homology of *ATG* genes is low, the *ATG* system is well conserved between yeast, mammals

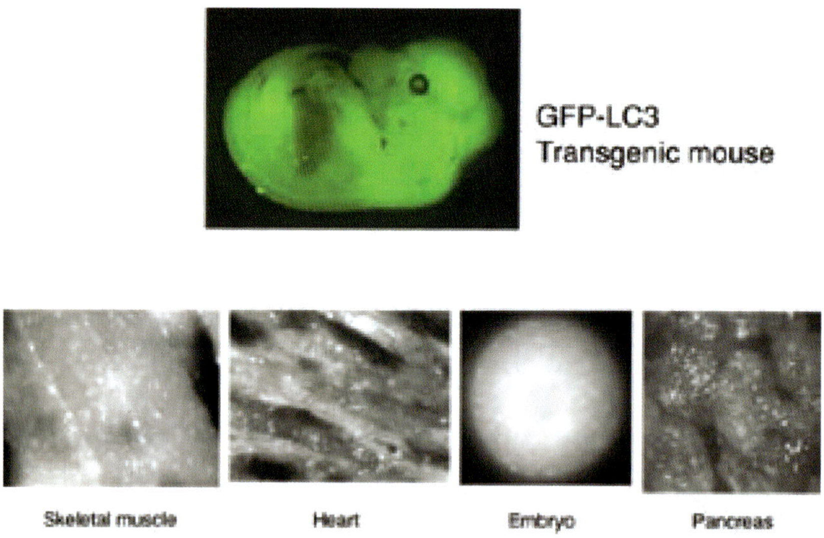

Figure 19. Expression of GFP-LC3 in tissues of a transgenic mouse. Shown are the localisation of autophagosomes, observed as puncta, in skeletal muscle (A, from a starved adult mouse), the heart (B, from a neonate), an embryo (C, post-fertilisation) and the pancreas (D, from a mouse with acute pancreatitis), as observed by fluorescence microscopy.

and plants, indicating that the core machinery of autophagy was acquired at an early stage of evolution.

The identification of *ATG* genes essential for autophagy changed the entire landscape of autophagy research. One aspect of this revolution is that we can now visualize the process of autophagy by fluorescence microscopy using Atg proteins as markers. Mizushima took advantage of this when he constructed a transgenic mouse expressing GFP-LC3 (Mizushima et al., 2004) (Figure 19). This mouse has allowed us to assess the degree to which autophagy occurs in every organ of the mouse's body and has been shared with labs all over the world. Mizushima also developed the first *ATG* gene knockout mouse, which allowed us to show that autophagy plays a crucial role for survival at birth (Kuma et al., 2004). Masaaki Komatsu was also able to demonstrate that conditional knockout of the ATG7 gene causes accumulation of ubiquitinated proteins and ultimately tumour in the liver (Takamura et al., 2011). From this point, manipulation of *ATG* genes in various organisms, different cell types, a variety of tissues and organs, and individuals started in many labs around the world and revealed a vast array of physiological functions of autophagy (Figure 20). The relevance of autophagy in various diseases is becoming clear. This rapid expansion of autophagy would not have been possible without the concerted efforts of autophagy researchers all over the world. But this expanding area falls outside the scope of this lecture, which focuses on my fundamental research in yeast.

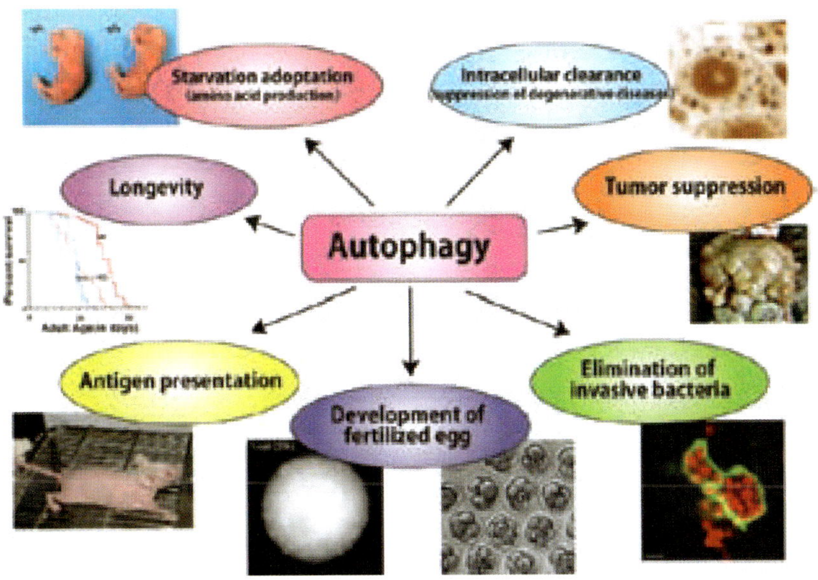

Figure 20. The diverse physiological roles of autophagy in mammalian cells.

STARTING MY FINAL LAB AT THE TOKYO INSTITUTE OF TECHNOLOGY

In 2009, another important change occurred in my career. I was offered a position as a specially appointed professor at the Tokyo Institute of Technology, where we were given access to excellent research facilities. I brought along many colleagues from the NIBB to this institute, and with Hitoshi Nakatogawa and Kuninori Suzuki as assistant professors initially, later joined by Assistant Professors Hayashi Yamamoto, Yasuhiro Araki and Tomoko Kawamata, we started a new laboratory. Here, we have been conducting increasingly advanced analyses of the Atg machinery and more recently have been considering the role of autophagy as one of many cellular processes.

Atg13 and its role in Atg protein assembly at the PAS

Yamamoto and Nobuo Noda worked together to uncover details of the early steps of PAS organisation. Atg13, a key player, has a unique structure that is made up of an N-terminal globular domain and a long, naturally disordered C-terminal region. This disordered region contains many amino acid residues that are phosphorylated by Tor1 kinase. As mentioned above, Atg13 is quickly dephosphorylated upon starvation, which allows it to bind to Atg1 and Atg17 via specific sites in the disordered region. Atg17 exists as a dimer of Atg17-Atg29-Atg31 complexes, the structure of which was recently solved (Ragusa et al., 2012). As a result, a dimeric complex of Atg1, Atg13, Atg17, Atg29 and Atg31 is formed, known as the Atg1 complex. However, work by Dan Klionsky's group had suggested that the actual number of Atg proteins assembling at the PAS is actually 30–40 (Geng et al., 2008). It was therefore clear that further assembly was occurring. Recently, we found that Atg13 has a second, weak Atg17 binding site that is also essential for PAS formation and autophagy (Yamamoto et al., 2016). Atg13 is not able to use both of the binding sites to associate with a single Atg17 protein, but rather links two Atg17 molecules, resulting in the formation of the PAS scaffold. In this way, the scaffold can form a large assembly (Figure 21).

Why is such a higher-order supramolecular structure necessary for autophagy? The answer lies in the activation of Atg1 kinase. Within the PAS scaffold, this higher-order organisation brings Atg1 kinases into proximity of each other, which results in auto-phosphorylation between adjacent Atg1 molecules (Figure 21). This activation of Atg1 kinase activity facilitates the next step of membrane formation, the recruitment of Atg9 vesicles to the PAS, contributing the initial membrane for isolation membrane expansion. Recently, Sho Suzuki found that the N-terminal HORMA domain of Atg13 binds to the N-terminal disordered region of Atg9 (Figure 22) (Suzuki et al., 2015). Assembly of many HORMA

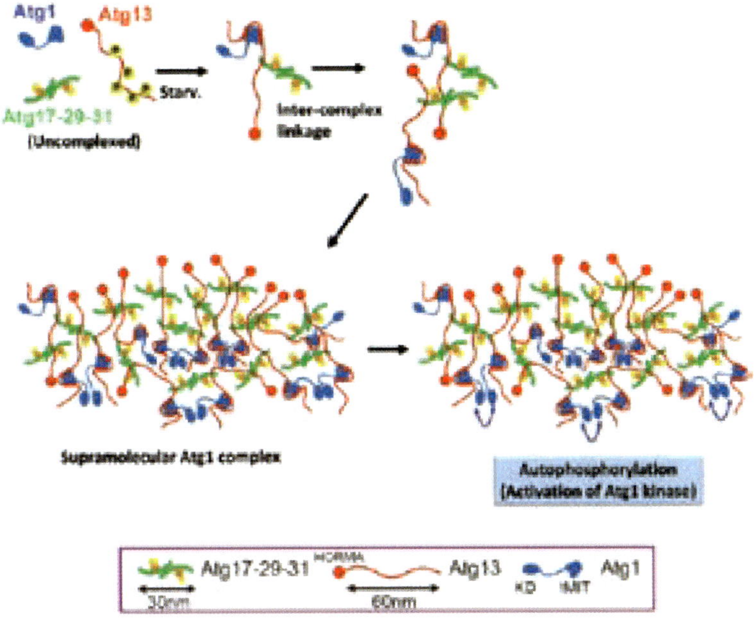

Figure 21. Formation of the supramolecular Atg1 complex by higher-order interactions of Atg13 molecules. Dephosphorylation of Atg13, causing Atg1 complex formation, is followed by inter-complex linkages that facilitate the activation of Atg1 kinase. Activation is achieved by autophosphorylation by adjacent Atg1 proteins enmeshed within the surpramolecular complex.

domains might be essential for interaction with Atg9 on the recruited vesicles. These findings together demonstrate that the PAS is not static, but is in fact a dynamic and flexible structure consisting of a large number of Atg proteins and membrane structures. During membrane formation, the PAS must be regulated spatiotemporally in a concerted manner by protein modification and the formation of a transient assembly, but we still need to extend our analyses to the later steps of autophagosome formation to get a clear picture of these events.

Alternate forms of autophagy

I mentioned earlier that nutrient starvation is the most potent signal for the induction of autophagy, representing a fundamental evolutionary adaptation to the persistent challenge of nutrient limitation, starvation. In our work, we have predominantly assessed autophagy in cells that have been subjected to starvation. But evolution is highly resourceful, and as with many highly-conserved core cellular processes, autophagy has found a range of diverse applications in the cell. In contrast to the canonical autophagy pathway, where induction is tightly

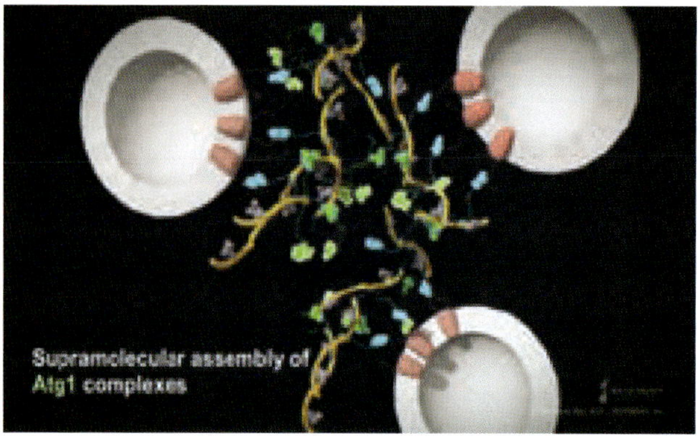

Figure 22. 3D model of the interaction between the supramolecular Atg1 complex (center) and Atg9 vesicles (membranes in pink, Atg9 in red). This interaction supports the delivery of the first membranes to the PAS, and is essential as the seed of the isolation membrane.

regulated and induced in response to starvation, constitutive forms of autophagy, which occur continuously at a basal level in the cell, have also been uncovered. Constitutive autophagy, the progress of which is independent of starvation, is important for the maintenance of cytoplasmic homeostasis. Dan Klionsky was central in the uncovering of one such form of constitutive autophagy, the Cvt pathway, which is an anabolic process that depends on the core autophagy machinery (Umekawa and Klionsky, 2012). In the Cvt pathway, the receptor protein Atg19 allows Ape1 complexes to interact with the PAS. Following this, the core autophagic machinery then very selectively packages the Ape1 complexes into a Cvt vesicle. This vesicle is then delivered to the vacuole in the same manner as the much larger autophagosome. Once delivered, the cargo of peptidases is released into the vacuolar lumen, where they are then processed to yield the mature, active hydrolases. This demonstrates that evolution has resulted both in selective applications of the autophagic machinery and anabolic processes in the cell, aside from its widespread induction in response to cellular crises such as nutrient deprivation.

This is not the only form of selective autophagy, which is now a burgeoning field in autophagy research. The targeted isolation and degradation of a range of cellular components, including specific proteins, supramolecular complexes, protein aggregates, organelles and even viruses and invasive bacteria (Amano et al., 2006) (Figure 23). As an example, Koji Okamoto, in our lab, identified Atg32, a receptor protein for the selective degradation of mitochondria that was concurrently identified by Klionsky's group (Okamoto et al., 2009; Kanki et al., 2009). Our analysis showed that Atg32 interacts with Atg11, adaptor protein for

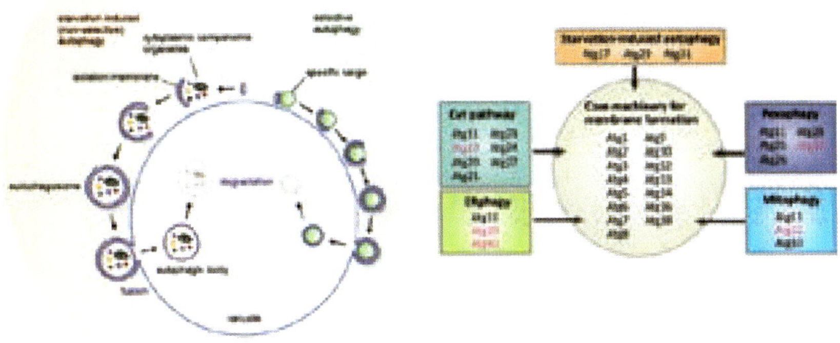

Figure 23. Selective autophagy in yeast. In contrast to starvation-induced autophagy, selective autophagy allows the degradation of specific cargo using the core autophagy mechanism. Proteins involved in the selective degradation of peroxisome (purple), mitochondria (blue), the ER (green) and the maturation of Ape1 (teal) are shown in the coloured boxes. Receptor proteins are indicated in red.

selective autophagy, which allows the canonical autophagy mechanism to isolate the targeted organelle for vacuolar degradation. In yeast, receptor proteins for the peroxisome (Atg28, Stasyk et al., 2006) and α-mannosidase (Atg34, Suzuki et al., 2010) have been uncovered in addition to Atg32, and Nakatogawa's group recently discovered receptors for selective autophagy of the ER and the nucleus (Atg39 and Atg40, Mochida et al., 2015). These receptor proteins, of which many more will be identified, carry an Atg8-interacting motif, as well as an Atg11 binding site. Atg11 is essential for all forms of selective autophagy, serving as a scaffold much like the Atg17 complex involved in starvation-induced autophagy. The mechanism by which individual organelles are marked for degradation is not clear and further research is essential, but it appears that selective autophagy plays a role in maintaining the efficient function and health of intracellular components, removing organelles as they age or accumulate damage. Autophagy therefore also plays a role in the maintenance of cellular homeostasis, regardless of environmental circumstances such as starvation.

Current work

Autophagy research is still at a very early stage, and our understanding of the physiological role of autophagy is only in its infancy. Over twenty years, our functional analysis of the Atg proteins allowed us to focus on one question: how the autophagosome is formed. To this end, we predominantly used cells that were shifted from nutrient rich conditions to a synthetic medium lacking nitrogen. This is because the induction of bulk autophagy is most striking using this experimental regime, and is thereby easiest to analyse. However, such nutrient rich conditions very rarely exist in nature, and I believe that we must endeavour to understand the contribution that autophagy makes during the natural life of yeast cells. From this perspective, I feel that we have returned to the starting point of autophagy research by going back to the original, fundamental questions—when, how, and what is degraded by autophagy? Yeast still offers striking advantages in comparison to mammalian cells, including the ease of vacuole purification and biochemical analyses. However, a major weakness in the study of autophagy has been the relatively limited repertoire of biochemical techniques available to researchers. Whilst developing new techniques, we must systematically address the conditions that induce autophagy, identify specific degradation targets, determine how degradation occurs within the vacuole and identify the final products of degradation and how they are exported back to the cytoplasm. We must extend our consideration of degradation to targets other than protein, which has been the focus until now, to include other cellular material, such as nucleic acids, lipids and polysaccharides. As an example, our group has recently shown bulk RNA degradation by autophagy, and that the nucleotides resulting from this process are secreted to the outside of the cell (Huang et al., 2015) (Figure 24). In order to capture such fascinating phenomena, the identification of the products that arise from autophagic degradation, the mechanism of their export from the vacuole as well as their effect on and implications for cellular metabolism have to be considered. I believe that this research will deepen our understanding of degradation and also help to characterise the ambiguous relationship between genetics, phenotype and disease.

Reflections on the nature of discovery in science

My autophagy research has always been driven by nothing more than intellectual curiosity and a thirst to get a better understanding of life through protein dynamics within the cell. When I started my work, I never thought it would become relevant in diseases as diverse as neuro-degeneration, infectious disease, cancer and others in such a short time. But now autophagy research has become a major field in biology. Figure 25 shows that the number of papers related to autophagy published in the literature continues to rapidly increase. It goes without saying

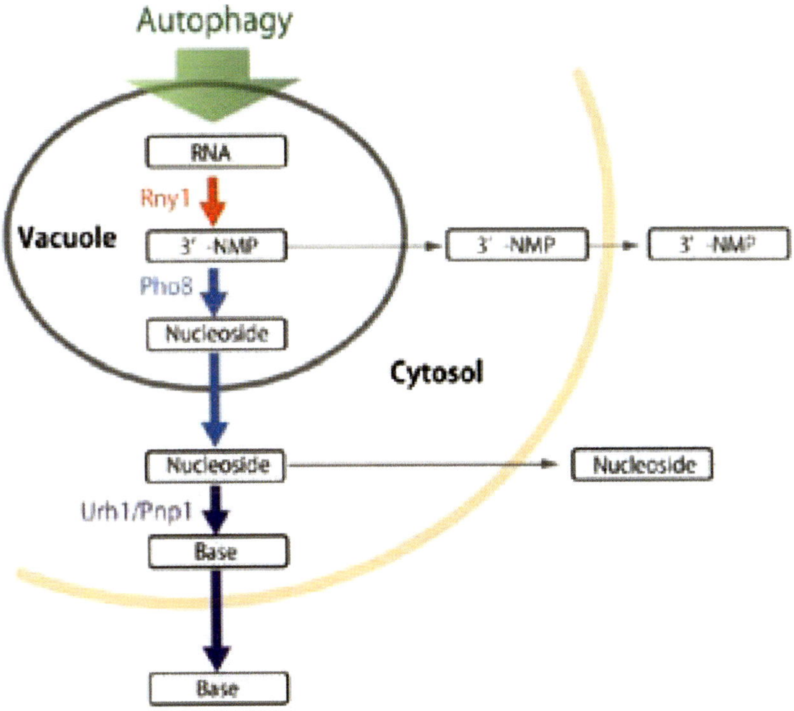

Figure 24. Diagram of autophagy-dependent RNA degradation in yeast. The standard process of RNA degradation by autophagy is indicated by thick arrows, whereas minor products are indicated by thin arrows.

that this is thanks in large part to the tremendous efforts of many researchers all over the world, so I would like to express my sincere thanks and share this honour with all of them.

But Figure 25 also tells us it takes a while to establish a new field. Truly original discoveries in science are often triggered by unpredictable and unforeseen small findings. Nowadays, the distance between basic discoveries and practical applications is getting closer. While this is exciting, scientists are increasingly required to provide evidence of immediate and tangible applications of their work. It is my sincere hope that society is able to nurture not only purpose-oriented science, but also science as a core cultural activity. If my small idea and decades of work have made a contribution to fundamental science through autophagy research, and the Nobel Foundation recognizes the basic nature of this work with this prize, it is honestly my greatest pleasure and satisfaction as a basic scientist.

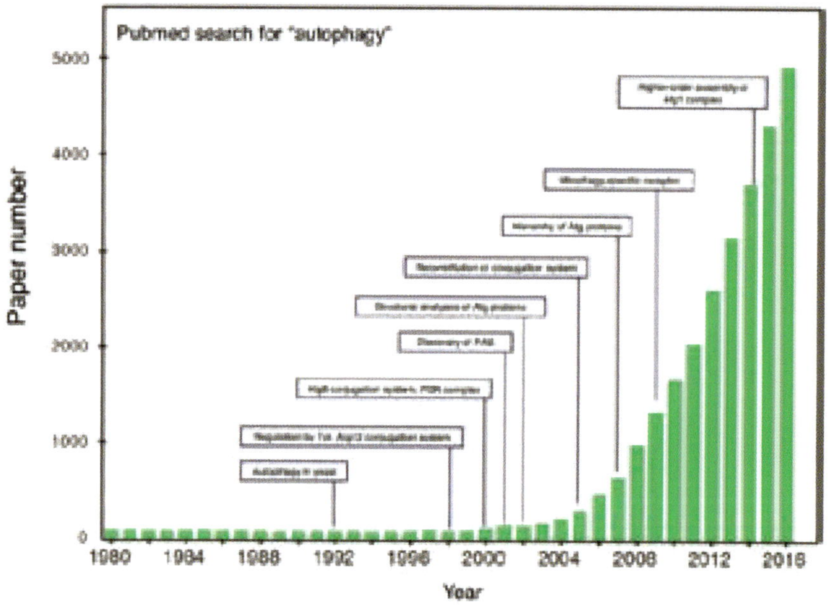

Figure 25. Number of papers published in the field of autophagy. Major developments in the Ohsumi lab are indicated in boxes for the relevant years.

Acknowledgements and closing remarks

I feel that I have been very fortunate over the 40 years of my research career. I am most appreciative for the many wonderful colleagues who I have worked with over the years. I have been lucky to work with postdoctoral researchers and graduate students who appreciate the importance of basic research. Many of them have shared my passion to study difficult projects that do not necessarily give immediate results. I am very pleased to note that many of these colleagues are now working at a range of universities as researchers. I have also been very fortunate to have met and worked with many collaborators over the years who have helped us along the way. In particular I want to note the contribution of two researchers who have collaborated closely with us for over a decade in the systematic analysis of the Atg proteins: Nobuo Noda, who now works at the Institute of Microbial Chemistry, and Fuyuhiko Inagaki, who very unfortunately passed away in 2016. I deeply appreciate their collaboration over the years.

I have only been able to make these achievements thanks to a good dose of fortune, the excellent colleagues who have joined me on this journey, the great efforts of many collaborators, good friends, the continuous support of many grants and my caring family, especially my wife Mariko. I am truly moved by your support over the years.

REFERENCES

AMANO, A., NAKAGAWA, I. & YOSHIMORI, T. 2006. Autophagy in innate immunity against intracellular bacteria. *J Biochem*, 140, 161–6.

ARAKI, Y., KU, W. C., AKIOKA, M., MAY, A. I., HAYASHI, Y., ARISAKA, F., ISHIHAMA, Y. & OHSUMI, Y. 2013. Atg38 is required for autophagy-specific phosphatidylinositol 3-kinase complex integrity. *J Cell Biol*, 203, 299–313.

BABA, M., OSUMI, M. & OHSUMI, Y. 1995. Analysis of the membrane structures involved in autophagy in yeast by freeze-replica method. *Cell Struct Funct*, 20, 465–71.

BABA, M., TAKESHIGE, K., BABA, N. & OHSUMI, Y. 1994. Ultrastructural analysis of the autophagic process in yeast: detection of autophagosomes and their characterization. *J Cell Biol*, 124, 903–13.

BARTH, H., MEILING-WESSE, K., EPPLE, U. D. & THUMM, M. 2001. Autophagy and the cytoplasm to vacuole targeting pathway both require Aut10p. *FEBS Lett*, 508, 23–8.

DE DUVE, C. 1963. The Lysosome Concept. *Ciba Foundation Symposium—Anterior Pituitary Secretion (Book I of Colloquia on Endocrinology)*. John Wiley & Sons, Ltd.

DE DUVE, C., PRESSMAN, B. C., GIANETTO, R., WATTIAUX, R. & APPELMANS, F. 1955. Tissue fractionation studies. 6. Intracellular distribution patterns of enzymes in rat-liver tissue. *Biochem J*, 60, 604–17.

FARADAY, M. 1861. *A Course of Six Lectures on the Chemical History of a Candle*, London, Griffin, Bohn and Company.

GAMOW, G. 1940. *Mr. Tompkins in Wonderland*, Cambridge, Cambridge University Press.

GENG, J., BABA, M., NAIR, U. & KLIONSKY, D. J. 2008. Quantitative analysis of autophagy-related protein stoichiometry by fluorescence microscopy. *J Cell Biol*, 182, 129–40.

HANADA, T., NODA, N. N., SATOMI, Y., ICHIMURA, Y., FUJIOKA, Y., TAKAO, T., INAGAKI, F. & OHSUMI, Y. 2007. The Atg12-Atg5 conjugate has a novel E3-like activity for protein lipidation in autophagy. *J Biol Chem*, 282, 37298–302.

HARTWELL, L. H. 1974. Saccharomyces cerevisiae cell cycle. *Bacteriol Rev*, 38, 164–98.

HUANG, H., KAWAMATA, T., HORIE, T., TSUGAWA, H., NAKAYAMA, Y., OHSUMI, Y. & FUKUSAKI, E. 2015. Bulk RNA degradation by nitrogen starvation-induced autophagy in yeast. *Embo j*, 34, 154–68.

ICHIMURA, Y., IMAMURA, Y., EMOTO, K., UMEDA, M., NODA, T. & OHSUMI, Y. 2004. In vivo and in vitro reconstitution of Atg8 conjugation essential for autophagy. *J Biol Chem*, 279, 40584–92.

ICHIMURA, Y., KIRISAKO, T., TAKAO, T., SATOMI, Y., SHIMONISHI, Y., ISHIHARA, N., MIZUSHIMA, N., TANIDA, I., KOMINAMI, E., OHSUMI, M., NODA, T. & OHSUMI, Y. 2000. A ubiquitin-like system mediates protein lipidation. *Nature*, 408, 488–92.

JOHNSTON, G. C., PRINGLE, J. R. & HARTWELL, L. H. 1977. Coordination of growth with cell division in the yeast Saccharomyces cerevisiae. *Exp Cell Res*, 105, 79–98.

KABEYA, Y., KAWAMATA, T., SUZUKI, K. & OHSUMI, Y. 2007. Cis1/Atg31 is required for autophagosome formation in Saccharomyces cerevisiae. *Biochem Biophys Res Commun*, 356, 405–10.

KABEYA, Y., MIZUSHIMA, N., UENO, T., YAMAMOTO, A., KIRISAKO, T., NODA, T., KOMINAMI, E., OHSUMI, Y. & YOSHIMORI, T. 2000. LC3, a mammalian homologue of yeast Apg8p, is localized in autophagosome membranes after processing. *Embo j*, 19, 5720–8.

KAKINUMA, Y., OHSUMI, Y. & ANRAKU, Y. 1981. Properties of H+-translocating adenosine triphosphatase in vacuolar membranes of SAccharomyces cerevisiae. *J Biol Chem*, 256, 10859–63.

KAMADA, Y., FUNAKOSHI, T., SHINTANI, T., NAGANO, K., OHSUMI, M. & OHSUMI, Y. 2000. Tor-mediated induction of autophagy via an Apg1 protein kinase complex. *J Cell Biol*, 150, 1507–13.

KAMETAKA, S., MATSUURA, A., WADA, Y. & OHSUMI, Y. 1996. Structural and functional analyses of APG5, a gene involved in autophagy in yeast. *Gene*, 178, 139–43.

KANKI, T., WANG, K., CAO, Y., BABA, M. & KLIONSKY, D. J. 2009. Atg32 is a mitochondrial protein that confers selectivity during mitophagy. *Dev Cell*, 17, 98–109.

KAWAMATA, T., KAMADA, Y., SUZUKI, K., KUBOSHIMA, N., AKIMATSU, H., OTA, S., OHSUMI, M. & OHSUMI, Y. 2005. Characterization of a novel autophagy-specific gene, ATG29. *Biochem Biophys Res Commun*, 338, 1884–9.

KIHARA, A., NODA, T., ISHIHARA, N. & OHSUMI, Y. 2001. Two distinct Vps34 phosphatidylinositol 3-kinase complexes function in autophagy and carboxypeptidase Y sorting in Saccharomyces cerevisiae. *J Cell Biol*, 152, 519–30.

KIRISAKO, T., ICHIMURA, Y., OKADA, H., KABEYA, Y., MIZUSHIMA, N., YOSHIMORI, T., OHSUMI, M., TAKAO, T., NODA, T. & OHSUMI, Y. 2000. The reversible modification regulates the membrane-binding state of Apg8/Aut7 essential for autophagy and the cytoplasm to vacuole targeting pathway. *J Cell Biol*, 151, 263–76.

KUMA, A., HATANO, M., MATSUI, M., YAMAMOTO, A., NAKAYA, H., YOSHIMORI, T., OHSUMI, Y., TOKUHISA, T. & MIZUSHIMA, N. 2004. The role of autophagy during the early neonatal starvation period. *Nature*, 432, 1032–6.

MATSUURA, A., TSUKADA, M., WADA, Y. & OHSUMI, Y. 1997. Apg1p, a novel protein kinase required for the autophagic process in Saccharomyces cerevisiae. *Gene*, 192, 245–50.

MIZUSHIMA, N., NODA, T., YOSHIMORI, T., TANAKA, Y., ISHII, T., GEORGE, M. D., KLIONSKY, D. J., OHSUMI, M. & OHSUMI, Y. 1998a. A protein conjugation system essential for autophagy. *Nature*, 395, 395–8.

MIZUSHIMA, N., SUGITA, H., YOSHIMORI, T. & OHSUMI, Y. 1998b. A new protein conjugation system in human. The counterpart of the yeast Apg12p conjugation system essential for autophagy. *J Biol Chem*, 273, 33889–92.

MIZUSHIMA, N., YAMAMOTO, A., MATSUI, M., YOSHIMORI, T. & OHSUMI, Y. 2004. In vivo analysis of autophagy in response to nutrient starvation using transgenic mice expressing a fluorescent autophagosome marker. *Mol Biol Cell*, 15, 1101–11.

MOCHIDA, K., OIKAWA, Y., KIMURA, Y., KIRISAKO, H., HIRANO, H., OHSUMI, Y. & NAKATOGAWA, H. 2015. Receptor-mediated selective autophagy degrades the endoplasmic reticulum and the nucleus. *Nature*, 522, 359–62.

MORTIMORE, G. E., HUTSON, N. J. & SURMACZ, C. A. 1983. Quantitative correlation between proteolysis and macro- and microautophagy in mouse hepatocytes during starvation and refeeding. *Proc Natl Acad Sci U S A*, 80, 2179–83.

MORTIMORE, G. E. & WARD, W. F. 1976. Behavior of the lysosomal system during organ perfusion. An inquiry into the mechanism of hepatic proteolysis. *Front Biol*, 45, 157–84.

NAKATOGAWA, H., ICHIMURA, Y. & OHSUMI, Y. 2007. Atg8, a ubiquitin-like protein required for autophagosome formation, mediates membrane tethering and hemifusion. *Cell*, 130, 165–78.

NODA, T. & KLIONSKY, D. J. 2008. The quantitative Pho8Delta60 assay of nonspecific autophagy. *Methods Enzymol*, 451, 33–42.

NODA, T. & OHSUMI, Y. 1998. Tor, a phosphatidylinositol kinase homologue, controls autophagy in yeast. *J Biol Chem*, 273, 3963–6.

NOVICK, P., FIELD, C. & SCHEKMAN, R. 1980. Identification of 23 complementation groups required for post-translational events in the yeast secretory pathway. *Cell*, 21, 205–15.

NOVIKOFF, A. B., BEAUFAY, H. & DE DUVE, C. 1956. Electron microscopy of lysosomerich fractions from rat liver. *J Biophys Biochem Cytol*, 2, 179–84.

OHSUMI, Y. & ANRAKU, Y. 1981. Active transport of basic amino acids driven by a proton motive force in vacuolar membrane vesicles of Saccharomyces cerevisiae. *J Biol Chem*, 256, 2079–82.

OHSUMI, Y. & ANRAKU, Y. 1983. Calcium transport driven by a proton motive force in vacuolar membrane vesicles of Saccharomyces cerevisiae. *J Biol Chem*, 258, 5614–7.

OKAMOTO, K., KONDO-OKAMOTO, N. & OHSUMI, Y. 2009. Mitochondria-anchored receptor Atg32 mediates degradation of mitochondria via selective autophagy. *Dev Cell*, 17, 87–97.

RAGUSA, M. J., STANLEY, R. E. & HURLEY, J. H. 2012. Architecture of the Atg17 complex as a scaffold for autophagosome biogenesis. *Cell*, 151, 1501–12.

SAKOH-NAKATOGAWA, M., MATOBA, K., ASAI, E., KIRISAKO, H., ISHII, J., NODA, N. N., INAGAKI, F., NAKATOGAWA, H. & OHSUMI, Y. 2013. Atg12-Atg5 conjugate enhances E2 activity of Atg3 by rearranging its catalytic site. *Nat Struct Mol Biol*, 20, 433–9.

SCHOENHEIMER, R. The Dynamic State of Body Constituents. The Edward K. Dunham Lectures for the Promotion of the Medical Sciences, 1942. Harvard University Press.

SCHOENHEIMER, R., RATNER, S. & RITTENBERG, D. 1939. The process of continuous deamination and reamination of amino acids in the proteins of normal animals. *Science*, 89, 272–3.

SEAMAN, M. N., MARCUSSON, E. G., CEREGHINO, J. L. & EMR, S. D. 1997. Endosome to Golgi retrieval of the vacuolar protein sorting receptor, Vps10p, requires the function of the VPS29, VPS30, and VPS35 gene products. *J Cell Biol*, 137, 79–92.

SEGLEN, P. O. & GORDON, P. B. 1982. 3-Methyladenine: specific inhibitor of autophagic/lysosomal protein degradation in isolated rat hepatocytes. *Proc Natl Acad Sci U S A*, 79, 1889–92.

SEGLEN, P. O., GORDON, P. B. & POLI, A. 1980. Amino acid inhibition of the autophagic/lysosomal pathway of protein degradation in isolated rat hepatocytes. *Biochim Biophys Acta*, 630, 103–18.

STASYK, O. V., STASYK, O. G., MATHEWSON, R. D., FARRE, J. C., NAZARKO, V. Y., KRASOVSKA, O. S., SUBRAMANI, S., CREGG, J. M. & SIBIRNY, A. A. 2006. Atg28, a novel coiled-coil protein involved in autophagic degradation of peroxisomes in the methylotrophic yeast Pichia pastoris. *Autophagy*, 2, 30–8.

SUZUKI, K., KIRISAKO, T., KAMADA, Y., MIZUSHIMA, N., NODA, T. & OHSUMI, Y. 2001. The pre-autophagosomal structure organized by concerted functions of APG genes is essential for autophagosome formation. *Embo j*, 20, 5971–81.

SUZUKI, K., KONDO, C., MORIMOTO, M. & OHSUMI, Y. 2010. Selective transport of alpha-mannosidase by autophagic pathways: identification of a novel receptor, Atg34p. *J Biol Chem*, 285, 30019–25.

SUZUKI, K., KUBOTA, Y., SEKITO, T. & OHSUMI, Y. 2007. Hierarchy of Atg proteins in pre-autophagosomal structure organization. *Genes Cells*, 12, 209–18.

SUZUKI, S. W., YAMAMOTO, H., OIKAWA, Y., KONDO-KAKUTA, C., KIMURA, Y., HIRANO, H. & OHSUMI, Y. 2015. Atg13 HORMA domain recruits Atg9 vesicles during autophagosome formation. *Proc Natl Acad Sci U S A*, 112, 3350–5.

TAKAMURA, A., KOMATSU, M., HARA, T., SAKAMOTO, A., KISHI, C., WAGURI, S., EISHI, Y., HINO, O., TANAKA, K. & MIZUSHIMA, N. 2011. Autophagy-deficient mice develop multiple liver tumors. *Genes Dev*, 25, 795–800.

TAKESHIGE, K., BABA, M., TSUBOI, S., NODA, T. & OHSUMI, Y. 1992. Autophagy in yeast demonstrated with proteinase-deficient mutants and conditions for its induction. *J Cell Biol*, 119, 301–11.

TSUKADA, M. & OHSUMI, Y. 1993. Isolation and characterization of autophagy-defective mutants of Saccharomyces cerevisiae. *FEBS Lett*, 333, 169–74.

UCHIDA, E., OHSUMI, Y. & ANRAKU, Y. 1985. Purification and properties of H+-translocating, Mg2+-adenosine triphosphatase from vacuolar membranes of Saccharomyces cerevisiae. *J Biol Chem*, 260, 1090–5.

UMEKAWA, M. & KLIONSKY, D. J. 2012. The Cytoplasm-to-Vacuole Targeting Pathway: A Historical Perspective. *Int J Cell Biol*, 2012, 142634.

YAMAMOTO, H., FUJIOKA, Y., SUZUKI, S. W., NOSHIRO, D., SUZUKI, H., KONDO-KAKUTA, C., KIMURA, Y., HIRANO, H., ANDO, T., NODA, N. N. & OHSUMI, Y. 2016. The Intrinsically Disordered Protein Atg13 Mediates Supramolecular Assembly of Autophagy Initiation Complexes. *Dev Cell*, 38, 86–99.

YAMAMOTO, H., KAKUTA, S., WATANABE, T. M., KITAMURA, A., SEKITO, T., KONDO-KAKUTA, C., ICHIKAWA, R., KINJO, M. & OHSUMI, Y. 2012. Atg9 vesicles are an important membrane source during early steps of autophagosome formation. *J Cell Biol*, 198, 219–33.

YOSHIMOTO, K., HANAOKA, H., SATO, S., KATO, T., TABATA, S., NODA, T. & OHSUMI, Y. 2004. Processing of ATG8s, ubiquitin-like proteins, and their deconjugation by ATG4s are essential for plant autophagy. *Plant Cell*, 16, 2967–83.

Physiology or Medicine 2017

Jeffrey C. Hall, Michael Rosbash and Michael W. Young

"for their discoveries of molecular mechanisms controlling the circadian rhythm"

The Nobel Prize in Physiology or Medicine, 2017

Presentation speech by Professor Carlos Ibáñez, Member of the Nobel Assembly at Karolinska Institutet, Member of the Nobel Committee for Physiology or Medicine, 10 December 2017.

Your Majesties, Your Royal Highnesses, Esteemed Nobel Laureates, Ladies and Gentlemen,

Phileas Fogg, the main character in Jules Verne's acclaimed novel *Around the World in Eighty Days*, could not have suffered from jet-lag during his trip, despite crossing multiple time zones. His body had plenty of time – more than 3 days per time zone – to get adjusted to the time differences encountered along his journey. Today, in the era of jet travel, we can cross several time zones in only a few hours; but our bodies suffer, as they struggle to adapt to the new time at our destination. Many of our foreign guests this evening are surely experiencing this now. Why can't our physiology adapt more rapidly? What keeps it behind?

Our physiology is regulated by an internal clock that generates daily rhythms known as "circadian", from the Latin *circadiem*, meaning "around one day". Circadian rhythms are ancient and exist in all forms of life. Life on Earth is adapted to the rotation of our planet, and the internal clock anticipates day/night cycles, helping organisms optimize their physiology and behavior. Although the existence of a biological clock has been known for nearly a century, only recently have we begun to understand what it is made of and how it keeps ticking.

Our story begins in 1729, when French astronomer Jean-Jacques de Mairan took a mimosa plant, which leaves are open during the day but close at night, and placed it in constant darkness. He observed that the leaves still opened and closed rhythmically at the appropriate time, suggesting an endogenous origin of the daily rhythm. Physiology is controlled by genes, and the biological clock is no exception. In 1971, Seymour Benzer and Ron Konopka isolated mutant flies that had alterations in their normal 24h cycle of activity. Fifteen years later, Jeffrey Hall and Michael Rosbash, working together at Brandeis University in Massachusetts, and Michael Young, at Rockefeller University in New York, isolated the mutated gene, called *period*.

As instrumental as this was, however, the isolation of the *period* gene did not tell very much about the mechanism of the biological clock. It was a remarkable series of discoveries made during the 1990s by this year's Nobel Laureates that finally elucidated how our biological clock ticks. The basic principle, first proposed by Jeffrey Hall and Michael Rosbash, is deceptively simple. The *period* gene produces a protein that accumulates in the cell and, after reaching a certain level, blocks the gene and hence its own production. As protein levels subside, the gene becomes active again and the cycle resumes. As many things in biology, however, the devil is in the details; as it was still unclear how the *period* protein can be stabilized long enough and then enter the cell nucleus to inhibit its own production. Michael Young discovered two additional genes, he named them *timeless* and *doubletime*, that partner with *period* and together contribute to the generation of robust oscillations of approximately 24hs.

The discomfort of jetlag is evidence of the strength of our biological clock, as it takes time for the machinery to readjust to a sudden change in environmental conditions. Although sunlight is scarce this time of year in Stockholm, the good news is that food is also a strong resetting stimulus, so the banquet that follows after this ceremony will surely help towards adjusting our internal clocks.

The 2017 Nobel Laureates have uncovered a mechanism controlling a truly fundamental process in physiology, how our cells and bodies keep time. Such time-keeping is essential for our adaptation, and has important implications for human health; not just jetlag, but also the incidence of chronic syndromes, such as cancer, metabolic and sleep disorders, and several neurological conditions.

Professors Hall, Rosbash and Young:

Your brilliant studies have solved one of the great puzzles in physiology. Your discoveries have unraveled the cogs and wheels of the biological clock, an essential mechanism for the survival of life on our planet.

On behalf of the Nobel Assembly at the Karolinska Institute, I wish to convey to you our warmest congratulations. May I now ask that you step forward to receive the Nobel Prize from the hands of His Majesty the King.

Jeffrey C. Hall. © Nobel Prize Outreach AB. Photo: A. Mahmoud

Jeffrey C. Hall

Biography

JEFFREY C. HALL WAS BORN in Brooklyn, New York, near the end of World War II (in Europe). His parents, fortunately for him, were among rare young adults in the U.S. who achieved college educations during the Depression. Hall's father used his higher education credentials to become a journalist, his mother a school teacher. These are arguably mindful vocations, and they promoted a mindful atmosphere in Hall's home when he was growing up, without there being any indoctrination and probably without Hall himself necessarily being aware of this salutary environment.

Eventually, what Hall's father achieved vocationally caused him to work in Washington, D.C., covering the United States Senate along with presidential campaigns. Thus his offspring were raised mostly in a Maryland suburb of Washington. The aforementioned intra-home atmosphere (enhanced by how interesting it was for Hall to absorb information from his father about politics, society, and their historical contexts) made it axiomatic that he and his two siblings would attend college. Hall did so, beginning in 1963, and became a Biology major at Amherst College (Amherst, Massachusetts). A key element of this experience stemmed from his desire to do "Senior Honors" research. This caused Hall to be assigned, as his Honors supervisor, Dr. Phillip T. Ives. The latter had, by then (mid-1960s), been a longstanding *Drosophila* geneticist. Though Hall was unaware of the following during his college stint: Ives was a distinguished such geneticist then and later, as Hall learned during his post-undergraduate time.

In any event, Ives was an excellent mentor, who not only instructed his small number of undergrad supervisees superbly, but also imbued them with a fervent interest in basic research generally and *Drosophila* genetics in particular. As Hall was performing a low-level genetics project at that college, his Biology Department superiors (including Ives) recommended that he try to become a graduate student at the University of Washington in Seattle. This advice came Hall's way in the context of an incipiently well-regarded Genetics Department having been established at "U-Dub" (W = double U). Hall took heed of those college-based recommendations and enrolled at U-Dub in 1967. Soon after joining the Genetics Department there, he joined (in turn) the laboratory of Prof. Larry Sandler. The latter was an excellent *Drosophila* geneticist, who like Ives happened to be a direct descendant of Thomas Hunt Morgan. Morgan, along with his students at Columbia University in New York, were pioneers who founded, sustained, and expanded the fruit-fly genetics "system," during the 19-teens and subsequent decades.

Within U-Dub's Genetics Department, Sandler's – and Hall's – leader was Professor Herschel Roman, founder and longstanding Chair of that department (1959–1980). Roman fostered departmental norms that promoted high-quality instruction, training, and mentoring. In this regard, "Hersch" was well acquainted, professionally and personally, with almost all members of his department. The interest he took accordingly caused him to pull Hall aside, albeit not by singling him out. In any case, Hersch recommend that Hall try for a postdoctoral position in the laboratory of Seymour Benzer, California Institute of Technology (CalTech, a.k.a. CIT) in Pasadena where, more than incidentally, Morgan's lab had moved from New York in the late 1920s. For Benzer's part, by the early 1970s, he had established a second career, after initiating his genetically-based vocation via "pure" genetic studies of microbes. Along with Seymour *also* moving to CalTech in the mid-1960s, he shifted his interests and activities into the nascent sub-field of behavioral-cum-neuro-genetics. Benzer chose *Drosophila*, possibly influenced by the famed "fly group" that had long been ensconced there. Morgan was dead – as was one of his famed students, Calvin Bridges (Columbia -> CalTech) – but the equally famed Alfred Sturtevant remained alive, as was the latter's student (then CIT Professor) Edward Lewis.

Back to Roman's intra-departmental office at U-Dub: That Chair's recommendation – in person to Hall *and* on the telephone to Benzer – about the former doing a postdoc in the latter's lab led to Hall joining that CIT group during the late summer of 1971.

For a few months thereafter, Hall was fortunate to overlap with one of Benzer's senior grad students, Ronald (Ron) J. Konopka. That investigator had made himself into a "chrono-geneticist," taking a genetic approach to study biological timing, viz. daily rhythms. Thus, as Konopka was completing his thesis project (PhD, 1972), Hall became vividly aware of what Ron had been doing and spectacularly accomplishing: Induce from scratch, via application of a chemical mutagen, novel mutants in *Drosophila melanogaster* that would potentially manifest abnormalities or anomalies of such rhythmicity (including that which is exhibited at the level of fruit-fly adult behavior: rest/activity cycles, normally manifested via ca. 24-hour cycle durations). Ron's mutant hunting, followed by high-quality analysis of the rhythm-based phenotypes *and* mutational genotypes, was stunningly successful. This was based in large part upon Konopka inducing and identifying a magnificent trio of novel mutants: one that displayed only 19-h daily cycles (in constant darkness), a 29-h mutant, and a third (the original) that was arrhythmic, a.k.a. aperiodic. Ron named these variants *period* mutants, formally justified, for he demonstrated that each of the three mutations involved newly induced changes within *one D.* melanogaster *gene* (famously abbreviated *per* and pronounced "purr" as opposed to "peer").

During these heady days of the early 1970s, Hall could not help become a fan of Konopka's research, owing to the findings themselves, and against a recent background that caused him previously to become aware of daily rhythmicity in *Drosophila*. Exposure to the relevant phenomena – including that fly cultures maintain daily rhythms in constant darkness, in which condition they display what are known as a "circadian" rhythm" – occurred when Hall was a student in a college course; its instructor happened to include a module about circadian rhythms manifested by "emerging flies" (metamorphosis -> adulthood = "eclosion"). That Assistant Professor at Amherst College had recently completed his PhD thesis research in the laboratory of Colin Pittendrigh (Princeton University, New Jersey), whereby the latter was becoming one of the "grand old men" of rhythm-related research. However, neither Pittendrigh nor anyone else had brought any definitive *Drosophila* genetics to bear on studies of these eclosion rhythms. At all events Hall was fortunate, by coincidence, to be attuned to the way that Ron Konopka pioneered a genetic approach to asking "what is a circadian clock?" That quoted phrase alludes to the notion that Konopka's novel variants seemed for all-the-world to be *clock* mutants, as the relevant publication, co-authored by Benzer, was entitled. Implicitly, if a mutation (actually two) can change the circadian-cycle duration in a constant environmental condition, that smacks of a "pacemaker" problem. The noun just quoted had been invoked in context of organisms of all kinds displaying daily rhythmicities of all kinds, including behavioral; and that such biological (physiological, biochemical, etc.) cycling persists in the absence of daily earth cycles, notably light:dark ones. Implication: Central pacemakers harbored *within organisms* underlie such rhythmicity. This kind of pacemaker can be regarded as synonymous with "the circadian clock," as it was initially inferred to exist via a plant experiment (-> animal studies, including of invertebrates and mammals, -> microbes as well).

Aside from the rhythm-related sub-enterprise extant within Benzer's lab, Hall's own research there from fall 1971 to the end of 1973 did not involve time-based phenomena. Instead: neurochemical ones as well as a project involving genetic "mosaics." The latter genotype entailed *Drosophila* that were each part-male (one X chromosome)//part-female (XX) and with the two kinds of chromosomal genotype marked phenotypically via "histochemical" genetic marking (one-X cells, including CNS neurons, unstained for an enzyme reaction; 2X ones stained). Hall's mentor, trainer, and co-worker for these projects was a fellow postdoc in the laboratory of Benzer, whose supervisory actions tended toward *laissez-faire,* named Douglas Kankel (Brown University, Providence, Rhode Island -> CIT, -> Yale University, New Haven, Connecticut). Doug taught his labmate a whole lot about *Drosophila* biology, including neurobiology, against a background of Hall having previously done genetic (qua genetic) studies alone. The latter did manage to bring to his pair of Benzer-lab projects genetic expertise, involving a heavy dose of *chromosome manipulations* in the context of many types of such being afforded by the "lore of *Drosophila,*" harking back to Morgan and his students.

Shortly after Hall's postdoc stint entered its third year, by which time his co-worker Kankel had moved to Yale as noted above, he received invitations to interview for Assistant Professorships at two U.S. universities: U. Missouri (Columbia) and Brandeis U. (Waltham, MA). These invites were promoted by none-other-than Herschel Roman, who had recently visited those two institutions and recommended that his former mentee be considered for faculty-level jobs. So Hall can never forget nor fail to appreciate how meaningful to him was Prof. Roman's career-sponsoring support (1967–1971; then intra-1973).

After traveling for the interviews just referred to, Hall received job offers from both of the universities in question, initially from Brandeis. That offer led him to begin an Assistant Professorship there, winter 1974. Hoping to get some research going there – at this near-Boston locale – he continued the pair of projects that had been maturing at CalTech around the time when Kankel and then Hall moved across the United States from Southern California (U.S.). This collaborative association continued for a while, "Back East," facilitated by both fledgling faculty members being located in New England. These two former Benzer lab members put forth publications presenting the neurochemical-genetic and nervous-system mosaic findings in 1976. Yet Hall and Kankel had published zero primary papers during their postdoc stints, during an era when that kind of ostensible non-productivity could nonetheless lead to faculty jobs; nowadays, applications for such must be accompanied by massive publication-ridden CVs.

One aspect of Hall's lab studies at Brandeis University involved *courtship* behavior in *Drosophila.* The starting point was to observe and quantify courtship capacities of the aforementioned sex mosaics: Which portions of the neuro-histochemically marked CNSs had to be genetically male (or female) if a given mosaic would perform one or more elements of the sex-specific, courtship-behavioral sequence? This category of behavioro-neuro analysis proceeded to a merger of genetic-mosaic principles and practices with neurochemically *disrupting* mutations, stemming from the other project Hall and Kankel had performed, starting at CalTech. The rather complex "mosaic dissection" experiment in question, based at Brandeis, was initiated in Hall's lab by him and a postdoc who joined it during the late 1970s: C.P. (Bambos) Kyriacou. A key reproductive-behavioral phenotype recorded during this study was a male-like courtship song (normally produced by a standard XY fly's wing vibrations, put forth when he follows a female, ramping up toward eventually attempting to mate with her). Bambos's and Hall's question: Which types of mosaics – set up to be each all-male, but with some tissues neurochemically mutated, others normal – might sing abnormally, putatively correlated within intra-CNS locations of the neurochemical deficit? But this project ended up dying on the vine, because of the following: Kyriacou and Hall also recorded (with microphones and magnetic tape) the singing behavior of *control* males: siblings of the mosaics, whereby such XY controls were uniformly normal for their neurochemistry. Analyzing visually appreciable renditions of the auditory recordings led to their inability to discern supposedly canonical "song parameters," previously but cavalierly reported by the early song recorders, who had been working mostly in the U.K.

By virtue of enhanced labor, Bambos broke down and analyzed the entirely of each several-minute recording; thus he pulled-out long series of relevant computations, concentrating on a key song parameter: rate of tone-pulse production (per series of 10-second bins). He therefore discerned that that singing element seemed to be fluctuating systematically. Further analyses revealed, indeed, that the rate in question oscillated rhythmically: speeding up, slowing down, speeding up again; with a cycle duration of about one minute for *D. melanogaster* free of behavioral/neural mutations. Next step, during the micro-era in question (late 1970s): Kyriacou and Hall wondered whether the only known rhythm-affecting mutations in "our" species might by-some-chance alter song rhythmicity. Hall, especially, knew that those variants were Ron Konopka's *circadian* mutants (n=3, harking back to the latter's PhD thesis project). So Hall wrote Dr. Konopka, asking for culture copies of Ron's mutants, even though the Brandeis researchers were aware that circadian rhythmicity and song such are defined by cycle durations three orders of magnitude different from each other. Yet, Konopka's *per* mutations – two of which caused altered cycle durations, the third causing arrhythmicity – were found to alter courtship-song cycling in ways paralleling effects of these genetic variants on daily rhythms.

So Hall's admiration for Konopka's chrono-genetic accomplishments and a coincidental matter of the former being instructed in a college course about biological rhythms in *Drosophila*, prompted that lab head at Brandeis, and others there, to enter an arena defined by actual chronobiological lab work. Meanwhile, a close colleague of Hall in Waltham, MA – then Associate Professor Michael Rosbash – had previously become aware of the Konopka mutants. After Hall and Rosbash met during 1974, the former could not help mention that banner study (emanating from his former postdoc lab) and speak highly of it; even though neither Hall nor Rosbash had any mid-'70s interest in doing anything about *per* mutants or the gene defined by them. Nonetheless, Rosbash also became well acquainted with Bambos Kyriacou, as the second half of the 1970s unfolded, enhancing the former's appreciation for the *period* gene's existence.

A few years later in the early 1980s, an experimental question suggested itself, whereby the *per* gene would have to be identified and isolated at the DNA level.

To make a medium-length story short as to how that question arose: We fantasized about "cloning *per*" from *D. melanogaster* (*mel*), using that DNA readily to do the same for *per* in the close interspecific relative *D. simulans* (*sim*), then transferring the latter into *mel* to ask whether such a single-gene infusion would bring with it regulation of *sim*-like singing rhythmicity. As of the early '80s, Kyraicou and Hall had found that *sim* and *mel* males generate species-specific song-cycle durations *and* that the genetic etiology of this difference mapped to the same chromosome on which *per* is located.

Therefore, and initially *not* based on a specific interest in *circadian* rhythmicity in *Drosophila*, the researchers at Brandeis with material help from Kyriacou and from Konopka himself, set out to isolate *period*-locus DNA and identify it via behavioral bioassays: introduction of putative *per* DNA into *per*-mutant "hosts" carrying the arrhythmia-inducing mutation. This new project was rooted in an incipient, close collaboration between the laboratories of Hall and of Rosbash at Brandeis, augmented by the two sending behaviorally-pertinent strains to professional locations where the "two K" guys had ended up as of the mid-'80s.

This multi-lab effort eventually ballooned *as* its interests shifted mainly into the *daily*-rhythm arena. The researchers began to so study via multi-pronged approaches: behavioral genetics, cyto-genetics (application of chromosome aberrations), molecular genetics, and neuro-genetics. For Hall's and Rosbash's part, it was meaningful that they had become close not only professionally but also personally: respectively, related to co-awareness that Hall was a generic fruit-fly geneticist; and Rosbash was a generic molecular biologist, initially studying vertebrates and yeast from the mid-'70s into the '80s. In addition, these two faculty members shared various personal interests, mostly revolving round low-culture stuff. Might their association – including separate, seemingly complementary, scientific backgrounds – promote a fruitful joining of forces: chrono-genetics deepening into the molecular-genetic area?

As this bi-lab collaboration at Brandeis got started and proceeded, the two were operating in competitive parallel with *per*-molecular studies performed in the lab of Mike Young (Rockefeller University, New York). Mike had become at least marginally interested in circadian rhythms affected by *period* mutations back when he was a graduate student in the 1970s, mainly studying other bio-genetic phenomena in *Drosophila*.

The *tri*-lab deal – a pair of collaborating groups plus one competing with the former two – began to work out. As of approximately the mid-'80s, a quartet of publications (two from Massachusetts, two from New York) reported *probable* cloning of *per*, followed by nailing that matter via DNA-mediated "rescue" of the relevant *per*-mutational effects, as alluded to above. Hall maintains that a companion piece in this volume, based on his Nobel Prize lecture in late 2017, usefully recounts sufficient additional features of these early-days stories, along with presenting the requisite background information from the 1970s and earlier.

No doubt the lecture-based articles by Profs. Rosbash and Young will flesh out all additionally relevant aspects of this history. Meanwhile, some quasi-editorial remarks, partly coming under the header of Hall tooting his own biographical horn; as well, what follows may usefully divulge some historical stuff, referring to the post-'80s era: In this respect, and in Hall's experience plus opinion, it is possible to get carried away with "the primacy of *per*": yes, the first clock gene cloned in any organism. Gene #2 of this type was identified at that level in *Neurospora* five years later. Could one clock factor, the PER protein, tell the whole fruit-fly story insofar as the circadian-pacemaker mechanism was concerned? A priori, no. Realizing this, Hall and associates such as Rosbash and Young "did Konopka's" of their own, spearheading renewed searches for rhythm-related genes in *Drosophila*, whose products could be gleaned to contribute materially to said mechanism. Such mutant hunting was set-up to involve chromosomes extending beyond the one where *period* is located.

As elaborated citationally within Hall's "lecture article," this tri-lab endeavor led to identification of Young's *timeless* (*tim*) gene, which encodes a PER companion; and *doubletime* (*dbt*), whose encoded enzyme influences the dynamics of PER protein "cycling." Now *per* and *tim* must be transcriptionally activated, of course, before their first-stage products (mRNAs) can go up and down each day. Indeed, although perhaps anti-climactically, the Hall/Rosbash crew – including several valued co-workers (grad students, post docs, and others) – induced behavior-arrhythmia-inducing mutations at loci named *Clock* (*Clk*) and *cycle* (*cyc*), which also lead to very low levels of *period* and *timeless* products. Neither *Clk* nor *cyc* mutations (loss-of-function "alleles") kill developing *Drosophila*; same for *per* or *tim* "nulls," but unlike *dbt* ones, the latter gene being a developmentally vital one with pleiotropic effects. Thus, *Clk* and *cyc* – whose transcrioton-factor products co-associate to turn-on both *per* and *tim* transcription -- can be regarded as semi-dedicated to rhythm-regulating processes.

Discovery of all these core clock factors in *Drosophila* was rooted in mutant hunting. Yet, extending beyond that core, circadian clocks must also receive inputs from the environment, e.g., so that these only *circa*-dian pacemakers can be subjected to daily re-sets, thus underpinning 24.0-hour cycle durations in natural conditions. One conspicuously acting input factor in *Drosophila* came to the fore thanks to yet another hunt for mutants performed by Hall and co-workers, which resulted in identification of a light-absorbing molecule called CRY. It is encoded by the mutationally defined *cry* gene, whose encoded protein – when activated by photic stimuli – "touches" TIM to promote the latter's degradation during the falling phase of *tim*-product cycling. Now the core clock also has to do more than spin its internal wheels, plus be sensitive to external stimuli. Therefore *output pathways* must project from central-pacemaking functions, ultimately to mediate revealed rhythmicity (behavior, physiology, and much more overall). Starting with performance of *molecular*-genetic tactics, many output-gene candidates in *Drosophila* were uncovered at Brandeis and Rockefeller, plus within farther-flung research groups. One such gene, which encodes a brain neuropeptide, comprised a fruit of such searching at Brandeis; contributed to gaining insights about "outputs from the clock;" then was exploited molecular-genetically, neuro-genetically, and behaviorally by Hall's research group and several others. The latter came to be composed of an ever-expanding venture, extending well beyond investigative activities occurring originally at only a small number of institutions. This swelling phenomenon included a variety of "other-directed" investigators, referring to how their careers started, dropping much of what they had been doing in order to start studying *Drosophila* chronobiology and that of other-organismal rhythms.

Hall could not help summarize elements of this post-*per* research, although the meaning of that seminal gene continued to be elucidated well past the mid-1980s. At a minimum, successful searching for further factors – notably at Brandeis and Rockefeller during the 1990s – signifies that these (dare we say) molecularly pioneering labs were serious about their hope to flesh-out rather robust understandings of the overall rhythm-related deal.

By analogy to entering what he just did, Hall comes near the end of this biog by inserting a partly personal coda: He has long regarded the *genetic* side of the overall enterprise to be extremely consequential, exemplified by what was just outlined within recent passages: so many "clock players" found in *Drosophila* via pheno-genetic screenings. So the attack on fruit-fly chronobiology has had a lot to do with variant genotypes and associated *pheno-genetics* (to re-invoke an arcane term, meaning rhythm-related effects of genic variants, chromosomally based ones, and molecular "clones"). Conversations between Hall and Rosbash, and among other associates, have stimulated Rosbash to say that "the molecular biology made all the difference."

Fair enough: Truly no one could have anticipated, for example, what *period*-gene products are about, absent "cloning *per*" then empirically analyzing the encoded RNA & protein (involving way more than sequencing the gene and describing PER protein on paper, hoping forlornly that that description alone would divulge much at first-blush). *In addition*, genotypic and phenotypic elements of the eventual extravaganza (momentarily factoring out molecular matters as a *gedanken* consideration) have been rate-limiting investigatively, in Hall's opinion. For his part, and in order potentially to "do anything" chrono-wise, he relentlessly sensed appreciation for genetically based interactions with and mentoring by his early-career associates: Ives, Sandler, Roman, Lewis, Kankel, and Konopka.

If Hall had not been fortunate enough to come straight out of the T.H. Morgan tradition of *Drosophila* genetics, he wonders in retrospect whether he could have contributed to the overall behavioral/neurobiological/chronobiological enterprise in some sort of meaningful manner. Prof. Young *might* sense something similar. Although he was not a molecular-geneticist in his early pre-postdoc days, he too is a direct descendant of Morgan, via Young's PhD research in a laboratory headed by an academic great-grandson of a fruit-fly-genetics pioneer. Hall has noticed, for instance, that Young merged with élan his molecular-genetic expertise with *Drosophila* cyto-genetics (matters revolving round the latter having been absorbed by that student of the fruit-fly in advance of Young beginning to focus heavily on biological rhythms and circadian clocks). Maybe Hall, as well, was able to tap into the aforementioned "lore of *Drosophila*," influencing ways that he helped sustain the chronobiological endeavor at hand, in part via some sort of genetic diligence.

The Little Flies and their Genes:

How Research Exploiting *Drosophila* has led to Prize-Winning Advances in Genetic and Biological Knowledge, all the way to uncovering Mysteries of the Circadian Clock

Nobel Lecture, December 8, 2017 by
Jeffrey C. Hall
Cambridge, Maine, ME, USA.

THIS ESSAY WILL PAY HOMAGE to basic research. Context: The 2017 Nobel Prize in Physiology or Medicine (P or M). For this, the word "or" alludes to the other noun (P) meaning essentially *anything* biological, irrespective of any apparent M relevance. All bio-phenomena, at base, are "physiological," which is so even for genetics (how cellular processes mediate replication of the genetic material and transmission of it from one generation to another: cellular and parent/offspring).

The subject matter that follows will have much to do with a subset of the basic-research enterprise: genetics, in part from historical perspectives, both generic and personal. The former adjective alludes to putting rhythm-related genetic studies in contexts extending, at times, beyond "the little flies;" the latter adjective points to 20th and 21st century developmental-, behavioral-, neural-, and molecular-genetic matters revolving round the fruit-fly *Drosophila melanogaster.*

With regard to the foregoing introductory passages, the 2017 trio of P or M awardees performed research that was deeply rooted in genetic and molecular-biological studies of *Drosophila*. The deep background of research in those related arenas led to knowledge and materials that were crucial for establishing and sustaining research enterprises in what came to be called "chrono-genetics:" studies of daily biological rhythms from genetic (mutational and molecular) perspectives.

I invoked the adjective "basic" up-top in order to claim that a Nobel Prize in the biological field need not necessarily have any practical significance, at least early on, such as impacts connected with "Medicine". That said, huge numbers of advances made on behalf of basic research (in scientific areas well beyond biology) turned out to be extra-interesting, in the sense that *applications* of fundamental findings can be exemplified and re-exemplified.

The immediately foregoing remark prompts a mention of "model systems," such as the *Drosophila* one. To decode the phrase just quoted: Potentially studying something or other in a "lower" organism, for example, might be a "model" for elucidating processes manifested other species, including "higher" ones such as mammals and us. In this light (or cynical dark), research-grant-proposal boilerplate often includes passages about hoped-for *relevance* of studying whatever in a given model system. Another point of view: Research findings can be appreciated for their own sake, involving new knowledge about an Earth-based life form in and of itself. This attitude belies any necessity that basic-research findings will somehow be applicable somewhere else, organismically. Thus, with regard to oh-so-many times I have been asked how and why is *Drosophila* a good model system, I eventually came to respond: "The genetics of *D. melanogaster* is a fine model system for studying hereditary phenomena in *Drosophila*."

Studies that purport, in contrast, to point toward practical applications come under the heading of what is called, nowadays, "translational research:" Can certain findings be rather quickly translated into scientific activities that promote human welfare, for example? I cannot claim that "non-translational" research is just as good, in terms of promoting even eventual applications. I do claim, with no compunction and high-mindedly, that essentially all categories of research have the positive potential to generate new knowledge per se, whether or not such is quickly (or ever) perceived to be especially interesting or useful beyond the newness itself.

Approaching matters revolving around daily biological rhythms in *Drosophila*, studied genetically, I have firm memories of the early days during which such research was initiated then limped along for a fair number of subsequent years. Few, perhaps none, of the relevant researchers could have predicted the broadly and deeply based implications of such chrono-genetic findings. Such expansions were based in part on ways that "non-drosophilists" went to school on the fruit fly factors and phenomena, thus extending the ostensible "insect-only" significance of what the 2017 awardees contributed, by applying *Drosophila*-based potential to analyze non-insect phenomena via genetic and molecular approaches.

Backing up to elements of my own career and how it connected, however loosely, with the history of genetic research performed using *Drosophila*, these endeavors turned out to generate biological findings with robust significance for understanding many ways that animals are formed, then function. Such accomplishments, harking back to the first half of the 20th century and unrelated to yours truly, were even deemed worthy of Prize awardings. (Stay tuned.) As for this one of the 2017 awardees, I began to get my feet wet in *Drosophila* genetics as an undergraduate. During about 40% of that stint, I performed a pair of low-level genetic projects. They had to with what-I-call "genetic genetics:" studying hereditary phenomena per se, with essentially zero general biological implications. I carried out these projects under the supervision of a *Drosophila* geneticist named Phillip T. Ives. As I learned later, Ives was a direct descendant of Thomas Hunt Morgan and Alfred Sturtevant, two of the giants in genetic history (Stay tuned.) Accordingly, as I registered in retrospect, Ives not only genetically instructed his handful of undergrad charges, he also imbued them with "the lore of *Drosophila*:" how that system was founded, expanded, and sustained over several decades, meaning from about the 1910s to what was then the mid/late 1960s.

As previewed, Ives performed his PhD-earning research at the California Institute of Technology (CalTech, a.k.a. C.I.T., Pasadena, CA, U.S.) under Prof. Alfred Sturtevant. The latter was an undergraduate then graduate student under the aforementioned Morgan, whose small lab – then at Columbia University, New York City, U.S. – had single-handedly founded *Drosophila* as a system in which one could elucidate genetic factors and events. This came down initially during the 1910s; then and later Morgan and his students, with them further functioning as alumni, generated vast amounts of generally important genetic knowledge. Not for nothing, they also discovered and created large numbers of genetic variants. Many of them have been maintained ever since, and several of them became crucially applicable by subsequent geneticists and bio-geneticists (For the latter, again stay tuned.) Owing to all this pioneering research, and supervision thereof, Morgan was awarded the Nobel Prize in 1933.

Several grad students working under him earned their PhDs via fruit-fly genetics. A trio of Morgan's "F1" (genetic jargon) became the best-remembered members of his group at Columbia and then CalTech: Sturtevant, along with Calvin Bridges and Hermann Muller. Who cares? Or why am I going on about these investigator-based histories? Well, elements of what Morgan and his F1 achieved turned out to be monumentally relevant to the performance of many subsequent studies, notably in the aforenoted area of bio-genetics (taking a genetic approach to study any number of biological processes), including *chrono*-genetics. Such instances of historical significance, catalyzed by Morgan and crew, extended well beyond the meaning of their genetic-genetic findings, hugely impactful though they were (e.g., Sturtevant, 1913; Bridges, 1916; Muller, 1927). It was an aspect of Muller's genetic career that caused him to be a Prize awardee (1946). Yet what Sturtevant and Bridges achieved was equally meaningful, at least in terms of ways that materials they established had whopping heuristic value down the decades. For example, the latter geneticist and *cyto*-geneticist observed and analyzed the chromosomes (viz. "cyto-") of D. *melanogaster* and thus established highly-resolved descriptions of those sub-cellular organelles. These findings (e.g., Bridges, 1935) held many drosophilists in good stead later on, as will be exemplified later in a chrono-molecular-genetic context.

Such chromosome-level observations often occurred in the context of Muller and his labmates generating genetic variants at, indeed, the chromosomal level. One category of such came to be called "balancer" chromosomes (Lindsley and Grell, 1968a). By definition – yet with apology for now diving into the cyto-genetic weeds, but importantly so I think – a balancer (here abbreviated *Bal*) harbors inverted (*In*) chromosomal segments along with a dominant (*Dom*) marker mutation. Now *In*'s, over an homologous chromosome in normal (*norm*) sequence, block meiotic recombination between such homologs. Thus, an offspring from a *Bal/norm* parent that does *not* exhibit effects of the *Dom* marker must carry the entire "norm" (normal) chromosome. The latter, however, might not an environmental agent having mutagenic effects (e.g., Muller, 1927; with other later, germane examples to follow).

Balancer chromosomes – which are essentially unique to *Drosophila* – are far from the only type of chromosome-level anomaly extant in the fruit fly system. Another example: the "attached-X," meaning two X chromosomes in a *D. melanogaster* female attached to one centromere, vs. standard "free Xs," each attached to their own centromere. The first case of an attached-X was discovered, as a spontaneously occurring sport (read: anomalous organism, possibly a genetic variant) by Thomas Hunt Morgan's better half (Morgan, 1922). Owing to principles & practices established previously by Bridges (1916), attached-X-over-Y females (the Y chromosome being *not* sex-determining), mated to standard males (XY), is a "true-breeding" situation: All live offspring are attached-X/Y females and XY males. The latter receive their one X from *fathers*, the reverse of "criss-cross" inheritance occurring in a standard free-Xs x XY mating (cf.

Morgan, 1922). Other types of zygotes produced from an attach-X x standard-male mating – triple-X and YY – carry *lethal* genotypes; more about those, more generally, later.

The attached-X deal established a most powerful cyto-genetic tool, applied again and again later to initiate various bio-genetic enterprises, including a chrono-genetic one. (Stay tuned.) In this regard, Muller himself designed and generated one of the now-classical and hugely-useful attached-X "strains." He published this accomplishment, aimed at establishing a particularly stable attached-X type and item, in a tiny research note (mid-1940's, albeit unrelated to the Prize that came his way then; see Lindsley and Grell, 1968b). The Mullerian strain in question – formally named *C(1)DX, y f* – got scattered all over the *Drosophila* community and held several subsequent investigators in good stead, as previewed briefly above.

An insert and disclaimer about *Drosophila*-genetic nomenclature: *C* (previous paragraph) always stands for Compound Chromosome, meaning 2 usually free chromosome "arms" attached to one centromere; *(1)* refers to the X, that is the first chromosome; *DX* means double-X; *y* and *f* are, respectively, abbreviations for the recessive X-chromosomal mutations *yellow* (body color) and *forked* (bristle morphology), with regard to which Muller's attached-X is homozygous (viz. *y f / y f*). It *might* be believed that there is some method to this nomenclature madness, whereby genotypic designators are invoked according to putatively comprehensible norms; those include designating inverted chromosomes in part with *In*, followed by the X or autosome number in question.

Back to me and early stages of my career. My grad student supervisor was Larry Sandler (LS), who like Ives was a superb mentor. Examples: (1) I learned from LS how well-regarded a geneticist was Ives. (2) Sandler taught his mentees about the deep history of *Drosophila* genetics, involving and specifically invoking the noun "lore", both in terms of *genic* variants and chromosome-level ones. (3) For the latter genotypes, we were informed about not only the properties of chromosome aberrations, but also how they were designed, constructed, and analyzed.

One thing that I learned from Sandler – the academic grandson of Sturtevant, by the way – was how well his "grandfather" had analyzed the meiotic behavior of inverted chromosomes. Elements of those analyses were co-performed by another to-become-famous geneticist. (See below.) It was notable that, when Morgan moved from New York to Southern California during the late 1920s, Sturtevant along with Bridges (but not Muller) moved with him. This move promoted founding of the "Division of Biology" at CalTech, whose ever-enhancing fame was to come, including with respect to pioneering chrono-genetics.

Speaking of the intellectual, research-related "line" at hand, another student under Sturtevant at CalTech in addition to Phil Ives was Edward (Ed) Lewis. He is long-remembered in bio-genetic and Nobel Prize winning history (Lewis, 1978, 1995). In order to recite elements of Lewis's estimable career, a piece of background information that is on-point (lore-wise): Back when I was "under" Ives (1960s), he mentioned that his CalTech labmate Bridges (1930s) had "the best eye" for recognizing novel sports. One that Bridges detected (1910s, back at Columbia) got named *bithorax* (*bx*), for a posterior-thorax structure was subtly modified in the direction of a more anteriorly located structure (Lindsley and Grell, 1968c). *bx* was demonstrated by Bridges and labmates to be a heritable mutant, hence the italicized name and abbreviation. And the *bx* single-locus mutation contributed to establishing an autosomal "linkage group" (cf. that for the X chromosome, e.g., Sturtevant, 1913). The autosome that harbors *bx* is one of the large non-sex chromosomes in *D. melanogaster.* This chromosome, number 3, was later realized to contain some of the all-time famous fruit fly genes (e.g., as a conceit, recounted by Villella and Hall, 2008). For *bithorax*'s part, instead of the mutation being applied as a genetic tool alone, how that mutation changes the fly's anatomy was regarded as potentially interesting *biologically*, as well being genetically useful.

After gaining his PhD at CalTech, where Dr. Lewis stayed ever after, Ed initiated a series of studies involving "complex loci." Those are chromosome sites at which several genetic variants were found to be clustered closely at and near a given chromosomal site. Typically, the various variants cause phenotypic changes that are related, e.g., all having to do with eye morphology. By digging into such loci, and the trio or more genotypic variants associated with each, Lewis was once again dealing with (I say again) *genetic* genetics: here meaning how various combinations of genetic changes at a given locus were apparently acting and interacting in – indeed – complex ways. Such Lewisian analyses seemed to consider the phenotypes – external anatomical changes caused by the heritable variants – as means to an end, allowing elucidations of the genetic interactions as such. Lewis himself recounted these elements of his career in the written version of his Nobel Prize Lecture [See Lewis (1995), if you are willing to grapple with the many genetic complexities he describes.]

Professor Lewis spent a fair proportion of his career investigating one of these complex loci, and his studies took on increasing degrees of *biological* significance (Lewis, 1995): how genotypic changes point to ways that genes materially influence animal *development*. This brings us back to the original mutant noticed by Bridges: *bithorax* (Lindsley and Grell, 1968c). Lewis established several additional genetic variants at the relevant autosomal locus, which became known as the BITHORAX COMPLEX (BX-C). A given BX-C-based variant – along with combinations of mutations – caused changes in the fly's external anatomy that were regarded as increasingly interesting, phenotypically. Without burdening readers with hard genotypic specificity, suffice it say that one of these combinations led to development of the famed "4-winged" fly (depicted, on behalf of Lewis, by Lakhotia, 1995). This highly anomalous animal had a pair of the usual wings, extending from the middle part of the fly's thorax, *and* an additional pair extending from the immediately posterior portion of this "major body region." Thus, this BX-C combination caused the highly-evolved dipteran insect in question (*Drosophila melanogaster*) to take on attributes like that of a more primitive insect, such as the normally 4-winged dragonfly. Another BX-C genotype caused legs to extend off an anterior segment of the fly's abdomen, à la primitive invertebrates that possess legs extending from nearly all body segments, e.g., a centipede or millipede. Such phenotypic changes are arguably more interesting than many others known in *D. melanogaster*, such as those involving altered pigmentation, e.g., the already introduced *yellow* mutant.

Now some of the mutations mapping to (within) the chromosomal locus in question (BX-C) have such severe effects on development as to kill the animal during a given pre-adult stage. These *lethal* BX-C genotypes can be exemplified in two ways: (1) *Ultrabithorax* (*Ubx*), whereby *Ubx/Ubx* kills developing *Drosophila*. (2) A deletion of the entire BX-C, which once again is a recessively lethal genotype. The latter type of genotypic change, a.k.a. *Deficiency* (*Df*), in general has lethal consequences, with the proviso that the *Df* in question removes an appreciable number of "chromosomal bands." To re-introduce the cyto-genetic matter: These intra-chromosomal entities, which will come heavily into play in a chrono-genetic context later, were described in magnificent detail by observing "giant chromosomes" contained within salivary gland cells of *D. melanogaster* larvae (e.g., Bridges, 1935). This genome – again at the level of the relevant larval cells – is composed of approximately 5,200 chromosomal bands. Each of the long chromosomal "arms" (one for the X, two each for each of the two large autosomes) harbors ca. 1000 such bands. As for deletions, their existence and attributes can be assessed by microscopic observations of larvae carrying *Df/+* genotypes. Yet if only one or two faint bands got removed by a particular *Df*, that genotypic alteration tends to be nearly impossible to resolve microscopically. A very short deletion of this type might be a "homozygous-viable" genotype, difficult to know if the *Df/+* or *Df/Df* type is "cytologically" unresolvable. Empirical fact: A deletion that removes a readily observable number of salivary-chromosomal bands (practically meaning a quartet, up to a dozen or more) invariably removes at least one v*ital gene*, such that those *Df/Df* genotypes are indeed lethal. One of many such examples: the cytologically demonstrable *Df* of BX-C (when homozygous) kills *D. melano-*

gaster at an early developmental stage. Such animals – to a first approximation – exhibit embryonic *segments* that appear to all alike among them (Lewis, 1995). It is as if this higher insect has been *transformed* to take on qualities of a most primitive invertebrate, viz. an annelid. This category of BX-C-based transformation is analogous to those described above, whereby viable genotypes change adult-segment qualities in the direction of evolutionary primitiveness. Implication (of a Nobel Prize winning variety): Genes of the BX-C, in their normal forms, evolved to *suppress the latent capacities* of *Drosophila* to develop in ways that more primitive invertebrates do. Yet BX-C genes turned out to be significant *way beyond* the invertebrate arena, as will be mentioned later.

The matter of lethal genotypes brings us back to reconsider multiply inverted *balancer* chromosomes (within this article, as a reminder, abbreviated *Bal*). X-chromosomal *Bal*'s usually do not carry recessive-lethal marker mutations, allowing X-*Bal/Y* males to live. *Bal/lethal-variant* females – expressing the effect of a non-lethal dominant marker mutation – are also alive, if the lethal factor carried "under" the *Bal* is recessive, à la Muller's discovery, alluded to above, of lethal genotypes newly induced by ionizing radiation. How about balancer chromosomes generated on behalf of the *D. melanogaster* autosomes? Consider the long "third chromosome," which harbors the BX-C. This developmentally (and evolutionarily) important genotype has long been maintained in a *Bal(3)/Df*-BX-C "stock." For this, the balancer (as for all such multiply-inverted autosomes in *D. melanogaster*) was deliberately constructed to contain a dominant marker, whereby *Dom/+* causes an conveniently recognizable anatomical change at the level of live flies. Such a stock is true-breeding, whereby only *Bal(3)/Df*-BX-C live flies are produced down the generations because the dominant mutation was deliberately chosen to cause, as well, *recessive lethality*. This "strain maintenance" situation allows developing animals homozygous for *Df*-BX-C to be "pulled out," whenever one wishes, for observations, etc. And the deletion genotype can never be lost in terms of routine stock preservation, because any live fly manifesting effects of the balancer's *Dom* marker must be *heterozygous* with *Df*-BX-C; this is because the multiple inversions disallow meiotic recombination, which otherwise would lead to *Dom*-with-*Df*-BX-C or "+ +" recombinant genotypes.

An insert, with a bit more about inverted chromosomes in *Drosophila*: I find it interesting that an eventual Nobel Prize winner – George Beadle, who came to study biochemical genetics in *Neurospora* (reviewed by Dronamraju, 1991) – was a geneticist studying fruit flies in his earlier days. For example, Beadle analyzed the properties and cellular behavior of *Drosophila* inversions, working with Sturtevant (e.g., Beadle and Sturtevant, 1935; historically summarized by Hawley and Ganetzky, 2016).

The Prize just referred to (1958) was co-awarded to the biochemical geneticist Edward Tatum, who began his career by studying the genetics of bacteria with Joshua Lederberg (e.g., Tatum and Lederberg, 1947). In fact, this late 50s Prize was awarded to Lederberg, along with Beadle and Tatum. It should go without saying that deconvoluting prokaryotic genetics – including genetically & molecularly characterized "vectors" that can be introduced into bacterial cells – turned out to be quintessential for essentially the entire enterprise that can be described as "genetic engineering" (DNA cloning and molecular manipulations of such macromolecules).

I should pause to mention that Beadle's and Tatum's exploitation of *Neurospora* genetics (e.g., Tatum and Beadle, 1942) was *not* designed to find out all that much about fungal genotypes alone (genetic genetics). By later analogy, *N. crassa* was magnificently utilized, from another *biological* perspective, to analyze bona fide circadian rhythms in that fungus. This microbial enterprise was pioneered and sustained by Jerry Feldman, Jay Dunlap, Jennifer Loros, and colleagues, who made a highly-successful genetic plus molecular assault on "what is the clock?" in this supposedly lower eukaryote (summarized, during the rather early days of these inquiries, by Dunlap, 1999).

With regard to a *higher* eukaryote, the principal *genetically* based power of inversion application as set-up to be possible by the likes of Beadle and Sturtevant has to do with – you guessed it – *recombination-eliminating balancer* chromosomes. I provided a primer about such chromosome aberrations of the multiply inverted variety and about other types of homozygous-lethal genotypes, for reasons extending beyond a tip of the hat pointed toward the estimable Professor Lewis. As one of the most knowledgeable and proficient drosophilists, ever, he was well aware of the power of balancer chromosomes, crucially applied in his case to maintain strains of developmentally interesting lethal factors. Now many other drosophilists, including bio-geneticists ever emerging as the 1970s unfolded, were also astute enough to be aware of (a) pertinent cytogenetic tools, e.g., *Bal*'s, or "the attached-X," which will re-surface yet again below, and (b) how lethal mutations can cause interesting biological changes as well as pointing to putatively interesting genes, in terms of what such factors are doing on behalf of wild-type development. Two such investigators, who began to work together during the 1970s, were Christiane (Janni) Nüsslein-Volhard and Eric Wiechaus. They initiated their developmental-genetic investigations by treating "normal" males from some wild-type stock or the other with a powerfully acting *chemical mutagen* (abbreviated EMS). Its properties and utility had been rattling round the fly world as of mid/late 1960s or so, and EMS has been applied to generate from-scratch a variety of interesting, novel mutants (see below). Lewis himself published a brief research note (Lewis and Bacher, 1968), which described EMS mutagenesis and how conveniently it can be effected by merely feeding flies on sugar water laced with the mutagen (thus no need to do something like inject the substance into treated, starting flies).

So Nüsslein-Volhard and Wieschaus, operating as third or fourth generation "mutagenizers" (late '70s, vis-à-vis a perceivable starting point for systematic developmental genetics approximately one decade earlier) proceeded roughly like this: treated the aforenoted starting flies and mated them to *Bal*-bearing females contained in a *Bal* + *Dom1/Dom2* stock, again of the true-breading variety, because each such *Dom* – here designated generically as containing 2 different dominant markers – was and is a recessive-lethal mutation, to repeat a previously introduced point. Next-generation flies – either *Bal/mutagenized autosome* (*mut*) females or *Dom2/mut* males – were backcrossed to *Bal* + *Dom1/Dom2* flies. Now any offspring of a *Dom2/mut* male that does not display the marker effect of *Dom2* must carry *mut*, because there is no recombination within meioses of males in *D. melanogaster* and thus no need for the *Dom2* marker to be contained within a multiply inverted autosome. Result: Nüsslein-Volhard's and Wiechaus's *mut* chromosomes were contained in next-generation offspring of *both* sexes (e.g., *Bal* + *Dom1/mut* females and *Dom2/mut* males). Question: Will any offspring in the *subsequent* (ultimate) generation *not* express effects of either *Dom* marker? If not, then *mut/mut* was a non-lethal genotype. Yet many of these ultimate matings, for the many "lines" established by this pair of developmental geneticists, each line originating from an individual *Bal* + *Dom1/mut* female or *Dom2/mut* male, displayed only *Dom* males and females, signifying that the *mut* in question carried at least one newly induced lethal mutation. So I hope the reader can discern how powerful are applications of balancer chromosomes: Observing the *Dom* marker *phenotypes*, as outlined briefly above, unequivocally reveals to the investigator the animal's *genotype*. Reality check: Nüsslein-Volhard and Wieschaus infused additional genetic "tricks," which made the genetic side of their screening even more proficiently performable (summarized by Wieschaus, 1995). As usual, these bio-geneticists tapped into the rich array of extant genetic variants and the "extra" opportunities afforded by them, beyond usage of standard *Dom* markers, for example.

The developmental(bio)-geneticists in question observed what amounted to dying embryos, as to *which* if any might manifest anatomical anomalies that could suggest the mutationally defined gene at hand to be involved in "pattern formation:" the insects formative body plan (e.g., nature of embryonic segments, segment identities). Thus, Nüsslein-Volhard and Wieschaus discovered mutants and genes – a small but appreciable subset of those pointed to by their newly induced mutations – that could mutate to lethality for reasons more bio-interesting than in cases where a normally appearing mutant might just grind to a developmental halt or begin to generically to disintegrate.

Here is an ostensibly stray aside (but not really), related to the outcomes of Nüsslein-Volhard's and Wieschaus's mutant hunting. The matter now at hand is known as "genetic saturation", which will also rear its handsome head later. Here, a not-always-appreciated element of these investigators' findings involved *numbers* of newly induced mutants. This quantitative matter was summarized, during the decade subsequent to the pertinent Prize winning paper (1980), by Nüsslein-Volhard (1995) and Wieschaus (1995). An editorial note: After this pair of biologists published their announcement report (1980), in a vanity journal not conducive to full disclosures (*re* page limitations), they followed up with "journal of record" reports (cited by Nüsslein-Volhard, 1995) that laid out *all* aspects of their successful hunting, including things like "just how many mutations" corresponded to a given genetic locus hit by the mutagen. Nowadays, not all out-of-breath "banner" papers put forth full accounts, of what was done and found; and usefully dense follow-ups tend minimally to appear in the later literature. In any case, the mutant hunters at hand documented a huge number of developmental-lethal mutations in *Drosophila*, only a subset of which caused apparent abnormalities of embryonic pattern formation. Yet that relatively small proportion of "extra interesting genes" was defined by appreciable more independently induced mutations than genetic-loci defined, even though some of the latter were pointed to by only "one hit" each (summarized by Wieschaus, 1995). Therefore, it was as if these developmental geneticists could have been approaching, but not-yet "at", a situation of genetic *saturation*. This scenario would have been altogether different if every new mutation turned out to define a previously *un*-identified gene. Certain of the *chrono*-genetic sub-enterprises revealed a similar state of affairs, as we shall see, related to number games played in context of # mutations vis-à-vis # genes.

A fair number of the developmental genes under consideration within this portion of the essay implied to the larger "developmental genetic community" that such factors could be molecularly operating on behalf of *regulating other genes*, ultimately those immediately underpinning the relevant anatomical structures. Indeed – and owing to the fact that certain mutations in fruit flies and in many other forms not only cause ostensibly interesting abnormalities, they also (I say again) identify genetic loci – when several of the genes "pointed to" by the Nüsslein-Volhard/Wieschaus mutations were identified based on where they were located chromosomally, a large proportion of the encoded products – determined post-DNA cloning – seemed as if their would be *transcription factors* (summarized by Wieschaus, 1995). Again, these outcomes were as gingerly predicted (e.g., Lawrence, 1992). As well, those types of gene products come out of Lewis's (originally Bridges's) BX-C. More broadly, a rather large percentage of "developmental genes" operating in all kinds of animals on behalf of embryonic pattern formation appeared – via first-blush "clone and sequence" data – to encode proteins with DNA-binding capacities. Nucleotide sequencing of the genes' coding regions divulged on-paper proteins, several of which harbor "motifs" known to promote physical associations between such polypeptides and DNA sequences surmised to possess gene-regulatory functions.

The ever-expanding significance of the purely bio-genetic discoveries put forth by the 1995 Prize winners was based in part on molecular geneticists, working in a variety of "non-insect" systems, identifying "cognates" of BX-C genes along with several of those pointed to by the Nüsslein-Volhard/Wieschaus mutations, thereby describing molecularly interpretable products to emanate from way more than the "fly genes" alone (reviewed, from a neuro-perspective, by Bellen et al., 2010). It bears mentioning that the "evo-devo" extravaganza – studying development from the perspective of homologous genes acting analogously among separate species – was pioneered by developmentally investigating drosophilists. They prayerfully wondered whether DNA sequences cloned from "homeotic" genes, such as within the BX-C, might be applicable to identify molecular relatives in other species of animal; several such prayers were answered (beginning with McGinnis et al., 1984a, b, for example; reviewed by Gehring et al., 2009).

How about the *chrono*-genetic enterprise? To ramp up toward recounting that currently important sub-story, it must be first noted that bio-genetics in *Drosophila* began to leak out a little into the literature several years before the Prize-wining reports of Lewis (1978) plus Nüsslein-Volhard and Wieschaus (1980). How systematic bio-genetics started can be exemplified by the *developmental*-genetic accomplishments of David Suzuki and co-workers. Back before he became a generic science publicizer (popularizer?) in his native Canada, he began his bio-career at a small level, doing a bit of *Drosophila* genetics at Amherst College in Amherst, Massachusetts, under the aforementioned Phil Ives (à la Hall, who was mentored by Ives later on). Suzuki went on to earn a PhD, studying aand investigating under a hard-core fruit-fly geneticist, leading the former on his own to perform early-career studies at the faculty level in Western Canada in that restricted investigative area. But as of the mid/late 1960s, Suzuki slipped sideways into performance of bio-genetic studies: He generated anew and identified developmental mutants that were *temperature-sensitive* (TS), e.g., as caused by heat-sensitive lethal mutations (reviewed by Suzuki et al., 1976).

Remarkably one of the "classic" *bithorax* mutants (Lindsley and Grell, 1968d) was found retrospectively to be a TS developmental mutant, e.g., Villee, 1945; Kaufman et al., 1973). Beginning in the early 1970s, Suzuki's group began also to induce and isolate *adult-functional* mutants. For this, these investigators exploited the famed attached-X type of chromosome aberration, as follows: Starting flies (genetically normal XY males) were treated with the powerful mutagen EMS (as introduced already) then crossed to attached-X/Y females. Individual male offspring who had inherited a treated X from their father were "backcrossed" to the same attached-X type, to created a series of "putant" lines (putative mutants). One type of adult-mutant hunting effected by the Suzuki group entailed ferreting out "patho-physiological" mutants, viz. post-natal flies that were paralyzed at high temperature but OK at lower ones (e.g., Suzuki et al., 1971; Poodry et al., 1973).

One might call the Suzukian enterprise just outline an aspect of "behavior genetics," although it is difficult to pit mutationally-caused TS paralysis against a normal phenotype (elementary well-being at relatively low temperatures). This issue, if it is one, introduces the matter of *bona fide* behavioral genetics, whereby measurable phenotypes of the normal variety can be quantitatively compared with mutationally caused abnormalities or anomalies. In this respect, a key behavioral genetic pioneer was Seymour Benzer, who searched for novel behavioral sports in a dedicated manner: induction from scratch via application of the same chemical mutagen under consideration here, followed by systematic screening for heritable variants manifesting adult defects (Benzer, 1967). What Benzer had opted to do career-wise was leave the microbial genetic field in which he had toiled, notably during the 1950s, to become a higher-eukaryotic biogeneticist. After treating starting flies with EMS and mating them to – you guessed it – Muller's attached-X, *y f*, the remainder of Benzer's genetic tactic was identical to that already described, Suzuki-wise. Ultimately, groups of males from a given putant line – each carrying the same originally mutagenized X – were tested for responses to light; this led to several sports that manifested subnormal or anomalous "phototaxis" (Benzer, 1967).

Shortly after the time of the report just cited, a student in Benzer's CalTech lab, Ronald (Ron) Konopka, carried out the precise same genetic tactic. In this case, two generations from the start, the lines were screened for males that might exhibit abnormalities of *daily rhythmicity*. Three such putant lines were identified, out of some 2000 tries (Konopka and Benzer, 1971). Usefully, it seemed, the *female* progeny in these late-60s/early-70s operations served as *internal controls* for the mutant hunting. By-eye observations of intra-line offspring were doable even at low resolution, because the *yellow* body-color phenotype is readily observable without recourse to a microscope.

Konopka's original trio of mutants, *each* involving mutagen "hits" at *one* X-chromosomal locus that got named *period* (*per*), arguably displayed interesting rhythm abnormalities (Konopka and Benzer, 1971): no rhythmicity, 19 hour cycle durations in constant darkness, or 29 h cycles in that condition (the latter two phenotypes being *way* off the ca. 24 h norm, for this species of insect and for animals in general). "Interesting" mutant phenotypes or not – your choice – this search for novel *circadian rhythm* mutants (referring to the altered cycle durations, or lack of rhythmicity, in a constant environmental condition) was clearly successful. It would not have been, had Konopka been "0 for 2000." Importantly, I think this mutant hunting was not *hyper*-successful, as it could have been perceived to be if the numerator had been something like 200 or 300. That kind of outcome could have prompted the following types of questions: *Which* among an avalanche of new mutants would one plan to pursue? *And* could such a large number (hundreds instead of a tractable trio) imply that a great many of separate genes feed somehow into rhythm-regulating processes, many or most of these genic factors causing such effects indirectly? Instead, the *one* gene defined by the seminal *period* mutations allowed Ron and his boss gingerly to surmise that the factor pointed to by the trio in questions is a "clock gene": hence the title of Konopka's and Benzer's seminal paper. The assumed product of such a gene would operate, again hypothetically, somewhere near the "core" of *Drosophila's* circadian clock.

In this regard – perhaps unwittingly allowing for eventual identification of his *period* gene at the molecular level – Konopka applied his *period* mutations to determine the gene locus's location at very high intra-chromosomal resolution. This nuts and bolts operation (genetic bookkeeping) was accomplished in conjunction with proving that all three of Ron's mutations mapped to one X-chromosomal site (Konopka and Benzer, 1971; Smith and Konopka, 1981; augmented by elements of Young and Judd, 1978). What Konopka, who was a fine chronobiologist *and* geneticist, accomplished accordingly was to "meiotically map" his *period* mutations, via application of Sturtevantian principles and practices (cf. Sturtevant, 1913). As well, Konopka applied various chromosome aberrations – notably deletions – that had to involve intra-X-chromosomal sites near *per's* map position, to refine the gene's location (Konopka and Benzer, 1971; Smith and Konopka, 1981). This chrono-investigator was fortunate that his *per* gene happened to be located near the classic *white* (*w*) one, mutational identification and mapping of which led to generating many other types of "near-*w*" variants (e.g., Judd et al., 1972; Young and Judd, 1978).

These genetic practices became the norm, of a sort, when subsequent rhythm-interested drosophilists went to school on the Konopka approach. Such successors of Konopka mutagenized starting flies, as usual with the workhorse chemical referred to already, or by what is known as "transposon mobilization"; then they worked through mating schemes strongly analogous to those effected by Nüsslein-Volhard and Wieschaus (making, you guessed it, crucial application of dominantly marked balancer autosomes). But in this case – three generations after the mutagenesis-based starting points – the investigators who set out to identify novel rhythm mutants with autosomal etiologies looked for "lines" in which *mut/mut* genotypes, as explained above, segregating out of the final-stage matings allowed for adult viability and daily-rhythm testings (behaviorally).

It follows that the post-Konopka chrono-mutant hunters exploiting *Drosophila*-based opportunities were hoping mutationally to define rhythm-related genes beyond the X chromosome (cf. Konopka and Benzer, 1971; Konopka et al., 1991). This "worked out," as summarized in the relatively early days of post-Konopka mutant hunting by Hall (2003). Two features of these outcomes: (1) At least a couple dozen novel rhythm-altering mutations were induced and identified, *but* they defined only about a half dozen novel genes; it follows that "repeat hits" kept being effected at previously identified loci, as if still further mutations with clear effects on daily rhythmicity were rather unlikely to expand the overall genetic horizon toward oblivion, e.g., three dozen mutations defining that many genes, implying that "mutation #37 or so" would point to yet another factor that would have warranted some sort of ever-broadening analysis; (2) As just implied, the genetic loci determined for the new rhythm mutations were homed-in-upon by genetic and *Df*-based mappings, so that these genes could be identified and physically isolated at the DNA level (Hall, 2003).

Once again, these investigative processes exemplify the "power of forward genetics" starting with mutants then genetic map positions and subsequent "positional cloning" (DNA-based identification of interesting genes based on their chromosomal locations). Yet it is fair to say that the chrono-molecular-genetic investigators coming up with this array of mutants and molecular clones could hardly ever predict what kinds of protein products would be encoded by a given rhythm-related genes; thus a different scenario than in the case of developmental-cum-molecular genetics (e.g., Lawrence, 1992; Wieschaus, 1995). As we shall see, "just what" a given chrono-molecular "is doing" on behalf of cellular and molecular processes – harking back to the initially mysterious PERIOD (PER) protein – had to be figured out by down-to-earth empiricism (histologically and biochemically based lab work).

I exaggerate a bit, for there were additional rhythm-related factors initially identified in their *normal* forms, by application of various "purely molecular" tactics (summarized, again, by Hall, 2003). For example, a gene inferred to encode a neuropeptide in *Drosophila*, eventually named "pigment dispersing factor" (albeit not operating on behalf of any pigment dispersal), was identified and isolated at the DNA level as a molecularly normal factor. This led to establishment of the *pdf* gene's chromosomal location, by effecting a cyto-molecular procedure requiring no genetic variant to be in hand (Park and Hall, 1998). These findings allowed for the possibility of *subsequent* identification of a *pdf*-locus mutation, whose effects on behavioral rhythmicity could be assessed and were (Renn et al., 1999). This kind of strategy is known as "reverse genetics:" start with a gene in its normal form, whose significance cannot necessarily be inferred by the encoded product's attribute "on paper," viz. "clone and sequence;" but then come across or deliberately generate mutations within the relevant loci and ask about their potential effects on the bio-process of interest.

I should pause further to exemplify what reverse genetic starting points have meant in the rhythm arena. One of the new "clock genes" found by the post-Konopka mutant hunting referred to above was named *Clock* (*Clk*). Why? This was because positional cloning of the *Clk*-identified locus in *D. melanogaster*, followed by DNA sequencing of the gene's coding subset (Allada et al., 1998), came up with an on-paper protein quite similar to that emanating from the *Clk* gene in mouse, which had hopefully been named *Clock*, even though the original variant solely signified "rhythm mutant," as opposed to a circadian-pacemaker variant, necessarily. After the murine factor originated by way of a chemically induced mutation (Vitaterna et al., 1994), à la Konopka, the responsible variant was fine-level mapped; this allowed for molecular cloning in mouse of *Clk* DNA (Antoch et al., 1997; King et al., 1997). The CLK protein's nature (on paper) suggested that it could have "something to do with" PER protein (cf. Allada et al., 1998), as if CLK were indeed operating on behalf of the circadian clock in mammals.

Findings of this sort made various investigators wonder whether might there be *per*-like genes in such animals. For several years, after *period* was cloned in *Drosophila*, molecular searches for mammalian *per*'s came up dry; until 1997, when two separate research groups identified molecular relatives of fruit-fly *per* in human and mouse (Sun et al., 1997; Tei et al., 1997). This suggested reverse-genetic possibilities, whereby the separate *per*-like genes in mouse (n=3 per mammalian species) were reverse-genetically mutated (e.g., Bae et al., 2001; reviewed in a broad context by Bućan and Abel, 2002).

How about going in the other direction interspecifically? Well, cloning of mouse *Clk* made certain chrono-drosophilists imagine that a molecular relative of mammalian *Clk* might be identified and isolable in *Drosophila*. Yes, as it turned out. This led to manipulations of fly-*Clk* DNA in its normal form, leading to inferences as to how CLK in flies functions as part of the insect's clockworks (e.g., Darlington et al., 1998). But pure reverse genetics as such was not necessary here, because fly-*Clk* was co-identified initially by an arrhythmia-inducing mutation (staring with EMS mutagenesis, as noted), which proceeded to *forward*-genetic elucidation of the gene and its product (Allada et al., 1998). This strategy had to include overt rhythm significance for *Clk* gene action in *Drosophila* (as in mouse), because the starting point was a rhythm-defective mutant: behaviorally arrhythmic *and* with the *Clk* mutation causing *per* product levels in *Drosophila* to be very low (Allada et al., 1998).

What was just summarized points, once again, to matters revolving round "genetic satutation" of the situation revolving round rhythm-regulation in animals. First, let it be noted the *Jerk* mutant (as fly-*Clock* was provisionally named pre-cloning) could have involved "anything" in terms of encoded-product quality. Yet what that gene makes in the fly is none other than a very similar rhythm-related protein in mammals. As the post-Konopka mutant hunting matured, Young's lab mutationally identified a locus (Price et al., 1998) called *doubletime* (*dbt*). It turned out to encode a known category of kinase enzyme (Kloss et al., 1998), which in *Drosophila* targets PER protein as one of its substrates (e.g., Syed et al., 2011). Incidentally, *dbt* – essentially at the same time it was discovered via rhythm mutants – could not have been surmised to be "chrono-specific," because the gene was contemporaneously identified via mutations at the self-same locus that cause embryonic lethality (Zillian et al., 1999). Thus *dbt* – *un*like *per* and the Young lab's famed *timeless* (*tim*) gene (originally Sehgal et al., 1991, 1994) – is developmentally vital. Therefore, in turn, the DBT enzyme must target substrates not limited to the inessential PER protein. Mutant fruit-flies with a ruined *period* gene, or lacking it altogether, are alive and "seem fine" until one tests them for daily rhythms, as will be certified later. So the case of *doubletime* caused that gene to take on expanded biological significance, even though the overall case of clock genes in flies and mammals caused the story *not* to get bloated beyond a relatively small number of key "players."

More about mammals, to drive home the point just claimed. Within a non-murine mammalian species, a dramatic rhythm-affecting mutation was run across: a spontaneous sport called *tau* which, when homozygous, causes *hamsters* to manifest ca. *20-hour* cycle durations in constant darkness (Ralph and Menaker, 1988). Even though forward-genetic possibilities are minimal for this type of mammal, Joe Takahashi's research group managed to home in upon the relevant genetic locus, in order to clone *tau* DNA from hamster (Lowrey et al., 2000) then reveal that the encoded product is none other than a close molecular relative of *Drosophila*'s DBT kinase (cf. Kloss et al., 1998). Once again, TAU could have been anything, barely predictable pre-cloning; but the outcome was to *re*-identify an already appreciated clock-functioning factor.

These molecular-genetic examples of saturation-signifying claims are analogous to "more rhythm-affecting mutations than genes pointed to." Here, the matter of *Clk* (née *Jerk*) in *Drosophila*, followed by the case of *tau* in hamster, could well have identified previously unknown (even unanticipated categories of) players materially contributing to animal clockworks. But both such cases got squeezed down to clock factors already on the "animal-rhythm table."

Another intra-fly example, whereby yours truly realizes he keeps getting ahead of the overall sweep of chrono-molecular history: The Konopka-like, but autosomally based, screening for new rhythm mutants resulted in afore-described *Clk* and a separate gene (*cycle*, a.k.a. *cyc*) whose category of protein-product co-functions with CLK (Rutila et al., 1998; cf. Allada et al., 1998). Two *additional* rhythm-altering mutations fell out of this screening, beyond those that which eventually were revealed to generate two independently induced mutations at *each* of the *Clk* and *cyc* loci.. The additional two – for a total of six – co-mapped with one another, at an autosomal locus well separated from those of *Clk* or *cyc* (Suri et al., 2000). Now the pair of "extra" mutants could have recognized a brand-new rhythm-related factor; instead the (non-*Clk*, non-*cyc*) mutations at hand turned out merely to re-identify *doubletime* (Suri et al., 2000), just as *tim* was hit by further mutageneses carried out in Young's research group (e.g., Rothenfluh et al., 2000) *and* independently within the Brandeis-based one (Rutila et al., 1996).

I have been soft-pedaling the matter of "cloning the first clock gene": yes, *period* in *Drosophila*. Owing to the 2017 Prize (Physiology or Medicine) being based in part on a trio of lab-heads whose research groups basically co-identified *per* at the molecular level, here is some requisite historical information. First, let it be known that the aforementioned matter of "genetic saturation" has to do with much more than "lots of *rhythm*-affecting mutations defining a much smaller number of genes." Thus the saturation issue was in play when reasons for Janni Nüsslein-Volhard's and Eric Wieschaus's anointment as Nobel laureates were outlined in previous passages. An earlier example from the sub-field of developmental genetics: A well-defined segment of *Drosophila*'s X chromosome – near the *white gene* as noted – was saturated by the research group of Burke Judd, in terms of loci that can mutate to lethality. Cleverly applying various genetic tricks (crucially, as usual, dependent on pre-existing chromosome aberrations), these investigators induced hundreds of novel mutations, all necessarily confined to a small, near-*w* region of the X chromosome, defined by the length of a certain X deletion. But that many mutations defined only a bit more than one dozen loci; it followed that *that* many "vital genes" were contained within the region (Judd et al., 1972). In other words, any further induced lethal mutation would almost certainly be yet another repeat hit at a locus already mutated. But these investigators wondered whether additional genes might be harbored within the region at hand, loci that could not mutate to lethality.

Now Mike Young (at least) was aware that Konopka had mapped his *per* gene to right smack in the middle of the intra-X segment in question (Judd et al., 1972; cf. Konopka and Benzer, 1971). A crucial concern here: All of Konopka's *per* mutants had to be alive, based on his chrono-screening tactic (Konopka and Benzer, 1971); yet this gene *could* have been a vital one, capable of mutating to lethality. If that were so, a set of allelic lethals from the "saturation screen" – lethal mutations independently mapped to be at least very near *per* – would fail to "complement" certain *per* mutations: e.g., arrhythmia-inducing per^{o} "over" given such lethal would cause locomotor arrhythmicity, just as Konopka had found *Df-per*/ per^{o} to cause. Yet Young found that all the relevant lethals complemented per^{o}: None of the pertinent heterozygous types was arrhythmic, just as was known for the strongly rhythmic attributes of per^{o}/+ or *Df-per*/+ types (Young and Judd, 1978; cf. Konopka and Benzer, 1971). Thus, Judd et al. (1972) had not really "saturated the region" in terms of identifying *all Drosophila* genes contained therein, only for the developmentally vital ones. It followed that the *period* gene in this insect cannot mutate to lethality – only, it seemed in terms of most severe mutational effects, causing biological arrhythmicity.

An obvious implication of the foregoing micro-history is that Dr. Young's interest in the *period* gene harked back to his formative years, when he was a graduate student in Judd's lab. When this investigator expanded his horizons on behalf of bio-*molecular* genetics, he began to study a classic "developmental gene" in *Drosophila* called *Notch* (*N*). Don't ask, but register that this factor had long been regarded as potentially interesting. And fly *N* is related to strongly analogous factors in other animals, of course (reviewed by Bellen et al., 2010), just as are several of the Prize-winning "Lewis/ Nüsslein-Volhard/Wieschaus" genes. Well, the *N* locus is on the X chromosome of *D. melanogaster*, not all that far from *w* and *per*. Thus Young's own research group – once he started his lab in New York City after his grad student days and a "molecular post doc" in California – aimed to "clone *Notch*." He did so (e.g., Kidd et al., 1983, 1986) and realized that this afforded the possibility of effecting a "molecular chromosomal jump" from the *Notch* locus to the *period* one. This, too, worked out thanks to clever application of the relevant molecular clones *and* of chromosome aberrations whose lesions crucially "touched" both of the pertinent intra-chromosomal loci (Bargiello and Young, 1984).

Contemporaneously, the neighboring Hall and Rosbash Labs: (within Waltham, Massachusetts) opted to "try and clone *per*" in competition with Young and co-workers. Why at Brandeis University? This motivation harked back to Hall, working with a postdoc C.P. (Bambos) Kyriacou in the late 1970s, stumbling upon a short-term behavioral rhythm in *Drosophila* connected with normal male courtship behavior (Kyriacou and Hall, 1980). Those two investigators were aware that the only extant rhythm mutants in this species were Konopka's *period* ones. Therefore Hall, who had long been in touch with his former labmate Konopka, asked the latter if he would send a trio of stock containers, representing the famed trio of *per* mutants. Konopka so transmitted, along with a snail-mailed note saying words to the effect of "that sounds like a nice experiment." Hall wondered why Ron would say such a thing: How could one expect mutations, which by definition affect daily rhythmicities, also to cause alterations of the one-minute courtship rhythm in question? Yet each of those three mutations *did* so affect (Kyriacou and Hall, 1980). Disclaimer: Other courtship-involved investigators are skeptical of that study and related ones that followed, as initially reported in the very early '80s. I will not burden you with this matter, but see the review of Hall and Kyriacou (1990); also Alt et al. (1998); Stern (2014); Stern et al., (2017); Kyriacou et al. (2017, 2019).

Because various elements of *Drosophila* courtship are species-specific, Kyriacou and Hall asked whether the "one-minute" rhythm displayed by *D. melanogaster* males would be the same, different, or non-existent in other species. They found and reported about five years later that males of *D. simulans* (here abbreviated *sim*) generate the relevant "courtship-song" rhythm with a cycle duration of only about 40 seconds, compared with about 60 sec for "*mel*" males (Kyriacou and Hall, 1986). As was kind of customary for species-diverging characters, these reproductive-behavioral investigators generated interspecific hybrids between *mel* and *sim*. Those two *Drosophila* types are close evolutionary relatives, long known to generate hybrid progeny when males of one species are coaxed to mate with females of the other. Yet, it was also known that hybrid males with a *mel* X (over a *sim* Y) are developmentally dead; whereas the sexually reciprocal type (*sim* X/*mel* Y) is alive and behaves OK or better. In this regard, Kyriacou and Hall were aware of – dare I say again? – the lore of *Drosophila*. Thus they applied a mutation in *D. simulans* known as *Lethal hybrid rescue*; contained in a *sim* male parent, it mated to a standard *mel* female, leading to viable *mel* X/*sim* Y sons. The reciprocal mating effected by these investigators was standard *sim* males x attached-X *mel* females; their sons are *sim* X/*mel* Y. Yet again (!) the "power of the attached-X," applied here to control for "maternal effects" on hybrid progeny properties, viz. species-specific egg cytoplasm influences thereon, putatively *not* operating in this case, for both female-parent types were *D. melanogaster*. Behavioral results: The reciprocally hybrid male types put forth courtship-song rhythms matching the type of X chromosome coming in from the dual-species parents (Kyriacou and Hall, 1986). It followed that the etiology of the species difference in rhythm singing mapped to only one chromosome, not the typical outcome inferable from the behavior of hybrid animals. So, could it be that said etiology was harbored within but *one genetic locus* on that chromosome? It was (and is) "the X," where the *period* gene is located.

Well, this question was in-principle answerable, as of the early 1980s: (A) Could the *period* gene be cloned from *D. melanogaster*? Yes, as exemplified above. This would allow for quick cloning of *per* from *D. simulans*, which would have to harbor a strongly homologous *period* gene. Now, a pair of *Drosophila* biologists-cum-molecular geneticists had burst forth contemporaneously with the wherewithal to generate routinely "DNA transformed" fruit flies (Spradling and Rubin, 1982; Rubin and Spradling, 1982). The potential for revolutionizing bio-molecular-genetic studies of this insect was realized by an endless number of DNA-mediated *transgenic* types designed and generated by drosophilists, thanks to proficient transgenesis doability established by Allan Spradling and Gerry Rubin, cited above. What *kinds of DNA engineerings* could be effected – followed by introduction into *Drosophila* and insertions of the introduced "constructs" into the flies' chromosomes – seemed limited only by investigators' imaginations. Here is but one example thereby, inserted here as something of a conceit. We at Brandeis, working with the research group of Steve Kay elsewhere, wondered whether "molecular cycling" of *period* gene products, dealt with historically below, might be monitorable at the level of *individual live flies*. Thus we fused a regulatory region cloned from the *per* locus to a DNA sequence (*luc*) encoding fire-fly luciferase; and when that transgenic type was fed the relevant enzyme substrate, "glow rhythms" could indeed be measured all day, and during the next and the next (Brandes et al., 1996). This ploy led, for instance, to establishment of "molecular-rhythm parameters" that (in contrast) can be mushily apprehendable via sacrificing and extracting-from large numbers of flies "per time point." This "*per-luc*" transgenic-type was exploited in several ways subsequently. For example, *Drosophila* carrying pre-mutagenized chromosomes *and* that transgene were screened for novel rhythm mutants, not based on behavioral phenotypes; instead upon mutational alterations of normal LUC-reported cycling (Stanewsky et al., 1998). This led to identification of a new mutant named *cry*, along with forward-genetic identification of the corresponding DNA. Not to burden readers with hardly any details: *Drosophila* CRY functions on an "input pathway" involving light-induced daily resets of the fly's circadian clock (e.g., Hall, 2000).

This enterprise within the enterprise exemplifies how certain fruit fly chrono-geneticists came to study not only core clock functions (via *period* and a handful of additional genes), but also the important matter of environmental *inputs to the clock*. Circadian clocks, in general, are not very accurate; but they are easily reset, daily, and notably by photic stimuli. We and various other research groups also hoped to tackle the aforementioned matter of *outputs* "from the clock" (e.g. Peng et al., 2003; cf. Park and Hall, 1998; Renn et al., 1999), realizing that a clock at its core does nothing for the organism, unless key elements of central pacemaker function "feed forward" into the regulation of revealed rhythmicity.

Back to the matter of courtship-song rhythmicity. Spradling's and Rubin's transgenically based breakthrough made us at Brandeis realize one more thing that became doable: Against a backdrop of the interspecific song-rhythm difference summarized earlier, and mapping the etiology of this species-specificity to the X chromosome, we could molecularly transfer *per*$^{+}$ DNA from *D. simulans* into *D. melanogaster*, whereby that latter type by itself would be genetically arrhythmic (*per*^{0}) for both locomotor activity (Konopka and Benzer, 1971) and male singing behavior (Kyriacou and Hall, 1980). Specific question: Would such a single-gene interspecific transfer bring with it regulation of courtship-song rhythmicity like that of the donor species (viz. ca. 40-sec cycle duration)? Many years later – however surprisingly, as just implied, or controversial (as referred to earlier) – the answer was "yes" (Wheeler et al., 1991).

Meanwhile, however, we had to get our hands on *per* DNA cloned from *D. melanogaster*. Taking a different tack from that of Young's lab (outlined above), I first realized that we at Brandeis could potentially get a running start with regard to "*per* region" DNA. How? Well this former Brandeisian – who routinely tracked the *Drosophila* literature, no matter how putatively obscure – remembered a then-recent paper, emanating in part from a university that happens to be located within the Prize awarding country: University of Lund, southern Sweden; not that important. What *was* important was that a research group headed there by Jan-Erik Edström had initiated a novel tactic for "positional cloning" by scraping DNA out of a well-defined region of the *D. melanogaster* genome (!). For this, those investigators looked at giant salivary-gland chromosomes (aforementioned) under the microscope and applied a fine-tipped glass needle to an intra-X segment near the *white* gene, thus also near the *period* one (Scalenghe et al., 1981). This subset of the genome was chosen because it had long been so well characterized as to mutationally defined and cytogenetically mapped loci, as already summarized above in various contexts. Scalenghe and co-workers may or may not have been aware that they could have "scraped out" *per* DNA among molecular clones connecting with nearby loci. In any case, when Hall was at a conference co-attended by one of the Edström lab collaborators (Vince, né Vincenzo, Pirrotta, cf. Scalenghe et al., 1981), the former asked the latter if a follow-up from the original report had identified any "*white*-region" clones corresponding to X subregion "3B1-2."

I burden you with that level of detail in context of (i) there being 102 major regions among the *D. melanogaster* chromosomes, going back to Bridge etc.'s original cytogenetic analyses; each such region is conventionally subdivided into about a half-dozen lettered sub-regions, which in turn are designated by further "sub-numbers" (each of which corresponds to an individualizable salivary-gland chromosomal band, amounting to a total of ca. 50 bands per major numbered region); (ii) Konopka himself, via his aforementioned fine-mapping endeavor, had localized his *period* gene "down to" 1 or 2 intra-X bands (within region 3), or perhaps to a tiny subsegment located between, indeed, bands 3B 1 & 2 (Smith and Konopka, 1981). Furthermore, Ron and his then co-worker had mated two deletion-bearing types to one another, creating *Df-1/Df-2* females (Smith and Konopka, 191), against a backdrop of each such *Df* (here generically designated) "uncovering" the arrhythmia-inducing effects of per^{o}; and one edge of *Df-1* (call it the righthand one) had been pre-designated to "just overlap" the relevant edge (lefthand) of *Df-2*; it was as if this two-deletion heterozygous type *removed* the *period* gene altogether (confirmed to be so in subsequent molecular analyses by Bargiello and Young, 1984; and Reddy et al., 1984). Relevant inferences of the pre-molecular variety were buttressed by the fact that the *Df-1/Df-2* females were found to be alive and arrhythmic (Smith and Konopka, 1981), same as for the behavior of per^{o}/per^{o} females. This cytogenetic ploy once again indicated that the *period* gene is not a vital one.

Against this background, and with apologies for including a rather genetically dense insert, the "DNA-scraper" collaborator whom Hall queried (Pirrotta) said he would check as to whether any of his and co-workers' X region-3 clones might have been found apparently to correspond with aforenoted 3B1-2 (owing to such molecular cytogeneticists having annealed in-situ their various scraped-out clones, of a labeled variety, to salivary-gland chromosomes). That cytogeneticist soon sent a pair of DNA clones to Brandeis, surmising that they could be, or be very near, the molecular starting-point for cloning *per* that we were hoping to acquire. This wish was realized, via various aspects of subsequent labwork performed in the collaborating groups of Hall plus Rosbash (Reddy et al., 1984). As already noted, Young's competing group reported similar, ultimately molecularly converging, findings, about *period* gene molecular genetics (Bargiello and Young, 1984; summarized by Young et al., 1985). Some ostensibly "banner" features of these findings, reported during a late stage of 1984, were "transgenic rescues" of per^{o}, effected by transforming-in putative *per*⁺ DNA (Zehring et al., 1984; Bargiello et al., 1984). It was these early-days findings that made us at Brandeis realize that doing essentially the same, but later, introduction of *per*⁺ DNA from *D. simulans*, had a chance of answering the question about "interspecific genetic etiology" for a behavioral variation, cavalierly posed in previous passages.

Before such an investigative prayer could be put forth (Wheeler et al., 1991), however, various nuts and bolts characterizations of the molecularly identified and transgenically confirmed *period* gene DNA were performed after 1984 and during a fair proportion of the mid/late 1980s. For instance, coding DNA within the *per* locus was sequenced (reviewed by Hall, 2003). But these descriptive findings led, for a while, to no meaningful insights as to how the on-paper *protein* encoded by this ostensible clock gene might function within cells and tissues.

Here is a final fillip to this early history, recounted on behalf of divulging how chronogenetics, then molecular-neurobiological genetics, was – dare I say – pioneered as a (then) cottage sub-industry. It was operating back then as a small part of the burgeoning industrial-level research that the "bio-genetic" system came to be, based conspicuously on *Drosophila* research. Speaking of low-level stuff, the Brandeis groups' "*per* cloning" goal was not really met at first, because Reddy et al. (1984) identified three separate RNA's to emanate from the tiny intra-X region defined by Konopka's two-deletion ploy. Which of that trio was *per* per se (maybe more than one transcription unit)? It occurred to us to track expression of three candidate genes by extracting RNA's at different timepoints during a given daytime, then nighttime. Might one such transcript type manifest systematic fluctuations in its concentration? If so, *that* RNA might be transcribed from *per*. It would have been registered as a gene that can *mutate* to alter daily *biological* rhythms, correlated with the *normal* form of this gene making its product to define a daily *molecular* rhythm. The answer to this preliminary inquiry was wrong, x 3 or so: The incorrect third of the gene trio at hand was reported to generate a daily-oscillating transcript type, whereas what turned out to be *per* RNA was said *not* to fluctuate via whole fly RNA extracts (Reddy et al., 1984; similarly reported in a summary article by Young et al., 1985). This "wrong-RNA" matter was corrected later (Lorenz et al., 1989).

More meaningful than asking at first-blush about *per* RNA oscillations were later demonstrations of PER protein cycling, via timed immuno-histochemical assessments of final product levels in "adult head" specimens (Siwicki et al., 1988; Zerr et al., 1990). Such assessments of temporally controlled expression of the gene were performed contemporaneously with re-examination of the "cycling RNA?" matter by Hardin et al. (1990), based on "head only" extractions of RNAs at a series of timepoints across 24-hour Earth-day cycles.

An investigative fillip, which refers to one of the overall goals that sustained the chrono-genetic enterprise: Siwicki (1988), along with Liu et al. (1988), took the first meaningful step toward describing *where* a clock gene is expressed *in the fly's nervous system*, potentially pointing toward the neural substrates of rhythmic behavior. Additional early-days findings in this arena were made by Brigitte Frisch, John Ewer, and Maki Kaneko (reviewed by Kaneko, 1998). Later, this *neuro*-genetic sub-enterprise expanded into studies made in various other research groups (including that headed by Michael Rosbash, e.g., Stoleru et al., 2004).

I should leave things there, assuming that Professors Rosbash and Young will better recount how *subsequent* investigations during the 1990s and 21st century decades have elucidated "what is PER protein, and what are co-acting or interacting clock gene products, doing to form and operate a circadian clock?" The single-letter adjective italicized shortly above alludes to the fact that figuring out a fair amount of "what's the clock in *Drosophila*?" turned out to be of fine heuristic value for doing so analogously in studies of many other life forms. At a minimum, daily oscillations of clock proteins – in terms of their concentrations, qualitative attributes, or both – became the Zeitgeist of many molecular-chrono-genetic analyses, performed on behalf of investigative systems ranging from microbes to mammals. The relevant investigative extravaganza has been nicely summarized – as these studies came to the fore and, at least for animals, quasi-converged – by Dunlap (1999); Reppert and Weaver (2000); Young and Kay (2001); Chang and Reppert (2001); Kondo, 2007); and by Cohen and Golden (2015).

ACKNOWLEDGMENTS

This author (and anointee) finds it impossible to acknowledge all the co-workers, labmates, colleagues, and other associates who contributed mightily to studying *Drosophila* rhythms and the clocks underlying them – most valued as these many companions were. There really had to be a large overall crew to perform and sustain theses studies, because they were multi-faceted: genetics, molecular biology, neurobiology, and behavior. Suffice it to say here, with regard to my rhythm-related associates, that I gratefully acknowledge associations with the other pair of "lab-head awardees:" Michael Rosbash and Mike Young *and* how that pair of professors fostered vast numbers of experimentally proficient – as well as interesting and enjoyable – interactions among supervisees working within the three research groups in question. Again, such students, post-doctorals, and other researching labmates – distributed between two groups in the Boston area and one located in New York City – have been too numerous to acknowledge individually. One exception to this, specifically with regard to the "Hall lab:" I can never forget, nor fail to appreciate retrospectively, the first professional-level associate (graduate student) within my laboratory at Brandeis University in Waltham, Massachusetts. That was now-Professor Ralph J. Greenspan (University of California at San Diego). Ralph showed the way with respect to exploiting *Drosophila* genetics in order to make that "system" a genuinely deep neuro-genetic one: hard-core cyto-genetics devoted to elucidating neurochemical, neuro-anatomical, neuro-physiological, and behavioral phenomena, forming something of a symphony of studies that Ralph made speak vigorously to one another. It was (now) Dr. Greenspan who, however unwittingly during the mid/late 1970s, showed the way whereby initially neuro- and molecular-neuro-genetic investigations of *Drosophila*'s "rhythm system" started us down a path along which *mechanistic* analyses of the flies' internal "clockworks" came to the fore in the late 1980s and beyond.

REFERENCES

Allada, R, White, N.E., So, W.V., Hall, J.C., and Rosbash, M. (1998). A mutant Drosophila homolog of mammalian *Clock* disrupts circadian rhythms and transcription of *period* and *timeless. Cell* **93**: 791–804.

Alt, S., Ringo, J., Talyn, B., Bray, W., and Dowse, H. (1998). The *period* gene controls courtship song cycles in *Drosophila melanogaster. Anim Behav.* **56**: 87–97.

Antoch, M.P., Song, E.-J., Change, A.-M., Vitaterna, M.H., Zhao, Y., Wilsbacher, L.D., Sangoram, A.M., King, D.P., Pinto, L.H., and Takahashi, J.S. (1997). Functional identification of the mouse circadian *Clock* gene by transgenic BAC rescue. *Cell* **89**: 655–667.

Bae, K., Jin, X., Maywood, E.S., Hastings, M.H., Reppert, S.M., and Weaver, D.R. (2001). Differential functions of *mPer1, mPer2*, and *mPer3* in the SCN circadian clock. *Neuron* **30**: 525–536.

Bargiello, T.A., and Young, M.W. (1984). Molecular genetics of a biological clock in *Drosophila. Proc. Natl. Acad. Sci. U.S.* **81**: 2142–2146.

Bargiello, T.A., Jackson, F.R., and Young, M.W. (1984). Restoration of circadian behavioural rhythms by gene transfer in *Drosophila. Nature* **312**: 752–754.

Beadle, G.W., and Sturtevant, A.H. (1935). X chromosome inversions and meiosis in *Drosophila melanogaster. Proc. Natl. Acad. Sci. U.S.* **21**: 384–390.

Bellen, H.J., Tong, C., and Tsuda, H. (2010). 100 years of *Drosophila* research and its impact on vertebrate neuroscience. *Nat. Rev. Neurosci.* **11**: 514–522.

Benzer, S. (1967). Behavioral mutants of *Drosophila* isolated by countercurrent distribution. *Proc. Natl. Acad. Sci. US* **58**: 1112–1119.

Brandes, C., Plautz, J.D., Stanewsky, R., Jamison, C.F., Straume, M., Wood, K.V., Kay, S.A., and Hall, J.C. (1996). Novel features of Drosophila *period* transcription revealed by real-time luciferase reporting. *Neuron* **16**: 687–692.

Bridges, C.B. (1916). Non-disjunction as proof of the chromosome theory of heredity. *Genetics* **1**: 1–52, concluded **1**: 107–163

Bridges, C.B. (1935). Salivary chromosome maps. *J. Hered.* **26**: 60–64.
Bućan M., and Abel, T. (2002). The mouse: genetics meets behaviour. *Nat. Rev. Genet.* **3**: 114–123.
Chang, D.C., and Reppert, S.M. (2001). The circadian clocks of mice and men. *Neuron* **29**: 555–558.
Cohen, S.E., and Golden, S.S. (2015). Circadian rhythms in cyanobacteria. *Microbiol. Molec. Biol. Rev* **79**: 373–385.
Darlington, T., Wager-Smith, K., Ceriani, M.F., Staknis, D., Gekakis, N., Steeves, T., Weitz, C., Takahashi, J.S., and Kay, S.A. (1998). Closing the circadian loop: CLOCK-induced transcription of its own inhibitors, *per* and *tim*. *Science* **280**: 1599–1603.
Dronamraju, K.R. (1991). Profiles in genetics: George Wells Beadle and the origins of the gene-enzyme concept. *J. Hered.* **82**: 443–446.
Dunlap, J.C. (1999). Molecular bases for circadian clocks. *Cell* **96**: 271–290.
Gehring, W.J., Kloter, U., and Suga, H. (2009). Evolution of the *Hox* gene complex from an evolutionary ground state. *Curr. Top. Devel. Biol.* **88**: 35–61.
Hall, J.C. (2000). Cryptochrome: sensory reception, transduction and clock functions subserving circadian systems. *Curr. Opin. Neurobiol.* **10**: 456–466.
Hall, J.C. (2003). Genetics and molecular biology of rhythms in *Drosophila* and other insects. *Adv. Genet.* **48**: 1–286.
Hall, J.C., and Kyriacou C.P. (1990). Genetics of biological rhythms in *Drosophila*. *Adv. Insect Physiol.* **22**: 221–298.
Hardin, P.E., Hall, J.C., and Rosbash, M. (1990). Feedback of the *Drosophila period* gene product on circadian cycling of its messenger RNA levels. *Nature* **343**: 536–540.
Hawley, R.S., and Ganetzky, B. (2016). Alfred Sturtevant and George Beadle untangle inversions. *Genetics* **203**: 1001–1003.
Judd, B.H., Shen, M.W., and Kaufman, T.C. (1972). The anatomy and function of a segment of the X chromosome of *Drosophila melanogaster. Genetics* **71**: 139–156.
Kaneko, M. (1998). Neural substrates of *Drosophila* rhythms revealed by mutants and molecular manipulations. *Curr. Opin. Neurobiol.* **8**: 652–658.
Kaufman, T.C., Tasaka, S.E., and Suzuki, D.T. (1973). The interaction of two complex loci, *zeste* and *bithorax* in *Drosophila melanogaster. Genetics* **75**: 299–321.
Kidd, S., Kelley, M.R., and Young, M.W. (1986). Sequence of the *Notch* locus of *Drosophila melanogaster*: relationship of the encoded protein to mammalian clotting and growth factors. *Molec. Cell. Biol.* **6**: 3094–3108.
Kidd, S., Lockett, T.J., and Young, M.W. (1983). The *Notch* locus of Drosophila melanogaster. *Cell* **34**: 421–433.
King, D.P., Zhao, Y., Sangoram, A.M., Wilsbacher, L.D., Tanaka, M., Antoch, M.P., Steeves, T.D.L., Vitaterna, M.H., Kornhauser, J.M., Lowrey, P.L., Turek, F.W., and Takahashi, J.S. (1997b). Positional cloning of the mouse circadian *Clock* gene. *Cell* **89**: 641–653.
Kloss, B., Price, J.L., Saez, L., Blau, J., Rothenflluh, A., and Young, M.W. (1998). The *Drosophila* clock gene *double-time* encodes a protein closely related to human casein kinase Iε. *Cell* **94**: 97–107.
Kondo, T. (2007). A cyanobacterial circadian clock based on the Kai oscillator. Cold Spring Harb. Symp. *Quant Biol.* **72**: 47–55.
Konopka, R.J., and Benzer, S. (1971). Clock mutants of *Drosophila melanogaster. Proc. Nat. Acad. Sci. U.S.* **68**: 2112–2116.
Konopka, R.J., Smith, R.F., and Orr, D. (1991). Characterization of *Andante*, a new

Drosophila clock mutant, and its interactions with other clock mutants. *J. Neurogenet.* 7: 103–114.
Kyriacou. C.P., Dowse, H.B., Zhang, L., and Green, E.W. (2019). Computational error and restricted use of time series analyses underlie failure to replicate *period*-dependent son rhythms in *Drosophila*. Tentatively accepted for publications in *J. Biol. Rhythms.*
Kyriacou. C.P., Green, E.W., Piffer, A., and Dowse, H.B. (2017). Failure to reproduce *period*-dependent song cycles in *Drosophila* is due to poor automated pulse-detection and low-intensity courtship. *Proc. Natl. Acad Sci U.S.* **114**: 1970–1975.
Kyriacou, C.P., and Hall, J.C. (1980). Circadian rhythm mutations in *Drosophila melanogaster* affect short-term fluctuations in the male's courtship song. *Proc. Natl. Acad. Sci. U.S.* **77**: 6929–6933.
Kyriacou, C.P., and Hall, J.C. (1986). Interspecific genetic control of courtship song production and reception in *Drosophila. Science* **232**: 494–497.
Lakhotia, S.C. (1995). The 1995 Nobel Prize: a tribute to the power of formal genetics. *Curr. Sci.* **69**: 797–801.
Lawrence, P.A. (1992). *The Making of a Fly: The Genetics of Animal Design*, Blackwell Scientific, Oxford, UK.
Lewis, E.B. (1978). A gene complex controlling segmentation in *Drosophia. Nature* **276**: 565–570.
Lewis, E.B. (1995). The Bithorax Complex: The first fifty years. *Nobel Lectures*, pp. 235–260.
Lewis, E.B., and Bacher, F. (1968). Methods of feeding ethyl methane sulfonate (EMS) to Drosophila males. *Drosophila Info. Serv.* **43**: 193.
Lindsley, D.L., and Grell, E.H. (1968a) *Genetic Variations of* Drosophila melanogaster. Special Chromosomes, *Carnegie Institution of Washington Publication* 627, p. 409.
Lindsley, D.L., and Grell, E.H. (1968b) *Genetic Variations of* Drosophila melanogaster. Mutations, *Carnegie Institution of Washington Publication* 627, pp. 43–44.
Lindsley, D.L., and Grell, E.H. (1968c) *Genetic Variations of* Drosophila melanogaster. Inversions, *Carnegie Institution of Washington Publication* 627, pp. 321–346.
Lindsley, D.L., and Grell, E.H. (1968d) *Genetic Variations of* Drosophila melanogaster. Mutations, *Carnegie Institution of Washington Publication* 627, p. 44.
Liu, X., Lorenz, L., Yu, Q., Hall, J.C., and Rosbash, M. (1988). Spatial and temporal expression of the *period* gene in *Drosophila melanogaster. Genes Devel.* **2**: 228–238.
Lorenz, L.J., Hall, J.C. and Rosbash, M. (1989). Expression of a *Drosophila* mRNA is under circadian clock control during pupation. *Development* **107**: 869–880.
Lowrey, P.L., Shimomura, K., Antoch, M.P., Yamazaki, S., Zemenidies, P.D., Ralph, M.R., Menaker, M., and Takahashi, J.S. (2000). Positional syntenic cloning and functional characterization of the mammalian circadian mutation *tau. Science* **288**: 483–491.
McGinnis, W., Garber, R.L., Wirz, J., Koroiwa, A. and Gehring, W.J. (1984a). A homologous protein-coding sequence in Drosophila homeotic genes and its conservation in other metazoans. *Cell* **37**: 403–408
McGinnis, W., Hart, C.P., Gehring, W.J., and Ruddle, F.H. (1984b). Molecular cloning and chromosome mapping of a mouse DNA sequence homologous to homeotic genes of Drosophila. *Cell* **38**: 675–680.
Morgan L.V. (1922). Non-criss-cross inheritance in *Drosophila melanogaster. Biol. Bull.* **42**: 267–274.

Muller, H.J. (1927). Artificial transmutation of the gene. *Science* **66**: 84–87.
Nüsslein-Volhard, C. (1995). The identification of genes controlling development in flies and fishes. *Nobel Lectures,* pp. 273–294.
Nüsslein-Volhard, C., and Wieschaus, E. (1980) Mutations affecting segment number and polarity in *Drosophila. Nature* **287**: 795–801.
Park, J.H., and Hall, J.C. (1998). Isolation and chronobiological analysis of a neuropeptide pigment-dispersing factor gene in *Drosophila melanogaster. J. Biol. Rhythms* **13**: 219–228.
Peng, Y., Stoleru, D., Levine, J.D., Hall, J.C., and Rosbash, M. (2003). *Drosophila* free-running rhythms require intercellular communication. *Pub. Lib. Sci. Biol.* **1**: 32–40.
Poodry, C.A., Hall. L., and Suzuki, D.T. (1973). Developmental properties of *shibire*: a pleiotropic mutation affecting larval and adult locomotion and development. *Devel. Biol.* **32**: 373–386.
Price, J.L., Blau, J., Rothenfluh, A., Abodeely, M., Kloss, B., and Young, M.W. (1998). *double-time* is a new Drosophila clock gene that regulates PERIOD protein accumulation. *Cell* **94**: 83–95.
Ralph, M.R., and Menaker, M. (1988). A mutation of the circadian system in golden hamsters. *Science* **241**: 1225–1227.
Reddy, P., Zehring, W.A., Wheeler, D.A., Pirrotta, V., Hadfield, C., Hall, J.C., and Rosbash, M. (1984). Molecular analysis of the *period* locus in Drosophila melanogaster and identification of a transcript involved in biological rhythms. *Cell* **38**: 701–710.
Renn, S.C.P., Park, J.H., Rosbash, M., Hall, J.C., and Taghert, P.H. (1999). A *pdf* neuropeptide gene mutation and ablation of PDF neurons each cause severe abnormalities of behavioral circadian rhythms in *Drosophila. Cell* **99**: 791–802.
Reppert, S.M., and Weaver, D.R. (2000). Comparing clockworks: mouse versus fly. *J. Biol. Rhythms* **15**: 357–364.
Rothenfluh, A., Abodeely, M., Price, J.L., and Young, M.W. (2000). Isolation and analysis of six *timeless* alleles that cause short- or long-period circadian rhythms in Drosophila. *Genetics* **156**: 665–675.
Rubin, G.M., and Spradling, A.C. (1982). Genetic transformation of *Drosophila* with transposable element vectors. *Science* **218**: 348–353.
Rutila, J.E., Suri, V., Le, M., So, W.V., Rosbash, M. and Hall, J.C. (1998). CYCLE is a second bHLH-PAS clock protein essential for circadian rhythmicity and transcription of Drosophila *period* and *timeless. Cell* **93**: 805–814.
Rutila, J.E., Zeng, H., Le, M., Curtin, K.D., Hall, J.C., and Rosbash, M. (1996). The tim^{SL} mutant of the Drosophila rhythm gene *timeless* manifests allele-specific interactions with *period* gene mutants. *Neuron* **17**: 921–929.
Scalenghe, F., Turco, E., Edström J.-E., Pirrotta, V., and Melli, M. (1981). Microdissection and cloning of DNA from a specific region of *Drosophila melanogaster* polytene chromosomes. *Chromosoma* **82**: 205–216.
Sehgal, A., Man, B., Price, J.L., Vosshall, L.B., and Young, M.W. (1991). New clock mutations in *Drosophila*. *Ann. N.Y. Acad. Sci.* **618**: 1–10.
Sehgal, A., Price, J.L., Man, B., and Young, M.W. (1994). Loss of circadian behavioral rhythms and *per* RNA oscillations in the *Drosophila* mutant *timeless. Science* **263**, 1603–1606.
Siwicki, K.K., Eastman, C., Petersen, G., Rosbash, M., and Hall, J.C. (1988). Antibodies to the *period* gene product of Drosophila reveal diverse tissue distribution and rhythm changes in the visual system. *Neuron* **1**: 141–150.

Smith, R.F., and Konopka, R.J. (1981). Circadian clock phenotypes of chromosome aberrations with a breakpoint at the *per* locus. *Molec. Gen. Genet.* **183**: 243–251.

Spradling ,A.C., and Rubin, G.M. (1982). Transposition of cloned P elements into *Drosophila* germ line chromosomes. *Science* **218**: 341–347.

Stanewsky, R., Kaneko, M., Emery, P., Beretta, B., Wager-Smith, K., Kay, S.A., Rosbash, M., and Hall, J.C. (1998). The cry^{b} mutation identifies cryptochrome as a circadian photoreceptor in Drosophila. *Cell* **95**: 681–692.

Stern, D.L. (2014). Reported *Drosophila* courtship song rhythms are artifacts of data analysis. *BMC Biol.***12**: 38 [citation format for online-only article]

Stern D.L., Clemens J., Coen P., Calhoun A.J., Hogenesch J.B., Arthur B.J., Murthy M. (2017). Experimental and statistical reevaluation provides no evidence for *Drosophila* courtship song rhythms. *Proc. Natl. Acad. Sci. U.S.* **114**: 9978–9983.

Stoleru, D., Peng, Y., Agosto, J., and Rosbash, M. (2004). Coupled oscillators control morning and evening locomotor behavior in *Drosophila. Nature* **431**: 862–868.

Sturtevant, A. H. (1913). The linear arrangement of six sex-linked factors in *Drosophila*, as shown by their mode of association. *J. Exp. Zool.* **14**: 43–59.

Sun, Z.S., Albrecht, U., Zhuchenko, O., Bailey, J., Eichele, G., and Lee, C.C. (1997). *RIGUI*, a putative mammalian ortholog of the *Drosophila period* gene. *Cell* **90**: 1003–1011.

Suri, V., Hall, J.C., and Rosbash, M. (2000). Two novel *doubletime* mutants alter circadian properties and eliminate the delay between RNA and protein in *Drosophila. J. Neurosci.* **20**: 7547–7555.

Suzuki, D.T., Grigliatti, T., and Williamson, R. (1971). Temperature-sensitive mutations in *Drosophila melanogaster.* VII. A mutation ($para^{ts}$) causing reversible adult paralysis. *Proc. Natl. Acad. Sci. U.S.* **68**: 890–893.

Suzuki, D.T., Kaufman, T., Falk, D., and the U.B.C. Drosophila Research Group (1976). Conditionally expressed mutations in *Drosophila melanognster.* In: *The Genetics and Biology of* Drosophila, Vol. la., M. Ashburner, and E. Novitski, Eds., Academic Press, New York, pp. 208–263.

Syed, S., Saez, L., and Young, M.W. (2011). Kinetics of *doubletime* kinase-dependent degradation of the *Drosophila* period protein. *J. Biol. Chem.* **286**: 27654–27662.

Tatum, E.L., and Beadle, G.W. (1942). Genetic control of biochemical reactions in *Neurospora*: an "aminobenzoicless" mutant. *Proc. Natl. Acad. Sci. U.S.* **28**: 234–243.

Tatum, E.L., and Lederberg, J. (1947). Gene recombination in the bacterium *Escherichia coli. J. Bacteriol.* **53**: 673–684.

Tei, H., Okamura, H., Shigeyoshi, Y., Fukuhara, C., Ozawa, R., Hirose, M., and Sakaki, Y. (1997). Circadian oscillation of a mammalian homologue of the *Drosophila period* gene. *Nature* **389**: 512–516.

Villee, C.A. (1945). Phenogenetic studies of the homeotic mutants of *Drosophila melanogaster.* III. The effects of temperature on the expression of $bithorax^{34e}$. *Amer. Nat.* **79**: 246–258.

Villella, A., and J.C. Hall. 2008. Neurogenetics of courtship and mating in *Drosophila. Adv. Genet.* **72**: 67–184.

Vitaterna, M.H., King, D.P., Chang, A.-M., Kornhauser, J.M., Lowrey, P.L., McDonald, J.D., Dove, W.F., Pinto, L.H., Turek, F.W., and Takahashi, J.S. (1994). Mutagenesis and mapping of a mouse gene, *Clock*, essential for circadian behavior. *Science* **264**: 719–725.

Wheeler, D.A., Kyriacou, C.P., Greenacre, M.L., Yu, Q., Rutila, J.E., Rosbash, M., and Hall, J.C. (1991). Molecular transfer of a species-specific behavior from *Drosophila simulans to Drosophila melanogaster. Science* **251**: 1082–1085.

Wieschaus, E. (1995). From molecular patterns to morphogenesis. The Lessons from Drosophila. *Nobel Lectures:* 302–314.

Young, M.W., and Judd, B.H. (1978). Nonessential sequences, genes, and the polytene chromosome bands of *Drosophila melanogaster. Genetics* **88**: 723–742.

Young, M.W., Jackson, F.R., Shin, H.S, and Bargiello, T.A. (1985). A biological clock in *Drosophila*. Cold Spring Harb. Symp. *Quant. Biol.* **50**: 865–875.

Young, M.W., and Kay, S.A. (2001). Time zones: a comparative genetics of circadian clocks. *Nat. Rev. Genet.* **2**: 702–715.

Zehring, W.A., Wheeler, D.A., Reddy, P., Konopka, R.J., Kyriacou, C.P., Rosbash, M., and Hall, J.C. (1984). P-element transformation with *period* locus DNA restores rhythmicity to mutant arrhythmic Drosophila melanogaster. *Cell* **39**: 369–376.

Zerr, D.M., Hall, J.C., Rosbash, M., and Siwicki, K.K. (1990). Circadian fluctuations of *period* protein immunoreactivity in the CNS and the visual system of *Drosophila. J. Neurosci.* **10**: 2749–2762.

Zilian, O., Frei, E., Burke, R., Brentrup, D., Gutjahr, T., Bryant, P.J., and Noll, M. (1999). *double-time* is identical to *discs overgrown*, which is required for cell survival, proliferation and growth arrest in *Drosophila* imaginal discs. *Development* **126**: 5409–5420.

Michael Rosbash. © Nobel Prize Outreach AB. Photo: A. Mahmoud

Michael Rosbash

Biography

My parents and Germany: tough times

My parents were born and raised in Germany. My mother Hilda was born in 1914 and came from a secular, quite comfortable family in Berlin. Her father, my grandfather, Magnus Sonntag had a pharmacy, which still exists today with the same name and in the same location (Marien-Apotheke, Wilhelmsaue 110, 10715 Berlin). As the story was told to me, Magnus was a soldier in WWI and home on weekend leave when he discovered his wife, my biological grandmother, with another man. Magnus kicked her out, she moved to Holland, which left her 3-year-old daughter with a father at the front and no other parent in Berlin. My mother was sent to live with her grandparents for a few years and never saw her biological mother again – although there were some subsequent communications (see below). My grandfather remarried, and my mother thankfully adored her step-mother. There were two much younger children from this marriage, and all 3 siblings have now passed away.

My father Alfred was born in 1912 and raised in Baden-Baden. My grandfather Joel Rosbasch (my parents dropped the c upon arriving in the US) went to Germany as a young man sometime around 1907 to escape from difficult economic times in the Ukraine. He left his family including 3 young children behind in Kremenchuk and got a job in a cigarette factory in Stuttgart, not very far from Baden-Baden. Cigarettes were rolled by hand in those days. My grandfather was skilled in the art and so had no problem securing a job in early 20th century Germany. He sent for his family, including my father's 3 older siblings, who were all then raised in Germany. According to my father's one younger sibling, my Aunt Lotte (she passed away at the age of 98 in 2013), my grandfather saw the handwriting on the wall and left cigarette-making shortly before that industry was mechanized. He moved his family to Baden-Baden and ran a small grocery-dry goods store there, together with my grandmother; they lived above the store. Although the Rosbasch family was quite religious and kept kosher, the children – my father and his siblings – were educated in the secular German system. My father was a top student at Gymnasium and was asked to tutor those fellow-students who were struggling with math – again according to my Aunt Lotte.

Both of my parents were in university, my father studying law and my mother medicine, until Jews were no longer allowed to pursue these professions -- in about 1933 I believe it was. They then both turned to Jewish professions. My mother studied some aspect of Jewish education, and my father – who had an excellent voice as well as a religious background -- became a cantor. My parents were married in 1937 and in that same year my father got his first real job as a cantor in a synagogue in Breslau, which is now Wroclaw in Poland. According to my mother and my Aunt, my dad loved that job as well as the Germany of his youth (see below). He didn't want to leave; like many German Jews he found it incomprehensible that the madness of the mid-late 30s would not somehow dissipate and then evaporate. My mother was more pessimistic, realistic one can say in hindsight, and insisted on going to the US – where there was some distant family who helped with visas. I vaguely recall that her biological mother sent her from Holland the money for passage. My mother was never one for negotiating or discussing something that she felt strongly about, and so just said to my father that she was going to the USA either with or without him. Being besotted about my mother (yes – story from Aunt Lotte once again), he left for the States with my mother in 1938 a few months before Kristallnacht. The next morning after this wicked night, the local Baden-Baden Gestapo came to my grandmother's home for my father, demanding "the tall one," to which she defiantly responded, "You're too late; he's gone to America."

At the risk of a tangent, it is interesting to note that this one year as a cantor was sufficient to make my father "beamte" or tenured in the German civil service. This status entitled my father to a pension, and my mother collected a widow's pension for decades based on this one year of service; this had nothing to do with being Jewish or restitution. Few people in the United States know that all clergy in Germany are state employees and receive their salary from the federal government. This is true for rabbis, priests, imams as well as protestant ministers, and it was true in the 30s and is still true today. What is remarkable, and a testament to German bureaucracy, that all of this continued to function like clockwork (pun intended) – even for Jews – at the same time as another wing of the German government was gearing up to efficiently carry out the final solution. As Dave Barry would say, "I am not making this up."

MOVING TO THE US AND MY CHILDHOOD

My parents arrived in New York with few resources and no job. While my father searched for a job, my mother cleaned hotel rooms – a fact she bitterly recounted to me for the rest of her life as if it were my doing. (I had/have lots of Jewish guilt but this was too much even for me.) My father joined the myriad of rabbis and cantors from Europe pounding the New York pavement and looking for work, to no avail. At this time my mother was looking through some National Jewish newspaper, the Forward perhaps, formerly the Jewish Daily Forward, and saw an advertisement for a cantor's job in a reform congregation in Kansas City Missouri. She suggested my dad apply, who said "But we're not Reform Jews." (To more religious Jews, Reform Judiasm at that time was an anathema, essentially indistinguishable from Christianity.) My mother was not to be deterred. "We are now; making a living comes before some ridiculous religious division," she replied. My dad went to Kansas City, interviewed for and was offered the job, and my parents moved there in late 1938. So my mother once again came to the rescue.

My parents loved Kansas City and had nothing but good things to say about their 8 years there. They made wonderful friends, learned lots about America including how to drive (taught by some of those friends) and had their first child; I was born there in 1944.

I have two family anecdotes about Kansas City. The first concerns a seminar that I gave perhaps 15 years ago in the Stowers Institute for Medical Research in that city. I was in the office of the Director Rob Krumlauf, when he told me that the Institute had been Menorah Hospital, a famous Kansas City hospital before the hospital building was gutted and turned into the Stowers. I vaguely recalled the name of the hospital where I was born; sure enough, the Stowers seminar room is at or very near the precise location of the Menorah Hospital maternity ward where I had been born.

The second anecdote comes from my Aunt Lotte, my father's younger sister whom I have already introduced. She was a 29 year old single woman when I was born and came to Kansas City in 1945 to help out my mother with her young child. One day in May during that stay, Aunt Lotte went to the cinema with my father. This was before TV and the only place where one could see a recent newsreel, which happened to show that day the surrender of the German generals. My father said to his sister after watching, "I am surprised to discover how sad I am, how ambivalent I apparently am about this allied victory." He was of course glad that the war had come to an end and with a victory for the US. Nonetheless and despite his religion and negative experiences in Germany during the 30s, he had been educated as a German patriot and to have respect for the German military; those sentiments had not entirely disappeared. The moral lesson: life is not simple.

My parents moved to Boston in the summer of 1946 when I was two and a half years old. Moving between synagogues is not unlike moving between academic institutions. A bigger, more prestigious synagogue offers a rabbi or cantor a job with better conditions and more salary, perhaps in a more interesting city. The offer was successful in moving my family.

We lived in an apartment in Brookline until 1950 at which point we moved to our own home in Newton, all part of the American dream. I was six at the time and began second grade at the Cabot School. As I have recently described in some detail ("Life is an N of 1"), I had some behavioral issues throughout school – probably throughout life. In hindsight, it is likely I had ADHD or some variant thereof, but the worlds of education and psychology were too naïve to treat these kinds of problems in the 50s. Perhaps this was a good thing, at least for the kids who were problematic but not too disruptive. Meds today may be given too liberally and too quickly, perhaps as much to help teachers and parents as to help the troubled kids. Certainly, I turned out OK without any meds or treatment.

My dad died of a heart attack in the fall of 1954 at the age of 42. He died in the synagogue on Yom Kippur eve, shortly after singing Kol Nidre, the liturgical chant which is sung by the cantor at the beginning of this most sacred holiday. It is said that only the holiest of men die on Yom Kippur; I would like to believe this is true. I should avoid any confusion at this point and state unequivocally that I am a devout atheist and quite anti-religion. Nonetheless, I do have an irrational connection to this event and to my family history, which is now obvious even to the most casual reader. Emotional issues are not easily dismissed, and my own health history is not unrelated. I had a heart attack at the age of 38 and am still here and in decent shape at the age of 74.

My father had had an initial heart attack a few weeks prior, for which he had been hospitalized. There was apparently a big family brouhaha that ensued over whether he should return to singing, which is physically quite stressful. My mother told him that he had to decide himself, and so with his physician's OK, he went back to his normal job. If hindsight is 20:20, foresight in this case was blind.

Our small nuclear family was destroyed by my father's death. My fragile mother, just 40 years old, never fully recovered; my brother and I – ten and six at the time – were left to fend for ourselves, at least emotionally. In this case perhaps and in contrast to my possible childhood behavioral issue mentioned above (ADHD), some professional intervention might have been a good thing, for my mother as well as for me and my brother.

My father's death was an additional blow for my mother in an entirely different way. She had matriculated at BU Medical School and was slated to begin classes in the fall of 1954. This was when my brother began first grade, when both of us would finally be in school full-time. This was therefore the first opportunity in more than 20 years that she had to fulfill her dream of becoming a physician, the path she had been on in 1933 when the Nazis had forced her to leave university. BU had even fast tracked her path by giving her two years of credit for her time in a German university studying medicine. However, we had no financial resources and so my mother had to go to work to support her two children. She took a six-month course at the Massachusetts General Hospital (MGH) and then went to work at the Beth Israel (BI) Hospital as a cytologist. Within a few years, she became the head of this new subdivision of the BI Pathology Department. She supervised a team of cytologists and ran this well-known and profitable department for many years. After being a widow and working at the BI for 20 years, she married another German Jew, a widower from Pittsburgh. She moved there in 1974, at about the same time I came to Brandeis, and lived in Pittsburgh for the last 34 years of her life, until she died in 2008 at the age of 94.

Although painful, this history still does not do justice to the difficult life my mother had. I have already described the extent to which her education and secure world as a comfortable Berlinerin were destroyed by the Nazis. She then had to begin life anew in the United States at the age of 24. In addition, however, her own Sonntag family was divided by thousands of miles; her parents and brothers went to Brazil in the late 30s when my mother and father emigrated to the US. My mother only saw her parents once after 1938, a visit she made alone to Brazil fourteen years later in 1952. Thank goodness she made that trip, because my grandparents both died one year later in 1953, a year before the death of my father. They never managed to see their daughter's children, and my brother and I never met our maternal grandparents. In addition to all this history, can you imagine losing both your parents and your husband within one year at the age of 40, and then having to go to work to support your two young children?

This short ode therefore serves to put into perspective some less than admirable personality features of my mother: she was emotionally distant, not very empathetic and also quite selfish. I do prefer nature (hard-wiring) over nurture (environment) as the principal explanation for personality, but the hard road my mother had to travel has made her family more forgiving about her shortcomings. Lastly, she was unbelievably proud of her children, including my academic accomplishments. She would always ask my wife, "Do you think he will win the big one?" What a shame that she died 9 years too early and did not live to see my Nobel Prize.

HIGH SCHOOL AND COLLEGE (CALTECH)

I was a rather indifferent high school student and went to Caltech to escape the unhappy home life briefly described above. Caltech was a good school and also as far away as I could get from Boston in the early 1960s. Going to college there was a stroke of good fortune, because it was academically very challenging. My fellow undergraduates were really smart, and most of them were also hard-working – a combination of diligence and fear of failure. I in contrast had no study habits from high school, a combination of the likely ADHD I referred to above and the fact that I could succeed reasonably well without any effort. This was impossible at Caltech, and after a year or 18 months of trying to succeed without working like in high school, I finally succumbed to the old adage, "If you can't beat 'em, join 'em." So I buckled down, started to study and did well academically.

I remember that the Caltech Dean of Students Paul Eaton told me at graduation in 1965 that I had the very lowest projected GPA of all the students in my class. (In those days before computers, Caltech had some primitive system for projecting the academic performance of their applicants.) He told me this to emphasize how proud he was of my performance, which had dramatically exceeded expectations. In hindsight, I had been accepted to Caltech almost certainly because of "geographical distribution;" west coast schools wanted east coast students as well as west coast students. My good academic performance at Caltech is an important lesson I try to remember: statistics are important, especially for making policy, but there are always individual outliers.

Not only did I do well academically, but the course material had also become MUCH more interesting. Skating along the surface, studying only just before exams, doesn't provide the positive feedback that comes with real learning, with thinking often about academic material. Once the old habits were broken, they were replaced by the positive feedback loops of understanding.

I was aided in this transition by a wonderful advisor and mentor, Norman Davidson. ND as he was called was a fantastic chemist, who was just transitioning in the early 60s from physical chemistry and statistical mechanics to nucleic acids. He was a no-nonsense guy, who challenged me to do well. He was also a fantastic role model with his complete joy in doing research and running a lab. I had never met a grown-up who loved his work so thoroughly. I remember the day I decided to do research for a living. ND was walking down the hall away from me and was wearing a t-shirt. On the back it said on top, "I'd rather be in the lab," and then on the bottom, "Or maybe playing tennis." (He really liked sports, skiing as well as tennis, and had been a basketball player as an undergraduate at the University of Chicago.). I didn't know what it was really like to run a lab, but I said to myself, "this has to be a good thing if it can bring such unbridled joy."

PARIS

I went to Paris for year after Caltech. It was highly unusual in the 60s to take time off, to not go directly to graduate or professional school, but I had a desire to see the world. (My mother was an inveterate traveler, so perhaps my wanderlust was inherited or perhaps culturally transmitted.) I had been working and saving money during my junior and senior years of college to travel for a year, when I decided on a lark to apply for a Fulbright Scholarship. I knew some French from high school and even one semester in college, and Paris was a romantic destination. I also had learned in a Caltech class about Jacob and Monod (the lac operon, gene regulation and allostery), and so Paris even seemed like a good scientific destination. My choice was prescient, because these two gentlemen – along with their French/Pasteur colleague Andre Lwoff – won the 1965 Nobel Prize in Physiology or Medicine, which was awarded a month or so after I arrived in Paris.

To my surprise, I was awarded that Fulbright Scholarship and was off to Paris by boat from New York together with the rest of my fellow Scholars in the late summer of 1965. The Fulbright organization used that trip and the first two weeks of our time in Paris for group bonding and orientation. My recollection is that I was the only scientist in the group, which was great for my general education. Most of my colleagues were Ph.D. students in French literature, from prestigious US universities. They were therefore not only older but also wiser than I was, especially in matters that concerned France, French language and French culture; my fellow Scholars therefore helped me acclimate to my new circumstances.

Despite my genuine praise above for Caltech, it was at the time an institution with no female undergraduates and more generally a rather narrow cultural bandwidth. It turned out that I was desperate for a different experience and so fully embraced what Paris had to offer. This was just about everything: in addition to the obvious – women, cuisine and wine – about which I knew virtually nothing at the age of 21, there was the remarkable cultural and political heterogeneity of mid 1960s Paris. I was stunned by all the refugees and students from all over the world, which reflected the genuinely cosmopolitan nature of the city as well as the influence of colonial France and the French language in the Middle East, Africa and Asia. It also reflected the parochial nature of my life in the US in the 50s and 60s; I don't think I had ever met an Arab before Paris, and I didn't know where the Maghreb was or what the word meant. (It is a major region of North Africa, including Morocco, Algeria, Tunisia and Libya; I also just learned from Wikipedia that it includes Mauritania.) There were also tons of Lebanese in Paris; I did know where Beirut and Lebanon were only because – like most Jews at the time – I knew they were just north of Israel.

1965 was at the end of France's colonial era, 9 years after Dien Bien Phu and only 3 years after the end of the Algerian war. This recent history explained many of the refugees and the left-wing politics that were thriving in the Paris streets. However, I also met charming and generous right-wing people. Most memorable were the Pieds-Noirs. (The term refers to Europeans, mostly French people like Albert Camus, who had lived in Algeria for generations and had now "returned" to France.) Their politics was understandably colored by their experience of having been uprooted from their homes and adopted land by revolution. I learned a lot from all of these people, from their politics, their languages, their experiences, their stories, their families, and even their home-made couscous.

You might be wondering: how did a young scientist have time for this cultural accretion? The short answer is that I didn't work much that year. The longer answer is as follows: I was assigned by the Fulbright organization to the lab of Marianne Grunberg Manago at the Institut de Biologie Physico Chimique. Marianne was a very famous 44-year-old scientist in 1965. She had been a post-doc of Severo Ochoa at NYU and famous for having done the work for which he won the Nobel Prize in 1959. In fact Marianne's enzyme, polynucleotide phosphorylase, won two Nobel Prizes as it was used to synthesize the oligonucleotides used by Marshall Nirenberg to crack the genetic code (1968 Nobel Prize).

Marianne was a very nice but somewhat imperious European professor and totally dumfounded by the assignment of a very assertive 21-year-old Caltech undergraduate to her laboratory. She put me to work with her wonderful technician Jacques Dondon to help make the stock of charged tRNAs for the lab. This preparative work took a full week and was interesting the first week, tolerable for the second but quite boring by the third. Every week I would ask Marianne to give me a proper research project, and she kept responding by saying she would do it "next week." So after 4-5 weeks of this back and forth, I changed my tack. I said, "Marianne, I am going back to graduate school in the United States without question, but I am now having the time of my life here in Paris in other ways. I am seeing and learning things I never imagined existed. I like the lab here and will work if I have a research project but not without one. So if you still don't give me a research project next week as I have been asking, I will stop coming to the lab except once a week to collect my paycheck and to peruse the journals." She never gave me a project, and I stopped working in the lab as previewed. I spent the year doing all these other things in Paris, including learning to speak French well, and I also traveled all across Europe. In the decades that passed, Marianne never mentioned this conflict to me – if I can call it that. She followed my career and proudly saw me as one of her scientific progeny.

In the immortal words of Edith Piaf, "I regret nothing" about my year in Paris. I made dear friends and acquired a life-long appreciation for French culture and the French language. I even benefited professionally because I had a long string of outstanding French students and post-docs who came to Brandeis years later and had a huge, positive impact on my career; it all began I believe with that year in France at the tender age of 21 and with my subsequent quite fluent facility with French. Moreover, I was offered big director jobs 20–25 years later. The offers were flattering and I was tempted, but I was recently married with my current wife, a Chilean who adored the States and was reluctant to change countries once again. Moreover, she convinced me – correctly in hindsight – that I would never be able to navigate institutional politics in a foreign culture like France. There is too much important that is left unsaid, in conversation and negotiation, that only a native can glean, despite excellent language skills. My wife said, "Let me put it this way; if you have had trouble at Brandeis, imagine what this would be like in France."

MIT

I went to graduate school at MIT, which was a great place for me. I decided to work on eukaryotic gene expression in the lab of Sheldon Penman. He was very smart, committed to the lab and a caring mentor. There was also a wonderful collection of students and post-docs in his lab. Notable from that period of time were Bob Weinberg, Hung Fang and Rob Singer. My work went well. I gained confidence and experience, and generated first-rate publications.

MIT was a much bigger, more cosmopolitan place than Caltech, and the late 60s was a more interesting period of time in the United States than the early 60s. Vietnam had polarized the country, MIT was a center of political activity, and marijuana had gone mainstream. I was very engaged in the anti-war movement and considered devoting more time to this political work. I spoke to Sheldon about it, and he told me I would have to change advisors if I wanted to do politics in a way that would interfere with my lab work. Moreover, he said that there were MIT faculty members who would be OK with a less than complete effort to the lab, who were themselves committing considerable time to anti-war activities. Because Sheldon was quite right-wing and in favor of the US military effort in Vietnam, I thought this might have contributed to his inflexibility. So I went to Sheldon's younger colleague David Baltimore for advice. David was anti-war and only six years older than I was, i.e., in some ways my peer. To my surprise at the time, he endorsed Sheldon's viewpoint by agreeing that a PI can insist on the effort that a graduate student should make in his/her lab. And he too suggested I switch labs or agree to pare down my political activities if I wanted to stay working for Sheldon. I respected and was grateful for David's candid opinion, which had a big influence on my decision to remain in Sheldon's lab. My scientific career has almost certainly had more influence than any political contribution I might have made, and I would not be writing this Nobel biography had I not made the decision to remain with Sheldon.

Two anecdotes stick in my mind from my 5 years at MIT; they are both illuminating I suspect. Sheldon and I had the identical old car. When he decided to upgrade and buy a new car, he generously offered me his old one for parts. He knew I had some experience fixing cars from my Caltech years and so might make use of the gift. I accepted but had another idea in mind. I spent an entire Sunday trying to remove intact his faculty parking sticker so I could affix it to my driver's side window. Failing that, I tried removing his window to replace mine, but that too was problematic. I finally settled on removing my driver's door and replacing it with his door. The only minor problem was that his car was blue and mine was white, so I drove for the next year a two tone car, white with a blue door. (No way was I going to pay to have my car painted to address this minor issue.) To my delight, the parking garage guards at MIT paid no attention to my new two-tone car, and so I parked in the close and prestigious faculty lot for a year. As a testament to habituation, I became more and more bold over the next year and was eventually caught by the police for parking illegally in that lot. Sheldon was ticketed because the sticker was registered to him. He read me the riot act but was secretly amused, I always thought, admiring perhaps of my chutzpa.

The second anecdote was when I – a graduate student – fired one of the technicians in Sheldon's lab while he was out of town. I was working at the hood (fume cupboard) and jostling for space with a technician, who had admittedly been there first. I was in a rush and told her in a rude way that my work was more important. She took offense, not unreasonably, and said something like "I can't work here any longer," to which I responded with an even more offensive remark. She then stormed out and did not return for the rest of the day. It took an hour or two for the reality to hit me, including the fact that Sheldon was due back at MIT the next morning. I had to drive to the technician's apartment that evening and beg forgiveness. She wouldn't open her door for me, and it took 30 minutes of pleading and throwing pebbles against her window to get her to let me in, finally accept my apology and promise not to tell Sheldon the next morning.

A simple, polite summary of both anecdotes is that I had a difficult character. Thank goodness for the permissive, tolerant environments of Caltech and MIT.

POST-DOC AND EDINBURGH

I had planned to go for my post-doc to the wonderful Hogness lab at Stanford, to study *Drosophila* chromosomes, and I wrote and received a Helen Hay Whitney Fellowship to do that. A few short months before leaving MIT however, I had a change of heart. It was catalyzed by meeting a couple of chromosome-nucleic acid researchers from Scotland, Mick Callan and John Bishop, both of whom spent some time in 1970–1971 in Boston. There was also my wanderlust; I thought this might be the last time I would be able to live in Europe, an opportunity I should not pass up. So I inquired about going to Callan's lab in St. Andrews and planned to collaborate with Bishop in Edinburgh at the same time. I wrote the Whitney foundation, which gave me permission to switch. I then wrote to Hogness and honestly explained my decision, including my desire to live for a while in Europe. He wrote me back a handwritten note and was a complete gentleman about the situation. After I became a PI myself, I appreciated even more his graciousness. Not going to his lab had multiple layers of irony: Mike Young went there a few years later, and many of my colleagues and friends were trained in the Hogness lab at the same time as I would have been there, e.g., Ray White, Gerry Rubin and Michael Grunstein to name just a few.

I was not very happy in St. Andrews. It was a lovely town but a bit sleepy after Boston and MIT, culturally as well as scientifically. Callan was an excellent scientist but not very communicative; he did his own work and kept to himself. So, I transferred to the Bishop lab in Edinburgh, which was a more dynamic city and scientific environment. John's lab was an excellent place, and I learned a lot from him as well as from my colleagues in his lab, for example Nick Hastie, Saveria Campo, and Stanley Perlman. There was the Birnstiel lab next door, which had Peter Ford as a post-doc and Adrian Bird and Michael Grunstein as students. I collaborated with Ford during my time in Edinburgh. A couple of hundred meters or so down the road was the Zoology Department with Ed Southern and his laboratory.

Edinburgh and the UK will always have a special place in my heart, and I have life-long friends from my time there, especially Michael Grunstein and his family. The Brits also have a special place in my head. Their educational system and general approach to science was not identical to the way things were done in the US. Simply put, the UK approach was more cerebral whereas Americans are more pragmatic. I would like to think my exposure to the UK way of doing science complemented my American education and natural instincts, which are almost pathologically pragmatic.

RECRUITMENT TO BRANDEIS

I was only about 15 months into my post-doc when Brandeis called me up and asked me to apply for a faculty position. I said it was too early and I wasn't interested, but Harlyn Halvorson, the Director of the Rosenstiel Center at Brandeis persisted. "Just come give us a seminar." Since my mother lived only a few miles from Brandeis, it was an opportunity to visit her too, so I left Edinburgh for a long weekend. I gave a seminar, met with faculty, and they ended up offering me the job I had said I didn't want. It turned out that they had a blue ribbon committee to recommend people for the two molecular biologists they wanted to hire. The chair of that committee was Jim Darnell. He was the mentor of my MIT mentor Sheldon, had known me since the beginning of my Ph.D. there and had kept his eye on me.

The upshot was that Brandeis leaned over backwards to get me to accept the job. They let me stay in Edinburgh for more than 18 months to complete my post-doc. The space, set-up money and teaching conditions were generous. And then to top it off, they offered to pay me my Brandeis salary for the year before I would arrive on campus. Although this turned out to be complicated (I couldn't keep my fellowship while keeping my salary), Brandeis supplemented my salary to bring my total compensation to match an assistant professor salary. This effectively doubled my 1973 salary, from about $10K to $20K. My post-doc salary of 10K was already extraordinarily generous by UK standards, especially with the then current dollar to pound exchange rate. To put my 1973 salary in perspective, it was considerably higher than that of my PI Bishop. Moreover, he was a Reader, the equivalent of an Associate Professor, with a mortgage and 3 children, and I was single. One important outcome of all this was that I banked the extra 10K, which was exactly the 25% down payment for the house I purchased in 1975–76 more than 40 years ago. We raised our children in that house and still live there. I therefore owe Brandeis the roof over my head in addition to the shirt off my back and a good fraction of my Nobel Prize.

There is of course more to say, but I am ending this brief biography to coincide with my arrival at Brandeis, when the Brandeis circadian rhythm story began.

Circadian Rhythms and the Transcriptional Feedback Loop

Nobel Lecture, 7 December 2017 by
Michael Rosbash
Howard Hughes Medical Institute, Brandeis University,
Waltham, MA, USA.

INTRODUCTION

Circadian clocks undoubtedly arose in response to the daily changes in illumination that are due to the rotation of the earth. Indeed, sunrise and sunset are very consistent external events, if we ignore the slight change in timing that usually occurs every day. Although this change can be substantial at some times and in some locations, in March near the north and south poles for example, life and biological clocks probably evolved in more temperate zones.

Clocks function to allow organisms to anticipate daily changes in their environment. When something happens every day at the same time, organisms "learn" that the event will occur. This anticipation, preparing for what is going to happen, is a superior strategy to merely reacting to that change. Animals use their clocks to maximize or minimize their encounters with what I like to call the big 3: finding food, finding mates, and avoiding predators.

Clocks may also serve to provide order to internal processes. The temporal separation of different biochemical reactions is an example and may provide an additional fitness advantage. These inexorable rhythms, due to the persistent ticking of a biological clock in our brains and in many if not most other tissues, serve to coordinate myriad features of our behavior, metabolism and physiology – even our sleep-wake cycle (Hardin and Panda, 2013).

Circadian rhythms are present in most if not all animals, plants, and some species of fungi like *Neurospora*. Even some photosynthetic bacterial species like cyanobacteria have well-studied circadian clocks (Kondo et al., 1993). Indeed, the first circadian clock may have belonged to photosynthetic cyanobacteria, which are probably responsible for oxygenation of the atmosphere. These cyanobacterial clocks as well as plant clocks are very different from those of animals, with no credible homology between the different clock proteins. Therefore circadian rhythms probably emerged multiple times in evolution, which underscores their importance (Rosbash, 2009).

The rotation of the earth on its axis is a source of daily light and temperature cycles in most locations on earth. These daily cycles are considered the oldest and most continuous environmental features to which life was originally exposed. Even the self-replicating molecules of the original RNA world that preceded cellular life probably arose in the presence of light-dark and temperature cycles, well before the atmosphere was anything like it is today. The rotation of the earth was also perhaps 20 percent shorter four billion years ago than it is today, which may help explain why circadian clocks are not precise twenty-four-hour timekeepers (Rosbash, 2009).

The fact that endogenous circadian rhythms have periods distinct from precisely 24hr is a key piece of evidence that the rhythms are entrained rather than driven by the solar cycle. It is generally acknowledged that the first person to recognize that circadian rhythms are not driven was an eighteenth-century French astronomer named de Mairan. He had been musing about the movements of plants, which famously extend their leaves during the daytime and retract them at night. It was assumed that this daily leaf movement rhythm is light-driven, but de Mairan tested this hypothesis in 1729. To perform this first "free-running" circadian experiment in the absence of a light-dark cycle or entraining cues, he took a plant down to the basement, probably his wine cellar. In principle, it had constant conditions, i.e., no daily variations in light. Miraculously, the leaf movements continued unabated on a circa twenty-four-hour schedule, indicating that they were driven by an endogenous circadian clock rather than by light (Daan, 2010).

Importantly however, de Mairan did not carefully measure the free-running period of the leaf movements, nor did other scientists addressing this problem for decades thereafter. Although there were no light fluctuations in the basement during the experiment, other oscillating stimuli were not measured or considered, e.g., humidity, temperature etc. Because the world lives on a 24hr schedule, any human interference could provoke an environmental oscillation of 24hr and drive the plant rhythms. Neither de Mairan nor other plant scientists who followed him considered that the key was not only the absence of light but also a careful measurement of non-24hr period length. Although a 24hr period could reflect an endogenous timekeeper with that precise period, it could also reflect a driven oscillator, one that requires external 24hr oscillations of some kind. A non-24hr period however must reflect an endogenous timekeeper. (I am taking a bit of license in making this statement so strongly.) For some reason, it was a very long time before this distinction was appreciated and leaf movement period measurements made with much more precision. Eventually, however, they were done and shown to be non-24hr, indicating unambiguously that leaf movements were not driven by a feature of the external world but reflected endogenous circadian timekeepers (Daan, 2010).

The circadian period of other organisms is similarly almost always measured under constant or "free-running" conditions, which usually means many days, weeks or even months in darkness; humans are commonly measured in another more elaborate way because of ethical issues with a constant darkness protocol. Endogenous periods usually vary between 22 and 26 hours. Mice for example have periods of about 23.5 hours and rats about 24.5 hours. Importantly, there are even differences between inbred strains of mice, indicating that quantitative features of the circadian system are under genetic control; I will have much more to say about genetics and mutants below. Importantly, these circadian or not exactly 24hr rhythms are normally reset or "entrained" by the light-dark cycle. This is how humans, rodents and other organisms have a cycle that is identical from one day to the next: a somewhat slow or fast clock is advanced or delayed every day by the 24hr light cycle (Daan, 2010).

In the early-mid 20th century, a similar endogenous clock assay was carried out for *Drosophila*. This organism shows a rhythm of eclosion, the emergence of adult flies from their pupal cases. This is a population assay as any one fly experiences this event only once. However, the population emerges in a circadian manner with a major peak after dawn. In other words, adult emergence is gated by the circadian clock; animals that have not emerged by late morning must wait until the following morning. This is presumably to reduce exposure of newly emerged adults to mid-day and late afternoon heat. As expected of a circadian phenomenon, this circadian gating of eclosion continues in constant darkness. Endogenous periods of *Drosophila* were measured under these conditions; although they vary somewhat from species to species and strain to strain, they are very close to 24hr (Bunning, 1935; Kalmus, 1935, 1940a, b).

The circadian pioneer Colin Pittendrigh introduced and championed many important aspects of circadian biology. He was among the first if not the first to appreciate the importance and universality of the following phenomenon: if clocks were to function properly, especially in poikilotherms like *Drosophila*, then there must be a mechanism to keep their timing system effectively insensitive to temperature changes (Pittendrigh, 1954). When the ambient temperature drops or increases substantially, say by 10°C, the insect still must be able to tell time, e.g., to properly anticipate dawn and dusk. Pittendrigh showed that the endogenous period of *Drosophila pseudoobscura* does not change much with temperature as measured with the eclosion assay in constant darkness (Pittendrigh, 1954), and the importance of temperature compensation was also appreciated by others (Hastings and Sweeney, 1957). The eclosion assay was used in the landmark 1971 paper by Konopka and Benzer, which arguably initiated the modern era of circadian molecular genetics and certainly initiated my circadian work (Konopka and Benzer, 1971).

KONOPKA AND BENZER

Seymour Benzer is an American hero, a one-of-a-kind scientist (Weiner, 1999). As a solid-state physics Ph.D. student at Purdue University in the 1940s, he came close to discovering the transistor, a discovery that won William Shockley, John Bardeen, and Walter Brattain of Bell Labs the 1956 Nobel Prize in Physics. He transferred his interests to molecular genetics and studied the fine structure of the gene as a Purdue faculty member. For this work during the '50s and early '60s, he shared many prizes including the 1971 Lasker Award with Sydney Brenner and Charles Yanofsky. Many people, including me, often wondered why Benzer did not win a Nobel Prize for this groundbreaking work. (Brenner won a Nobel Prize in 2002 for his later work on the worm, *C. elegans*.)

Benzer moved to Caltech in the mid-'60s and changed his research focus, from a study of phage genes to the foundations of behavior using fly genetics. Caltech had been a major center of *Drosophila* genetic work since T.H. Morgan moved there from Columbia in 1928. (Morgan won the 1933 Nobel Prize in Physiology or Medicine; this was the first American Nobel Prize in Physiology or Medicine and the first *Drosophila* Nobel Prize.) Benzer's strategy followed what had been done in other areas of fly biology, namely associate single mutations and their underlying genes with a phenotype. The strategy was simple: the phenotypes in this case would be behavioral and the identified genes would therefore be – by definition – behavioral genes.

Ron Konopka was Benzer's first student (Fig. 1). Importantly, Konopka had become interested in circadian rhythms in college and brought the

Figure 1. Ronald J. Konopka, around the time he began graduate school at Caltech (Rosbash, 2015).

circadian problem to Benzer, contrary to what one might imagine from first principles. Konopka also designed as well as carried out the first eclosion screen to search for credible circadian mutants (Rosbash, 2015).

Konopka exploited the remarkable genetics toolkit of *Drosophila* to simplify the eclosion screen. He used the classic attached X strategy so that the X chromosome of a mutagenized male would be inherited by his male progeny; like in humans, the Y chromosome is normally inherited by males from a male parent. However in *Drosophila* and now unlike in mammals, it is the X:autosome ratio that determines phenotypic sex; the Y chromosome predominantly contributes to sperm motility. With this attached X strategy therefore, the phenotypes of mutated genes on the X could be assayed in males of the first (F1) generation with no need for

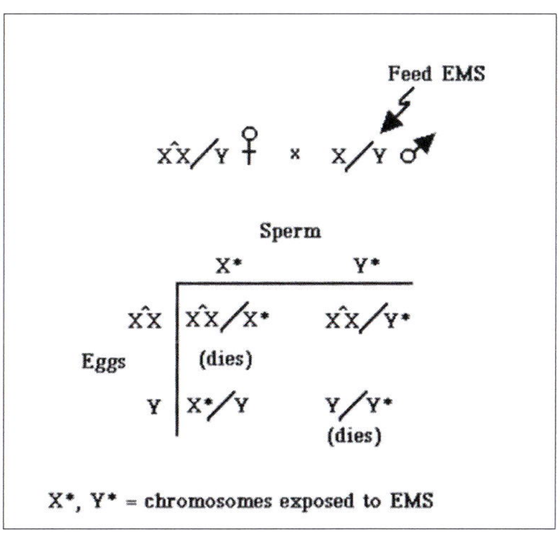

Figure 2. Identification of X-linked mutations in D. melanogaster. EMS-mutagenized sperm is used to fertilize eggs containing either an attached X chromosome or a Y chromosome. Recessive mutations on the X chromosome from mutagenized sperm are phenotypically visible in the F1 male offspring (Baker and Woodard, 2007).

backcrossing. As there is only a single mutagenized X chromosome in these F1 males, even recessive mutations would be visible. The disadvantage was that only about 20% of fly genes are present on the X chromosome, so potentially interesting genes present on the autosomes would be invisible (Fig. 2).

The screen was remarkably successful, and the Konopka-Benzer paper identified 3 mutant strains, each with a very different circadian phenotype from the normal wild-type fly with a ca. 24hr period. One mutant had a short period, about 19–20 hours, the second had a long period, about 29 hours, and the third had no apparent rhythm at all; it appeared arrhythmic. Remarkably, the phenotypes were not only assayed with the traditional fly eclosion assay (Fig. 3) but also with a locomotor activity assay (Fig. 4). Even more remarkably, the mutations would not complement, i.e., they appeared to be alleles of the same gene, which they named *period*. More extensive genetic analyses in the paper were consistent with this single gene conclusion. In addition, the phenotypes were semi-dominant, a result that has been extended to many circadian genes in different systems in the subsequent literature. These properties – including the mind-boggling fact that a single gene could apparently mutate to speed

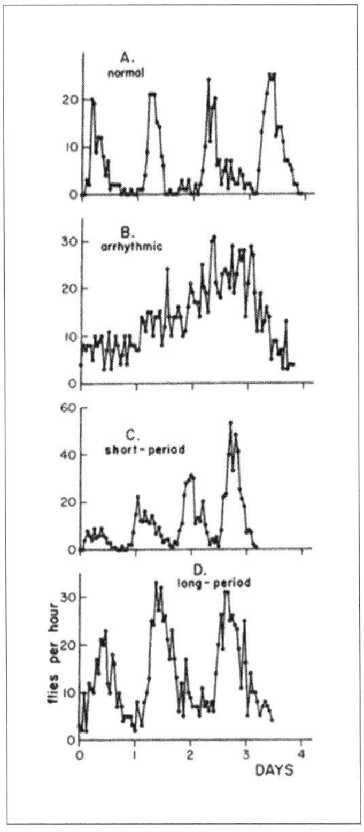

Figure 3. Eclosion rhythms of wild-type and Konopka and Benzer's period mutants (Konopka and Benzer, 1971).

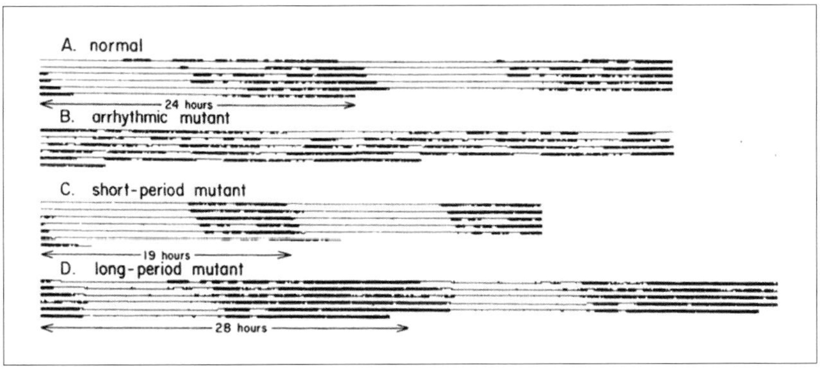

Figure 4. Locomotor activity patterns of wild-type and period mutants (Konopka and Benzer, 1971).

up, slow down or stop the clock – suggested that the rhythm-generating machinery reads out *period* gene expression in some semi-quantitative manner to influence circadian time-keeping in multiple directions (Konopka and Benzer, 1971).

The field made very little progress for the next decade or so. Although I am not aware of another laboratory other than Konopka's that was working on *Drosophila melanogaster* circadian rhythms in the early-mid '70s, I suspect that the major reason for the lack of progress was not the lack of manpower but the lack of viable approaches other than genetics. The insect was too small for most physiological approaches, and the strictly genetic approach, or phenogenetic approach as it is sometimes called, was inherently limited.

BRANDEIS AND RECOMBINANT DNA

One new lab that did enter the field of *Drosophila* circadian biology in the '70s was that of my friend and Brandeis colleague Jeff Hall. He also trained with Seymour Benzer but as a post-doc. Jeff knew Ron Konopka well from their time together at Caltech; Ron was doing his Ph.D. in the Benzer lab and then returned to Caltech as an assistant professor after a brief post-doc with Colin Pittendrigh at Stanford. Jeff's research interests in the late '70s were focused on neurogenetics writ large and also more narrowly on fruit fly courtship. His post-doc Bambos Kyriacou made a finding in the late '70s that merged the field of courtship with that of circadian rhythms. He discovered that the courtship song that the male fly sings to a female has about a 1 min rhythm. Moreover, the *period* mutants affect this ultradian rhythm similar to the way that these mutants affect circadian rhythms, i.e., shorter, longer and no rhythm at all (Kyriacou and

Hall, 1980). Although this finding lacks any mechanistic underpinning and has recently become quite controversial (Kyriacou et al., 2017; Stern, 2014; Stern et al., 2017), the reason for mentioning it here is historical: it enhanced Jeff's interest in *Drosophila* circadian biology and by extension my own.

Despite all this fascinating phenomenology, it was my contention in the early '80s that one could not strike off in a new direction, with new methods and ideas, without being able to get at the molecules – the RNAs and proteins – that were participating in timekeeping; they included the *period* RNAs and proteins. I must confess that this view also reflected a certain subjective bias; I was not comfortable with strictly genetic arguments, regardless of their strength in hindsight. This means that I was still far from certain that *period* and its gene products played a central role in circadian timing.

Enter recombinant DNA technology into the story in the late '70s-early '80s; it had transformed genetics, including my own yeast work. We were actively cloning yeast genes and using recombinant DNA methods to study their expression. It goes almost without saying that this new approach also solved my "comfort problem"; one could finally imagine cloning and sequencing enigmatic genes like *Drosophila period*. The application of recombinant DNA technology would therefore not only enable an entirely new generation of experiments but also might provide some insight into the function of this gene and by extension the molecular underpinnings of circadian rhythms.

Jeff and I discussed the possibility of collaborating in this new direction for at least a year but were unable to begin for a number of reasons, including most importantly the lack of personnel. This obstacle was overcome when Pranitha Reddy, a Brandeis Biochemistry graduate student, transferred into my lab. Her young thesis advisor Vivian Ernst had tragically passed away, and Pranitha needed a new home. I offered her this new project, the cloning of the *period* gene, and she enthusiastically accepted. The loose collaboration plan was that Jeff's lab would provide the genetics and behavior expertise, whereas Pranitha and my lab would do the cloning and provide the recombinant DNA expertise.

Importantly, this technology was making rapid progress in ways that were highly relevant to the identification of period. It became feasible by the early '80s to pursue the cloning of *Drosophila period* by chromosome walking, a strategy pioneered by the Hogness lab at Stanford (Bender et al., 1983a; Bender et al., 1983b; Spierer et al., 1983). Moreover, gene identification could then be done by gene rescue, which had just been accomplished by Rubin and Spradling (Rubin and Spradling, 1982; Spradling and Rubin, 1982). Our dream was to rescue the per^{o} strain and its arrhythmic phenotype with a piece of cloned DNA containing the wild-type (normal) *period* gene. Although chromosome walking and gene identification by rescue were not relevant to the yeast recombinant DNA research ongoing in my lab, we were deep into other recombinant DNA studies as mentioned above. The gap between what was ongoing in my lab and these new genetic strategies was therefore not big.

Pranitha led the successful Brandeis effort to clone this DNA, and she was aided by Will Zehring and Dave Wheller from Jeff's lab (Reddy et al., 1984). We had collaborative help in the important initial phases of this work from Christopher Hadfield and Vince Pirrotta then at the EMBO labs in Heidelberg, Germany. They had expertise in making phage libraries from DNA microdissected from *Drosophila* giant salivary glands. We knew from the genetic mapping work of Konopka that the *Drosophila* period gene was located at position 3B 1-2, between the white and zeste genes on the X chromosome (Konopka and Benzer, 1971). Our European collaborators made a DNA library from a piece of the salivary gland X chromosome surrounding region 3B 1-2, which reduced the amount of chromosome walking we would have to undertake.

Our rescue experiments were spearheaded by Zehring and Wheeler in the Hall lab with help from Reddy as well as Ron Konopka and Bambos Kyriacou (Zehring et al., 1984). Bambos had by this time returned to a faculty position in the UK but remained a frequent collaborator. He was responsible for the courtship song rhythm assay, which was also measured in this first gene rescue publication. Konopka had also moved, from Caltech to Clarkson College in upstate New York. Jeff was in frequent

a.

b.

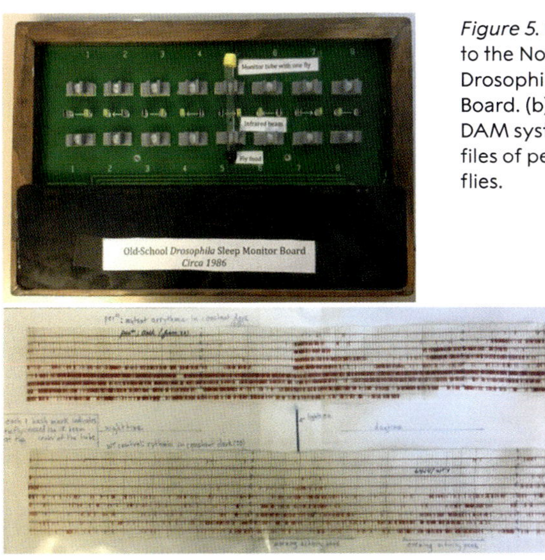

Figure 5. Gift of Hall and Rosbash to the Nobel Museum. (a) Original Drosophila Activity Monitor (DAM) Board. (b) Recording traces from DAM system showing activity profiles of per01 mutant and wild-type flies.

communication with him, and Ron helped us set up our initial Brandeis locomotor activity monitoring system. It became our gift, from me and Jeff, to the Nobel Museum (Fig. 5). Konopka also acted as an advisor, to make sure that we were carrying out and interpreting these locomotor activity rhythm assays – effectively sleep-wake assays – properly. This was our principal assay (Fig. 6); we only rarely assayed eclosion. My graduate student Qiang Yu also helped with these experiments, with cloning and especially with DNA injections into embryos. Qiang had spent several years in the Chinese countryside during the cultural revolution. Some of that time was spent repairing watches, which made him an expert at fine motor movements under magnification. These were precisely the skills required for the embryo injections, which generated the transgenic fly strains.

These two first papers, the cloning and characterization of the putative period DNA as well as the gene rescue experiments, relied for identification on classical genetic rearrangements used by Konopka in his characterization of the *period* gene (Konopka and Benzer, 1971). They dated from an earlier characterization of the white-zeste region of the fly genome within which period maps. This region and its chromosomal rearrangements were studied in detail by Bert Judd and his collaborators at the University of Texas. It is not coincidental that they included his

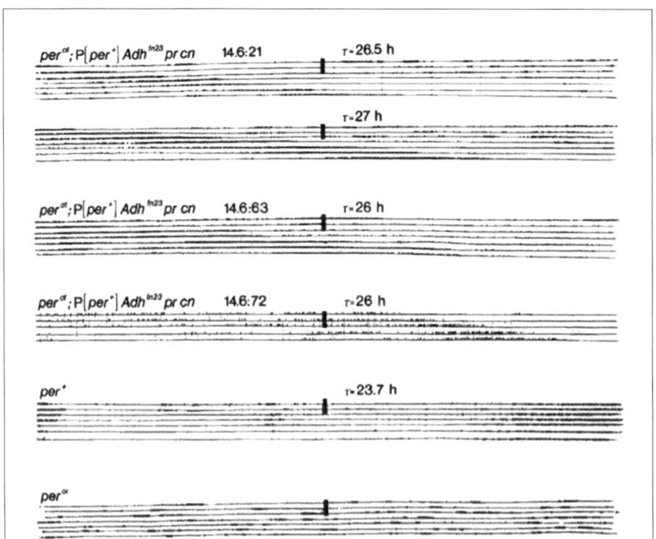

Figure 6. Activity grams showing rescue of per01 arrythmicity by segments of DNA from the per region. Each line represents a subjective day, while the midline mark in each actogram denotes the timing of lights turning on (Zehring et al., 1984).

Ph.D. student Mike Young, with whom Jeff Hall and I have shared this Nobel Prize (Young and Judd, 1978). Importantly in the context of our initial cloning and gene rescue efforts, our combined group at Brandeis did not know that the Young lab at Rockefeller University was engaged in very similar experiments (Bargiello et al., 1984; Bargiello and Young, 1984). Both groups soon learned about the other one, and we were locked in an intense battle for primacy during the first few years of this work. Importantly, the cloning and rescue of *period* was done independently and at the same time in both places. Although unpleasant at the time, the competition and focus probably contributed to some of these early successes. I note that the functional identification of a behavioral gene by transgenic rescue by both groups, Rosbash-Hall and Young, was accomplished 13 years before the same goal was achieved in mouse with the *Clock* gene by Takahashi and colleagues. Admittedly, this was a much more challenging task given the size of the mouse genome and other issues.

Mike Young and his colleagues deserve high marks for these early accomplishments. Although many of the conclusions were similar to ours, the early Young papers did a better job than we did of focusing in on a 4.5 kb transcript as the correct *period* mRNA. We were distracted by a 0.9 kb transcript, which was encoded just downstream of period and underwent apparent circadian oscillations in level (Reddy et al., 1984)) This transcript turned out to be a red herring: it is strongly expressed during pupation, just prior to eclosion. The daily synchronized emergence of young adults combined with the rapid decay of the transcript after eclosion gave rise to "apparent" circadian oscillations during adulthood (Lorenz et al., 1989).

SEARCHING FOR *PERIOD* PROTEIN FUNCTION

The next big question was, what is the function of the *period* protein or PER? This was approached in the obvious way, i.e., by DNA sequencing and done once again in parallel by the Young group and our Brandeis group (Baylies et al., 1987; Citri et al., 1987; Jackson et al., 1986; Yu et al., 1987b). The single nucleotide changes of the 3 mutations handsomely conformed to the simplest interpretation of the Konopka and Benzer data, but we were disappointed that the protein sequence itself was not instantly illuminating. These were, however, the early days of sequencing with rather few proteins in the database. This explains why the PER sequence did not link to its now known family of proteins; it was the first member of this family, i.e., a "pioneer protein." However, PER did link to proteoglycans, which often have repeated sequences that resemble the glycine-threonine (GT) repeat sequences of PER (Jackson et al., 1986; Reddy et al., 1986). The serines of proteoglycan GS repeats are sites of O-linked glycosylation.

We pursued this relationship and published a paper indicating that PER is indeed a proteoglycan (Reddy et al., 1986). In hindsight, the biochemical characterization of PER was simply too superficial to draw such a strong conclusion. The Young lab competition issue (I feared – correctly it turned out – that they too were working on this proteoglycan hypothesis) and a lack of prudence led me to push forward and publish these modest biochemical experiments. I also thought that the GT repeat sequence had to be significant and made the classical error of letting a potential exciting conclusion erase what should have been vigilant skepticism about my own data. I still don't know to this day if those experiments were in error or just their interpretation. I suspect the latter, for which I am 100% responsible, i.e., the results were probably correct but simply due to protein aggregation. In any case, the "PER is a proteoglycan" hypothesis stood tall in 1986 with no other proposed functions for PER. Moreover, this hypothesis soon stood even taller because of a high profile paper published in 1987 through a collaborative effort from the Young laboratory and that of David Spray at Albert Einstein (Bargiello et al., 1987). This paper argued that the per mutations affect intercellular communication in *Drosophila* larval salivary glands. PER would be resident in the extracellular matrix and quantitatively affect gap junction function, a perfect role for a proteoglycan. Importantly, there were no strong data to the contrary nor even another competing hypothesis for PER function in 1987.

The first hint that this proteoglycan hypothesis might be on the wrong track was also published in 1987 and came from my outstanding graduate student Qiang Yu (Yu et al., 1987a). In one of the first of many behavior-transformation papers in which per DNA was manipulated in vitro and then assayed functionally in vivo, Qiang cleanly excised the GT repeat region from the PER coding region and assayed the function of this deleted gene and protein. We assayed as usual two behaviors, circadian locomotor activity rhythms and the courtship song rhythms. The deletion had a strong effect on the more enigmatic courtship rhythms, but the canonical ca. 24hr locomotor activity rhythm assay was unaffected. Although the paper title featured the positive result, we recognized that the lack of an effect on the circadian assay was important. If the GT repeat region is the cardinal feature that defines proteoglycan properties, e.g., O-linked glycosylation, might this mean that the protein was not a proteoglycan? Might the biochemistry and the salivary gland gap junction results be misleading?

The results of a study we published in the following year was also relevant to the proteoglycan hypothesis (Colot et al., 1988). We cloned and sequenced period from D. *pseudoobscura* and D. *virilis*. The coding sequences of both genes are missing the GT repeat region but can rescue behavioral rhythms of *per*^{o} *D. melanogaster* flies, consistent with the deletion experiment described above. Because the coding regions of the two genes still contain several pairs of TG and SG amino acids and because single SG dipeptides can serve as sites for the O-linked attachment of glycosaminoglycan side chains, the data could not eliminate the proteoglycan possibility. They did, however, add to a growing sense of doubt.

Unrelated to proteoglycans, the assay of behavior with manipulated *period* DNA was used in other papers published at about the same time. We characterized an unusual collection of spliced transcripts encoded by *per* and showed that two of them have biological activity in transformation-rescue experiments (Citri et al., 1987). The first author, Yoav Citri, became fascinated with fly rhythms and began these experiments when he was a post-doc in the Baltimore lab at MIT; he then moved over to my lab to continue the work. He left Brandeis to become an assistant professor at the Weizmann Institute around 1989 and died tragically in 1995 in a traffic accident.

We also used this sequence and rescue strategy to address the molecular identity of the famous Konopka mutations, with Qiang Yu once again principally responsible for these experiments in my laboratory (Yu et al., 1987b). Our sequence data largely agreed with that published the previous year from the Young group (Jackson et al., 1986), but our per gene contained an extra exon and was also missing an intron present in the Young publication, which interrupted a long contiguous coding exon. As a consequence, our predicted sequence was 97 amino acids longer than the one published by Young. We identified the *per*^{o1} and *per*^{S} mutations, and in both cases identified only a single nucleotide change from wild-type per DNA. The work did not just rely on these sequence differences but also used chimeric DNA fragments to show that altered biological activity tracked with the mutation. Remarkably and consistent with the original hypothesis of Konopka that *per*^{o1} is a null mutant with no biological activity, this mutation creates a stop codon; there is therefore no wild-type PER encoded and perhaps no protein at all. The *per*^{S} nucleotide change creates a serine to asparagine missense mutation at position 589. Both changes were in locations that underscored the fact that the 4.5 kb transcript encodes PER as originally suggested by Young and colleagues.

Although our original sequencing data made it almost certain that the mutation at position 589 is responsible for the *per*^{S} short period phenotype, we recapitulated this change in a subsequent study in which position 589 of wild-type DNA was changed to encode a number of different amino acids including asparagine (Rutila et al., 1992). The results not only verified that the serine to asparagine change was indeed responsible for the short period phenotype but also indicated that short period is likely to be the default phenotype of most amino acids at position 589, i.e., only serine and threonine have a ca. 24 hr period. The data in this paper also indicated that the *per*[S] protein is a hypomorph (has decreased function) rather than a hypermorph, with an increase in gene/protein activity as originally postulated by Konopka and Benzer (Konopka and Benzer, 1971). The Young lab identified all 3 Konopka and Benzer mutations and also showed that more copies of *period* transgenes led to shorter periods (Baylies et al., 1987).

At the risk of making a slight tangent to discuss much more recent data, we now have a quite sophisticated view of at least part of the PER molecular cycle, and these data underscore the hypomorphic or loss-of-function character of the *per*[S] protein. PER amino acid 589 is phosphorylated by the kinase Nemo, which probably delays subsequent PER phosphorylation events by the important circadian kinase doubletime (DBT); they lead to the degradation of PER, which happens too rapidly in the absence of PER 589 phosphorylation (Chiu et al., 2011). This rapid degradation of PERS occurs after the disappearance of TIM in the late night or early day, a scenario that is also consistent with more recent data on the effects of the *perS* mutation (Li et al., 2014; Li and Rosbash, 2013). In contrast to our understanding of the PERS effect, there is still no good mechanistic understanding of the PERL amino acid change and why it slows

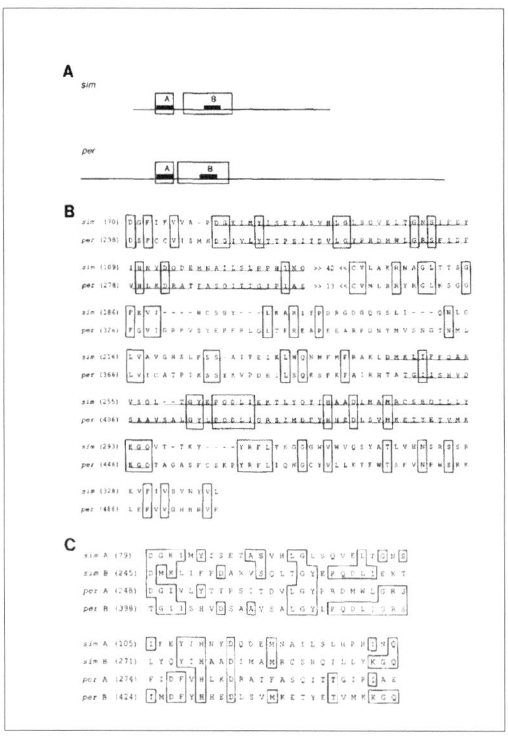

Figure 7. Comparison of SIM and PER sequences (Crews et al., 1998).

the circadian cycle (Kidd et al., 2015). Although PERL may interact more weakly with TIM (Gekakis et al., 1995), it is still uncertain which feature(s) of PER activity is specifically affected by PERL.

The 1987 mechanistic speculations about *perS* protein function were difficult to reconcile with a proteoglycan function. Moreover, we soon learned of an impending publication that further eroded my confidence in this hypothesis. Steve Crews, working in Corey Goodman's lab had cloned and sequenced a *Drosophila* gene called *single-minded* with protein coding homology to PER. Their 1988 paper showed that Single-minded (SIM) is a nuclear protein and strongly suggested it was a transcription factor (Crews et al., 1988). Especially difficult to square with proteoglycan theory was SIM nuclear localization. We now know that the homology reflects a protein-protein interaction domain called PAS and present in a family of transcription factors; the three founding members are PER, ARNT and SIM; ARNT is a mammalian protein, the aryl hydrocarbon receptor nuclear translocator (Huang et al., 1993). Although the homology was quite modest and could have reflected a motif shared by proteins with completely different functions (Fig. 7), it offered an important competing possibility: perhaps *period* encodes a transcription factor.

BACK ON TRACK: GENE EXPRESSION AND THE FEEDBACK LOOP

The possibility of a gene expression function was dramatically enhanced by the first relevant data from my post-doc Paul Hardin, who was examining *period* mRNA from fly heads at different circadian times. The paper that was ultimately published made two findings (Hardin et al., 1990). The first is that these mRNA levels undergo robust circadian oscillations in fly heads. The second is that the phase of RNA cycling was affected by the *per* missense mutations identically to their effects on behavioral phase (Fig. 8). Paul gets full credit for making these seminal findings, including the important idea to use fly heads rather than whole flies for the source of RNA. The lack of cycling in certain body tissues diluted the cycling signal and therefore reduced the cycling amplitude in RNA from whole flies so that it was not reliable.

Paul's experimental success was helped by his local environment, namely, the fact that a good fraction of his colleagues in my lab studied yeast RNA and were facile with state-of-the-art RNA methods. As PCR did not exist at this time, Paul had to use RNase protection to assay specific per mRNA levels. This method was not terribly difficult but also not trivial, so this aspect of the Brandeis circadian work benefited from being adjacent to – in effect embedded within – a nucleic acid-centric research program. Circadian labs elsewhere had a difficult time establishing these kinds of nucleic acid-centric approaches (personal communication).

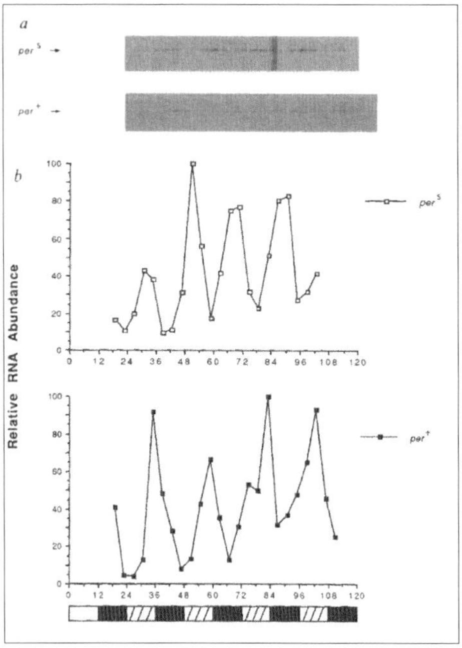

Figure 8. Both Per^{+} and per^{s} mRNA levels undergo circadian oscillations. pers mRNA oscillations correspond to the shortened period exhibited by these mutant flies (Hardin et al., 1990).

We now know that the feedback effects of the *per* missense mutations on *per* RNA cycling operate predominantly at the transcriptional level and are upstream of their effects on behavior, i.e., the protein directly affects its own transcription as part of the timekeeping mechanism. (I should note that this is still inferred rather than directly shown, because we do not have a *per* mutation that is known only to impair the feedback effect on its own transcription with a period effect as measured behaviorally.) At the time of this first paper, however (Hardin et al., 1990), there were many possibilities, and we were very conservative in interpreting these results. For example, there were no transcriptional or pseudo-transcriptional assays, so feedback could have operated at a post-transcriptional level. We also had no evidence that PER functions directly to affect its own gene expression, for example by binding to chromatin or to pre-mRNAs, and our model considered the possible role of intermediate effector proteins. We even considered that circadian behavior would be required to generate oscillating mRNA levels, i.e., the feedback effects on gene expression might then be downstream of a PER effect on behavior. In this case, PER would be affecting circadian behavior and perhaps the circadian clock independent of its feedback effect on its own gene expression (Fig. 9). Nonetheless, the Hardin et al. data taken together with a simple and positive interpretation of the modest homology to the Single-minded transcription factor suggested a simple parsimonious hypothesis: PER

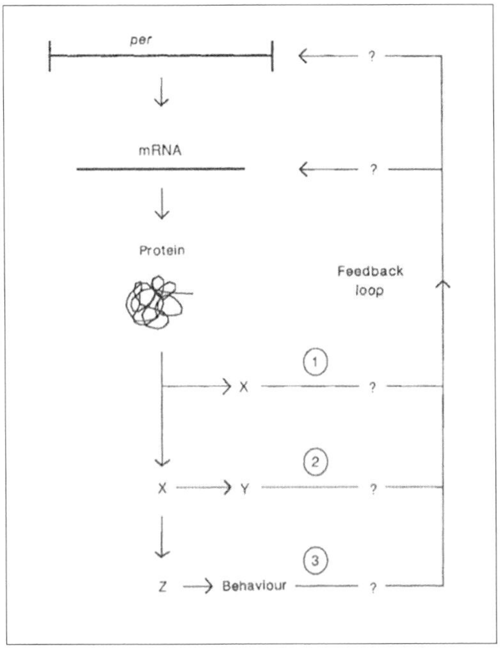

Figure 9. Original feedback loop model showing broadly how per might regulate its own expression (Hardin et al., 1990).

was engaged in a negative feedback loop that affects its own expression and that this loop was "the clock" or at least central to timekeeping. The negative feedback loop would therefore be upstream of the behavioral rhythms that are usually measured by circadian biologists.

This new working model of how PER functions, namely at the level of gene expression regulation, was very attractive, and my lab pushed hard to support and extend this new view of timekeeping. First was to address transcriptional vs. post-transcriptional regulation. To this end, Paul Hardin showed that feedback regulation almost certainly occurred at the transcriptional level (Hardin et al., 1992). This paper had two principal experiments. i) It showed that pre-mRNA oscillations were similar if not identical to mRNA oscillations. ii) It showed that the feedback could be recapitulated with *per* 5' flanking sequences, i.e., they would drive cycling of a reporter gene (Fig. 10).

To my surprise, we had serious trouble publishing this important paper. Perhaps this was because transcriptional regulation was taken for granted by the scientific community; post-transcriptional regulation, splicing for example, was not yet broadly appreciated as a viable regulatory mechanism.

The second paper addressed PER subcellular localization. Is it a nuclear protein like SIM or is it cytoplasmic? The latter would favor an indirect effect on transcription and possibly even other functions. To

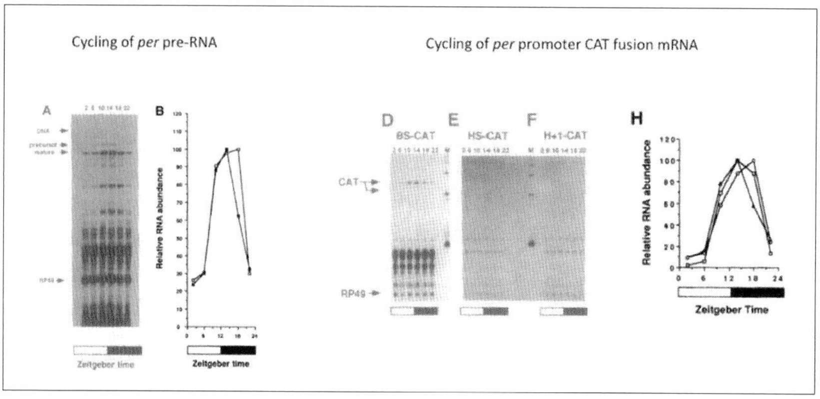

Figure 10. per mRNA oscillations are transcriptionally regulated (Hardin et al., 1992).

address this issue, my graduate student Xin Liu went to Seymour Benzer's lab for a few months to do immunoelectron microscopy. (Xin recently retired after many years as a UCLA professor.) PER was a nuclear protein (Liu et al., 1992). This was consistent with a more direct role on transcription and made the proteoglycan hypothesis even less likely.

The third paper was carried out by an excellent graduate student, Honkui Zeng, who addressed whether PER functioned in a direct, cell-autonomous manner. (Honkui has been with the Allen Brain Institute for more than a decade and is currently the Executive Director of its Structured Science Division.) There was at this time no evidence that PER inhibited transcription within the same cells in which it was expressed. To this end, Honkui showed that the photoreceptor cells of the adult fly eye contain a circadian oscillator system similar to the brain. She also showed that constitutively overexpressing PER in these cells prevented endogenous *per* RNA cycling and caused low *per* RNA levels with no effects on locomotor activity rhythms, consistent with cell-autonomous negative feedback (Zeng et al., 1994). A similar strategy was subsequently applied to brain circadian neurons with similar results (Yang and Sehgal, 2001).

The fourth paper came from yet another outstanding graduate student, Josh Huang. (Josh is current a Professor of Neuroscience at the Cold Spring Harbor Laboratory). He addressed the role of the PAS domain, the protein sequence link between PER and known transcription factors (Huang et al., 1993). At the time of this work in 1994, there were 3 known PAS-containing (PAS-bHLH, basic-helix-loop-helix) transcription factors: In addition to PER there were Single-minded as mentioned above and the two subunits of the aryl hydrocarbon (AH) receptor. Josh's work showed that PAS could function as a novel protein dimerization motif and mediate associations between different members of the PAS protein family. As PER had and still has no known DNA interaction motifs, we proposed that it regulates circadian transcription by interacting with the PAS domain of an important positively-acting circadian transcription factor. This prediction turned out correct with the identification of the PAS-containing positive circadian transcription factors, the heterodimeric Clock and BMAL1 in mammals and their orthologs Clock and Cycle in flies (see below).

Taken together with the original Hardin feedback loop paper, these four subsequent papers laid the foundation of a coherent transcription-centric view of circadian timekeeping; it seemed incompatible with gap junctions. Indeed, I already had doubts about this hypothesis in 1989; at this time the work for the first Hardin feedback loop was largely done (Hardin et al., 1990), and we had shown that the GT region – the linchpin of the proteoglycan relationship – is not conserved in related *Drosophila* species (Colot et al., 1988) and that this region is also not necessary for PER circadian function (Yu et al., 1987a). Although I had no idea how to advance or challenge our biochemistry paper (Reddy et al., 1986) in an efficient way, I thought we should repeat the simpler Young-Spray salivary gland-gap junction paper to see if it was correct (Bargiello et al., 1987). Also, for this experiment, I really had no idea what to do; it was not only completely beyond my expertise but also outside that of my circle of friends and colleagues. Nonetheless I was quite obsessed and for the next year or two would ask scientific strangers, "Do you know anything about gap junctions?" "Can you help us out with this experiment?" One day at a Brandeis cocktail party I was introduced to Dan Goodenough. I asked him naively, "Where do you work and what do you do?" He said, "Harvard and on gap junctions." I latched onto his ankle: "Can you PLEASE help us repeat a gap junction experiment in Drosophila larval salivary glands?" He said sure, and we were on our way. I had no available personnel, but Jeff had a technician with time to spare. So, Kimberly Flint spent several months hosted by the Goodenough lab and showed that those key gap junction experiments would not repeat. Dan Goodenough was a prince, not only to host Kimberly and let her work in his lab but also because he refused to coauthor any paper that might emerge from the work. We communicated our failure to Young before publication, and Jeff also told his former post-doc Kathy Siwicki about our results. She was at the Marine Biological Laboratory (MBL) at Woods Hole at the time and was apparently doing the same experiments in collaboration with David Spray, the senior author of the Bargiello et al. paper who was also there during that summer of 1992. They wrote a short paper for the *Biological Bulletin*, the local MBL journal. It was published super rapidly, in October, and they included Kimberly, Jeff and me as coauthors (Siwicki et al., 1992). Young,

Spray and all coauthors then acted properly and retracted their *Nature* paper in December of 1992 (Saez et al., 1992), and we published an independent and properly documented paper the following year in the *Journal of Membrane Biology* (Flint et al., 1993). This was the death knell for the proteoglycan hypothesis, and we forged ahead with gene expression.

THE LINK TO MAMMALS

The identification of mammalian PER in 1997 was incredibly important. This is because for the first time it seemed likely that what we had learned in flies would be applicable to mammals including humans. This finding, that mammals contain one or more *period* genes, was independently accomplished in two different laboratories and done before the existence of the human genome sequence, meaning one could not just search the data base for a mammalian protein sequence with a credible relationship to PER (Sun et al., 1997; Tei et al., 1997). It was done by PCR and required a judicious selection of primer sequences. My laboratory and others also tried to do this but unsuccessfully (data not shown; personal communication).

NEW CIRCADIAN MUTANTS

The Young laboratory also made a connection to mammals, with their modern (post-Konopka and Benzer) efforts to identify *Drosophila* circadian mutants. They discovered two very important fly clock genes, one of which linked in an important and simple way to mammals. That gene was *doubletime*, which encodes the *Drosophila* version of the important mammalian kinase CKI (Kloss et al., 1998; Price et al., 1998). It modifies PER and also CLK during the circadian cycle. Well after the publications on *doubletime* from the Young lab, we also identified and studied two novel mutant versions of *doubletime*, which emerged from our own efforts to isolated new circadian mutants (Suri et al., 2000). CKI not only plays an important role in the mammalian clock but a mutation in this gene also underlies the phenotype of the famous 20hr *tau* hamster (Lowrey et al., 2000).

The other Young laboratory clock gene from this early period is *timeless* (Gekakis et al., 1995; Myers et al., 1995; Price et al., 1995; Sehgal et al., 1994; Sehgal et al., 1995). Although its relationship to mammals was not immediately apparent and is considerably weaker even today than that of *period* and *doubletime* (Gotter, 2006), *timeless* played an incredibly important role in the fly circadian system. In addition, its identification and cloning in 1995 showed that *period* was not unique and that scientists other than Konopka and Benzer could identify clock mutants. This fear was not entirely unfounded, as almost 25 years had passed between the publication of Konopka and Benzer on *period* and these Young lab publications on the second *Drosophila* clock gene *timeless* (*tim*).

Importantly, the characterization of *tim* connected back to PER and to the feedback loop, i.e., *tim* mRNA undergoes RNA cycling like *per* mRNA, the *tim* mutants affect *per* as well as cycling of their own mRNA, i.e., *tim* mRNA (Sehgal et al., 1995) and the *timeless* protein (TIM) interacts with PER (Gekakis et al., 1995). In fact, TIM and PER associate together in the cytoplasm and are probably transported together into the nucleus (Jang et al., 2015; Meyer et al., 2006; Vosshall et al., 1994). However, the regulatory role of this transport event – the extent to which it contributes to circadian timing (Meyer et al., 2006) – is less certain in my view. All proteins are synthesized on cytoplasmic ribosomes. So nuclear proteins must be transported from the cytoplasm to the nucleus, but this transport event need not be under clock control. Temporal control of nuclear and cytoplasmic localization (Shafer et al., 2002) could equally well reflect regulation of cytoplasmic or nuclear retention rather than nuclear transport per se. Indeed, we have some evidence that PER-TIM retention on chromatin is temporally regulated (data not shown).

Our own principal contribution to the *tim* story came from biochemistry (Zeng et al., 1996). My graduate student Hongkui Zeng was trying to identify PER partner proteins, by characterizing proteins associated with PER in fly head extracts with mass spectometry. Because we realized that there might be different partners at different times of day, Hongkui began by characterizing the size of the protein complex within which PER was located at different times. PER appeared to be monomeric in the morning during its declining/degrading phase but in a larger complex later in the day during its increasing/accumulating phase. Moreover, that complex appeared to have one additional major protein, with an apparent molecular weight somewhat greater than PER (Zeng et al., 1996). That complex and the additional protein disappeared in the late night or rapidly after a light pulse. We were gearing up to identify that seemingly light-sensitive protein by mass spectrometry when we heard about the Young lab success in identifying TIM. Because the TIM molecular weight was somewhat greater than that of PER and the two proteins associate in a yeast two hybrid assay (Gekakis et al., 1995), TIM was an excellent candidate for the PER partner protein we had identified in fly head extracts. The Young laboratory was very gracious in providing us with an anti-TIM antibody with which we were able to definitively identify it as the higher molecular weight PER partner identified in our gels (Zeng et al., 1996).

The rapid degradation of TIM at the end of the night or in response to a phase-shifting pulse of light was independently observed and proposed to be important for light-mediated phase-shifting and entrainment in 3 labs at about the same time (Hunter-Ensor et al., 1996; Myers et al., 1996; Zeng et al., 1996). We also provided evidence that TIM degradation and phase-shifting do not just occur in parallel. This study (Suri et al., 1998) made use of a new *tim* allele. In an attempt to shed light on the enigmatic *perL* mutation, we had screened for allele-specific suppressors or enhancers of that mutation, namely, additional mutations elsewhere in the fly genome that would have a different phenotypic effect on that *perL* background than on a wild-type background. It was to my knowledge the first genetic screen of its kind, a behavioral suppressor-enhancer screen. It identified a new *tim* allele, which we named *tim suppressor of long* or *timSL* (Rutila et al., 1996). In this context of TIM degradation and light, *timSL* enhanced the light-sensitivity of TIM degradation and phase-shifting in parallel (Suri et al., 1998), indicating that TIM destabilization indeed connects the core central clock to entrainment by the light-dark cycle, the principal zeitgeiber (temporal cue) of the environment in which we live.

What is the molecular connection between light and TIM degradation? Patrick Emery in my lab and Ralf Stanewsky in the Hall lab worked closely together and identified Cryptochrome as a major circadian photoreceptor; they also cloned the gene and identified a mutant strain that was significantly impaired in circadian light perception (Emery et al., 1998; Stanewsky et al., 1998). The work also relied on important previous work from mammals and constituted another strong link between the fly and mammalian circadian systems.

The story begins with the important DNA repair enzyme photolyase, which is present from E. coli to man. This enzyme had been studied for a long time by the biochemist Aziz Sancar, who shared the 2015 Nobel Prize in chemistry for his outstanding work. The plant blue light photoreceptor cryptochrome is a close structural relative of photolyase. There were mammalian gene sequences that were related to cryptochrome, and Sancar was convinced that their encoded proteins, cryptochrome 1 and 2 (CRY1 and CRY2) must function in circadian photoreception. Indeed, the initial properties of these gene and proteins, including phenotypes of mouse mutants missing these cryptochromes, was consistent with a circadian photoreceptor function for cryptochromes in mammals (Vitaterna et al., 1999). Moreover, there was strong evidence that there was an unknown photoreceptor in the mammalian eye that accounts for persistent circadian photoreception in rodless-coneless mice (Lucas et al., 1999); this is in addition to the known photoreceptor function of cryptochromes in plants. Importantly, melanopsin was soon discovered as that missing eye photoreceptor (Provencio et al., 1998) and mammalian cryptochromes currently have no function in circadian photoreception.

Nonetheless, these Sancar ideas about a circadian photoreceptor function for cryptochrome and the known mammalian cryptochrome genes resonated with us at Brandeis. The fly circadian system was known to be maximally sensitive to blue light, and there was no known eye pigment or other identified photoreceptor that might play this role. So, Patrick Emery in my lab set out to clone and characterize *Drosophila cryptochrome* (*cry*; the sequence information indicated there was only gene as is commonly the case in flies). The work showed that its protein product CRY was light-sensitive, i.e., degraded in response to light exposure, and that CRY overexpression enhanced circadian light sensitivity (Emery et al., 1998).

At the same time, Ralf Stanewsky in the Hall lab had identified a new circadian mutant that was light-insensitive. It had very interesting properties, which were consistent with the absence of a circadian photoreceptor. With Emery's map position and the identification of a deletion that uncovered the locus, Stanewsky could quickly assign the mutation to the *cry* gene. Characterization of the mutant strain, called *cryb*, indicated that the gene *cry* indeed encoded a circadian photoreceptor (Stanewsky et al., 1998) as was also indicated by the Emery et al. paper described directly above (Emery et al., 1998).

Subsequent work over the next couple of years, predominantly from Emery, extended this initial work by showing quite remarkably that *cryb* flies remain rhythmic in constant light (Emery et al., 2000a). This is unlike wild-type flies as constant light destroys their rhythms – and those of many organisms. This simple observation indicates that CRY acts as the only principal circadian photoreceptor in *Drosophila*.

Emery and colleagues also showed at about the same time that expressing CRY only in a small number of key circadian neurons in the brain could rescue the circadian deficits of the *cryb* mutant background. This indicates that CRY functions within the brain and is therefore a deep brain photoreceptor (Emery et al., 2000b).

How does CRY work, i.e., how does it interface with the circadian clockworks? CRY and TIM interact in a light-dependent fashion, even in yeast (Ceriani et al., 1999). Amita Sehgal and colleagues then discovered that the F-box protein Jetlag interacts with CRY and TIM after light exposure and helps deliver them both to the proteosome. As predicted, phase-shifts are reduced in jetlag mutants (Koh et al., 2006).

This CRY-TIM mechanism is not the only way that the *Drosophila* clock system can achieve phase shifts. Clock neuron firing, specifically PDF neuron firing, can also elicit phase shifts and in a CRY-independent manner (Guo et al., 2014). This mode of firing also results in TIM degradation and presumably reflects in part the connections of the fly eye to the PDF neurons and/or the role of other clock neurons (Fig. 11).

The common feature of these two modes of phase-shifting, light-CRY and light-firing, is TIM degradation or post-transcriptional regulation. This contrasts with the enhanced clock gene transcription that is the pri-

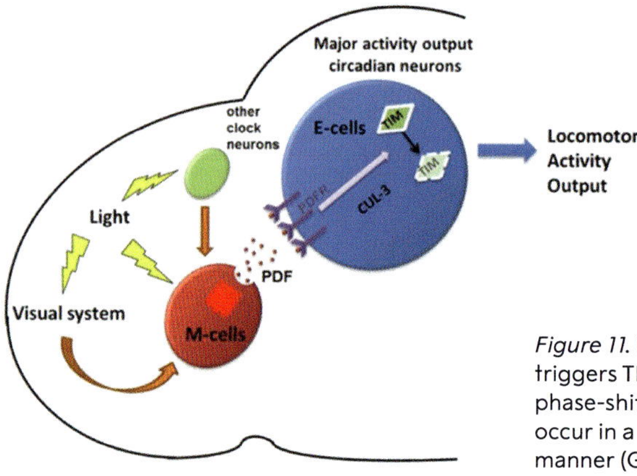

Figure 11. Firing of PDF cells triggers TIM degradation and a phase-shift, both of which can occur in a CRY-independent manner (Guo et al., 2014).

mary means of achieving light-mediated phase-shifts in mammalian systems (Albrecht et al., 1997; Shearman et al., 1997; Shigeyoshi et al., 1997). Nonetheless, it remains uncertain whether this difference will remain so striking, i.e., whether future results will reveal some important light-mediated post-transcriptional regulation of clock gene expression in mammals or some light-mediated clock gene transcriptional regulation in flies.

However, one strong argument in favor of this difference between flies and mammals is the different roles of cryptochrome in the two systems. In contrast to its photoreceptor role in flies, cryptochrome is the major transcriptional repressor in mammals (Reppert and Weaver, 2002). These two roles reflect two major features of the cryptochrome ancestor photolyase, namely, photon capture and interaction with DNA (Emery et al., 1998; Gehring and Rosbash, 2003). It is striking that butterflies have two cryptochromes: one that resembles more closely *Drosophila* cryptochrome and functions as a photoreceptor; and another that resembles more closely mammalian cryptochrome and functions as a transcriptional repressor (Zhu et al., 2005). The butterfly situation appears to be the insect "rule," with the *Drosophila* situation the "exception" (Yuan et al., 2007).

The role of cryptochrome in transcriptional repression was preceded by the discovery of CLK-BMAL1 in mammals and their orthologs CLK-CYC in flies. Joe Takahashi and colleagues spearheaded the cloning of CLK through a forward genetic screen in mice. A robust behavioral phenotype of the CLKD17 allele was first identified by behavioral screening (Vitaterna et al., 1994). The gene was identified three years later as a member of the bHLH-PAS-containing protein family (Antoch et al., 1997; King et al., 1997). The bHLH domain suggested DNA binding and other bHLH-PAS family members had been identified as bona fide transcription factors, making this function an excellent candidate for CLK. Moreover, given the presence of a PAS domain in PER and the circadian behavioral phenotype of the mutant, CLK was probably also a major circadian transcription factor as described above. Importantly however, the two 1997 papers did not connect CLK to the transcription of PER (Antoch et al., 1997; King et al., 1997), i.e., they did not clarify whether CLK was the positive transcription factor that drives the synthesis of the negative regulator PER. This was done a year later by Takahashi and coworkers, in papers that also identified the CLK partner BMAL1 (Darlington et al., 1998; Gekakis et al., 1998).

We came to the identical conclusion at almost the same time, by identifying the CLK-BMAL1 orthologs Clock-cycle or CLK-CYC (Allada et al., 1998; Rutila et al., 1998). This was the major result of a genetic screen we had been carrying out over several years. It had the goal of finding the positive transcription factor that drives the transcription of *per* and *tim*. To this end, we decided to focus on a large number of mutant strains, which were totally arrhythmic. The logic was that a loss of function mutation in this positive transcription factor would not be able to synthesize PER and TIM and therefore should be arrhythmic. We also adopted this strategy to do something different than what we knew was well underway in the Young laboratory, i.e., competently searching for mutant strains with altered circadian periods. Some of these inevitably emerged from our screen as well.

It turned out to be much easier to find arrhythmic strains than strains with altered periods, presumably because functional rhythms rely on many different functions and pathways, e.g., output pathways from circadian neurons to motor neurons, in addition to the core timekeeping mechanism. A subscreen was therefore required to avoid these output pathways and focus in on that subset of the arrhythmic strains that might be candidates for encoding the positive transcription factor(s). So about 50 arrhythmic strains were assayed for cycling of *per* and *tim* mRNA. The vast majority of these mutant strains still manifested robust cycling of *per* and *tim* mRNAs, consistent with the notion that they did not impact core circadian timekeeping. (As a byproduct, these results indicated that robust behavioral oscillations are not necessary for molecular feedback, e.g., for the effect of PER on the cycling of its own mRNA (Fig. 9).) However, 3 mutant strains were exceptional and manifested non-cycling, low levels of *per* mRNA even in a light-dark cycle. We then identified and cloned the 3 mutant genes. Remarkably, two of them were alleles of *cycle*, and one was an allele of *Clock*, exactly the phenotype one might expect from a mutant in the key positive circadian transcription (Allada et al., 1998; Rutila et al., 1998). Given the results of Takahashi and colleagues (Darlington et al., 1998; Gekakis et al., 1998), the field was really on its way by 1998 with powerful, parallel systems in flies and mammals.

The final coda to the feedback loop chapter was the demonstration that PER interacts directly with CLK-CYC in flies (Lee et al., 1998; Menet et al., 2010) and with CLK-BMAL1 in mammals (Lee et al., 2001).

A FAMILY STORY RESULTING IN TRIKINETICS LOCOMOTOR ACTIVITY MONITORS

I went on sabbatical to Paris with my family in 1988–1989. In the spring of 1989, shortly before returning to Brandeis, I received a phone call in Paris from Max Cowan, the Chief Scientific Officer of the Howard Hughes Medical Institute (HHMI). He informed me that I would be appointed as an Investigator of HHMI as of July 1, 1989. He asked me what I wanted for "set-up funds." I told him I had no idea because I already had a functional lab. He then suggested $100,000.00, and I said "sure." As there were no pressing needs, I began to consider what I might do with those funds.

Shortly after returning to Brandeis in late June, I went to a family (wife's family) barbecue in Newton, Massachusetts where I live. My wife is Chilean, and her brother-in-law was visiting with us. His first cousin, Sergio Simunovic, is also Chilean by birth and lives in Newton, had gone to MIT, and was the host of the barbecue. I asked Sergio what he did for a living. He answered that he was an engineer and owned a small robotics company in Waltham, Massachusetts. Waltham is next to Newton and also where Brandeis University is situated. I immediately asked Sergio if he might be interested in designing and building *Drosophila* locomotor activity monitors for us. He introduced me to his partner Mark Spencer, and we decided to begin a collaboration. It used up half my HHMI set-up funds, i.e., I paid $50,000.00 to TriKinetics to fund the R&D costs of the *Drosophila* locomotor activity monitors. (Perhaps some monitors were included in this initial cost; I can't recall.) For several months, Mark worked almost exclusively with my post-doc/research assistant Joan Rutila, the talented individual in my lab who supervised our genetic screening efforts, to design functional activity monitors. The challenge was to adjust the detection sensitivity so beam crossing detection was reliable – with minimal false positives and false negatives; this obstacle was eventually surmounted by Mark and Joan. In summary, success was achieved through a 3-way collaboration: my HHMI funds and initiative, Mark, and Joan. These same monitors are still sold today (Fig. 12).

I never imagined that these monitors would be used all over the world, i.e., I had assumed that they would be used by our two labs at Brandeis, perhaps also the Young lab at Rockefeller and then eventually by a few of our students who would establish independent labs. One would expect that I or my lab received some money or discount from TriKinetics but this was not the case.

Figure 12. Drosophila Activity Monitors.

EVOLUTIONARY CONSIDERATIONS

Somewhat after the first transcriptional feedback loop papers appeared from Brandeis, the Dunlaps at Dartmouth published an outstanding mechanistic paper describing a rather similar feedback loop scheme for *Neurospora* circadian rhythms (Aronson et al., 1994). This story started with Jerry Feldman, who was at UC Santa Cruz. Feldman and his colleagues isolated *Neurospora* circadian mutants not too long after the Konopka story was published (Feldman and Hoyle, 1973; Feldman and Hoyle, 1976). These studies identified the *frequency* gene, which had genetic properties remarkably similar to those described by Konopka and Benzer for *Drosophila* and its *period* gene (Gardner and Feldman, 1980; Loros and Feldman, 1986; Loros et al., 1986). Jay Dunlap was a post-doc in the Feldman lab at Santa Cruz and moved along with his now wife Jennifer Loros to Dartmouth and began a molecular attack on the circadian problem in *Neurospora*. Over the subsequent almost 25 years since the first molecular paper (Aronson et al., 1994), the *Neurospora* circadian clock has developed into a wonderful scientific tale, which has striking similarities to the fly and mammalian clock (Dunlap and Loros, 2017). Dunlap and Loros have even argued that these clocks have a single origin (Lee et al., 2000). However, I find these data less than totally convincing, i.e., the similarities either reflect convergent evolution or the evolutionary relationship between animal and *Neurospora* rhythms has not yet been clarified.

From a more general perspective and as mentioned near the beginning of this manuscript, circadian clocks almost certainly arose multiple times in evolution; there is essentially no molecular bridge between the cyanobacterial clock and eukaryotic clocks, and only the most speculative connection between plant clocks and the mammalian-fly clock (Rosbash, 2009). Fly cryptochrome does suggest, however, that this particular branch of the circadian world arose in response to the DNA damage caused by sunlight. Moreover, this "flight from light" hypothesis also explains the blue light sensitivity of cryptochrome. This is the wavelength that penetrates substantial depths in the ocean, perhaps reflecting the origin of life in the oceans as well as the diel migration that still occurs there today (Emery et al., 1998; Gehring and Rosbash, 2003; Pittendrigh, 1993; Rosbash, 2009). Responding to light would have preceded the origin of clocks, e.g., the ability to anticipate the daily appearance of light at dawn. In this view, the circadian kinases and other molecules that participate in the DNA damage response might even bridge plant and animal clocks (Rosbash, 2009). There is also another competing hypothesis to "flight from light" for the origins of clocks, namely the temporal organization of metabolism (Tu and McKnight, 2006).

WHAT REMAINS INTERESTING TO STUDY IN CIRCADIAN BIOLOGY?

Metabolism segues conveniently to medicine and future practical advances: metabolism is one of the myriad aspects of human physiology that are under circadian control. However, this Nobel Prize was given predominantly for our contributions to physiology rather than to medicine. (The Prize is after all for Physiology OR Medicine.). We now know about the dangers of shift work, the benefits of good circadian and sleep hygiene as well as the possible benefits of restricted feeding (Melkani and Panda, 2017), but major medical applications remain a future goal. Although this sentiment is similar to the one expressed by T.H. Morgan about genetics and medicine in his Nobel manuscript (The Nobel Lecture 1934), I am more optimistic about future medical advances from circadian biology than he was about medical advances from genetics in 1933. For understandable reasons, Morgan could not foresee the remarkable future that lay ahead for genetics, a practical as well as a conceptual future. The future is now more tangible in the case of clocks, because so much of the mammalian genome and therefore so much of physiology is known to be under circadian control (Zhang et al., 2014). There are also a number of possible drug targets implicated in the regulation of circadian physiology.

Additional future challenges are the precise timing mechanisms of animal circadian clocks and temperature compensation. What is really keeping time, i.e., what are the rate-limiting steps and why is there such little change with temperature? Although some progress has been made in our understanding of temperature compensation (Mehra et al., 2009), I suspect there is more to learn.

Another frontier is the relationship between neuroscience, clocks and sleep. Circadian clocks have been known for a long time to impact sleep (Borbely, 1982; Borbely et al., 2016; Daan et al., 1984). Moreover, sleep remains a major mystery. There are many hypotheses about how and why we sleep, but in my opinion these issues – especially why we sleep – remain major mysteries. Moreover, the neuronal circuitry that underlies the "how" of sleep and perhaps also the "why" is quite complex in mammals (Saper and Fuller, 2017). I am optimistic that the rather simple organization of the fly brain and especially its circadian circuitry will provide an entrée into this fascinating topic; my own current research is beginning to focus in this direction (Guo et al., 2016). I look forward to a future that will reveal the mysteries of sleep comparable to the progress I have witnessed in circadian biology since I began in this field 35 years ago.

FINAL THOUGHTS

What began in 1982 as a foray into cloning a clock gene became a grand adventure. Much to my surprise, a remarkable landscape was eventually revealed. I say this not because I was particularly pessimistic or uninterested in circadian biology 35 years ago but because most research problems don't work out in such remarkable fashion.

I have been asked by numerous students since October 2 something like, "How can I do something important and significant like you have done?" The answer is, "I have no idea, not only for you but also for me: I don't know how I arrived at this place today. This answer is a bit tongue in cheek of course, so here's a somewhat more detailed answer.

Luck played a huge role in my success, indeed in anyone's success. A substantial fraction of this luck was just being in the right place at the right time. Another fraction, perhaps not unrelated to the "right place-right time" comment, was knowing about nucleic acids and macromolecular metabolism when this turned out to be a key to thinking about the problem. Having an operating yeast RNA lab interspersed among my circadian researchers was an enormous advantage when it was time to put in place these methods.

It is also important to stress that almost all aspects of luck are just "one-off" and cannot be repeated. Like the pinball that seemingly traces the same path, it doesn't come out exactly the same the second time 'round. For this reason, trying to do something great – shooting for prizes and fame let's call it – is a fool's errand. It is good to pick an important problem, but even more important is to love your work, the craft of being a scientist. My partner in crime Jeff Hall adored the craft of fly genetics; he was damn good at it too. A second piece of advice is to enjoy the people. Life science research is truly a social activity – or at least it was during my more than 50 years since I began graduate school at MIT. I have made remarkable friends through science, people I truly admire and in some cases love. I am also fond of almost all the fantastic people who joined my circadian lab at key points in time. It goes without saying that I wouldn't be writing this document without their efforts, intellectual as well as physical. There is also family to thank, wife as well as children. They have been long-suffering and an unending source of support and joy.

I will close by saying that I never saw myself in historical context when work was ongoing (I still don't) but as the scientific equivalent of a working class guy; I was always trying to put one foot in front of the other so that my research was moving forward more often than it was moving backwards – at least some of the time. Like almost every other active scientist, I was trying to maintain my funding, make sure my students and post-docs had decent papers and fulfill my obligations to Brandeis and to HHMI. Even when we published the feedback loop paper (Hardin et al., 1990), there was no dramatic sense of accomplishment. We simply moved on to the next stage of the work, the next stage of worry I might call it: Is the paper correct? How could we resolve the ambiguities in the final figure/model? How does PER work? This is truly how I see myself; I am just a regular guy, a scientist who got incredibly lucky.

REFERENCES

Albrecht, U., Sun, Z.S., Eichele, G., and Lee, C.C. (1997). A differential response of two putative mammalian circadian regulators, *mper1* and *mper2*, to light. *Cell* **91**, 1055–1064.

Allada, R., White, N.E., So, W.V., Hall, J.C., and Rosbash, M. (1998). A mutant Drosophila homolog of mammalian Clock disrupts circadian rhythms and transcription of period and timeless. *Cell* **93**, 791–804.

Antoch, M.P., Song, E.-J., Chang, A.-M., Vitaterna, M.H., Zhao, Y., Wilsbacher, L.D., Sangoram, A.M., King, D.P., Pinto, L.H., and Takahashi, J.S. (1997). Functional identification of the mouse circadian *clock* gene by transgenic BAC rescue. *Cell* **89**, 655–667.

Aronson, B.D., Johnson, K.A., Loros, J.J., and Dunlap, J.C. (1994). Negative feedback defines a circadian clock: Autoregulation of the clock gene *frequency*. *Science* **263**, 1578–1584.

Bargiello, T.A., Jackson, F.R., and Young, M.W. (1984). Restoration of circadian behavioural rhythms by gene transfer in *Drosophila*. *Nature* **312**, 752–754.

Bargiello, T.A., Saez, L., Baylies, M.K., Gasic, G., Young, M.W., and Spray, D.C. (1987). The *Drosophila* clock gene per affects intercellular junctional communication. *Nature* **328**, 686–691.

Bargiello, T.A., and Young, M.W. (1984). Molecular genetics of a biological clock in *Drosophila. Proc Natl Acad Sci USA* **81**, 2142–2146.
Baylies, M.K., Bargiello, T.A., Jackson, F.R., and Young, M.W. (1987). Changes in abundance or structure of the per gene product can alter periodicity of the *Drosophila* clock. *Nature* **326**, 390–392.
Bender, W., Akam, M., Karch, F., Beachy, P.A., Peifer, M., Spierer, P., Lewis, E.B., and Hogness, D.S. (1983a). Molecular Genetics of the Bithorax Complex in *Drosophila melanogaster. Science* **221**, 23–29.
Bender, W., Spierer, P., and Hogness, D.S. (1983b). Chromosomal walking and jumping to isolate DNA from the Ace and rosy loci and the bithorax complex in *Drosophila melanogaster. J Mol Biol* **168**, 17–33.
Borbely, A.A. (1982). A two process model of sleep regulation. *Hum Neurobiol* **1**, 195–204.
Borbely, A.A., Daan, S., Wirz-Justice, A., and Deboer, T. (2016). The two-process model of sleep regulation: a reappraisal. *J Sleep Res* **25**, 131–143.
Bunning, E. (1935). Zur Kenntniss der endergonen Tagesrhythmik bei Insekten und Pflanzen. *Ber Deut botan Ges* **53**, 594–623.
Ceriani, M.F., Darlington, T.K., Staknis, D., Mas, P., Petti, A.A., Weitz, C.J., and Kay, S.A. (1999). Light-dependent sequestration of TIMELESS by CRYPTOCHROME. *Science* **285**, 553–556.
Chiu, J.C., Ko, H.W., and Edery, I. (2011). NEMO/NLK phosphorylates PERIOD to initiate a time-delay phosphorylation circuit that sets circadian clock speed. *Cell* **145**, 357–370.
Citri, Y., Colot, H.V., Jacquier, A.C., Yu, Q., Hall, J.C., Baltimore, D., and Rosbash, M. (1987). A family of unusually spliced biologically active transcripts encoded by a *Drosophila* clock gene. *Nature* **326**, 42–47.
Colot, H.V., Hall, J.C., and Rosbash, M. (1988). Interspecific comparison of the period gene of *Drosophila* reveals large blocks of non-conserved coding DNA. *EMBO J* **7**, 3929–3937.
Crews, S.T., Thomas, J.B., and Goodman, C.S. (1988). The *Drosophila single-minded* gene encodes a nuclear protein with sequence similarity to the *per* gene product. *Cell* **52**, 143–152.
Daan, S. (2010). *A History of Chronobiological Concepts: In The Circadian Clock*, U. Albrecht, ed. (New York, NY: Springer), pp. 1–35.
Daan, S., Beersma, D.G., and Borbely, A.A. (1984). Timing of human sleep: recovery process gated by a circadian pacemaker. *Am J Physiol* **246**, R161–183.
Darlington, T.K., Wager-Smith, K., Ceriani, M.F., Staknis, D., Gekakis, N., Steeves, T.D.L., Weitz, C.J., Takahashi, J.S., and Kay, S.A. (1998). Closing the Circadian Loop: CLOCK-induced Transcription of Its Own Inhibitors *per* and *tim*. *Science* **280**, 1599–1603.
Dunlap, J.C., and Loros, J.J. (2017). Making Time: Conservation of Biological Clocks from Fungi to Animals. *Microbiol Spectr* **5**.
Emery, P., So, W.V., Kaneko, M., Hall, J.C., and Rosbash, M. (1998). CRY, a *Drosophila* clock and light-regulated cryptochrome, is a major contributor to circadian rhythm resetting and photosensitivity. *Cell* **95**, 669–679.
Emery, P., Stanewsky, R., Hall, J.C., and Rosbash, M. (2000a). A unique circadian-rhythm photoreceptor. *Nature* **404**, 456–457.
Emery, P., Stanewsky, R., Helfrich-Forster, C., Emery-Le, M., Hall, J.C., and Rosbash, M. (2000b). Drosophila CRY is a deep brain circadian photoreceptor. *Neuron* **26**, 493–504.

Feldman, J.F., and Hoyle, M. (1973). Isolation of circadian clock mutants of *Neurospora crassa. Genetics* **755**, 605–613.

Feldman, J.F., and Hoyle, M.N. (1976). Complementation analysis of linked circadian clock mutants of *Neurospora crassa. Genetics* **82**, 9–17.

Flint, K.K., Rosbash, M., and Hall, J.C. (1993). Transfer of dye among salivary gland cells is not affected by genetic variations of the period clock gene in *Drosophila* melanogaster. *JMembrane Biol* **136**, 333–342.

Gardner, G.F., and Feldman, J.F. (1980). The frq locus in Neurospora crassa: A key element in circadian clock organization. *Genetics* **96**, 877–886.

Gehring, W., and Rosbash, M. (2003). The coevolution of blue-light photoreception and circadian rhythms. *J Mol Evol* **57** Suppl 1, S286–289.

Gekakis, N., Saez, L., Delahaye-Brown, A.-M., Myers, M.P., Sehgal, A., Young, M.W., and Weitz, C.J. (1995). Isolation of timeless by PER protein interactions: Defective interaction between timeless protein and long-period mutant PERL. *Science* **270**, 811–815.

Gekakis, N., Staknis, D., Nguyen, H.B., Davis, C.F., Wilsbacher, L.D., King, D.P., Takahashi, J.S., and Weitz, C.J. (1998). Role of the CLOCK protein in the mammalian circadian mechanism. *Science* **280**, 1564–1569.

Gotter, A.L. (2006). A Timeless debate: resolving TIM's noncircadian roles with possible clock function. *Neuroreport* **17**, 1229–1233.

Guo, F., Cerullo, I., Chen, X., and Rosbash, M. (2014). PDF neuron firing phase-shifts key circadian activity neurons in *Drosophila. Elife* **3**.

Guo, F., Yu, J., Jung, H.J., Abruzzi, K.C., Luo, W., Griffith, L.C., and Rosbash, M. (2016). Circadian neuron feedback controls the *Drosophila* sleep–activity profile. *Nature* **536**, 292–297.

Hardin, P.E., Hall, J.C., and Rosbash, M. (1990). Feedback of the *Drosophila* period gene product on circadian cycling of its messenger RNA levels. *Nature* **343**, 536–540.

Hardin, P.E., Hall, J.C., and Rosbash, M. (1992). Circadian oscillations in period gene mRNA levels are transcriptionally regulated. *Proc Natl Acad Sci USA* **89**, 11711–11715.

Hardin, P.E., and Panda, S. (2013). Circadian timekeeping and output mechanisms in animals. *Curr Opin Neurobiol* **23**, 724–731.

Hastings, J.W., and Sweeney, B.M. (1957). On the mechanism of temperature independence in a biological clock. *Proc Natl Acad Sci USA* **43**, 804–811.

Huang, Z.J., Edery, I., and Rosbash, M. (1993). PAS is a dimerization domain common to *Drosophila* Period and several transcription factors. *Nature* **364**, 259–262.

Hunter-Ensor, M., Ousley, A., and Sehgal, A. (1996). Regulation of the *Drosophila* protein timeless suggests a mechanism for resetting the circadian clock by light. *Cell* **84**, 677–685.

Jackson, R.F., Bargiello, T.A., Yun, S.-H., and Young, M.W. (1986). Product of the per locus of *Drosophila* shares homology with proteoglycans. *Nature* **320**, 185–188.

Jang, A.R., Moravcevic, K., Saez, L., Young, M.W., and Sehgal, A. (2015). *Drosophila* TIM binds importin alpha1, and acts as an adapter to transport PER to the nucleus. *PLoS Genet* **11**, e1004974.

Kalmus, H. (1935). Periodizität und Autochronie (= Ideochronie) als zeitregelnde Eigenschaften des Organismus. *Biologia Generalis* **11**, 93–114.

Kalmus, H. (1940a). Diurnal Rhythms in the Axolotl Larva and in *Drosophila. Nature* **145**, 72–73.

Kalmus, H. (1940b). New research in the diurnal periodicity of animals. *Acta Med Scand*, suppl **108**, 227–233.
Kidd, P.B., Young, M.W., and Siggia, E.D. (2015). Temperature compensation and temperature sensation in the circadian clock. *Proc Natl Acad Sci USA* **112**, E6284–6292.
King, D.P., Zhao, Y., Sangoram, A.M., Wilsbacher, L.D., Tanaka, M., Antoch, M.P., Steeves, T.D.L., Vitaterna, M.H., Kornhauser, J.M., Lowrey, P.L., et al. (1997). Positional Cloning of the mouse circadian clock gene. *Cell* **89**, 641–653.
Kloss, B., Price, J.L., Saez, L., Blau, J., Rothenfluh, A., Wesley, C.S., and Young, M.W. (1998). The *Drosophila* clock gene double-time encodes a protein closely related to human casein kinase Iepsilon. *Cell* **94**, 97–107.
Koh, K., Zheng, X., and Sehgal, A. (2006). JETLAG resets the *Drosophila* circadian clock by promoting light-induced degradation of TIMELESS. *Science* **312**, 1809–1812.
Kondo, T., Strayer, C.A., Kulkarni, R.D., Taylor, W., Ishiura, M., Golden, S.S., and Johnson, C.H. (1993). Circadian rhythms in prokaryotes: Luciferase as a reporter of circadian gene expression in cyanobacteria. *Proc Natl Acad Sci USA* **90**, 5672–5676.
Konopka, R.J., and Benzer, S. (1971). Clock mutants of *Drosophila melanogaster. Proc Natl Acad Sci USA* **68**, 2112–2116.
Kyriacou, C.P., Green, E.W., Piffer, A., and Dowse, H.B. (2017). Failure to reproduce period-dependent song cycles in *Drosophila* is due to poor automated pulse-detection and low-intensity courtship. *Proc Natl Acad Sci USA.*
Kyriacou, C.P., and Hall, J.C. (1980). Circadian rhythm mutations in *Drosophila melanogaster* affect short-term fluctuatons in the male's courtship song. *Proc Natl Acad Sci USA* **77**, 6729–6733.
Lee, C., Bae, K., and Edery, I. (1998). The Drosophila CLOCK protein undergoes daily rhythms in abundance, phosphorylation and interactions with the PER-TIM complex. *Neuron* **4**, 857–867.
Lee, C., Etchegaray, J.P., Cagampang, F.R., Loudon, A.S., and Reppert, S.M. (2001). Posttranslational mechanisms regulate the mammalian circadian clock. *Cell* **107**, 855–867.
Lee, K., Loros, J.J., and Dunlap, J.C. (2000). Interconnected feedback loops in the Neurospora circadian system. *Science* **289**, 107–110.
Li, Y., Guo, F., Shen, J., and Rosbash, M. (2014). PDF and cAMP enhance PER stability in *Drosophila* clock neurons. *Proc Natl Acad Sci USA* **111**, E1284–1290.
Li, Y., and Rosbash, M. (2013). Accelerated degradation of perS protein provides insight into light-mediated phase shifting. *J Biol Rhythms* **28**, 171–182.
Liu, X., Zwiebel, L.J., Hinton, D., Benzer, S., Hall, J.C., and Rosbash, M. (1992). The period gene encodes a predominantly nuclear protein in adult *Drosophila. J Neurosci* **12**, 2735–2744.
Lorenz, L.J., Hall, J.C., and Rosbash, M. (1989). Expression of a *Drosophila* mRNA is under circadian clock control during pupation. *Development* **107**, 869–880.
Loros, J.J., and Feldman, J.F. (1986). Loss of temperature compensation of circadian period length in the frq-9 mutant of Neurospora crassa. *J Biol Rhythms* **1**, 18767–18198.
Loros, J.J., Richman, A., and Feldman, J.F. (1986). A recessive circadian clock mutant at the frq locus in Neurospora crassa. *Genetics* **114**, 1095–1110.
Lowrey, P.L., Shimomura, K., Antoch, M.P., Yamazaki, S., Zemenides, P.D., Ralph, M.R., Menaker, M., and Takahashi, J.S. (2000). Positional syntenic cloning and functional characterization of the mammalian circadian mutation tau. *Science* **288**, 483–492.

Lucas, R.J., Freedman, M.S., Munoz, M., Garcia-Fernandez, J.M., and Foster, R.G. (1999). Regulation of the mammalian pineal by non-rod, non-cone, ocular photoreceptors. *Science* **284**, 505–507.

Mehra, A., Shi, M., Baker, C.L., Colot, H.V., Loros, J.J., and Dunlap, J.C. (2009). A role for casein kinase 2 in the mechanism underlying circadian temperature compensation. *Cell* **137**, 749–760.

Melkani, G.C., and Panda, S. (2017). Time-restricted feeding for prevention and treatment of cardiometabolic disorders. *J Physiol* **595**, 3691–3700.

Menet, J.S., Abruzzi, K.C., Desrochers, J., Rodriguez, J., and Rosbash, M. (2010). Dynamic PER repression mechanisms in the Drosophila circadian clock: from on-DNA to off-DNA. *Genes Dev* **24**, 358–367.

Meyer, P., Saez, L., and Young, M.W. (2006). PER-TIM interactions in living *Drosophila* cells: an interval timer for the circadian clock. *Science* **311**, 226–229.

Myers, M.P., Wager-Smith, K., Rothenfluh-Hilfiker, A., and Young, M.W. (1996). Light-induced degradation of TIMELESS and entrainment of the *Drosophila* circadian clock. *Science* **271**, 1736–1740.

Myers, M.P., Wager-Smith, K., Wesley, C.S., Young, M.W., and Sehgal, A. (1995). Positional cloning and sequence analysis of the *Drosophila* clock gene timeless. *Science* **270**, 805–808.

Pittendrigh, C.S. (1954). On temperature independence in the clock-system controlling emergence time in *Drosophila. Proc Natl Acad Sci USA* **40**, 1018–1029.

Pittendrigh, C.S. (1993). Temporal organization: Reflections of a Darwinian clock-watcher. *Annu Rev Physiol* **55**, 17–54.

Price, J.L., Blau, J., Rothenfluh, A., Abodeely, M., Kloss, B., and Young, M.W. (1998). double-time is a novel *Drosophila* clock gene that regulates PERIOD protein accumulation. *Cell* **94**, 83–95.

Price, J.L., Dembinska, M.E., Young, M.W., and Rosbash, M. (1995). Suppression of PERIOD protein abundance and circadian cycling by the *Drosophila* clock mutation timeless. *EMBO J* **14**, 4044–4049.

Provencio, I., Jiang, G., De Grip, W.J., Hayes, W.P., and Rollag, M.D. (1998). Melanopsin: An opsin in melanophores, brain, and eye. *Proc Natl Acad Sci USA* **95**, 340–345.

Reddy, P., Jacquier, A.C., Abovich, N., Petersen, G., and Rosbash, M. (1986). The period clock locus of *D. melanogaster* codes for a proteoglycan. *Cell* **46**, 53–61.

Reddy, P., Zehring, W.A., Wheeler, D.A., Pirrotta, V., Hadfield, C., Hall, J.C., and Rosbash, M. (1984). Molecular analysis of the period locus in *Drosophila melanogaster* and identification of a transcript involved in biological rhythms. Cell **38**, 701–710.

Reppert, S.M., and Weaver, D.R. (2002). Coordination of circadian timing in mammals. *Nature* **418**, 935–941.

Rosbash, M. (2009). The implications of multiple circadian clock origins. *PLoS Biol* **7**, e62.

Rosbash, M. (2015). Ronald J. Konopka (1947–2015). *Cell* **161**, 187–188.

Rubin, G.M., and Spradling, A.C. (1982). Genetic transformation of *Drosophila* with transposable element vectors. *Science* **218**, 348–353.

Rutila, J.E., Edery, I., Hall, J.C., and Rosbash, M. (1992). The analysis of new short-period circadian rhythm mutants suggests features of D. *melanogaster* period gene function. *J Neurogenet* **8**, 101–113.

Rutila, J.E., Suri, V., Le, M., So, W.V., Rosbash, M., and Hall, J.C. (1998). CYCLE Is a Second bHLH-PAS Clock Protein Essential for Circadian Rhythmicity and Transcription of *Drosophila* period and timeless. *Cell* 93, 805–814.

Rutila, J.E., Zeng, H., Le, M., Curtin, K.D., Hall, J.C., and Rosbash, M. (1996). The timSL mutant of the *Drosophila* rhythm gene timeless manifests allele-specific interactions with period gene mutants. *Neuron* **17**, 921–929.

Saez, L., Young, M.W., Baylies, M.K., Gasic, G., Bargiello, T.A., and Spray, D.C. (1992). Per – no link to gap junctions. *Nature* **360**, 542–542.

Saper, C.B., and Fuller, P.M. (2017). Wake-sleep circuitry: An overview. *Curr Opin Neurobiol* **44**, 186–192.

Sehgal, A., Price, J., Man, B., and Young, M. (1994). Loss of circadian behavioral rhythms and per RNA oscillations in the *Drosophila* mutant timeless. *Science* **263**, 1603–1606.

Sehgal, A., Rothenfluh-Hilfiker, A., Hunter-Ensor, M., Chen, Y., Myers, M., and Young, M.W. (1995). Rhythmic expression of timeless: A basis for promoting circadian cycles in period gene autoregulation. *Science* **270**, 808–810.

Shafer, O.T., Rosbash, M., and Truman, J.W. (2002). Sequential Nuclear Accumulation of the Clock Proteins Period and Timeless in the Pacemaker Neurons of *Drosophila melanogaster. J Neurosci* 22, 5946–5954.

Shearman, L.P., Zylka, M.J., Weaver, D.R., Kolakowski, L.F.J., and Reppert, S.M. (1997). Two period homologs: circadian expression and photic regulation in the suprachiasmatic nuclei. *Neuron* **19**, 1261–1269.

Shigeyoshi, Y., Taguchi, K., Yamamoto, S., Takekida, S., Yan, L., Tei, H., Moriya, T., Shibata, S., Loros, J.J., Dunlap, J.C., et al. (1997). Light-induced resetting of a mammalian circadian clock is associated with rapid induction of the mPer1 transcript. *Cell* **91**, 1043–1053.

Siwicki, K.K., Flint, K.F., Hall, J.C., Rosbash, M., and Spray, D.C. (1992). The *Drosophila* period gene and dye coupling in larval salivary glands: a re-evaluation. *Biol Bull* **183**, 340–341.

Spierer, P., Spierer, A., Bender, W., and Hogness, D.S. (1983). Molecular mapping of genetic and chromomeric units in *Drosophila melanogaster. J Mol Biol* **168**, 35–50.

Spradling, A.C., and Rubin, G.M. (1982). Transposition of cloned P elements into *Drosophila* germ line chromosomes. *Science* **218**, 341–347.

Stanewsky, R., Kaneko, M., Emery, P., Beretta, M., Wager-Smith, K., Kay, S.A., Rosbash, M., and Hall, J.C. (1998). The cryb mutation identifies cryptochrome as a circadian photoreceptor in *Drosophila. Cell* **95**, 681–692.

Stern, D.L. (2014). Reported *Drosophila* courtship song rhythms are artifacts of data analysis. *BMC Biol* **12**, 38.

Stern, D.L., Clemens, J., Coen, P., Calhoun, A.J., Hogenesch, J.B., Arthur, B.J., and Murthy, M. (2017). Experimental and statistical reevaluation provides no evidence for *Drosophila* courtship song rhythms. *Proc Natl Acad Sci USA* **114**, 9978–9983.

Sun, Z.S., Albrecht, U., Zhuchenko, O., Bailey, J., Eichele, G., and Lee, C.C. (1997). RIGUI, a putative mammalian ortholog of the *Drosophila* period gene. *Cell* **90**, 1003–1011.

Suri, V., Hall, J.C., and Rosbash, M. (2000). Two novel doubletime mutants alter circadian properties and eliminate the delay between RNA and protein in *Drosophila. J Neurosci* **20**, 7547–7555.

Suri, V., Qian, Z., Hall, J.C., and Rosbash, M. (1998). Evidence that the TIM light response is relevant to light-induced phase shifts in *Drosophila melanogaster. Neuron* **21**, 225–234.

Tei, H., Okamura, H., Shigeyoshi, Y., Fukuhara, C., Ozawa, R., Hirose, M., and Sakaki, Y. (1997). Circadian oscillation of a mammalian homologue of the *Drosophila* period gene. *Nature* **389**, 512–516.

Tu, B.P., and McKnight, S.L. (2006). Metabolic cycles as an underlying basis of biological oscillations. *Nat Rev Mol Cell Biol* **7**, 696–701.
Vitaterna, M.H., King, D.P., Chang, A.-M., Kornhauser, J.M., Lowrey, P.L., McDonald, J.D., Dove, W.F., Pinto, L.H., Turek, F.W., and Takahashi, J.S. (1994). Mutagenesis and mapping of a mouse gene, Clock, essential for circadian behavior. *Science* **264**, 719–725.
Vitaterna, M.H., Selby, C.P., Todo, T., Niwa, H., Thompson, C., Fruechte, E.M., Hitomi, K., Thresher, R.J., Ishikawa, T., Myazaki, J., et al. (1999). Differential regulation of mammalian period genes and circadian rhythmicity by cryptochromes 1 and 2. *Proc Natl Acad Sci USA* **96**, 12114–12119.
Vosshall, L.B., Price, J.L., Sehgal, A., Saez, L., and Young, M.W. (1994). Block in nuclear localization of period protein by a second clock mutation, timeless. *Science* **263**, 1606–1609.
Weiner, J. (1999). *Time, love, memory: A great biologist and his quest for the origins of behavior,* 1st ed. (New York: New York: Knopf).
Yang, Z., and Sehgal, A. (2001). Role of molecular oscillations in generating behavioral rhythms in *Drosophila. Neuron* **29**, 453–467.
Young, M.W., and Judd, B.H. (1978). Nonessential Sequences, Genes, and the Polytene Chromosome Bands of DROSOPHILA MELANOGASTER. *Genetics* **88**, 723–742.
Yu, Q., Colot, H.V., Kyriacou, C.P., Hall, J.C., and Rosbash, M. (1987a). Behaviour modification by *in vitro* mutagenesis of a variable region within the period gene of *Drosophila. Nature* **326**, 765–769.
Yu, Q., Jacquier, A.C., Citri, Y., Hamblen, M., Hall, J.C., and Rosbash, M. (1987b). Molecular mapping of point mutations in the period gene that stop or speed up biological clocks in *Drosophila melanogaster. Proc Natl Acad Sci USA* **84**, 784–788.
Yuan, Q., Metterville, D., Briscoe, A.D., and Reppert, S.M. (2007). Insect cryptochromes: Gene duplication and loss define diverse ways to construct insect circadian clocks. *Mol Biol Evol* **24**, 948–955.
Zehring, W.A., Wheeler, D.A., Reddy, P., Konopka, R.J., Kyriacou, C.P., Rosbash, M., and Hall, J.C. (1984). P-element transformation with period locus DNA restores rhythmicity to mutant, arrhythmic *Drosophila melanogaster. Cell* **39**, 369–376.
Zeng, H., Hardin, P.E., and Rosbash, M. (1994). Constitutive overexpression of the *Drosophila* period protein inhibits period mRNA cycling. *EMBO J* **13**, 3590–3598.
Zeng, H., Qian, Z., Myers, M.P., and Rosbash, M. (1996). A light-entrainment mechanism for the *Drosophila* circadian clock. *Nature* **380**, 129–135.
Zhang, R., Lahens, N.F., Ballance, H.I., Hughes, M.E., and Hogenesch, J.B. (2014). A circadian gene expression atlas in mammals: Implications for biology and medicine. *Proc Natl Acad Sci USA* **111**, 16219–16224.
Zhu, H., Yuan, Q., Briscoe, A.D., Froy, O., Casselman, A., and Reppert, S.M. (2005). The two CRYs of the butterfly. *Curr Biol* **15**, R953–954.

Michael W. Young. © Nobel Prize Outreach AB. Photo: A. Mahmoud

Michael W. Young

Biography

I remember reading this big book, *The Wonders of Life on Earth*, when I was eleven or twelve. It said bird migration was controlled by some kind of internal timer, some kind of clock. Some birds, it explained, use the position of the sun to orient their migration, and the clocks in their heads are able to track time in order to assume the same direction of flight even as the position of the sun moves throughout the day. I thought that was both mysterious and fascinating. It was my first exposure to the notion of biological clocks.

I was born in Miami, Florida on March 28, 1949, into an environment where biology was all around. The neighborhood kids spent a lot of time running around each other's backyards together. Many of the families had greenhouses with an array of exotic plants that always interested me. We were close to tourist attractions: the Parrot Jungle, the Monkey Jungle, the Serpentarium, the Orchid Jungle. The Parrot Jungle was especially close and a lot of the birds weren't caged, so toucans and parrots would sometimes arrive in our yard. You'd see native wildlife, too. One time, my sister, Denise, and I caught a small alligator in the creek behind our grandmother's house. The local newspaper published a photograph. But my favorite place was the "Rockpit". This covered an area of about 50 acres and was the source of the landfill under our elementary school, David Fairchild. The excavations left scattered hills of coral rock that we would climb to survey the ponds below. The place had plenty of reptiles and the ponds were full of tadpoles. There was a "No Trespassing" sign next to the narrow gravel road into the Rockpit that said "Violators will be Prosecuted", which to my eight year old mind meant electrocuted.

At the local hobby shop they sold models of all sorts of things. You could build model airplanes or cars, but you could also build model biological systems, like of the brain or something called the Visible Man. These were a mix of colored and transparent plastic that came in boxes just like those for model cars and airplanes and with similar price tags.

Figure 1. Michael and his sister Denise (1954).

Figure 2. Michael's parents, Lloyd and Mickey, vacationing in the Florida Keys (~1947).

You'd get these things and put them together, and they were a pretty reasonable representation of real anatomy. The other kids, especially my sister's friends, thought I was really weird for putting people together, taking them apart, and have them sitting on the desk. But for me, although inert representations, the models were ways I could explore biology a bit more formally, beyond just observing nature as it was around me.

I was interested in chemistry, too. I got a chemistry set at one point and set up a lab in the back room of our house. We had this beautiful terracotta tile floor. You'd do things like use potassium chlorate to generate some oxygen and hold a match to it and get it to explode. But I also would just mix and match things kind of carelessly. I never blew myself up, but I did create something that bubbled over and shot small droplets onto the floor that left lots of little white stains. After that, my parents confined my chemistry lab to the garage.

There was a kid down the street whose father worked at Porsche, and he was always bringing home these sports cars and dissecting them while the neighborhood kids watched. Mechanics interested me and we started building our own vehicles – the most typical starting point was a lawnmower engine – for running around in the neighborhood. I built a go-kart. It's amazing that I survived that, but it taught me a lot about the way things can be put together to work. Biology and motorized machinery are obviously not equivalent, but it was useful to think about what makes something work and to see what vehicles look like on the inside.

When I was in the fourth or fifth grade, I remember putting my books together in my desk with the science book on top and thinking somehow, I'm going to do something that involves the sciences. That evolved over the next couple of years into medicine. I had no idea how you could become a scientist. That was something else completely. But if you had an interest in the sciences, then becoming a doctor seemed like a reasonable path.

No one in my family shared my interest in science, but my parents, each with a high school education, were very supportive of my interests. My father was a bomber pilot in World War II. His plane was shot down in 1944 (he crash-landed in a potato field in Holland) and he was in a prisoner-of-war camp for the rest of the war. I have a copy of a letter a woman in the mid-west sent to my grandmother, saying that she had heard my father on a short-wave radio, asking that someone contact his mother to say he was alive and ok.

My parents met when my father came home to Knoxville, Tennessee. I think because of his war experience, he was not averse to taking risks, so when they went to Miami for their honeymoon in 1946, they just decided to stay. We went to church until I was about 10, but it was a very soft touch. And I don't remember a particular political bent to any of the discussions in our family. I think an advantage of this was that I didn't come out of a mold with a set of ideas that I'd eventually have to undo, because I became very irreligious in high school. I was much more enchanted by what you could prove to yourself than by having to put your faith into something.

In 1966 my father's job changed and we moved to Euless, Texas, a small town near Dallas. I went to LD Bell, the local public high school, and later to the University of Texas at Austin, where the tuition was $50 a semester, still expecting to study medicine eventually. Having grown up in the Great Depression, my father was always concerned that I needed to work at developing a career, by which he meant something tangible, a well-established profession. Medical doctor was something he understood.

Figure 3. My little league baseball team. I'm kneeling on the lower right. We made it to the Miami-wide playoffs, but lost when I struck out (~1962).

But my path quickly took a lasting turn, first with my father's sudden death (he had a heart attack on a business trip during my first year of college) and then when I took a course in genetics with Burke Judd in my senior year. Burke was very contemporary in the way he thought about genetics – his course looked ahead to the questions on the horizon – and I went to speak with him on several occasions because I really liked the class. I learned from him that I could sign up to do a summer research project in his lab and see how I liked research. I hadn't even known that this kind of thing was possible. That's when I first began to realize how you could train to be a scientist.

It's also how I ended up with my first mentor. Until then, the notion of having an advisor was foreign to me. Neither of my parents had gone to college, so although I took a lot of interesting courses, I didn't have a way of thinking about how to make decisions about the future. Burke's support and positive feedback very early on were very important in showing me this whole new world.

And I met my wife, Laurel Eckhardt, in Burke's course. I had tried to introduce myself to her, but she paid no attention to me at all. Later, when we met again in Burke's office I guess she realized maybe I was legitimate. I later learned that Burke and his course had also introduced her to the idea of a scientific research career. Like me, she was clueless about where scientific discoveries were made.

So I did spend that summer in the lab and it was a great time. The lab was a classical genetics lab and was focused on cytogenetics, that is, looking at chromosomes. There were two postdocs in the lab, Lenny Robbins and Ron Woodruff, who took me under their wings and seemed to think that I was serious enough to invest some time in. Lenny, especially, helped me learn more about biology, particularly molecular biology.

This was the early 1970s, and everybody was thinking about the new molecular biology. Burke's lab was very interested in understanding the genetic complexity of eukaryotes using *Drosophila*, the fruit fly, as a model. Fruit flies have these enormous chromosomes in their salivary glands and it's very easy to tell one chromosome from the other and even to tell specific regions apart. They have these striations that were thought at the time to be visible markings of the genes. You could count off the striations and you were just counting genes. But the tools were still too dull to go much further, to bring that down to a real understanding of the organization of these genes or other information in the chromosomes.

That summer, my project was to use genetic screens to look for sterile mutations in the X chromosome region on which Burke's lab was focusing. But I could do other things. If I wanted to use the microscope to look at my own chromosome preparations, I could do that. It was like stepping all the way back to my chemistry and biology sets, but here was a real laboratory at my disposal. I'd spend hours talking with the postdocs in the lab and those in adjoining laboratories, just dreaming up experiments that we might want to do. I began pushing to introduce molecular biology into Burke's lab so we could do some of the experiments that might get us to a better understanding of how eukaryotic genes and chromosomes are put together. By the end of the summer of 1971, I had made a career choice – to pursue genetics – and I had decided to stay on as a graduate student in Burke's lab.

The lab was focused on a short part of the X chromosome and how many genes there were in that region, which included enough DNA to include about 400–500 genes the size of those in a bacterium like E. coli. The first sweep through had given us something like twenty genes, so what's all the rest of that DNA doing? If knocking a gene out resulted in lethality, you wouldn't miss that gene. But how do you know what you've missed? We knew that knocking some genes out led to a much more subtle phenotype, so that was a real question.

I was in the process of mapping mutations I had found that affected female fertility when Burke came in waving around a paper from Ron Konopka and Seymour Benzer, published in *Proceedings of the National Academy of Sciences (PNAS)* in September 1971, describing *Drosophila* circadian clock mutants that they had discovered. The gene they had found, which they named *period*, was in an area that seemed to be within, or very close to, the region we were studying in Burke's lab. I thought these were pretty interesting mutations: in addition to their relevance to the general question Burke's lab had about how many genes there were in the region, they affected wake-sleep behavior, a behavior with very solid properties.

So, I wrote to Ron and Seymour to ask for the mutations. One of the things about the *Drosophila* community is the unspoken agreement that if you discover something, once you publish it, you should give any mutations, any materials that you've produced, freely to anyone who asks for them. Ron sent the mutants right away.

I did experiments that proved that *period* was, in fact, a new gene and that it lived between two genes we already knew about. Missing *period* raised big questions about how we were counting genes, but there was only so far anyone could push with the tools we had. We needed to be able to really look at a gene, to actually have the DNA corresponding to a gene and map the mutations affecting that gene so we could understand how those mutations affected the gene's activity and/or the gene's protein product.

We weren't in a position to do that work yet, but it was thinking ahead – where will this field be in five or ten years – that was exciting to me. That's what you want to be thinking about, those are the questions you want to have in mind.

Figure 4. I had just arrived at Stanford in the fall of 1975. My two fellow postdocs, Gerry Rubin and David Finnegan, chained me to a microscope until I mapped the chromosomal locations of their backlog of in situ hybridizations.

While I was working on locating *period*, I found two deficiencies that seemed to overlap, each of which eliminated the *period* gene and appeared to be missing only the *period* gene. That was a lucky find. But an even luckier find was a translocation – an inter-chromosomal exchange – that occurred right in the *period* gene, knocking out some of the gene's activity. Because deleting *period* eliminated circadian rhythmicity without producing any other visible consequences, I began to think *period* was central to whatever larger process was involved in biological timekeeping. It could be a mainspring or a gear, by analogy, to a mechanical clock.

The question of the relationship between genes and behavior was completely unresolved at the time, and I saw that *Drosophila's* clock and the *period* gene would be in many ways an ideal starting point to study that relationship. I saw that a periodic, quantifiable behavior like the circadian wake-sleep cycle would be a great starting point to begin digging into the genetic mechanisms that make a behavior work.

As I was finishing my graduate work in 1975 and thinking about where to do my postdoc, I learned from Burke that David Hogness at Stanford was going to use *Drosophila* to isolate pieces of cloned DNA and study single genes from all over the genome, in detail, at the molecular level. I realized that if I ever wanted to come back to the circadian problem, the translocation I had found would be a very powerful entry point to find the *period* gene within cloned DNA. But no matter what my questions were going to be, I knew I was going to have unsatisfying answers without the ability to isolate a gene and map mutations along that gene. With that, you could define everything: phenotype and genotype, chromosomes and DNA. This was only going on at Stanford and only in Dave's lab. The big questions, as far as I was concerned, could only be answered by going there. Dave was a little bit like Benzer in the sense that he wanted to ask new questions and set off on a new frontier that presented risks, but also opportunities for some real excitement. I thought his taste in problems was fantastic.

Laurel was ready for graduate school at the same time that I was ready to begin my postdoc. She had been accepted into a Ph.D. program at Stanford and was looking forward to studying with Len Herzenberg in the Genetics Department. This meant I would be right around the corner from her in the Biochemistry Department. We packed up her car and a U-Haul truck and drove to California. We remember coming into San Francisco at night and all of a sudden seeing these lights up on the hills. We both thought it is like going to Mars or something. It was so different from anyplace we'd been before.

The big breakthrough that had happened at Stanford was the development of a method for cloning DNA – what we call recombinant DNA – in order to isolate large amounts of a single gene, or even a single segment of a gene. They had protocols for assembling these DNAs, for propagating them in bacteria or viruses, and for making maps of these isolated DNAs once they were present. And all of these things were very new at the time. Dave's project was really the first genome project, although it wasn't called that. It was an attempt to use *Drosophila* to fully define what the flies' chromosomes carried. There was a chromosome map outside Dave's office with tacks pressed into place everywhere a cloned piece of DNA had been located. Eventually these landmark DNAs would begin connecting with their neighbors. We all were enchanted by the tools and by the possibilities they represented. My first project was to isolate breaks in the DNA, using a method that I'd come up with at Texas, so I would have the complete picture of what a gene was doing and what phenotypes it was affecting.

I never lost my love for the outdoors and got pretty deeply invested in technical rock-climbing with a group of biologists at Stanford. We used to set up our experiments on a Friday night, knowing that they would have to incubate through the weekend, and we'd then take off for Yosemite with ropes and pitons and carabiners and all this equipment to climb rocks. We'd talk about work when we were out together, things that we wanted to find out, but there also was this excitement and terror of being on a big rock face. While I don't climb rocks anymore, I remember those times as great fun that rounded out our experience in California.

When it came time to look for a faculty position, Laurel and I realized that I had to go to a metropolitan area with more than one academic institution, so she would not be limited when looking for post-doctoral positions once she finished her Ph.D. We settled pretty quickly on Rockefeller. It had tremendous advantages and was just very different from all the other institutions I had considered. I remember my first visit and talking with Norton Zinder. He was in Smith Hall – where I would soon have my first lab next to his. It was a dusty, dark, cavernous building, but there was a charm to the classical laboratories because of the deep history they held.

When I came to Rockefeller in 1978, I had a series of things I thought I could accomplish in five or six years. I was still asking, what's odd or interesting and new about a gene from a multicellular animal? My thought was to isolate one or more genes from a genetically well-defined region and to see if we could hammer at them at the genetic and the physical levels to bring the functional and physical maps together and produce a complete picture of a piece of a chromosome. We had the new technology, recombinant DNA. We had libraries of cloned DNAs. That should mean that every gene in the fly was somewhere in that library; it was just a matter of fishing the right ones out.

And pretty quickly, we isolated two genes – *period* and *Notch*, a developmental gene – that we worked on in parallel for several years.

With *period*, the goal was to try to understand why or how it was making a contribution to behavior, to the flies' sleep-wake rhythm. Laboratories had been studying circadian biology for decades, but they were really just guessing, hypothesizing, about what the underlying mechanisms might be.

But what *Drosophila* had were mutations in these genes, like those that Ron Konopka and Seymour Benzer had found, that suggested you didn't have to rely on hypotheses and you didn't have to guess about the basis for the biology. You could use the fly and find out answers. You could create mutants that have an interesting change in behavior and then ask what the underlying gene looked like. By exploring the genes that were critical to behavior, you could go in without the baggage of a model. You could simply let the fly and its genetics tell you what path to take. Before molecular approaches were in place, naming the *period* gene and having mutations that could be mapped by classical genetics to a given location on a chromosome provided no clear path for learning more about circadian biology.

So, we thought we had a great problem to work on and that we could attack it on our own. Initially, it was just a postdoc, Ted Bargiello, and I who started a "chromosomal walk" to get to the *period* gene. From my graduate school days in Texas, I had the translocation that broke the gene and told us where in this chromosomal walk it could be found. So, we were able to take a series of cloned DNAs to make a map, to put them together, and then to see where this break was. Then we asked which parts of the chromosomal walk were transcribed to make an RNA. When we discovered an RNA that was broken by this translocation, we knew this must be the *period* gene.

It was an interesting and stressful time because we started this work in the early 1980s without knowing anyone else was interested enough to invest time and energy in it. Ron and Seymour's paper was not heavily cited. But as we were somewhere approaching this level of analysis, we learned that Jeff Hall and Michael Rosbash were on a similar mission to try to isolate the *period* gene and that they also had made some pretty good progress.

When we finished the first round of work, we had very good genetic and molecular evidence that we had found a single transcription unit that was *period*. We didn't know any more about it. We didn't have a gene sequence, but we could say that it was about 7,000 base pairs long. We published that in *PNAS* in April of 1984. It was the first publication to come out on the isolation of *period*, and the first molecular study to address circadian rhythms in any organism.

The next step, we realized, was to see if we could confirm that everything was limited to that one transcription unit. We needed a device that could record the sleep-wake activity of individual flies, so the in-house shop at Rockefeller made it for us. It was a combination of old microscope slide boxes, some sheet metal, and some primitive electronics. We generated DNA composed of just the wild type version of our transcription unit and we microinjected it back into embryos that were *period* null mutants, that is, they were arrhythmic. Then we put the flies in the machine, five flies at a time. The event recorder wiggled when a fly moved up or down. Ted, Rob Jackson (another postdoc who had joined the lab), and I camped out most of the time in the lab just to see what would happen. A few feet of chart paper would come out every day and after a few days, we could see that putting the transgene back into the null mutants had restored the flies' rhythms!

That was exciting because it confirmed that we had the gene. It also had another level of import because this was the first time anyone had transplanted a gene to generate behavior. We had a simple animal and an easily measured, ubiquitous behavior that was well-understood but only at the level of phenomenology. The fly went from having none of that behavior to having the full range of that behavior just because this one gene was put back into it.

We published the report, which confirmed that the gene was the only thing that was necessary to restore the behavior, in the 1984 year-end issue of *Nature*. Hall and Rosbash published in 1984, too, and we could all see that *period* was the gene to move in on. A couple of years later we reported the sequence of the gene and the changes associated with Konopka and Benzer's long-period, short-period and arrhythmic mutations. All three were single nucleotide changes, and the two that adjusted period length each changed a single amino acid.

Because we had competition, we thought carefully about how we might carry the project in a new direction, and give ourselves more room to operate independently. We wanted to avoid having two groups just doing the same things. And that's when we decided to initiate a big screening effort to find additional genes involved in the clock. We thought we could learn more about the sleep-wake rhythm by looking for other genes that affected it. Meanwhile, Hall and Rosbash continued to work exclusively on the molecular biology of *period*. And I think both plans worked well because it kept us from making duplicative discoveries.

Initially our genetic screen went on and on without giving us anything meaningful, until finally, in the early 1990s, after seven thousand assays, two postdocs in the lab, Amita Sehgal and Jeff Price, found a new mutation. We named the gene *timeless*, and it had many of the same behavioral properties as *period*. Most importantly, in 1995 we discovered it encoded a protein that was a physical partner for the Period protein.

What excited us most is that we'd taken an agnostic approach to how circadian rhythms worked. We weren't asking for genes that work with *period*. We weren't saying let's collect the things with which *period* interacts. We just said let's find another gene that affects circadian rhythms. Incredibly, that took us to another piece of the same molecular system that depended on *period*. This was our first hint that there would be a single mechanism for keeping time in the fly.

Isolating *timeless* and unpacking its relationship to *period* made us realize that following the genetics could perhaps take us to the heart of what was controlling circadian rhythms. It was a huge boost – and it convinced us that we should keep screening and get every mutation in every gene that we could to see if we could repeat our success. Would every new gene that we isolated point us back to the same system, to a single machine responsible for circadian biology? In fact, it did, and that was extremely gratifying.

When we found *timeless*, Jeff and Michael immediately saw the benefits of driving genetics of the fly forward – that is, searching for all genes relevant to the clock system – as rapidly as possible. Our labs began a joint screen for new mutants that was supported by the National Science Foundation. With any luck at all, we thought, there would be several new genes to work on and no one group could handle all of them. It would be more productive to have complementary assessments of different genes going on in our labs.

What we now understand is that all of the genes our labs ultimately found participate in the same mechanism: there are nine or ten key components that interact with one another to produce an oscillating molecular system that has a natural cycle time of about 24 hours. All of this was revealed by genetics. It wasn't by anyone coming up with ideas to test; it was by admitting that we didn't have the foggiest idea how the clock worked. We sought to collect all the mutations that are important to circadian rhythms and to understand them one by one, and then we looked at their relationships to each other. This is a story not of a single big discovery, but of the accumulation of smaller discoveries over many years that together enabled us to see the underlying biology of how cells track time.

We now know that the same clock system that applies to the fruit fly applies all the way to humans. We have also learned, by following gene expression, that flies and humans are a collection of clocks. We did not expect in the beginning that we would have cells outside the nervous system that use these clocks. And we didn't anticipate the degree to which cells in different organs find it useful to independently determine the time of day. In retrospect, it makes sense, for example, that liver cells are regulated with a 24-hour periodicity that is synchronized with feeding times, and that tissue-autonomous clocks are involved in that regulation. There is some coordination of what happens in different organs, but it is beginning to look like there is as much signaling among organs as there is between the brain and these systems. You can produce an animal that has its head set on New York time and its liver set on Tokyo time, thereby revealing a really surprising degree of autonomy between clock systems in the body.

I came to Rockefeller with a five-year plan and am still here after 40 years. I don't think there's any place in the world that would have provided me with the way forward on this adventure in the way that Rockefeller did. It's such an unusual community and the unwavering interest and encouragement from colleagues early on – particularly Norton Zinder, Jim Darnell, Torsten Wiesel, and Günter Blobel (who to my great sadness died as I was writing this) – made all the difference.

In 1998 Seymour Benzer wrote me – I have saved the letter in my copy of Jonathan Weiner's biography of him – to congratulate me on our progress. He was thrilled, he said, to see how the work on the clock was going. Seymour had pioneered the notion that you could study genetics connected to behavior, but he met with a lot of resistance. When he started his work, many people just didn't believe that working on single genes would reveal important things about behavior. And the work couldn't move very rapidly before the molecular tools became available. It has been incredibly gratifying to bring molecular biology to this field and to prove, with Michael and Jeff, that a gene-based approach could solve a deep problem about behavior and reveal this beautiful circadian mechanism.

For the past several years, research in my lab has focused on both *Drosophila* and humans. The *Drosophila* work now centers on the related problem of sleep: what controls its duration and how is that driven by genes? We know sleep is fundamentally important – when we collect mutants that get 60 percent of the sleep of a typical wild-type fly, their

Figure 5. Hiking a desert canyon with Arissa, Laurel and Natalie in the mid-1990s.

lifespan is cut in half – yet we still don't know what sleep is actually for, how it is regulated, or what it accomplishes. I think we can use *Drosophila* genetics to get at these questions in as meaningful a way as we did with the circadian work.

On the human side, we think that most clinically significant circadian disorders are probably represented in the very large genetic databases that are available now. The challenge is to figure out which genetic variations are important and why.

Sleep, as I see it, has as much mystery to it as circadian biology once had – and deep progress is still ahead. And so we have a very fundamental set of questions on the *Drosophila* side and a more medically oriented set of questions now on the human side, and I think that's a nice mix at this point in the lab's history.

* * *

Life isn't all research, of course. Laurel, who is a biology professor at Hunter College, The City University of New York and I have been navigating family life and our careers together for a lucky, long time. We had two daughters in the 1980s: Natalie in 1986 and Arissa in 1989. Today, Natalie has a Ph.D. in sociology and Arissa is a medical resident. We built a house in New Mexico, where we retreat to hike, write papers, and explore the natural mysteries that have intrigued us from the beginning.

Editor's note: Michael Young delivered his Nobel Lecture "Time Travels: A 40 Year Journey from Drosophila's Clock Mutants to Human Circadian Disorders" on 7 December 2017 at Aula Medica, Karolinska Institutet in Stockholm. He was introduced by Professor Juleen Zierath. See https://www.nobelprize.org/prizes/medicine/2017/young/lecture/

Physiology or Medicine 2018

James P. Allison and Tasuku Honjo

"*for their discovery of cancer therapy by inhibition of negative immune regulation*"

The Nobel Prize in Physiology or Medicine, 2018

Presentation speech by Professor Klas Kärre, Member of the Nobel Assembly at Karolinska Institutet, Member of the Nobel Committee for Physiology or Medicine, 10 December 2018.

Your Majesties, Your Royal Highnesses, Honoured Laureates, Ladies and Gentlemen,

Cancer, one of the greatest scourges of humanity, is caused by the uncontrolled proliferation of diseased cells that can spread in the body and form metastases. Cancer therapy has been based on three pillars: surgery, radiotherapy and drugs that attack cancer cells. Today more than two thirds of cancers are cured, but the disease still claims millions of lives every year. There is a great need for new forms of therapy, and for a long time researchers have pinned their hopes on the possibility of utilising our immune system, which normally protects us against infections.

The immune system is based on a diversified array of instruments, in the form of various cells and molecules. Each of them has its own special sound and technical requirements to get them to work, a bit like the instruments in the orchestra here today.

The fundamental task of the immune system is to distinguish between foreign cells – and react against them – and the body's own cells, which it must leave in peace.

T cells – a kind of white blood cells that act, among other things, as killer cells – use a special instrument called a receptor to identify cancer cells as foreign. By the 1990s, cancer immunology researchers had shown that T cells will then react, but unfortunately in a much too timid way. Let's ask the orchestra to illustrate.

Beautiful and clear, but alas – so short, slow and weak! Playing "andante and pianissimo" was not enough to eliminate cancer cells. This year's Nobel Prize in Physiology or Medicine goes to two researchers who discovered how we can release the brakes that hold back the immune system, thereby mobilising it for cancer therapy. Both Laureates are immunologists, but neither was actually a cancer researcher from the beginning. This story therefore illustrates something important: the unexpected benefits of basic research.

In the mid-1990s, Jim Allison was working in a field of research that identified "gas pedals" that amplified the reaction of T cells, but also a "brake pedal" that weakened it. He tested the very bold idea of releasing the brake pedal by using antibodies to trigger T cell reactions against cancer. He began by curing cancer in mice. Then he worked step by step to develop this therapy for humans. He called it "checkpoint inhibition". In the first clinical studies of malignant melanoma, some patients responded dramatically – even in patients with the disease all over their bodies, the metastases shrank away and disappeared.

Tasuko Honjo had already discovered a new molecule on T cells in the early 1990s. Through many years of systematic work, he was able to show that it serves as another braking mechanism in the immune system. Inspired by Allison, he showed that blocking this brake also triggered attacks on cancer cells in mice, and that it worked in a new way. Honjo suggested that this might be developed into a powerful cancer therapy. Clinical researchers were later able to confirm his hypotheses: this form of checkpoint inhibition leads to responses in more patients and also works against additional forms of cancer.

The two therapies have especially strong effects when used together. Let's get a preview by returning to our immune system orchestra. You remember how pathetic it sounded last time. Now let's hear how the same reaction looks when we release the brakes. Maestro?

Allegro e Fortissimo! That was a bit different, the way it should sound! By orchestrating the immune system in the right way, it has proved possible to control or eliminate the disease in tens of thousands of patients. Many are still tumour-free after more than five years. This new pillar of cancer therapy is already solid, and the Laureates' discoveries have inspired a whole new field of research. Like the Carmen Overture, it promises an exciting future.

Professors Allison and Honjo:

Your ground-breaking research has added a new pillar in cancer therapy. It represents a new paradigm for treatment, not directly targeting the cancer cells, but rather releasing the brakes of the immune system. Your seminal discoveries constitute a landmark in the fight against cancer, for the benefit of numerous patients and all humankind.

On behalf of the Nobel Assembly at Karolinska Institutet, I wish to convey to you our warmest congratulations. May I now ask you to step forward to receive the Nobel Prize from the hands of His Majesty the King.

James P. Allison. © Nobel Prize Outreach AB. Photo: A. Mahmoud

James P. Allison

Biography

AS A BASIC SCIENTIST, I have been fortunate to see my research findings translate into a powerful new potentially curative treatment strategy for cancer. The first patient I met was Sharon Belvin. I met her in 2006 at Memorial Sloan-Kettering Cancer Center (MSKCC). Jedd Wolchok, an oncologist at MSKCC, called and asked me to meet him in his clinic. There, he introduced me to Sharon Belvin, who was 24 years old and had been battling metastatic melanoma. Melanoma had invaded her brain, lungs, and liver. She was diagnosed at the age of 22 and had received multiple prior therapies but her cancer continued to grow and, when she was 23, she was told that she only had a few months left to live. But, as a last-ditch effort, she participated in a clinical trial of a then experimental drug, anti-CTLA-4. Within 3 months of starting treatment, her tumors shrank in size and then disappeared. In 2006, when I met her, it was her one-year anniversary of having completed the treatment. Her disease had responded to the treatment and she was considered in remission and possibly cured.

Sharon and I have become good friends. When her first child was born a few years later, she sent me pictures. Then pictures of her second child. She is now 13 years out from her battle with cancer and enjoying life with a vibrant family. I can't help but cry whenever I tell this story. My meeting with Sharon was my first experience of how years of research as a basic scientist could have an impact on patients. Her experience, and those of many other patients with many types of cancer who have benefitted from this work, has provided inspiration to work to continue improving this therapy for the benefit of many more patients.

EARLY LIFE, SOUTH TEXAS

I was born in 1948 in Alice, a small farming and oil town in the brush country of the Rio Grande Valley in South Texas. Those early years in Alice shaped who I would later become. Being in Texas, the whole town, including my two older brothers, was obsessed with football. But, I learned from a young age that being crushed under a pile of big sweaty boys on a hot Texas afternoon was not my idea of fun. My interests lay more in knowledge and playing at chemical and biological experimentation. If you couldn't find me curled up with a book, I was probably in the garage dissecting frogs or making homemade bombs to test in the nearby woods.

My Dad was the old-fashioned kind of doctor that made house calls, and I would sometimes accompany him on his rounds around town. I think he was the first immunologist I met because, in the days before vaccines for measles, mumps, and other childhood diseases, he would take me with him so that I could be exposed to the other children who had these illnesses. It was his way of exposing me to the infectious agents while I was young so that I could develop immunity against these infections and avoid the dangers of contracting these illnesses as an adult.

My Mom died when I was 11 years old. This was my first loss to cancer. At the time I didn't know it was cancer because people simply did not speak of such things during those days. All I knew was that my Mom was getting sicker and spending increasing amounts of time in bed. She would go to the hospital for treatment and come back with burns on her neck. I later learned that she was suffering from lymphoma and was receiving radiation therapy, the standard-of-care treatment at the time. One morning in summer of 1960, as I was leaving with family friends to go swimming, I was told to go back inside and see my Mom. I sat at her bedside and held her hand as she died. That was a defining moment in my early life. And then, not long after my Mom died, my uncle died from melanoma and another uncle later died from lung cancer.

Cancer treatment has always been in the back of my mind. Still, I can't really say that the impact of cancer on my family motivated all that I would later do, but years later I knew that should the opportunity ever arise I would do all that I could to apply my work to curing cancer.

In high school in Alice, I was fortunate to have a few great teachers who went the extra mile to encourage me to do my best and to take advantage of available resources, some outside the reach of the standard curriculum. A counselor arranged for me to participate in special summer programs for talented students at University High School in Austin after the 8th, 9th, and 10th grades. The topics for the programs varied, but all focused on individual or small team conducted projects. These summers were invaluable in immersing me in new situations and broadening my perspectives. The summer after my junior year, I participated in a biology course sponsored by the National Science Foundation at UT Austin. The mornings during that summer were spent in lectures presented by Irwin Spear, a truly exceptional teacher who taught freshman biology at the University of Texas at Austin (UT Austin) during the academic year. And the afternoons during that summer were spent working in research laboratories. This was a life changing experience, and I began to seriously consider a career in science.

I had an unusual experience when I returned to Alice for my final year in high school, when biology was typically taught. I knew of Darwin and the Origin of Species from my reading, and Dr. Spear's lectures had made it clear how essential the ideas of selection were to the understanding of many aspects of biology. So, when I feared that biology as taught in Alice was going to be completely devoid of any mention of these ideas due to the religious beliefs of the faculty, I refused to take the class. I could see no value in such a course. I told the school officials that teaching biology without Darwin was like teaching physics without Newton and would be a waste of time, and I would not enroll in the course. The response was that it was a required course and, without it, I would not be allowed to graduate (among other threats). Finally, a wise counselor proposed that I fulfill the requirement by taking a correspondence course from UT Austin. I successfully completed it and graduated high school in 1965.

THE UNIVERSITY OF TEXAS AT AUSTIN: UNDERGRADUATE AND GRADUATE TRAINING

I enrolled at UT Austin in the summer session immediately after high school graduation. Due to my summers in Austin, I never thought of going anywhere else. In accordance with my father's hopes, I began as a pre-med student. However, I became dissatisfied with the rote memorization required in some of the pre-med courses. At the beginning of my second year, in order to make a little money and hopefully gain entrance into a research lab, I became a dishwasher for G. Barrie Kitto, a new assistant professor of biochemistry at UT Austin. Over time I was allowed to help with experiments, then do my own, and eventually have my own projects as an undergraduate researcher.

I gradually learned of a key difference between medical practice and laboratory research. I realized that a physician must have a brain full of facts which can be accessed rapidly to deal with emergency situations that may arise during patient care. Physicians have to be able to quickly analyze symptoms and implement a treatment plan. And physicians cannot be wrong. They must be accurate in order to help patients or, at least, avoid harming them. A scientist's job is very different. A scientist is usually focused on interesting, and hopefully important questions, and generating experiments to test hypotheses. As a scientist, it is equally valid to prove the hypothesis true or false. That's fortunate, because many of our hypotheses are wrong; in fact, if you are asking interesting questions, most of them are wrong. Being wrong can actually be a good thing, because the answers generated in disproving an incorrect hypothesis will help you and others to propose alternate hypotheses. Then, you go back to the lab to do more – hopefully better – experiments. I did not have the discipline to be a physician, so I chose to be a scientist and have more fun. My initial projects involved the biochemical taxonomy of sea urchins, sea cucumbers and starfish. These studies were interesting and taught me the importance of precision and rigor. My next major project focused on biochemical and serological characterization of bacterial asparaginases, which at the time were showing promise in the treatment of childhood leukemias.

In addition to my courses and laboratory studies, I truly enjoyed the music scene in Austin. These were wonderful years in Austin, especially with respect to the great music from Jerry Jeff Walker, Willie Nelson, and many others as well as the eclectic venues for hearing live music such as the Broken Spoke, Soap Creek Saloon, and, of course, the Armadillo World headquarters. Life was good!

It was also during my time at UT Austin that I took an undergraduate course in immunology taught by the late Professor Bill Mandy. Professor Mandy gave one lecture on T cells and I was hooked. T cells travel throughout the body seeking out cells that shouldn't be there, such as virus-infected cells or even tumor cells, and T cells then eradicate these aberrant cells. Nothing was known about how T cells recognized foreign cells, or what regulated their proliferation and function, or virtually anything else. I was fascinated, and decided to devote my career to solving their secrets.

SCRIPPS CLINIC AND RESEARCH FOUNDATION: 1974–77

Professors Mandy and Kitto helped me get a post-doctoral appointment in Ralph Reisfeld's laboratory at Scipps Clinic in La Jolla, California. I felt that this was a great choice because Scripps was a hotbed of research in immunology, and because Reisfeld had recently reported the isolation and initial characterization of human histocompatibility (HLA) antigens. These molecules had been shown to be important in rejection of allografts in humans, and the mouse homologs, MHC antigens, were suspected as having a more general role in T cell recognition of other antigens. I was reasonably productive, and succeeded in obtaining some of the very first primary amino acid sequence information from both HLA class I and class II antigens.

But, once again it was not all work. One day we met Clay and Ailene Blaker, two expatriate Texans who moved to San Diego to surf and start a band, "Clay Blaker and the Texas Honky Band." Clay was a singer/songwriter who wrote much of his own material. (One of his songs, "Lonesome Rodeo Cowboy" later became a hit for the country superstar George Strait). The music filled a void and reminded me of Austin. I had been playing harmonica for years, either tooting by myself or trying to play along with records by blues masters like Muddy Waters and Jimmie Reed, or with county outlaws like Willie Nelson and Waylon Jennings. I started sitting in with Clay's band occasionally, and ended up becoming a regular band member, playing with the band at a country dive bar called the Stingaree every Tuesday night for more than a year. At one point the band decided to move back to Texas to play in the dance halls around Gruene and New Braunfels in the hill country of central Texas. I decided to keep my day job as a post-doctoral fellow, but Clay and Aileen were very successful, earning a strong following.

THE UNIVERSITY OF TEXAS SYSTEM CANCER CENTER – SCIENCE PARK, SMITHVILLE: 1977–1984

I heard from a friend that the M. D. Anderson Cancer Center in Houston was opening up a small research institute in Smithville, which was about an hour away from Austin, and I secured a position as Assistant Professor. It was my chance to seriously devote my efforts to immunology, and also to get back to Austin.

It was a wonderful situation. We were in a bucolic setting in the Lost Pines area of central Texas, in the midst of a state park complete with a lake and hiking paths. The lab was about 7 miles from town, and we regularly saw deer, armadillos, and other wildlife on the laboratory grounds. There were only six faculty members and perhaps 50 staff in all. There was a grand esprit de corps, and we were accustomed to working together when extra hands were needed. We worked very hard, but also enjoyed the Austin music, exciting as ever, with many of the original artists still there, but also many new faces, like Stevie Ray Vaughn, Beto and the Fairlanes, and too many others to name.

Nominally the theme of the lab was carcinogenesis research, and my main project was to make monoclonal antibodies and use them to detect cell surface changes as liver cells become neoplastic. My colleague Douglas Hixson and I made good progress, but I was also free to begin my own efforts in T cell biology.

It was during my time at Smithville that I heard a lecture that Irv Weissman presented at the main campus in Houston that gave me some ideas, and led me to think of a series of experiments that might lead to identification of the T cell antigen receptor (TCR). The TCR was considered the "holy grail" of immunology at the time. Many high-profile labs and scientists were feverishly trying to find the TCR as its identification was key to unlocking the mysteries of T cells. T cells recognized MHC plus antigen that was found on antigen presenting cells (APCs), but the question remained: what was the specific molecule on T cells that enabled this interaction? I made a series of monoclonal antibodies that recognized clonotypic antibodies on mouse T cell lymphomas and I designed a set of biochemical experiments that, with the help of graduate student Bradley McIntyre, led to the identification of the protein structure of the TCR. This discovery led to a lot of notoriety and invitations to high level conferences.

After that, the race was on to clone the genes encoding the TCR. I spent a sabbatical as a visiting scientist in Irv Weissman's lab in order to conduct experiments aimed at finding the genes for the TCR but, other brilliant scientists, including Mark Davis and Steve Hedrick, as well as Tak Mak, were the first to identify the genes for the TCR.

Nonetheless, my time in Irv's lab opened a whole new world to me. I was invited to the University of California Berkeley to give a seminar and eventually I was offered a tenured-track position as a full Professor.

THE UNIVERSITY OF CALIFORNIA, BERKELEY: 1984–2004

UC Berkeley is a marvelous place, teeming with bright, inventive people seeking knowledge. My time at UC Berkeley was notable for my interactions with many wonderful colleagues and students. It was a time of discovery, collegiality, long days working in the lab, and long nights partying with everyone in the lab. Max Krummel described it best by saying that the lab was filled with many personalities who enjoyed each other's company so much that they worked together and had fun together, like a family, or like a pirate ship filled with great people having a great time, even as they continued to steer the ship in the right direction. We all worked hard to understand fundamental aspects of T cell activation and its regulation. These were exciting and heady times. I remember many scientific meetings, especially conferences at Asilomar where my lab members would lead intense scientific discussions during the day, and then celebrate at night by building a bonfire on the beach and dancing and singing until sunrise.

I also took on leadership roles for the first time at UC Berkeley, including chairing a fledgling new Division of Immunology in the Department of Molecular and Cell Biology, recruiting faculty as the department continued to grow, and building a supportive environment for junior scientists to succeed. I should point out that my strategy for mentoring consisted of protecting faculty from bureaucracy, providing resources, such as lab space, funding, and collaborations, encouraging people to pursue their scientific passion, and basically leaving them alone to pursue their goals, with some guidance when needed. I am very proud of the outstanding faculty that I hired, most of whom quickly rose to the top ranks of their fields. It was a joy to share data and interpretations of our research finding.

I also tried to inspire a culture of "work hard, play hard" in my lab and the department. New ideas and projects were always encouraged and drinks after work were also encouraged to discuss successes and disappointments. The bonds that were formed with people at Berkeley are lifelong because they were forged with a sense of camaraderie and shared vision.

My own scientific vision was clear at Berkeley: decipher the way in which T cell responses are regulated. The entire process of T cell responses starts with the TCR. The TCR is signal one and can be compared to the ignition switch of a car. It is needed to turn the car on and start the process of T cell activation, but we knew that it was not enough to get it going. A second, costimulatory signal, which could only be provided by specialized APCs such as dendritic cells was required. Without the second signal, T cells failed to proliferate and failed to respond to antigen, a state defined as anergy. In 1992, Fiona Harding in my lab showed that CD28, another protein on the surface of T cells, was sufficient and necessary to provide the second signal for full activation of naïve T cells and to prevent induction of anergy in T cell clones. CD28 can be compared to the gas pedal in a car, which allows the car to start moving.

Still, there was another important piece of the puzzle that had yet to be solved. Another receptor on the surface of T cells, named CTLA-4, had been found that bore significant homology to CD28. There was no scientific consensus on the potential function of CTLA-4 as most thought it was another positive costimulator. In 1993 and 1994, studies from my lab and Jeff Bluestone's conclusively demonstrated that CTLA-4 is a negative costimulator that directly opposes CD28. CTLA-4 can be compared to the brake in a car, which acts to stop responses before they cause any damage.

Max Krummel worked tirelessly to conduct experiments focused on defining the role of CTLA-4. After we showed that CTLA-4 acts as an inhibitory receptor to control T cell responses, I immediately designed the experiments to test the idea that antibody blockade of CTLA-4 would lead to tumor eradication. I outlined the experiments to Dana Leach and he injected mice with tumors and then treated them with anti-CTLA-4. He showed me results from the first experiment in 1994 and I was absolutely blown away. The treated mice had all rejected the tumors. It was astounding. By blocking a single molecule, CTLA-4, we had reversed tumor growth and death! I knew that we had to immediately confirm the data by repeating the experiment in a blind fashion.

But, Dana was leaving on Christmas vacation to visit his family, and he did not have time to repeat the experiment. We agreed to do a blinded experiment where he injected a new set of mice with tumors and to treat half of them with anti-CTLA-4 and leave the other half untreated, without my knowing which group was which. He set up the experiment and left on his trip. I monitored the mice daily and recorded tumor growth on all of them. Initially, I was disappointed, since all of the mice demonstrated tumor growth equally. But then, around day 14, I noticed that some of mice had tumors that appeared to have stopped growing. Over the next few days it became apparent that the tumors were shrinking, and eventually disappeared, from some of the mice. It, of course, turned out that all the mice that eventually rejected tumor had received the anti-CTLA-4 antibody. It was another eureka moment! I knew that we had just figured out a way to treat cancer in patients. We conducted many other experiments in mice and confirmed our data over and over again with a variety of tumor types. We did not find any tumors that we could not cure with CTLA-4 blockade as monotherapy, or in combination with other agents, such as chemotherapy, radiation, vaccines, or local ablation.

We clearly demonstrated that CTLA-4 was an inhibitory signal on T cells, which was the first time that an inhibitory pathway had been described for T cell responses, and we showed that blockade of CTLA-4 could lead to tumor eradication. We tried to convince many different pharmaceutical companies to work with us over the next 5 years but had little success due to skepticism in the field after many failed clinical trials with other immunotherapy agents such as vaccines and cytokine therapies. Many people did not believe that it was possible to treat cancer with a drug that did not target tumor cells but instead targeted T cells. The field of oncology was focused on identifying genetic mutations in cancer cells and then targeting these specific mutations with drugs, even though such studies indicated that tumor mutations would continuously change and accumulate, which would lead to escape from the genetically-targeted drugs. I argued that the mutations would elicit T cell responses; therefore, we needed to target T cells to improve their responses against these tumor mutations. I repeated my claims that the immune system was a living and evolving system and if we allowed T cells to do their job, by removing the brake with anti-CTLA-4, the T cells would keep up with the evolving mutations in tumors so that T cells would eradicate the tumors and cure patients. My claims were met with skepticism over and over again. Luckily, my friend Alan Korman, convinced Nils Lonberg at the biotech company Medarex that it was a worthwhile endeavor to help with our project. Medarex had a technology well-suited to the task: mice whose antibody genes had been replaced by human genes and could directly produce human antibodies. Alan joined Medarex and with Nils quickly produced an antibody (MDX-CTLA-4 or MDX-010), which was renamed ipilimumab by the US Food and Drug Administration when it entered clinical trials.

As the inventor of CTLA-4 blockade I was not, of course, involved in the trials, but was kept informed of the main efforts. The early trials were promising, with objective responses in a Phase I trial of ipilimumab in melanoma, and in some small trials in prostate, kidney, and a few other types of cancer. While this generated considerable excitement, I heard troubling reports that ipilimumab failed with more stringent trials with progression-free survival (PFS) as the endpoint. I began to worry that the clinical investigators might not fully appreciate the biology of CTLA-4 blockade, and were treating it as a conventional cancer drug that kills cancer cells, rather than a drug that activates immune responses by removing inhibitory signals.

As much as I loved UC Berkeley, I began to feel the need to learn more about cancer clinical trials, which was not possible there. In 2003 Harold Varmus and Thomas Kelly offered me the opportunity to lead the Immunology program at the MSKCC in Manhattan, and I accepted, not only to develop a world-class immunology program, but also to be close to the anti-CTLA-4 trials, many of which were being conducted at MSKCC.

MEMORIAL SLOAN-KETTERING CANCER CENTER: 2004–2012

I arrived in Manhattan during the summer of 2004 with most of my laboratory intact. Over the next 8 years, I did my best to help Thomas Kelly hire a cadre of the best immunologists that we could, and I feel that we succeeded in building a truly spectacular group of basic immunologists involved in a variety of studies relevant to the cancer problem. My own lab continued to study mechanisms of checkpoint blockade, new combinations that enhanced efficacy, and to identify and characterize new checkpoints that might also be targeted to improve the efficacy of checkpoint blockade.

One of the biggest pleasures of being at MSKCC was getting to know Dr. Lloyd Old, considered by many to be the father of modern tumor immunology and immunotherapy. I was very lucky to have him as a friend and mentor. One of Lloyd's mantras that still resonates with me today: "We must learn from each and every patient!" Lloyd and Dr. Padmanee (Pam) Sharma, an oncologist who was near completion of her postdoctoral studies as an immunologist with Lloyd, had been consulting with the Medarex team in designing and conducting trials with ipilimumab. Sharma's experience in patient care, clinical trials, and immunology were impressive, and she left MSKCC to begin her independent career at the

University of Texas MD Anderson Cancer Center in 2004. We agreed to stay in touch and try to work to better understand how CTLA-4 blockade impacted patients' immune responses.

At MSKCC I worked closely with Dr. Jedd Wolchok, who was the clinical investigator leading the anti-CTLA-4 trials in patients with metastatic melanoma. Jedd and I worked with Bristol-Myers Squibb (BMS), including Dr. Rachel Humphrey, who was leading the clinical trial development of anti-CTLA-4 with the Medarex team. Rachel and I became colleagues and good friends, and even band members in a new band that we established with Tom Gakjewski (University of Chicago) and Patrick Hwu (MDACC), both of whom were working in the emerging field of immuno-oncology. Naturally we agreed to name the new band "The Checkpoints". The band has grown to include a full horn section with Jedd on tuba and has a fairly large repertoire of rock and roll, blues and pop songs. We have entirely too much fun playing at scientific meetings! Again, the theme of "work hard, play hard" permeated my years at MSKCC.

During this time, BMS faced a critical decision that would determine the fate of immune checkpoint blockade as a strategy for cancer therapy. A Phase III registration trial of ipilimumab in metastatic melanoma with the endpoint of PFS had begun. While this endpoint was reasonable for cytotoxic drugs, it was not clear that it was appropriate for CTLA-4 blockade. Our work in mouse models indicated that anti-CTLA-4 works at the time of T cell priming, and allows the expansion of T cells to large numbers to deal with the tumor mass. In our preclinical studies in mice, the tumors almost always grew before they regressed. Of course there are many reasons why the details of mouse results might not reflect those of clinical studies. A more serious concern was that PFS had been one of the endpoints evaluated in many of the earlier trials, and many of the earlier trials had failed to identify clinical benefit using the PFS endpoint. Also, a trial by Pfizer with a different anti-CTLA-4 antibody had failed. The choice for BMS was to shut down their trial with anti-CTLA-4, or incur the tremendous additional costs of a trial with the endpoint of overall survival, which would likely take many more years to complete. The pressure to stop the BMS Phase III anti-CTLA-4 trial was tremendous, but clinical investigators, including Pam Sharma and Jedd Wolchok, and some of the BMS team members, including Rachel Humphrey and Elliott Sigal, championed keeping it open, and prevailed.

While waiting for the next few years for the outcome of the Phase III trial, we all continued our studies. We were invited to many scientific meetings where we would meet with other researchers in the field, including Pam, to share our data at podium presentations during the day and over drinks at the bar in the evenings. It was an exciting time as we all shared our experiences from different institutions and different clinical trials. Those involved in research related to CTLA-4 became a growing community of investigators. Anti-CTLA-4 was working for patients with metastatic melanoma in early clinical trials and, in 2006, Pam shared with us that she was conducting pre-surgical studies in patients with bladder cancer and she observed tumors disappearing in those patients as well. She also performed novel immune monitoring studies on the tumor and blood samples that she took from patients and identified CD-4 T cells expressing inducible costimulator (ICOS) as a critical component of the anti-tumor responses. It was impressive to me that she could design studies that provided both significant clinical data and novel insights into the scientific mechanisms underlying responses with anti-CTLA-4. I asked her to collaborate with Jedd and myself on the involvement of the ICOS+ T cells in melanoma, and we found in those studies, and in others, that the novel ICOS+ T cells play a causal role in the therapeutic effect of anti-CTLA-4 both in patients and animal models.

Our team of investigators expanded to include many more people over the years. The field of cancer immunotherapy exploded as a result of the clinical success of anti-CTLA-4, which opened a new field termed immune checkpoint therapy. Scientific meetings had more and more talks related to immune checkpoint therapy. Immunotherapy sessions at meetings, which previously had less than 50 people in attendance, required larger and larger rooms as hundreds and then thousands of people attended. Cancer immunotherapy went from a fringe science to events that were compared to sold-out rock concerts, which was fitting since our band played after some of the scientific meetings to an equally growing fan base.

The results of the Phase III trial of ipilimumab were finally presented in the Plenary Session of the annual meeting of the American Society of Clinical Oncology (ASCO) in June of 2010. The audience, used to many years of failure of immunotherapy trials presented to tiny audiences, was electrified: ipilimumab was successful in extending the median overall survival of melanoma patients. This had never before been reported in trials of melanoma patients for any other agent. Several years later, when sufficient data were available, a retrospective study showed that thousands of patients with metastatic melanoma who had undergone a single course of treatment, typically consisting of 4 doses of therapy over 3 months, were alive 10 years later.

In March of 2011, I was attending a small Banbury meeting on melanoma at the Cold Spring Harbor Laboratories and I was nervously checking my email because it was the day that the US Food and Drug Administration was scheduled to announce its decision on ipilimumab. At about 11:30AM I received an email from Alan Korman with a picture of of him and Nils Lonberg sharing celebratory drinks of whiskey! This was my first indication that the news from the FDA had been positive and anti-CTLA-4 had become an approved standard-of-care treatment for patients with melanoma.

I've met many patients who were successfully treated with anti-CTLA-4, including Sharon Belvin who became a close friend over the years. Unfortunately, during my time at MSKCC, I also experienced many losses, including the death of my brother Mike in 2005, and my friend, and mentor Lloyd Old in 2011, from metastatic prostate cancer. Anti-CTLA-4 led to tumor eradication and even cures in some patients but not in all patients. And, for patients with certain tumor types, such as prostate cancer and pancreatic cancer, anti-CTLA-4 did not seem to work at all. I was familiar with Pam's work focused on evaluating patients' samples from clinical trials to identify mechanisms of response and resistance to anti-CTLA-4 therapy, including a trial in prostate cancer patients that she designed in 2009, and I was convinced that we needed to expand these studies to build a larger program.

It is safe for me to say that the collaborations with Pam and I had over the years had clearly become a highlight in both my scientific career and my personal life. With time we developed a remarkable level of mutual respect and admiration and love. Eventually we became partners not just in science, but also in life.

UNIVERSITY OF TEXAS MD ANDERSON CANCER CENTER: 2012–PRESENT

At MSKCC, my lab had continued exploring ways in which to improve anti-tumor responses with anti-CTLA-4 treatment. We performed many experiments including combinations with chemotherapy, radiation therapy, cryoablation and with other immune checkpoint agents. After anti-CTLA-4 was shown to have some clinical success in the early clinical trials, many investigators searched for other potential inhibitory molecules. These studies led to identification of the PD-1/PD-L1 pathway as another inhibitory pathway that regulated T cell responses. Clinical trials also led to FDA-approval of anti-PD-1 and anti-PD-L1 antibodies as treatments for patients with cancer. Our lab also showed that combination therapy with anti-CTLA-4 plus anti-PD-1 improved anti-tumor responses in mice. These data were also translated to the clinic and the US FDA approved combination therapy with anti-CTLA-4 plus anti-PD-1 for patients with melanoma and patients with kidney cancer. Again, these studies demonstrated that some patients responded to therapy but, some patients did not. I knew that we were at the tip of the iceberg in terms of unleashing the full power of the immune system to treat cancer and I wanted to delve deeper by building a translational research program to quickly move between the lab and the large number of clinical trials with immune checkpoint therapy that had emerged as a result of the success with anti-CTLA-4. I decided to move to MDACC to build the Immunotherapy Platform with Pam.

In 2013, I left MSKCC to partner with Pam, who was already conducting mechanism-based clinical trials to study immune responses in patients who receive the immune checkpoint agents. I am Chair of Immunology at MDACC. I'm also continuing to work on fundamental immunologic mechanisms in my lab and we recently demonstrated significant differences between CTLA-4 blockade and PD-1 blockade. In addition, based on Pam's clinical and laboratory studies, which had already established a translational research program for genitourinary malignancies, I worked with institutional leadership to expand this translational model to the entire institution, encompassing 18 different departments and multiple tumor types. Pam and I established and lead the Immunotherapy Platform. Partnering with physicians from around the institution, we obtain and analyze tumor samples from patients on treatment in clinical trials with immune checkpoint inhibitors or other combination immunotherapy studies. We're currently involved with immune monitoring studies from over 100 clinical trials. Our goal is to gain mechanistic insight into the therapies that will allow us to develop rationally based combination therapies to bring the benefits of immuno-oncology to more patients with more types of cancer with the goal of obtaining cures.

Of course, my time at MDACC also reverberates with my motto, "work hard, play hard". I have a second band in Houston called The CheckMates. I continue to play harmonica at events with both The Checkpoints and The CheckMates. I've also had the privilege to play with Willie Nelson and Mickey Raphael, who I consider to be the greatest harmonica player on the planet. I continue to follow my passion for science and for life, and I'm truly grateful for the privilege of having a lifetime spent studying fundamental issues of biology that interested me, and that turned out to be of benefit to people with cancer.

Immune Checkpoint Blockade in Cancer Therapy

Nobel Lecture, December 7, 2018 by
James P. Allison
University of Texas MD Anderson Cancer Center, Houston, TX;
Parker Institute for Cancer Immunotherapy, San Francisco, CA, USA

INTRODUCTION

The idea of using the immune system to fight cancer has been around for decades, with multiple false starts along the way. The field of cancer immunotherapy had largely been focused entirely on ways in which to turn T cells on, with implementation of therapies such as cytokines, for example IL-2, or antigenic vaccines. However, our understanding of T cells at the time was limited. Our work, together with that of a cadre of immunologists dedicated to understanding the fundamental mechanisms involved in regulating T cell responses, showed that the process was more complicated. These fundamental studies also revealed inhibitory pathways that could be modulated to generate powerful T cell responses with subsequent tumor elimination. This led to the successful development of immune checkpoint blockade as a potential curative cancer therapy.

My love for T cells took root during an undergraduate immunology course in the 1960s. I was majoring in biochemistry with my thesis project being biochemical and serologic characterization of anti-neoplastic bacterial asparaginases, which at the time were showing promise in treating childhood leukemias. Professor Bill Mandy, who was primarily a B cell/antibody guy, taught the course, and he gave one lecture on T cells, which in 1965 had been shown by Max Cooper and Robert Good to be a lineage distinct from antibody generating B cells (1). While B cells and antibodies are certainly fascinating, T cells were new and little was known about them beyond the fact that they had the ability to travel all over the body and recognize and eliminate cells with foreign features, including viral, transplantation, and possible tumor antigens. I was hooked and determined to spend my career delving into the secrets of T cells.

With the help of Bill Mandy and Barrie Kitto, my dissertation advisors in 1974, I secured a postdoctoral position in Ralph Reisfeld's laboratory at the (then) Scripps Clinic and Research Foundation in La Jolla, California. Scripps was a hotbed of immunological research. My main project in the Reisfeld lab was to bring my background in protein biochemistry to bear on structural studies of human histocompatibility antigens. While I was fairly productive and obtained some of the first amino acid sequence data on both Class I and Class II antigens, I was somewhat frustrated in that I was not able to take full advantage of the opportunities available in T cell biology at Scripps.

THE SCIENCE PARK YEARS

In 1978, as a newly minted assistant professor with a small lab in a pine forest near Smithville, Texas, at the Science Park campus of the University of Texas M. D. Anderson Cancer Center, I remained fascinated by the complexity of the T cell response, which involved moving through the body to sample different antigens and then making decisions about when to proliferate in order to amass an army of cells that would eradicate any foreign entity. The super hero of our immune system! But, how did T cells make such complicated decisions? What were the signals that regulated T cell responses?

Throughout the 1970s and early 1980s there were remarkable advances in our understanding of T cell biology, including the clonal nature and MHC restriction of antigen recognition, and the role of a variety of cell surface molecules and cytokines in regulation of T cell responses. However, although many theories existed regarding the nature of the T cell receptor (TCR), its identification remained elusive and began to be regarded as the "holy grail" of immunology.

My first set of experiments to identify the TCR stemmed from a lecture I heard by Irv Weissman and led me to use a biochemical approach in identification of the protein structure of the TCR. I should point out that I was abetted by my relative ignorance of the field and the relative isolation of the Science Park.

The basic assumptions were: (1) the T cell antigen receptor, like the individual idiotypes associated with the B cell receptor, should be detectable in a clonotypic fashion as individually specific antigens, which could be identified by monoclonal antibodies; (2) the structure detected by these antibodies should be on all T cell tumors and normal T cells but not on B cells; and (3) the structure from different clones should have both tumor specific and shared peptide structures. Thus, I made a monoclonal antibody, 124-40, against C6XL, which was in the panel of lymphomas that I used in my thesis project on anti-tumor activity of asparaginase. The activity of the antibody with C6XL could be detected by flow cytometry and radioimmune assay. None of the other lymphomas in the panel reacted with 124-40, nor did normal splenic T cells. Working with Brad McIntyre, a summer undergraduate intern, then a research technician and ultimately a doctoral student, we carried out a series of immunochemical tests of the assumptions. Radioimunoprecipitation using the 124-40 monoclonal on the original C6XL lymphoma yielded a disulfide-bonded heterodimer with 39 and 41 kDa components. The antibody did not precipitate anything from other T cell lymphomas, again indicating that the antigen was clonally-specific (2). Taking the antigen eluted from that antibody, I generated a rabbit polyclonal antibody that interacted with all T cells. After reduction, the protein eluted from the antibody could be seen clearly on a 2D blot and consisted of α and β heterodimers. Peptide mapping showed that each consisted of both shared and clonally distinct peptides. We ran the whole cell lysate out on a gel and saw that the rabbit polyclonal reacted with the same protein on all T cells. The clonally-specific candidate receptor was also common to all T cells (3).

We published these data and proposed that the protein we had identified was the elusive TCR. Unfortunately, we had no functional data and there was initially skepticism. Luckily, at about the same time, John Kappler and Pippa Marrack, at the National Jewish Hospital in Denver, had developed a clonotypic monoclonal antibody that caused IL-2 release by an antigen-specific hybridoma, which was an indication of activation. I sent them our antibody and they confirmed that the protein identified by both our labs had the same heterodimeric structure (4).

There were other laboratories with similar antibodies, and others with molecular cloning techniques, so there was an intense race to clone the genes for the TCR. I decided to go to Irv Weissman's lab and work with Tom St. John to clone the genes, but we were ultimately beaten on that front by excellent work from Mark Davis and Steve Hedrick, along with Tak Mak, who cloned the beta chain of mouse and human TCR (5–7).

THE BERKELEY YEARS

While I was at Stanford trying to clone the TCR genes, I was invited to present a seminar on my work at the University of California, Berkeley and meet some of the faculty. A few weeks later, I was asked if I was interested in a faculty position. In 1984, after much encouragement by Marion "Bunny" Koshland, who was the leader of Immunology at Berkeley, and who also became a close friend and mentor, I decided to accept the offer and moved to Berkeley. After eight years in the bucolic environment of Science Park, UC Berkeley was a revelation and initially somewhat scary. The campus teemed with intellectual and scientific fervor, and new ideas could be welcomed, but subjected to a healthy and helpful degree of skepticism.

By the late 1980s, it became clear that the TCR was not the whole story in T cell activation. For T cell tumors and hybridomas, engagement of the TCR alone was sufficient for activation. However, as it became possible to grow and study clones of normal T cells, it was found that TCR engagement alone was not sufficient for T cell activation. Several labs, including notably that of Ronald Schwartz at the National Institute for Allergy and Immunological Diseases, showed that additional signals were required (8). The Schwartz group showed that while antigen presenting cells (APCs) pulsed with the appropriate antigenic peptide could fully activate T cell clones, treatment of the pulsed APCs with certain chemical fixatives resulted in loss of T cell activation. Furthermore, T cells exposed to antigen in this manner were unresponsive when re-exposed to antigen on unfixed APCs. Thus, the T cells had been rendered anergic. However, if T cells were exposed to chemically fixed peptide-pulsed APCs in the presence of unfixed APCs that had not been pulsed with antigen, the T cells were fully activated. This led to the 2-signal model for T cell activation by APCs. Signal 1 is provided upon recognition of antigen/MHC complexes by the T cell antigen receptor. Signal 2, the costimulatory signal, was provided by an antigen-independent signal engaging an unknown costimulatory receptor on the T cell. The race was on to determine the identity of the receptor/ligand pair that mediated costimulation.

In the mid 1980s, Hansen, Ledbetter, June, and others in Robert Nowinski's group had produced an antibody to human T cells named 9.3 and showed that it elevated IL-2 production from T cells (9). The target of 9.3 was later shown to be CD28 (10). Enhancement of IL-2 production seemed to be a property consistent with costimulation but, unfortunately, the mouse homolog was not known. With the help of Brian Seed at MIT and Jane Gross, a graduate student in my lab, we succeeded in cloning the mouse CD28 gene and went on to produce a monoclonal antibody to mouse CD28 (11). Fiona Harding, a postdoc in the lab, in collaboration with my colleague, David Raulet, at Berkeley, used this antibody to show that engagement of CD28 by the antibody was sufficient and necessary to prevent the induction of anergy in T cell clones by fixed antigen-pulsed APCs. In addition, we showed that the CD28 antibody, in combination with antibodies to the T cell antigen receptor, allowed for full activation of naïve T cells (12). Thus, CD28 met the criteria for an essential costimulatory receptor.

Peter Linsley demonstrated that a molecule called B7, which was originally found on B cells but also expressed by dendritic cells and other hematopoietic cells, was a ligand for CD28 (13). A few years later, several groups showed that a second molecule, B7-2, was also a ligand for CD28. Thus, it appeared that when a T cell received contemporaneous signals through the TCR and CD28 from antigen/MHC complexes and B7 molecules on an APC, such as a dendritic cell, a program initiated in the T cell allows rapid proliferation and acquisition of effector function, which results in generation of a T cell army from a small number of relevant T cells.

It was at this point that that there was an unexpected twist to the story. In 1988 Pierre Golstein (Marseille) cloned a gene for a T cell molecule he named cytotoxic lymphocyte antigen-4 (CTLA-4) (14). Little was known of it except that it was highly homologous to CD28, but was only found on T cells after activation. Peter Linsley made a recombinant form of CTLA-4 and showed that, like CD28, it bound to both B7-1 and B7-2, albeit with much higher avidities. While CTLA-4 was initially thought to be another costimulatory molecule, the evidence remained inconclusive. Finally, in the mid 1990s both Jeff Bluestone's lab and Max Krummel, an excellent postdoc in my lab, using an antibody that blocked CTLA-4, showed conclusively that CTLA-4 negatively regulates T cell activation by opposing CD28-mediated costimulation (15, 16). It was a paradigm-shifting finding, revealing CTLA-4 as the first immune checkpoint. Subsequently, it was shown that mice in which the *ctla-4* gene had been inactivated suffer from a rapid T cell lymphadenopathy and die about 3 weeks after birth (17–19). Basically, the T cells in CTLA-4 knockout mice cannot stop dividing. This suggests that CTLA-4 is a crucial downregulator of the proliferative phase that follows activation of T cells by antigen regulation and costimulation. Thus, activation of T cells results in not only the proliferative phase, but also induction of expression the CTLA-4 gene and subsequent events which will terminate proliferation.

One additional observation that we had made contributed to the development of immune checkpoint therapy. In a series of experiments designed with the goal of developing a new strategy for therapeutic vaccination, we engineered mouse tumor cell lines to express the CD28 ligand B7. Sarah Townsend, a postdoc in my lab, showed that tumor cell lines that normally grew in mice did not grow after transfection with the B7-1 gene unless T cells were depleted (20). Thus, the tumors expressed sufficient antigens for rejection, but were unable to initiate an effective T cell response because of the lack of ability to provide the B7 costimulatory signal required by CD28. This suggested that arising tumor cells might be invisible to T cells, allowing them to grow uninhibited by immune responses until reaching a size that results in tumor cell death to cause inflammation and accumulation of antigen presenting which can phagocytize dying tumor cells and presenting tumor antigens in the context of costimulation and cross-prime naïve T cells. This would lead to T cell activation and proliferation. This would result in induction of the CTLA-4 pathway, which would perform its normal function of limiting the T cell response.

Based on our new data regarding the signals that regulate T cell responses, I wondered how we could use the information to treat cancer. I surmised that tumor cells, which do not express B7 molecules, have a head start against the immune system because T cells could not recognize the tumor cells initially; however, as tumors got bigger and a few tumor cells died, APCs would take up the dead tumor cells by phagocytosis and then present mutated tumor antigens via MHC to T cells. T cells would interact with APCs, in the context of T cell receptor plus MHC with antigen (signal 1) and CD28 plus B7, which would lead to T cell activation consisting of proliferation and cytokine production, and forming an army of activated T cells focused on eradicating tumors. But, this process would be limited by CTLA-4 (Figure 1). CTLA-4 would halt the T cell response in order to prevent T cells from continuous proliferation and cytokine production, which would have pathologic consequences including autoimmunity and death. In my mind, it became a race; if CTLA-4 halted T cell responses too soon, before all tumor cells were eradicated, the tumor would win and the cancer would persist. I considered the possibility that an antibody to block CTLA-4 would enable T cell responses to persist long enough to eradicate all tumor cells and perhaps cure at least some cancers (Figure 2).

Then, along with Max Krummel and Dana Leach in my lab, I designed experiments in tumor-bearing mice to test whether anti-CTLA-4 antibody would lead to tumor rejection. These studies were initially conducted in 1994 and then reproduced in multiple other tumor models before we published our findings in 1996. The data were clear and convincing, in the presence of a CTLA-4-blocking antibody, after a short delay while the immune system ramped up, tumors simply melted away (21) (Figure 3). Not only that, those animals then mounted a strong immune response to rechallenge the same tumors, often completely rejecting them.

My lab conducted many successful experiments in murine models over the next several years; however, I desperately wanted to translate my idea to the clinic. I approached many different pharmaceutical companies but they were resistant to the idea of developing an immunotherapy agent to treat cancer. The field of cancer immunotherapy had many failures, and companies were not interested in pursuing another potential failure.

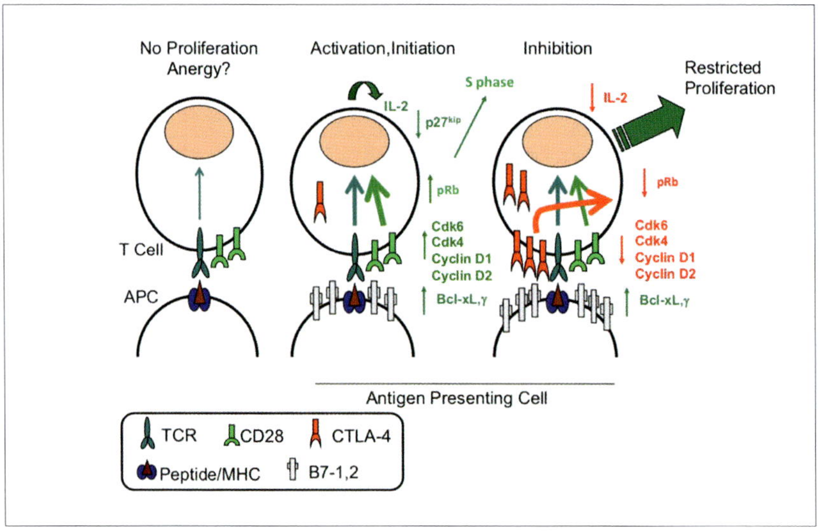

Figure 1. Dynamic Integration of TCR and Costimulatory Signals (circa 1996). Cartoon illustrates the primary cell surface signaling receptors involved in antigen-specific T cell activation and cognate ligands on APCs. T cells that are primed through the interaction of the TCR with antigen-bound MHCs enter a state of anergy in the absence of co-stimulation. Interaction of CD28 with B7 on the APC surface fully activates the T cell and initiates proliferation, while CTLA-4 on the T cell surface disrupts co-stimulation and inhibits T cell proliferation. (represents work performed in my laboratory by Jane Gross, Fiona Harding, Max Krummel, Cynthia Chambers, Monika Brunner, Egen and Kuhns).

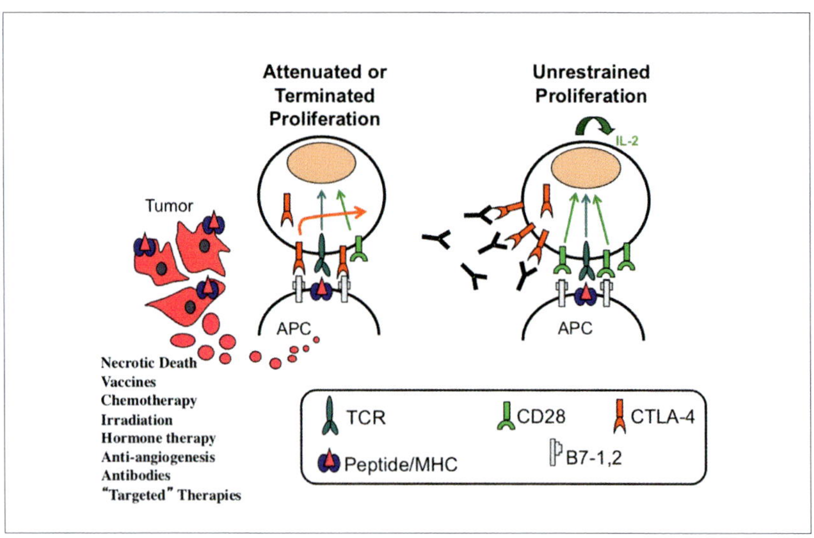

Figure 2. CTLA-4 Blockade Enhances Tumor-Specific Immune Responses. CTLA-4 on the T cell surface inhibits effective tumor-immune response by reducing T cell proliferation. Blockade of CTLA-4 removes inhibition of T cell proliferation and increases the antitumor response.

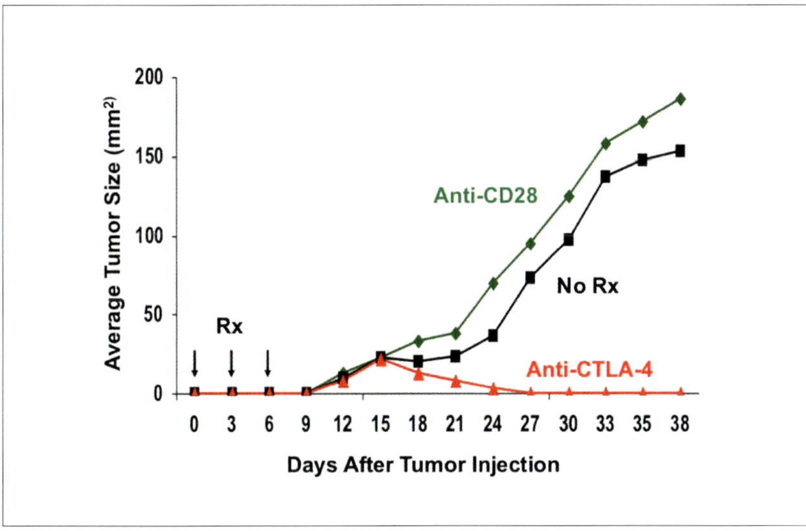

Figure 3. Anti-CTLA-4 Induces Regression of Transplantable Murine Tumor. Addition of CTLA-4 inhibitory antibody results in complete tumor rejection in BALB/c mice injected with V51BLim10 colon carcinoma cells (21).

Luckily, my friend Alan Korman was interested in my idea and, with the encouragement of Nils Lonberg and others, his new job at Medarex provided him with the opportunity to develop a humanized anti-CTLA-4 antibody in a novel murine model that had been genetically engineered to express human immune genes.

The new humanized anti-CTLA-4 antibody was initially called MDX-CTLA-4 and then was referred to as MDX-010. Medarex designed a Phase I clinical trial with MDX-CTLA-4 based on input from multiple investigators, including world-renowned immunologist, Dr. Lloyd Old, and his post-doctoral fellow at the time, Dr. Padmanee (Pam) Sharma. I had very little knowledge regarding clinical trials at the time, but I was determined to learn everything possible.

A pilot clinical trial with the anti-CTLA-4 antibody at a single dose of 3mg/kg was initiated in patients with metastatic prostate cancer and reported for the first time at the American Society of Clinical Oncology (ASCO) in May 2002, with subsequent publication in 2007, which demonstrated decline in the prostate-specific antigen (PSA) tumor marker by greater than or equal to 50% in 2 of 14 treated patients (22). In another clinical trial, MDX-CTLA-4 antibody was given as a single dose of 3mg/kg to patients with different tumor types, including patients with metastatic non-small cell lung cancer, patients with metastatic ovarian cancer, patients with acute myelogenous leukemia, and patients with metastatic melanoma. Preliminary data from the patients with metastatic melanoma, who were enrolled on this clinical trial between November 2000 and October 2002, were encouraging based on tumor biopsies demonstrating tumor necrosis and T cell infiltration into tumors (23). These data quickly led to multiple clinical trials for patients with metastatic melanoma. Another clinical trial with MDX-010 in 14 patients with metastatic melanoma reported 3 patients, who either had complete (2 patients) or partial regression (1 patient) of their tumors, including regression of brain metastasis (24).

Prior to anti-CTLA-4 therapy, patients with metastatic melanoma had poor long-term prognosis, with approximately 75% of patients surviving less than one year and 5-year mortality rate of 90% (25). The median overall survival at the time was approximately 8 months for patients with metastatic melanoma (26), and there was typically shorter life span for patients with brain metastases (27). At the time, approved therapies for patients with metastatic melanoma consisted of the chemotherapeutic agent dacarbazine (DTIC) or its oral analogue temozolamide or fotemustine chemotherapy or biochemotherapy regimens that included cytokine therapies with IL-2 or IFN-α, which led to reported responses but, the increased response rates observed with these agents did not translate to improved overall survival for patients (26–31). In fact, prior to anti-CTLA-4 therapy, no agent approved for the treatment of metastatic melanoma had been shown to improve the overall survival of patients in a randomized, phase III clinical trial.

The initial studies with anti-CTLA-4 in patients with metastatic melanoma led to multiple phase I and phase II clinical trials with various monotherapy or combination therapy studies, including combination studies with IL-2, gp-100 peptide vaccine, or chemotherapy (32). These studies identified toxicities, termed immune-related adverse events (irAEs), which were associated with anti-CTLA-4 therapy and confirmed to be inflammatory in nature, with increased infiltration of immune cells into affected tissues, which was documented by histopathology studies. The irAEs were found to be reversible in most cases, especially with cessation of anti-CTLA-4 therapy and administration of steroid therapy. Another key finding from these clinical trials was related to the unique kinetics of responses observed after treatment with anti-CTLA-4. Some patients had clinical responses that could be categorized as responding to therapy as per the previously established Response Evaluation Criteria in Solid Tumors (RECIST) for evaluation of tumor response to therapy, which required tumor shrinkage (at least a 30% decrease) without any associated growth (defined as greater than or equal to 20% increase in the sum of diameters and at least 5 mm increase) in the existing tumor lesions designated as target lesions or any development of new tumor lesions (33). However, some patients had disease progression (with increase in size of the tumors or development of new tumor lesions) before they eventually had response to anti-CTLA-4 therapy. According to RECIST criteria, these patients would not be counted as having had a response to anti-CTLA-4 therapy, which would imply that the drug did not work. The clinical terminology and endpoints were foreign to me. I needed to learn more.

Like in the animal models, it takes some time for patients to initiate a strong immune response after anti-CTLA-4 checkpoint blockade, and as such, even in the best responders, tumors often continued to progress for some time before shrinking and disappearing, which we published in our animal studies in 1996 (21). Because of this, the traditional RECIST criteria, which rely on definitions that were used for evaluating responses to chemotherapy agents that directly target tumor cells, as opposed to immunotherapy agents, which target immune cells to enable activation and proliferation for subsequent attack on tumor cells, many early studies underestimated the clinical benefit of anti-CTLA-4 therapy.

THE MSKCC YEARS

Our goal, based on our animal studies, was to eliminate the disease. When examining overall survival (a more appropriate metric for immunotherapy), we saw that patients treated with anti-CTLA-4 were surviving longer than expected. These observations led me to become involved with the clinical development of anti-CTLA-4, which meant that I needed to be working with the clinical teams that were developing the clinical trials. In 2004, I moved from the University of California at Berkeley to Memorial Sloan-Kettering Cancer Center (MSKCC), which was leading the phase III clinical trials for anti-CTLA-4 therapy. I took a faculty position as Chair of Immunology at MSKCC and became collaborators with the clinical investigators and the pharmaceutical company, Bristol-Myers Squibb (BMS), which had entered into collaboration with Medarex to lead the phase III clinical trials with anti-CTLA-4.

I quickly learned that clinical trials comprised 3 phases: phase I trials represented safety studies to evaluate how a drug affected humans, with various doses and/or schedules of a drug being tested; phase II trials represented efficacy studies to determine whether an established dose and schedule of drug could elicit clinical responses in patients; and phase III trials represented studies comparing the new drug to an established standard-of-care treatment in order to determine whether the new drug will become the new standard-of-care treatment. The phase I and phase II studies highlighted the safety issues and potential clinical benefit of anti-CTLA-4 but, the reported response rates as per defined RECIST criteria were less than 10%, which was not sufficient to lead to FDA approval. In addition, a second pharmaceutical company, Pfizer, had launched their own anti-CTLA-4 program and their phase III clinical trial was declared a failure in 2008.

BMS started their first phase III clinical trial with anti-CTLA-4 (MDX-010, which was subsequently named ipilimumab) in 2003. Based on available data at the time, many of us, including myself, Jedd Wolchok, Padmanee Sharma and Lloyd Old, felt strongly that the primary endpoint of the trial should be changed to overall survival (OS). We worked closely with Rachel Humphrey at BMS and she relayed our concerns to the leadership team at BMS, including Elliott Sigal, who was then Chief Scientific Officer and President of Research and Development. The phase III clinical trial was then amended to change the primary endpoint to OS. In 2010, the data was reported at the ASCO annual meeting and, for the first time, a randomized phase III clinical trial for patients with unresectable stage III or IV metastatic melanoma demonstrated an overall survival benefit, with ~45% of patients who received anti-CTLA-4 reported to be alive at 1 year and ~23% of patients who received anti-CTLA-4 reported to be alive at 2 years (34). Subsequent data reported from a cohort of about 5000 patients treated with anti-CTLA-4 (ipilimumab) on multiple clinical trials demonstrated 5-year survival for about 20% of patients (35) (Figure 4). The US FDA approved ipilimumab for the treatment of patients with metastatic melanoma in 2011.

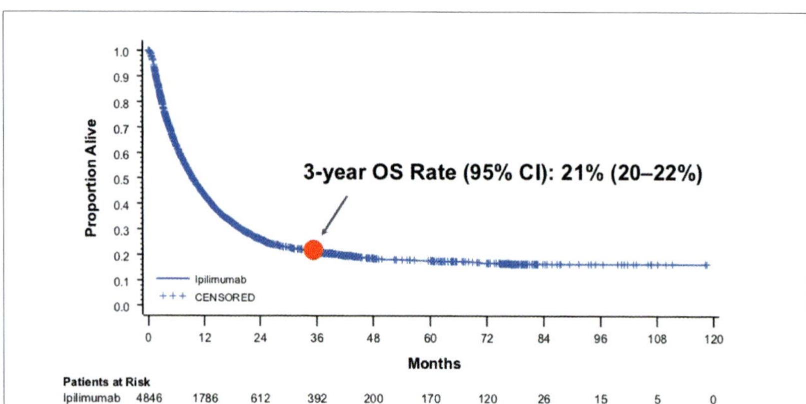

Figure 4. Ipilimumab induces lasting responses in 21% of patients with metastatic melanoma. Graph shows pooled data from 4846 patients with metastatic melanoma treated with ipilimumab. The group showed a 3-year overall survival (OS), and patients surviving 3 years remained tumor-free for 10 years and longer (35).

Clearly, the discovery of immune checkpoint inhibition had set into motion a line of work with the potential to cure all cancers. The success of anti-CTLA-4 in the clinic led to research work to identify other immune inhibitory pathways, with elegant work by other investigators, including Gordon Freeman, Arlene Sharpe, Lieping Chen and Tasuku Honjo in deciphering the inhibitory role of the PD-1/PD-L1 pathway and development of antibodies that block these inhibitory pathways to improve anti-tumor T cell responses and clinical outcomes (reviewed in Dr. Honjo's lecture). More recently, combination therapy with anti-CTLA-4 (ipilimumab) plus anti-PD-1 antibody (nivolumab) was US FDA-approved for the treatment of patients with metastatic melanoma (36) and for patients with metastatic renal cell carcinoma (37).

Since immune checkpoint therapy (anti-CTLA-4 and anti-PD-1/PD-L1 antibodies) targets immune cells, and specifically T cells, these treatments may theoretically provide clinical benefit for all cancer patients, regardless of tumor type. The clinical data supports this notion since multiple clinical trials have reported benefit for many different cancer types, which led the US FDA to approve immune checkpoint therapies for patients with diagnoses of: melanoma, renal cell carcinoma, bladder cancer, gastric cancer, head and neck cancer, non-small cell lung cancer, Hodgkin's disease, and even tumors defined by genetic mutations known as microsatellite instability (MSI) as opposed to tumors defined by anatomical location.

MDACC AND THE FUTURE

Thousands of patients have benefitted from immune checkpoint therapy, but there are thousands of patients who do not respond to the treatments. We need dedicated research efforts to evaluate patients' immune responses within the tumor microenvironment, both before and while on treatment, in order to have a better understanding of why some patients respond to treatment while others do not. The first pre-surgical study to address this issue was conducted by Dr. Padmanee Sharma and published in 2006. She designed a clinical trial with anti-CTLA-4 in order to obtain entire cystectomy samples from patients with bladder cancer who received anti-CTLA-4 prior to surgery. She designed novel immune monitoring assays to evaluate the immune responses in the bladder tumors and identified T cells expressing inducible costimulator (ICOS) as a critical component of successful anti-tumor responses (38-41). Clinical trials are now underway to test the concept of targeting ICOS plus anti-CTLA-4 as a novel combination strategy to improve clinical responses for more patients. Dr. Sharma pioneered these types of reverse translational studies, which consist of innovative clinical trials focused on studying human

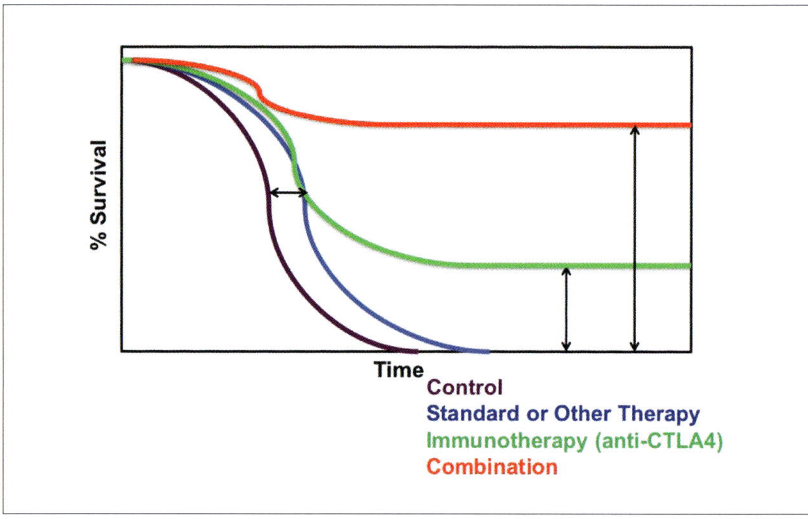

Figure 5. Improving Survival with Combination Therapy. Traditional cytotoxic therapies, seek to improve survival incrementally (horizontal arrow). Immunotherapies such as ipilimumab produce long-term responses, resulting in a 'tail' (above 1st vertical line) of patients that remain cancer-free. Combination therapies seek to 'raise the tail' (2nd vertical line), provided lasting responses to a larger number of patients.

immune responses with subsequent development of hypotheses that can be tested in the laboratory to generate new data for the next set of clinical studies. These studies are the foundation on which we built the Immunotherapy Platform at M. D. Anderson Cancer Center. The Immunotherapy Platform integrates both clinical trials and laboratory studies to develop novel treatments for cancer patients, with the hope of increasing overall survival for even more patients (Figure 5).

ACKNOWLEDGEMENTS

In the course of the work described in this document I have benefited from interactions with a large number of individuals, both from the four Institutions with which I have had the privilege to be associated, and even more that I have encountered in my journey. It is not possible to overestimate my gratitude for these interactions, whether they be discussions, collaborations, arguments, or social interactions. Many of these, too many to name, have led to long lasting friendships. These include faculty, trainees, and laboratory staff. I have tried to mention those primarily responsible for aspects of the work, but it is not possible to mention all, and I apologize for any omissions.

I would like also to thank the National Cancer Institute, the Howard Hughes Medical Institute, and the Cancer Prevention and Research Institute of Texas, as well the Cancer Research Institute, the Prostate Cancer Foundation, the Melanoma Alliance, and the Parker Institute for Cancer Immunotherapy for supporting my research over the years.

I would also like to express my gratitude for Medarex and Bristol Myers-Squibb for having the courage in making it possible to bring my dream to the clinic, and most of all the patients involved in the clinical trials for the courage to take a chance on a new modality of cancer therapy.

Finally, I dedicate this Nobel Lecture and my Nobel Prize to my love and soulmate, Padmanee Sharma, my partner in life and science.

REFERENCES

1. Cooper MD, Peterson RD, & Good RA (1965) Delineation of the Thymic and Bursal Lymphoid Systems in the Chicken. *Nature* **205**:143–146.
2. Allison JP, McIntyre BW, & Bloch D (1982) Tumor-specific antigen of murine T-lymphoma defined with monoclonal antibody. *Journal of immunology (Baltimore, Md.: 1950)* **129**(5):2293–2300.
3. McIntyre BW & Allison JP (1983) The mouse T cell receptor: structural heterogeneity of molecules of normal T cells defined by xenoantiserum. *Cell* **34**(3):739–746.
4. Kappler J, *et al.* (1983) The major histocompatibility complex-restricted antigen receptor on T cells in mouse and man: identification of constant and variable peptides. *Cell* **35**(1):295–302.
5. Yanagi Y, *et al.* (1984) A human T cell-specific cDNA clone encodes a protein having extensive homology to immunoglobulin chains. *Nature* **308**(5955):145–149.
6. Hedrick SM, Nielsen EA, Kavaler J, Cohen DI, & Davis MM (1984) Sequence relationships between putative T-cell receptor polypeptides and immunoglobulins. *Nature* **308**(5955):153–158.
7. Hedrick SM, Cohen DI, Nielsen EA, & Davis MM (1984) Isolation of cDNA clones encoding T cell-specific membrane-associated proteins. Nature 308(5955):149–153.
8. Jenkins MK & Schwartz RH (1987) Antigen presentation by chemically modified splenocytes induces antigen-specific T cell unresponsiveness in vitro and in vivo. *J Exp Med* **165**(2):302–319.
9. Martin PJ, *et al.* (1986) A 44 kilodalton cell surface homodimer regulates interleukin 2 production by activated human T lymphocytes. *Journal of immunology (Baltimore, Md.: 1950)* **136**(9):3282–3287.
10. Linsley PS & Ledbetter JA (1993) The role of the CD28 receptor during T cell responses to antigen. *Annu Rev Immunol* **11**:191–212.
11. Gross JA, Callas E, & Allison JP (1992) Identification and distribution of the costimulatory receptor CD28 in the mouse. Journal of immunology (Baltimore, Md.: 1950) **149**(2):380–388.
12. Harding FA, McArthur JG, Gross JA, Raulet DH, & Allison JP (1992) CD28-mediated signalling co-stimulates murine T cells and prevents induction of anergy in T-cell clones. *Nature* **356**(6370):607–609.

13. Linsley PS, Clark EA, & Ledbetter JA (1990) T-cell antigen CD28 mediates adhesion with B cells by interacting with activation antigen B7/BB-1. *Proc Natl Acad Sci U S A* **87**(13):5031–5035.
14. Brunet JF, et al. (1987) A new member of the immunoglobulin superfamily-CTLA-4. *Nature* **328**(6127):267–270.
15. Krummel MF & Allison JP (1995) CD28 and CTLA-4 have opposing effects on the response of T cells to stimulation. *J Exp Med* **182**(2):459–465.
16. Walunas TL, *et al.* (1994) CTLA-4 can function as a negative regulator of T cell activation. *Immunity* **1**(5):405–413.
17. Waterhouse P, et al. (1995) Lymphoproliferative disorders with early lethality in mice deficient in Ctla-4. *Science* **270**(5238):985–988.
18. Tivol EA, *et al.* (1995) Loss of CTLA-4 leads to massive lymphoproliferation and fatal multiorgan tissue destruction, revealing a critical negative regulatory role of CTLA-4. *Immunity* **3**(5):541–547.
19. Chambers CA, Sullivan TJ, & Allison JP (1997) Lymphoproliferation in CTLA-4-deficient mice is mediated by costimulation-dependent activation of CD4+ T cells. *Immunity* **7**(6):885–895.
20. Townsend SE & Allison JP (1993) Tumor rejection after direct costimulation of CD8+ T cells by B7-transfected melanoma cells. *Science* **259**(5093):368–370.
21. Leach DR, Krummel MF, & Allison JP (1996) Enhancement of antitumor immunity by CTLA-4 blockade. *Science* **271**(5256):1734–1736.
22. Small EJ, *et al.* (2007) A pilot trial of CTLA-4 blockade with human anti-CTLA-4 in patients with hormone-refractory prostate cancer. *Clin Cancer Res* **13**(6):1810–1815.
23. Hodi FS, *et al. (*2003) Biologic activity of cytotoxic T lymphocyte-associated antigen 4 antibody blockade in previously vaccinated metastatic melanoma and ovarian carcinoma patients. *Proc Natl Acad Sci U S A* **100**(8):4712–4717.
24. Phan GQ, et al. (2003) Cancer regression and autoimmunity induced by cytotoxic T lymphocyte-associated antigen 4 blockade in patients with metastatic melanoma. *Proc Natl Acad Sci U S A* **100**(14):8372–8377.
25. Korn EL, *et al.* (2008) Meta-analysis of phase II cooperative group trials in metastatic stage IV melanoma to determine progression-free and overall survival benchmarks for future phase II trials. *J Clin Oncol* **26**(4):527–534.
26. Garbe C, Eigentler TK, Keilholz U, Hauschild A, & Kirkwood JM (2011) Systematic review of medical treatment in melanoma: current status and future prospects. *Oncologist* **16**(1):5–24.
27. Sloan AE, Nock CJ, & Einstein DB (2009) Diagnosis and treatment of melanoma brain metastasis: a literature review. *Cancer Control* **16**(3):248–255.
28. Agarwala SS (2010) Novel immunotherapies as potential therapeutic partners for traditional or targeted agents: cytotoxic T-lymphocyte antigen-4 blockade in advanced melanoma. *Melanoma Res* **20**(1):1–10.
29. Eggermont AM & Kirkwood JM (2004) Re-evaluating the role of dacarbazine in metastatic melanoma: what have we learned in 30 years? *Eur J Cancer* **40**(12):1825–1836.
30. Rosenberg SA (2012) Raising the bar: the curative potential of human cancer immunotherapy. *Sci Transl Med* **4**(127):127ps128.
31. Eggermont AM (2010) Advances in systemic treatment of melanoma. *Ann Oncol* **21** Suppl 7:vii339–344.
32. Wolchok JD, *et al.* (2013) Development of ipilimumab: a novel immunotherapeutic approach for the treatment of advanced melanoma. *Ann N Y Acad Sci* **1291**:1–13.

33. Eisenhauer EA, *et al.* (2009) New response evaluation criteria in solid tumours: revised RECIST guideline (version 1.1). *Eur J Cancer* **45**(2):228–247.
34. Hodi FS, et al. (2010) Improved survival with ipilimumab in patients with metastatic melanoma. *N Engl J Med* **363**(8):711–723.
35. Schadendorf D, *et al.* (2015) Pooled Analysis of Long-Term Survival Data From Phase II and Phasc III Trials of Ipilimumab in Unresectable or Metastatic Melanoma. *J Clin Oncol* **33**(17):1889–1894.
36. Wolchok JD, *et al.* (2017) Overall Survival with Combined Nivolumab and Ipilimumab in Advanced Melanoma. *N Engl J Med* **377**(14):1345–1356.
37. Motzer RJ, *et al.* (2018) Nivolumab plus Ipilimumab versus Sunitinib in Advanced Renal-Cell Carcinoma. *N Engl J Med* **378**(14):1277–1290.
38. Liakou CI, *et al.* (2008) CTLA-4 blockade increases IFNgamma-producing CD4+ICOShi cells to shift the ratio of effector to regulatory T cells in cancer patients. *Proc Natl Acad Sci U S A* 105(39):14987–14992.
39. Carthon BC, *et al.* (2010) Preoperative CTLA-4 blockade: tolerability and immune monitoring in the setting of a presurgical clinical trial. *Clin Cancer Res* **16**(10):2861–2871.
40. Fu T, He Q, & Sharma P (2011) The ICOS/ICOSL pathway is required for optimal antitumor responses mediated by anti-CTLA-4 therapy. *Cancer Res* **71**(16):5445–5454.
41. Fan X, Quezada SA, Sepulveda MA, Sharma P, & Allison JP (2014) Engagement of the ICOS pathway markedly enhances efficacy of CTLA-4 blockade in cancer immunotherapy. *J Exp Med* **211**(4):715–725.

Tasuku Honjo. © Nobel Prize Outreach AB. Photo: A. Mahmoud

Tasuku Honjo

Biography

PROLOGUE

Burning wooden walls collapsed in front of us. I was on the back of my mother, who was running away from our burning house, gasping for air. Although I was just three and a half years old, I vividly remember this scene that seems to be engraved forever in my memory. It was one of the most devastating American bombing raids targeting civilians, which took place in Toyama City on August 1, 1945, just two weeks before the end of the war. We escaped the fire by jumping into a ditch beside a rice field to avoid the heat from the nearby burning houses. Just minutes before, an incendiary bomb had hit the bomb shelter in our garden where we had been hiding. Luckily, that bomb did not explode. I believe I was extremely fortunate from this early stage of my life.

I was born in Kyoto on January 27, 1942, less than two months after World War II started in the Pacific region. My family traces its ancestry back to a priest from a temple called Honjo-san Sensho-ji. Many years ago, when visiting the temple for the first time, I immediately recognized my relatives among others without any introduction. Their familiar eyes, eyebrows, noses and cheeks resembled those of my immediate family members.

We were welcomed by the head priest, Kuniyuki Honjo, who showed us the Sensho-ji Temple and writings about its origins. We were told that the Temple had been established sometime between 1077 to 1080 by Goromaru Fujiwara, a guard who helped the third son of the Emperor Go-Sanjo escape fighting in Kyoto, then Japan's capital.

My father worked as a surgeon at the Kyoto University Hospital. He later became the head of the Otorhinolaryngology Department in the Yamaguchi University Medical School in Ube City, where he served until his retirement in 1976. Having spent several years in Montreal's McGill University and Northwestern University in Chicago learning advanced surgical operations, he was fluent in English and perceived it was the new

Figure 1. Honjo family in 1955. From left to right Nobuko, Sho-ichi, Ryuko and Tasuku.

lingua franca one needed to master. Thus, my father introduced me to Ms. Kato, a Japanese woman born in Hawaii who taught me English two years ahead of my classmates, which helped my communication skills later in my career. My father enjoyed painting and was also an excellent golf player. I seem to have inherited his athletic but not artistic traits, although I do appreciate and enjoy paintings and music. While my father educated me rather strictly, my mother was protective and warm, forever my guardian goddess. She not only saved my life but shaped the person I was to become with her love (Fig. 1).

My first moment of enchantment with the natural sciences took place when I observed the tiny rings around Saturn in a clear sky using a portable telescope in the playground of Kamihara Elementary School in Ube during a summer vacation. Afterwards, I began to read many astronomy books and was determined to be an astronomer. My passion for astronomy continued for a few years until my mother gave me the biography of Hideyo Noguchi, a world-famous yet unconventional Japanese microbiologist. Noguchi obtained his medical doctor's license while struggling with incredible poverty and also a physical handicap after seriously burning his hand. He went to Philadelphia at the age of 24 to meet the great pioneer in the infectious disease research Simon Flexner, who later hired Noguchi at the Rockefeller Medical Institute. Noguchi is mostly known for his finding that progressive paralytic disease is caused by infection of the brain with syphilis, and for his pursuit of the cause of yellow fever. Ironically, the latter was to cause his sudden death in Ghana.

When I first had to make decisions about my future path at the end of senior high school, I was still uncertain. I knew I did not want to be a clerk or a civil servant, because I wanted to do something I was passionate about. I considered becoming a lawyer or embracing a diplomatic career, because I thought I could communicate effectively. However, the medical profession was very attractive to me, as not only my father but many of our relatives were clinical doctors. In the end, I chose to continue this legacy and applied for entry to the Kyoto University Medical School.

During the years around 1960, as I entered medical school, I felt I was witnessing a revolution in biology, with discoveries such as the double helix structure of DNA in 1953 and the decoding of the triplet nature of the genetic code in 1964 by Marshall Nirenberg and Philip Leder. I recall being extremely impressed by a Japanese book called *Revolution in Biology* by Atsuhiro Shibatani, who happened to be a colleague of my father at Yamaguchi University. Published in 1960, the book contained amazing speculations: that cancer is a manifestation of mutations in our DNA; that automated DNA sequencing will enable us to define these mutations; that the development of 'molecular surgery' will allow us to repair defects in DNA and may ultimately cure cancer. The night after I was exposed to Shibatani's extraordinary imagination, I could not sleep. I was totally immersed and fascinated by biochemistry and molecular biology, which I believed would lead to a better understanding of what life is.

EARLY TRAINING

Fortunately for my scientific career, I became bored with memorizing Latin names for anatomy and the symptoms of many diseases with unknown causes in medical school and began to spend most of my time in the laboratory of Professor Osamu Hayaishi. Hayaishi was a great biochemist – the discoverer of oxygenase enzymes – who had just returned to Kyoto University from a nine-year stint at the National Institutes of Health (NIH) and Washington University in St. Louis. He worked with the biochemist and Nobel laureate Arthur Kornberg, with whom he kept a close friendship throughout his life. I eagerly joined Hayaishi's department for graduate studies between 1967–1971. During this time, and also later in my life, I learned much from Hayaishi – not only about science but also other enjoyable pastimes like golf and wine tasting, which later gave me much pleasure. In Hayaishi's lab, I was mentored by Yasutomi Nishizuka, who was then famous for elucidating how the molecule NAD, essential for energy production in all living cells, is synthesized from the amino acid tryptophan, and later discovered protein kinase C and its regulation by phorbol ester.

I believe it was in the summer of 1967 when I found a very interesting paper by John Collier demonstrating the requirement for NAD by diphtheria toxin for inactivation of protein synthesis enzyme EF-2. This work was striking, as at that time toxins were thought to function by directly binding components of cellular machinery to block their activity. I had a strong desire to find out why NAD is essential for diphtheria toxin's function. I was, fortunately in the best place to answer this question, due to the availability of labeled forms of NAD generated by Nishizuka's research on the NAD pathway. I quickly established that diphtheria toxin is an enzyme that catalyzes transfer of the ADP-ribose moiety of NAD to EF-2 and published this finding in the prestigious *Journal of Biological Chemistry* (Honjo et al., 1968). Publishing this work while still a graduate student gave me a boost of confidence. The remaining two years of my graduate course were, sadly, scientifically lost, as student riots forced the closure of the university campus. I was already thinking of the next step. Following the example of my mentors and my father before me, I was determined to challenge myself in an international scientific environment.

I was fortunate to be hired as a postdoctoral fellow at the Carnegie Institution of Washington in Baltimore. The beautiful Carnegie building was located next to the northwest corner of the Homewood campus of Johns Hopkins University. Retrospectively, an incident which changed my life occurred about one year after my arrival. Donald Brown gave a seminar in Carnegie about the organization of the immunoglobulin (Ig) gene. At that time, not only immunologists but all biologists were keen to find how animals generate specific antibodies against almost every foreign antigen they encounter. Two major hypotheses were put forward to explain the enormous diversity of antibodies produced by B lymphocytes. The first so-called 'germline' hypothesis claimed that we may have as many as one thousand genes responsible for the antibody diversity phenomenon. The second theory, called 'somatic hypermutation' originally proposed by Sir Frank Macfarlane Burnet, claimed that DNA mutations in immune cells could generate enormous lymphocyte diversification from a limited number of inherited genes. A major problem with the germline hypothesis was explaining how individual copies of the Ig variable (V) gene could differ so greatly while maintaining the important structural information contained in the Ig constant (C) region gene. Based on his studies on the organization of ribosomal RNA genes, Don proposed that multiple copies of the C gene might be conserved by frequent homologous recombination between other copies of C genes to repair errors and suggested that this long-standing mystery of biology could be approached with modern molecular biological technology. I could not stop asking

Don where the best place to tackle this fundamental problem was. He suggested Philip Leder, whom I met three months later when he gave a seminar in Carnegie. My long journey in molecular immunology began when Phil accepted me to his lab at NIH.

The work in Phil's lab was exciting every day. There was a movable arrow on the wall pointing either to the germline or somatic hypermutation theory. The strategy Phil proposed was to count the number of the light (L)-chain C genes by hybridization kinetics, using radiolabeled cDNA derived from purified Ig L chain messenger RNA. The speed with which these labeled genetic probes were reannealed after denaturation, together with total cellular DNA, would reflect the number of L-chain C genes present in the genome. Our conclusion was that there is only one copy or very few copies of the Ig L_{κ} or L_{λ} chain gene. Ironically, these results ran contrary to Don's preferred germline theory of diversity.

BACK TO JAPAN

With this result, after almost four years in America I felt drawn back to Japan. Phil advised me to stay, as in 1974, it was apparent that the Japanese universities had neither money nor the sophisticated facilities to carry out advanced molecular biological studies. I faced the second big choice in my life. I finally made up my mind to go back to Japan. Firstly, I wanted my small children (my daughter, Yasuko who had recently been born in Baltimore, and my five-year-old son, Hajime) to experience aspects of Japanese culture that might be lost to them if we continued to live abroad, and secondly, I wanted to nurture a strong culture of molecular biology research into Japan, as my mentor Hayaishi had done for biochemistry. I am happy to have seen my children thrive back in Japan, Hajime as a gastroenterologist and Yasuko as an embryologist, although for their success I owe a debt of gratitude to my wife Shigeko as I was a typical, workaholic Japanese husband. Phil accepted my decision and kindly recommended me for a small grant from the Jane Coffin Childs Memorial Fund, which was instrumental in launching my career at the University of Tokyo. He also generously sent precious RNA reverse transcriptase, an essential reagent to generate cDNA that was then difficult to obtain.

The University of Tokyo was considered the best university in Japan. However, I found that not only the infrastructure but also the spirit of science was far behind what I had experienced in Hayaishi's laboratory. To avoid direct competition with Phil, I decided to reorient my work on the Ig heavy (H)- chain gene structures that had their own unique biological question, i.e., how different classes of antibody like IgG, IgE or IgA were made during immune responses. Although the question was clear, I faced many technical hurdles as even a basic piece of equipment like a gel dryer was difficult to find. I remember going down to the Akihabara area – where many shops sell electronic components – to buy thick rubber plates, hose and mesh to build my own vacuum drying apparatus for electrophoresis gels, which were essential for visualizing proteins. A few brilliant medical students – who later took the graduate course in my lab – voluntarily worked after classes with me.

It took almost three years until we began to obtain interesting data. Surprisingly, we found that the gamma chain constant (C_γ) gene was deleted from DNA in myelomas. To examine whether C_H gene deletion was an artifact of the myeloma cancer we were studying, I asked Michael Potter – an expert in plasma cell cancers working at NIH – to send more myeloma cells producing different Ig classes. Tohru Kataoka and I carried out extensive Cot Curve analyses of many myeloma DNAs (Fig. 2). At that time, I was commuting from Yokohama to Tokyo, which took 90 minutes each way. One night, while poring over results on the train on my way back home, I found that the C_H gene deletion always takes place upstream the C_H gene expressed in that particular myeloma, assuming a specific order of C_H genes on the chromosome. That was the moment when I reached the DNA deletion hypothesis model for class switch recombination (CSR) (Honjo and Kataoka, 1978). I was extremely delighted when our research was highlighted by the *Nature* editor Miranda Robertson, in that journal's News and Views.

In 1977, I returned to Phil's lab for three months when they restarted

Figure 2. Students in Tokyo around 1979. From left to right, Tohru Kataoka, Yoshio Yao-ita, Naoki Takahashi, Toshiaki Kawakami and Akira Shimizu.

experiments after the lifting of a long, frustrating moratorium on gene cloning technology. I quickly succeeded in isolating $C_{\gamma 1}$ cDNA, starting from purified mRNA using molecular cloning techniques, with the help of John Seitman. I was fascinated when we established the sequence of the $C_{\gamma 1}$ gene, and finally discovered that each domain of the $C_{\gamma 1}$ chain is separated by noncoding spacers. Akira Shimizu and others went on to map all the C_H genes on the mouse chromosome (Fig. 2). We were very much delighted to prove that the proposed DNA deletion model of antibody diversity was absolutely correct. The next clear goal was to elucidate the mechanism in detail, and especially to identify the enzyme(s) that catalyze(s) CSR.

JOURNEY TO AID

In 1979, five years after I came back from Phil's laboratory, I was invited to become the head of the Department of Genetics at Osaka University School of Medicine. Elevating a relatively early-stage researcher to such a prominent position was something unprecedented in Japan. However, my stay in Osaka was short, as in 1984 I was invited back to Kyoto University to succeed Hayaishi after his retirement. A highlight during my time in Osaka was being invited to the Nobel symposium organized by Erna Möller and Göran Möller in 1982. There I met Eva Severinson, who later proposed a collaboration to clone cytokines which could enhance class switching, as she had established a mouse helper T cell line and a sensitive experimental assay system for class switching. By expressing a full-length cDNA library in amphibian oocytes, we discovered IL-4 and IL-5, which turned out to be critical for not only CSR but also T cell differentiation.

In 1991, I was invited to NIH as a Fogarty Scholar, a very valuable sabbatical opportunity. It provided me with a year's salary and residence within the NIH campus, to encourage interactions with its many scientists. I enjoyed staying at the NIH campus and talking with many immunologists, including William Paul, who provided an education in cellular immunology, especially regulation of the immune response. One day I met Warren Strober and his associate Yoshio Wakatsuki from Kyoto University, who told me they had a cell line that could be stimulated to switch from producing IgM to IgA antibodies *in vitro*. Instantly, I realized this cell line could be what I was looking for in order to identify the enzyme(s) responsible for induction of CSR. After repeatedly selecting efficiently switching clones for two years, we finally isolated a cell line (CH12F3) that robustly expressed IgA after 48-hour stimulation with CD40 ligand, IL-4 and TGF-β1.

Masamichi Muramatsu began looking for differences in cDNAs

Figure 3. From left to right, Sidonia Fagarasan, Tasuku Honjo and Masamichi Muramatsu in 2013.

expressed by CH12F3 cells before and after stimulation (Fig. 3). One gene that caught his attention was expressed only in activated B cells but not T cells. In addition, the gene was turned on in germinal centers, the active sites in the lymphoid tissues where CSR was known to take place. Surprisingly, the clone closely resembled genes in the cytidine deaminase family, including RNA editing enzyme APOBEC-1. Indeed, Muramatsu could show it had noticeable if weak cytidine deaminase activity on free cytidine. We published a paper on the isolation of the activation induced cytidine deaminase (AID) gene in 1999 in the *Journal of Biological Chemistry.* Privately, I had chosen to name the gene AID because both aid and Tasuku can mean "help", which I did not tell Masamichi. We continued to investigate the AID function by knocking out the gene in mice. By the end of January 2000, we could show that AID-deficient animals lacked both CSR and SHM (Muramatsu *et al,* 2000).

We had been asked for collaboration by Anne Durandy and Alain Fischer from Hôpital Necker-Enfants Malades in Paris, who were trying to identify the gene responsible for a severe immune deficiency in patients with hyper-IgM syndrome type 2 (HIGM II). These patients had very high concentrations of IgM in their blood, strongly suggesting a genetic defect in class switching. We sent primers to Paris to detect the AID gene, which their group used to identify AID mutations in all HIGM II patients. They also found HIGM II patients were defective in SHM. We quickly confirmed that this process was also lacking in AID-deficient animals. The result was completely unexpected, as the scientific consensus at the time was that SHM and CSR were completely different modifications carried out by different genes. However, these findings showed that the AID protein alone was responsible for these two critical genetic alterations in Ab genes during immune responses. It was a eureka moment for me and the entire group working on AID. At this time, serving as a dean of the Medical School of Kyoto University, I was afflicted with very severe back pain, to the point of not being able to walk or sit comfortably. I worked on the AID paper lying on the floor in my office while colleagues kneeled next to me to discuss plans and ongoing experiments. I remember the year 2000 for both the physical challenge and some of the most exciting scientific breakthroughs in my life.

I presented this result at the Annual Meeting of the American Association of Immunologists in Seattle in May 2000, after we sent the two manuscripts to *Cell* describing AID deficiency in mice and humans. The lecture was well accepted by a large audience. Although it was clear that the game was over in the hunt for the enzymes responsible for CSR and SHM, for us this was the beginning of another journey to find out how such a small protein can mediate two complicated genetic changes in immune cells.

DISCOVERY OF PD-1, ITS FUNCTION AND APPLICATION

The mechanisms of clonal selection, namely, how immature T cells are selected during development in the thymus, was another big question in immunology. In my lab, Yasumasa Ishida was striving to identify the molecule(s) responsible for the thymic T cells selection, and proposed comparing the gene expression of growing and dying cells from the thymus (Fig. 4). In May 1991, he used this technique to identify a gene expressed during this process which we named PD(programmed cell death)-1 (Ishida *et al.*, 1992). The structure predicted from the gene sequence indicated that PD-1 is a surface receptor protein, related to but distinct from known receptor molecules involved in lymphocyte activation. Therefore, I thought we should focus on finding the function of this protein in immune cells. Hiroyuki Nishimura generated knock-out (KO) mice on a mixed genetic background in June 1994, but to our disappointment the KO mice were identical to controls (Fig. 4). Fortunately, Nagahiro Minato, a very knowledgeable and experienced tumor immunologist in our faculty, suggested that we should breed the PD-KO mice to create an inbred strain and continue to monitor the mice before drawing any conclusion (Fig. 5). Thanks to Minato's guidance, by November 1997, we observed that when introduced to the autoimmune-prone lpr/lpr strain, PD-1 deficiency led to arthritis and nephritis at five months of age. Continuing our patient monitoring, we found that PD-1 deficiency on the C57BL/6N background caused mice to develop similar autoimmune symptoms at 14 months of age. These findings convinced us that PD-1 is a negative regulator of the immune system (Nishimura *et al,* 1999). Disease manifestation in PD-1 KO mice was milder than that of CTLA-4 KO mice, all of which were reported to die within 4–5 weeks. The CTLA-4 gene, originally discovered by Pierre Golstein, was shown to encode a negative immune-receptor by Jeff Bluestone, Craig Thomson, Jim Allison, Tak Mak and Arlene Sharpe. The comparatively mild and chronic symptoms in PD-1 KO mice suggested to us that PD-1 could safely be targeted to treat diseases. Around March 1998, we began a series of experiments targeting PD-1 as a therapy for various disorders characterized by defects in immune regulations, such as cancer, infection, autoimmunity, and rejection of transplanted organs. Reasoning that blocking PD-1 would be simpler than enhancing its function, boosting the immune response against cancer became our first goal. Yoshiko Iwai began work looking ways to use PD-1 in therapeutic applications and generated many reagents and experimental systems (Fig. 4).

Ever since discovering PD-1, we had predicted the presence of a spe-

Figure 4. From left to right, Yasutoshi Agata, Yasumasa Ishida, Yoshiko Iwai, Taku Okazaki, Hiroyuki Nishimura at the celebration party of the CHAOO award by the Japanese Cancer Association in 2014.

cific binding partner or ligand (Ishida *et al.*, 1992). On September 17, 1998, we spoke with Steve Clark, the head of the Genetic Institute, about our problem with identification of ligands for PD-1. He proposed using their Biacore instrument to screen for PD-1 ligands in the supernatants from a variety of cultured cells. I agreed to collaborate and provided not only the reagents but all our knowledge about PD-1 because this molecule was completely unknown to them. Meantime, we tried to identify the PD-1 ligand by fishing for molecules which bound to an engineered PD-1-Ig fusion protein. Although we clearly detected binding of PD-1-Ig protein to various cell types, including some tumor cells, we were unable to purify or identify the miniscule amount of binding proteins. After a year of silence, on October 5, 1999, Clive Wood, who was responsible for this project under Steve, sent word out of the blue that they might have identified the PD-1 ligand. On October 25, 1999, in Boston, Clive explained that their collaborator Gordon Freeman at Dana Farber identified several B7 family cDNA in a deposited data base and the Clive group found the protein encoded by one of them (clone 292) had bound to the PD-1 expressing cells we had sent from our laboratory. In Japan, we confirmed binding of the 292 protein to PD-1, and performed experiments showing its immune inhibitory function. Together, we published our identification of a PD-1 ligand (PD-L1) (Freeman *et al.*, 2000). Just before the submission of our paper, I learned that Lieping Chen had published a paper describing the identical cDNA as a B7 family member with costimulatory activity, without knowing its receptor (Dong *et al.*, 1999).

Meanwhile, we continued our studies aimed at treating cancer by PD-1 blockade. By September 2000, Iwai had found that myeloma expressing the PD-L1 grew more slowly in PD-1 deficient mice compared with wild-type mice. I was very excited and convinced that PD-1 would make an excellent target for cancer treatment. As I needed good blocking antibodies against PD-1 or PD-L1, I talked with Minato and urged him to hurry in our collaborative work towards this goal. Minato`s group had already started searching for antibodies which could effectively block either PD-1 or PD-L1. Yoshimasa Tanaka in Minato's laboratory worked diligently on the project, and by the end of 2001 we had gathered enough data to demonstrate that PD-1 blockade by PD-L1 antibodies can prevent different types of tumor in mouse models (Iwai *et al.*, 2002) (Fig. 5). In my own group, Iwai showed that the spread of B16 melanoma cancer from spleen to liver could also be prevented by anti-PD-1 antibody treatment (Iwai *et al.*, 2005).

As we prepared our publication about PD-L1 isolation, I asked Wood how we might approach patent applications, and was informed that Wood and Gordon had already applied for a patent covering PD-L1 and its usage in November 1999 without informing me. I was surprised and registered a

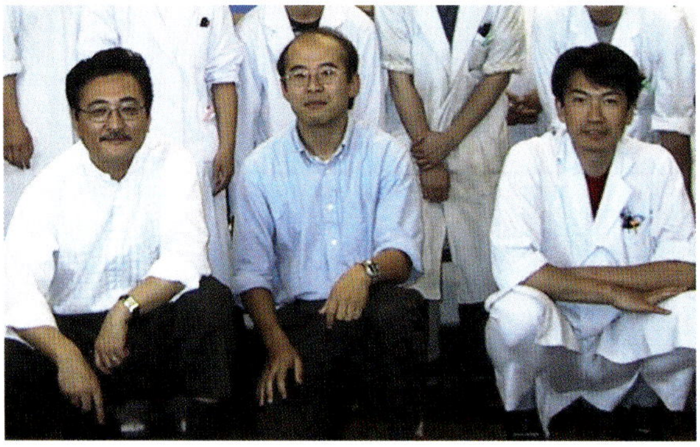

Figure 5. From left to right, Nagahiro Minato, Masakazu Hattori and Yoshimasa Tanaka in 2000.

complaint through a lawyer, but to no avail. However, Wood's patent was declined, perhaps lacking sufficient data to support its usage. More unpleasant surprises were to come, as more than 10 years after our PD-1 blockade paper was published, Freeman and Wood sued me by claiming co-inventorship of the PD-1 cancer immunotherapy patent. The precise timelines in my story thus arise from a desire to publicly clarify how we came to discover the therapeutic value of PD-1, due to my draining and still unresolved legal battle regarding the intellectual property rights.

CLINICAL APPLICATION

As I anticipated the clinical application of our finding on tumor treatment by PD-1 blockade, I approached Kyoto University to apply for the intellectual property right for usage of PD-1 blockade for cancer treatment. Frustratingly, they lacked the human and financial resources to support my application. At this point I turned to a small Japanese company, Ono Pharmaceutical, because we had been involved in other collaborations. Unfortunately, Ono also had no capacity to organize clinical trials for cancer drug development and searched for a bigger partner. They spent almost a year visiting more than a dozen companies without success. It was obvious that, at this time, neither the pharmaceutical industry, clinicians, nor even many biologists believed that trying to enhance the immune system could help cure cancer. I was terribly disappointed when Ono officials told me they had decided to abandon the project. I decided to turn to my network of friends in the USA, many of whom had started their own companies.

I visited one such venture capital company in Seattle and they surprisingly agreed to invest in my proposal almost instantly, on the condition that Ono would be excluded from future developments. I waited for several months for Ono to confirm they had no intention to use our patent independently before committing to the Seattle partner. Ono then stunned me with the news that they now had the resources to invest in developing a PD-1 treatment. Later, I learned that Medarex, the biopharmaceutical company led by Nils Lonberg had found our published PCT patent application and directly approached Ono with a collaboration proposal. I was extremely lucky again. Without this chance encounter, the PD-1 cancer drug development might have been delayed by many years.

The paper summarizing the first clinical trial (Phase I) on PD-1 antibody published by Suzanne Topalian's group in 2012, showed very promising complete or partial responses in roughly 20 to 30 percent of patients with terminal melanoma, lung, or renal cancer. In addition, patients continued in unprecedented good health for more than a year and, half after treatment. In Japan, we carried out a smaller clinical trial on drug-resistant ovarian cancer patients in collaboration with the Gynecology Department at Kyoto University. I was immensely pleased to hear about a patient whose large tumor had been completely cured by PD-1 blockade therapy, and to watch her enjoying a round of golf on a TV program about the breakthrough drug. Two patients treated for a year with PD-1 antibody are still free from tumors almost five years later. The PD-1 antibody was approved for melanoma treatment by PMDA in Japan in 2014, quickly followed by the FDA in the USA.

SCIENCE MEDIA

In 2013, *Science* magazine selected cancer immunotherapy as the number one scientific breakthrough of the year. The author of that article described the development of therapy targeting CTLA-4 by Jim Allison in detail. However, I was shocked as the article continued: "In the early 1990s, a biologist in Japan discovered a molecule expressed in dying T cells, which he called programmed death 1, or PD-1, and which he recognized as another brake on T cells. *He wasn't thinking of cancer, but others did.*" I was speechless! Clearly, this *Science* editor reporting cancer immunotherapy had either not read our papers describing cancer therapy by PD-1 blockade, or had intentionally neglected them (Iwai *et al.*, 2002, 2005).

This article was a symbol to me of deficits in our current system of science journalism. Scientists are eager to publish their papers in the so-called top journals like *Science, Nature* or *Cell*, as they are judged by their peers and funders based on such publications. When I first became a researcher, editors of these journals such as Miranda Robertson, Peter Newmark, Ben Lewin and Linda Miller, were very knowledgeable, reliable and sincerely interested in research. I was happy to send my manuscripts to their journals as I felt confident these editors had the knowledge and training to allow them to perceive the importance of our science. However, more recently, perhaps due to the enormous expansion in the number of journals, this no longer seems to be the case. When in-house editors rely on the majority vote of outside reviewers with different opinions rather than their own judgment, paradigm-shifting papers have difficulty being published, as by definition they run against the concepts held by the majority. It is notable that none of my papers cited by the Nobel Committee were published in these so-called top journals. The current peer review system tends to reject newly emerging concepts or new findings running against the established dogma in the scientific community. Since scientists feel they have at some point received unfair statements from our colleagues, I believe revealing the names of peer reviewers is one way we might mitigate some of these problems. Certainly the *status quo* needs to change to make our current peer review system more constructive and efficient.

SIX "C"S IN SCIENCE

To carry out science, I always say to my students; if they do not have *C*uriosity, then they should not choose a career in science. There are many types of scientists. It is always important to find out what you really want to know. Imagine research as a journey deep into mountains to find the origin of the mysterious river. Sometimes on the path you may discover an interesting stone which you bring back and find it very precious upon careful examination. My style of doing science is to find something totally unknown, or to map a totally new route to an unknown cave from which water is coming out. You should keep asking what you wish to know rather than what you can do. You need to embrace *C*hallenge and have the *C*ourage to invest your effort and time. These three "*C*"s were the basis of my research career. Once you fix on a project, three more "*C*"s are required. Of course, you have to *C*ontinue your research with lots of *C*oncentration. *C*oncentration means placing research as a central activity in your life, sometimes sacrificing other important calls on your time to focus on your goal. To *C*ontinue and *C*oncentrate, you may need *C*onfidence that you can do it. Or it could be the other way around. *C*oncentration and *C*ontinuation may eventually build up *C*onfidence. The last three "**C**"s are as important as the first.

DREAMS

Aside from scientific activities, I still have two dreams at the age of 76. As I have watched younger generations struggling for support, I have long thought Japanese universities should create their own funds to support young scientists. Compared to American universities with large endowments, Japanese universities rely more on government funds. Since the future prospects of the science budget from the Japanese government are not so promising, I have proposed that Kyoto University set up a fund using royalties from our PD-1 patent. Fortunately, this fund has already started accepting donations. I sincerely hope others will join this initiative to help instill scientific enthusiasm in the next generations of researchers. This is especially critical in Japan, where human creativity is our most important natural resource.

Another dream of mine is age-shoot. I started playing golf while living in the United States. Over there, the green fee was only five dollars, which even an impoverished post-doc could easily afford. Perhaps I love golf for its similarities to science. For every shot, the given conditions are never the same. You have to think how far you are going to hit, in which direction and whether to add spin to stop on the green and so on. Most importantly, one must avoid big mistakes. It is totally different from team sports or those where you must score points. In golf, you have to always choose from many ways to approach a shot and this decision lies with you alone. Probably, this challenging game sparked my curiosity to which I am so attached. I dream of achieving a score equal to my age before I turn 80.

CODA

The curiosity surrounding my two favorite molecules, PD-1 and AID, will continue thanks to my long-time collaborator, the Romanian scientist Sidonia Fagarasan (Fig. 3). Her work in my lab and beyond revealed that PD-1 and AID are essential molecules for the generation and selection of intestinal IgA that regulates our microbiota. Microbiota, in turn, control the immune system by promoting expression of AID, PD-1 and IgA, which are essential for maintaining immune tolerance to microbiota and self, metabolic balance and even brain function. Using AID-deficient mice, Sidonia was the first scientist to show that dysregulated gut microbiota greatly change the host immune system (Fragasan *et al.* 2002). How serendipitous that my two favorite molecules should collaborate in such a fascinating way to control the immune system, the microbiome and to fine-tune anti-tumor activity! It feels impossible to learn everything even just about these two molecules, as they have a far-reaching impact on other major physiological systems of the body. Indeed, despite so much progress, I believe we are just beginning our exploration of life sciences, because we understand so little about biology and physiology. After all, medicine has made great leaps in fixing acute problems, but when it comes to physiology we are humbled by the body's complexity, as we are decoding the continuous evolutionary experiments of the past four hundred and sixty million years. Every day I come to the lab enthusiastic to learn more about immune regulation in the context of physiological systems, still dreaming that curing cancer patients everywhere on this planet by immunotherapy will become possible someday.

ACKNOWLEDGEMENTS

I am deeply grateful for the precious support for the preparation of this article by Sidonia Fagarasan, Alexis Vogelzang, Blake Thompson and Maki Kobayashi.

REFERENCES

Dong, H., Zhu, G., Tamada, K., and Chen, L. (1999) B7-H1, a third member of the B7 family, co-stimulates T-cell proliferation and interleukin-10 secretion. *Nature Medicine* **5,** 1365–1369.

Fagarasan, S., Muramatsu, M., Suzuki, K., Nagaoka, H., Hiai, H., and Honjo, T. (2002) Critical roles of activation-induced cytidine deaminase in the homeostasis of gut flora. *Science* **298**, 1424–1427.

Freeman, G. J., Long, A. J., Iwai, Y., Bourque, K., Chernova, T., Nishimura, H., Fitz, L. J., Malenkovich, N., Okazaki, T., Byrne, M. C., Horton, H. F., Fouser, L., Carter, L., Ling, V., Bowman, M. R., Carreno, B. M., Collins, M., Wood, C. R., and Honjo, T. (2000) Engagement of the PD-1 immunoinhibitory receptor by a novel B7 family member leads to negative regulation of lymphocyte activation. *The Journal of Experimental Medicine* **192**, 1027–1034.

Honjo, T., Nishizuka, Y., Hayaishi, O., and Kato, I. (1968) Diphtheria toxin-dependent adenosine diphosphate ribosylation of aminoacyl transferase II and inhibition of protein synthesis. *Journal of Biological Chemistry* **243**, 3553–3555.

Honjo, T., and Kataoka, T. (1978) Organization of immunoglobulin heavy chain genes and allelic deletion model. *Proceedings of National Academy of Sciences of the United States of America* **75**, 2140–2144.

Ishida, Y., Agata, Y., Shibahara, K., and Honjo, T. (1992) Induced expression of PD-1, a novel member of the immunoglobulin gene superfamily, upon programmed cell death. *The EMBO Journal* **11**, 3887–3895.

Iwai, Y., Ishida, M., Tanaka, Y., Okazaki, T., Honjo, T., and Minato, N. (2002) Involvement of PD-L1 on tumor cells in the escape from host immune system and tumor immunotherapy by PD-L1 blockade. *Proceedings of National Academy of Sciences of the United States of America* **99**, 12293–12297.

Iwai, Y., Terawaki, S., and Honjo, T. (2005) PD-1 blockade inhibits hematogenous spread of poorly immunogenic tumor cells by enhanced recruitment of effector T cells. *International Immunology* **17**, 133–144.

Muramatsu, M., Kinoshita, K., Fagarasan, S., Yamada, S., Shinkai, Y., and Honjo, T. (2000) Class switch recombination and hypermutation require activation-induced cytidine deaminase (AID), a potential RNA editing enzyme. *Cell* **102**, 553–563.

Nishimura, H., Nose, M., Hiai, H., Minato, N., and Honjo, T. (1999) Development of lupus-like autoimmune diseases by disruption of the PD-1 gene encoding an ITIM motif-carrying immunoreceptor. *Immunity* **11**, 141–151.

Serendipities of Acquired Immunity

Nobel Lecture, December 7, 2018 by
Tasuku Honjo
Kyoto University, Kyoto, Japan.

INTRODUCTION

For a long time, biology was perceived as the lesser of the natural sciences because, unlike physics, deductive reasoning could not be used to solve biological problems. Biology has been full of mystery since I started my career in biological sciences almost half a century ago. Although the basic principles of biology stem from the rules of physics, biological systems have such an extraordinary, layered complexity derived from a tremendous number of parameters, beautifully and magically intertwined and controlled to achieve what we call "life". Paradoxically, we start with a rather limited number (about 20,000) of coding region genes. However, many transcripts can be generated from a single coding gene locus, indicating that a single gene can produce many proteins. Furthermore, there are a much higher number of non-coding transcripts that may affect the expression of the coding genes. DNA, RNA and proteins can be chemically modified by methylation, phosphorylation and acetylation. In addition, at least 20,000 metabolites circulate in our blood. These can also be sensed by cells, interacting with various proteins and influencing gene expression, thus generating enormously complicated regulatory mechanisms to achieve homeostasis. The origin of metabolites can be traced not only to the biochemistry of our own cells but also to the diverse communities of microbes inhabiting every surface of the body. If we imagine roughly 10^{13} order of our own cells, each expressing different proteins and containing different metabolites, in constant dialog with 10^{14} order of microbial cells, also in different metabolic states, the complexity of our biological system exceeds by far the physical and chemical complexity of the universe. The task of a medical scientist is deciphering this overwhelming complexity in order to understand the healthy state and diseases. We have yet to get a clear vision of exactly how gene expression is regulated. To make a long story short, the two major questions that we need to answer in biology are, what the mechanisms are for diverse yet strictly specific systems, and how these systems, are regulated. Biological regulation is robust, with many layers of redundancy as a safeguard. While we still do not have a thorough understanding of single-celled organisms like *E. coli* or yeast, multicellular organisms like human beings have far more complex regulatory systems with multiple layers of functional redundancy as well as tremendous specificities.

The immune system is perhaps one of the best-studied integrative regulatory systems in mammals. A large number of immune cells in different activation stages play specific roles, either through direct cell interactions or communication via soluble molecules like cytokines and their receptors. Although immunology is interesting and fascinating for a large audience, the immunologists' language is often misunderstood by other biologists, probably because of their use of scientific jargon. The best example is the cluster of differentiation (CD) annotation of lymphocyte surface molecules, which is necessary but at the same time very limiting and confusing. Even more difficult for non-aficionados is to understand sophisticated hypotheses and conceptual frameworks of the regulatory mechanisms of the immune system, including its innate and acquired immunity branches. The acquired immunity emerged during the late stage of our evolution. Therefore, this branch of the immune system utilizes all pre-existing regulatory systems employed by other systems, such as the nervous and endocrine systems. As such, the immune system is an ideal biological system to study the enormous specificity and highly sophisticated regulatory networks acquired by the biological systems during evolution.

When I started working on the immune system after my training in biochemistry and molecular biology, I often felt uneasy because immunologists were concerned mostly with functional aspects of the immune system rather than the molecules responsible. To give just one example, many immunologists used the term "antibody" for serum containing hundreds of proteins! Nevertheless, I witnessed a drastic change in the field with the advent of molecular biological techniques that could be easily applied to address immunological questions. I was very fortunate to begin my career at a time when molecular biology facilitated the elucidation of molecular mechanisms for the immune functions using their valuable assay systems. The most important conceptual questions in immunology were: a) how does the acquired immune system generate the enormous diversity required to distinguish myriads of microorganisms and substances foreign to our body and b) how does this enormously diverse recognition system discriminate between self *versus* non-self? Basically, these major immunological questions also represent the core questions in biology: how organisms can have enormous numbers of reactions with highly specific yet robust regulations.

IMMUNOGLOBULIN DIVERSITY

Karl Landsteiner was astonished by the observations that almost all synthesized organic compounds bound to proteins could induce generation of specific antibodies (Ab) or immunoglobulins (Ig) by B cells in animals. This brought the question of how many Ig genes might exist there to generate such a diverse repertoire of Ab. Around 1970, the Ig structure had been elucidated by the efforts of many scientists, including Rodney Porter and Gerald Edelman, by finding that a typical Ab like IgG, consists of two identical light (L) and two heavy (H) chains linked by disulfide bonds (Edelman, 1959; Porter, 1959). Frank Putnam (Titani *et al.*, 1965) and two of Edelman's colleagues, Norbert Hilschman and Lyman Craig (Hilschmann *et al.*, 1965), discovered that the human L chain has two distinct regions, the N-terminal variable (V) and the C-terminal constant (C) region. The next obvious question was related to the number of V and C genes. How many V genes or C genes are there? Do we have the same number of V and C genes, or perhaps a large number of V genes and a small number of C genes? Two major hypotheses were put forward to explain the enormous Ab diversity. The 'germline hypothesis' claimed that there might be as many as one thousand genes responsible for the Ab diversity phenomenon. Another hypothesis proposed by Sir Frank Macfarlane Burnet, called 'somatic hypermutation theory', claimed that somatic mutations could be responsible for the generation of enormous diversity of the lymphocyte repertoire starting from a rather limited number of inherited genes. This debate caught the attention of not only immunologists but also biologists in general.

After I obtained my Ph. D. in Osamu Hayaishi's laboratory, I went to the United States of America to carry out my postdoctoral training (Fig. 1). In 1972, I listened to a lecture by Donald Brown from the Department of Embryology of the Carnegie Institution of Washington in Baltimore (Fig. 2). He proposed a very brave idea in support of the germline hypothesis, which postulates that animals have a large number of V and C genes.

Figure 1. Osamu Hayaishi and myself around 1996.

Figure 2. Donald Brown at Carnegie Institution in Baltimore, 1971.

Assuming the V and C sequences to be reiterated more than a hundred times, the major problem of the germline hypothesis was finding a mechanism for keeping many copies of the C gene unchanged while many copies of V genes varied. The evolutionary selection pressure on each copy of the C gene would be very weak if there were many copies of the same gene. Don proposed that the frequent homologous recombination could repair genetic mutations and maintain a large number of copies of the Ig C genes. The idea came from his own study on the structure of the ribosomal RNA or the 5S RNA genes, both of which consist of the spacer and RNA coding regions which correspond to alternating V and C region sequences, respectively (Brown *et al.*, 1968). A set of the spacer and coding regions is tandemly repeated almost a thousand times in the ribosomal RNA gene loci, which are often located close to the telomere of the

Figure 3. Philip Leder at NIH, 1973.

chromosome where the homologous recombination frequency is expected to be high.

That seminar struck me like a lightning bolt. I knew that the question was very important, but only then did I realize that emerging molecular biological techniques could be used to find an answer. I was fortunate that Philip Leder at NIH accepted me into his laboratory to work on this question (Fig. 3). Our results indicated that the number of copies of the IgL (C_κ or C_λ) chain gene is low, one or a few at most, ironically disproving Don's germline hypothesis (Honjo *et al.*, 1974). The results also suggested that some gene alterations, either somatic mutations, recombination events or both are required to generate diverse Ab from a small number of C genes. Indeed, soon afterwards Susumu Tonegawa discovered that recombination of a small number of variable (V), diversity (D) and joining (J) segments generates a repertoire of millions of unique receptor genes during lymphocyte differentiation (Brack *et al.*, 1978).

CLASS SWITCH RECOMBINATION AND AID

Upon returning to Japan in 1974 to the University of Tokyo, I was thinking long and hard about what fundamental problem I could tackle while setting up my own small laboratory. I finally decided to work on the IgH chain genes to address the specific question related to molecular mechanism of Ig isotype or class switching, thus avoiding direct competition for the mechanism of Ab repertoire generation on which Phil, Susumu and many others were working. It was known that antigen stimulation of mature B cells induces isotype class switching, a process that replaces the H chain C (C_H) region gene without changing antigen-binding specificity of the V region gene. The C_H region determines the Ig classes such as IgM,

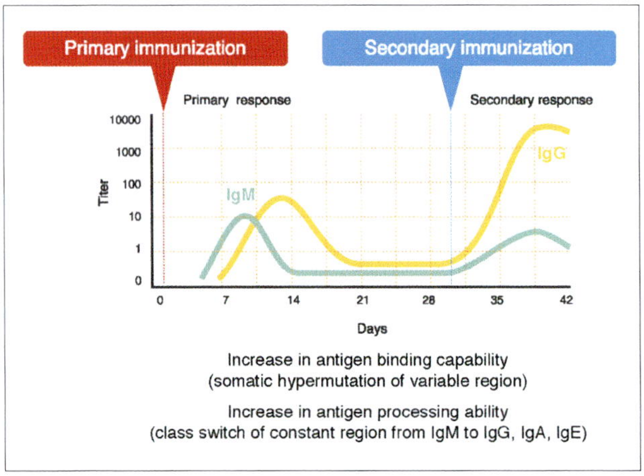

Figure 4. Antibody memory generation by vaccine (antigen) administration.

IgG, IgE and IgA. Another genetic alteration associated with antigenic stimulation of B cells is somatic hypermutation, a process known to facilitate generation of high-affinity Ab. These two phenomena represent the basis of immunological antibody memory and are critical to successful vaccination (Fig. 4).

We were very fortunate to find a collaborator, Masao Ono, who established the method for immunoprecipitating the Ig protein-mRNA complexes in polysomes, from which we could purify H chain messenger RNA rather quickly (Ono *et al.*, 1977). Using the cDNA from this purified mRNA, Tohru Kataoka and I were surprised to find the copy number of the C_H gene differed among DNA obtained from various myelomas producing different classes of Ig. By comparing the copy number of C_γ genes in DNAs of more than 10 myelomas expressing different C_H genes, I realized that a simple deletion between the expressed V_H gene and the C_H gene to be expressed could explain this phenomenon if the C_H gene were aligned in a specific order. We proposed the allelic deletion model of the Ig C_H locus (Honjo *et al.*, 1978).

In 1997, I isolated the first important C_H gene, $C_\gamma1$ from purified $C_\gamma1$ mRNA while on three-month stay in Phil's lab. Once $C_\gamma1$ cDNA was cloned, it did not take long to clone other C_H genes and to convincingly prove that DNA recombination occurs by the isolation of actively transcribed Ig H chain genes (Kataoka *et al.*, 1980). This discovery was made amid incredible competition, as in addition to us, four other groups reached the same conclusion almost simultaneously (Cory, 1981; Davis *et al.*, 1980; Maki *et al.*, 1980; Rabbitts et al., 1981). Finally, four years after we first conceived the allelic deletion, we were able to elucidate how all C_H genes are aligned on the mouse chromosome (Shimizu *et al.*, 1982).

We then proved that the deleted DNA segments are looped out from the chromosome (Iwasato *et al.*, 1990), a useful finding allowing us later to measure ongoing class switch recombination (CSR). Importantly, CSR takes place in the intergenic regions in the C_H locus, characterized by highly repetitive sequences. By studying how the Ig H chain locus is arranged in the mouse chromosome, we saw that each C_H region gene is proceeded by a region (a few to 10 Kb) of highly repetitive DNA which we named the switch (S) region (Kataoka *et al.*, 1981) (Fig. 5). Biologically, CSR must be very efficient because it takes place immediately after antigen stimulation. The structure we revealed fulfills this requirement because CSR can take place anywhere within the large S regions, unlike the site-specificity of V(D)J recombination.

Retrospectively, these studies were milestones in understanding immune diversity, yet I was determined to identify the enzyme(s) responsible for this efficient type of recombination. For this purpose, it was essential to set up a good cellular system to measure class switching *in vitro*. It was fortunate that Eva Severinson proposed collaboration for isolation of as-yet-unknown soluble factor that induces class switching. Using a factor-secreting T cell line that she developed, we isolated cDNA encoding interleukin 4 (IL-4) and IL-5, which turned out to be critical cytokines to not only class switching of B cells but also other important aspects of lymphocyte differentiation (Azuma *et al.*, 1986; Noma *et al.*, 1986).

Another important tool that helped elucidating enzymatic mechanisms of CSR was the B1 cell tumor-derived line CH12, obtained from Warren Strobel at NIH in the early '90s. Although the original cell line had a very high background and low switching efficiency, we repeatedly recloned the CH12 cells and finally obtained the CH12F3 line. This subclone contained

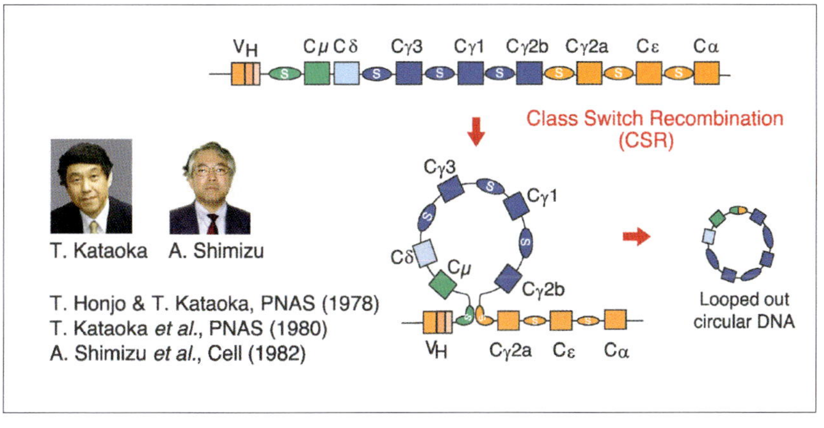

Figure 5. Class switch recombination takes place by deletion of a large DNA segment.

less than 1 percent IgA positive cells under non-stimulated condition but switched to more than 50 percent IgA positive cells after stimulation with IL-4, TGF-β1 and CD40 ligand (Nakamura *et al.*, 1996). The system was used by Masamichi Muramatsu to carry out subtractive hybridization between nonstimulated and stimulated cells, which led to identification of the 23c9 cDNA clone, specifically expressed by activated CH12F3 cells, and also uniquely detected in germinal center B cells but not in other cells. Its sequence revealed the strongest homology with the RNA editing enzyme APOBEC1, and had weak cytidine deaminase activity *in vitro.* We named it Activation-Induced cytidine Deaminase (AID) (Muramatsu *et al.*, 1999). AID-deficient mice completely lost not only class switching but also SHM, a finding which was unexpected as these two genetic alterations were thought to be mediated by different enzymes (Muramatsu *et al.*, 2000) (Fig. 6). AID was thus shown to be the enzyme I had dreamed of finding for so long. Using the human AID primers, we designed, and our collaborators Anne Durandy and Allan Fisher identified, that every one of their Hyper-IgM syndrome type 2 patients had mutations in the AID gene (Revy *et al.*, 2000). The phenotypes of AID deficiency in human and mouse were remarkably similar. Thus, we obtained clear evidence that AID is the enzyme that engraves antigen memory in the antibody genes, which serves as the mechanistic basis of vaccination.

However, the molecular mechanism of AID is still not fully resolved. Although we have accumulated evidence supporting the idea that RNA is directly edited by AID (Begum *et al.*, 2015), most scientists in this field

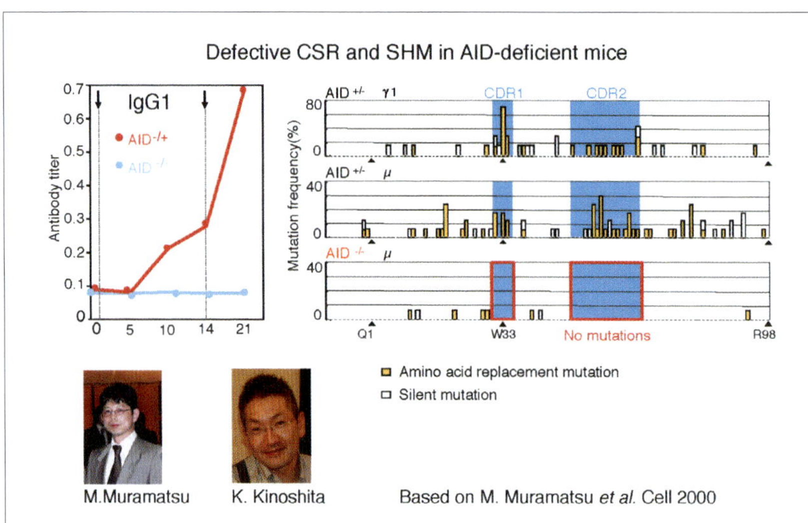

Figure 6. AID-deficient mice lost both CSR and SHM.

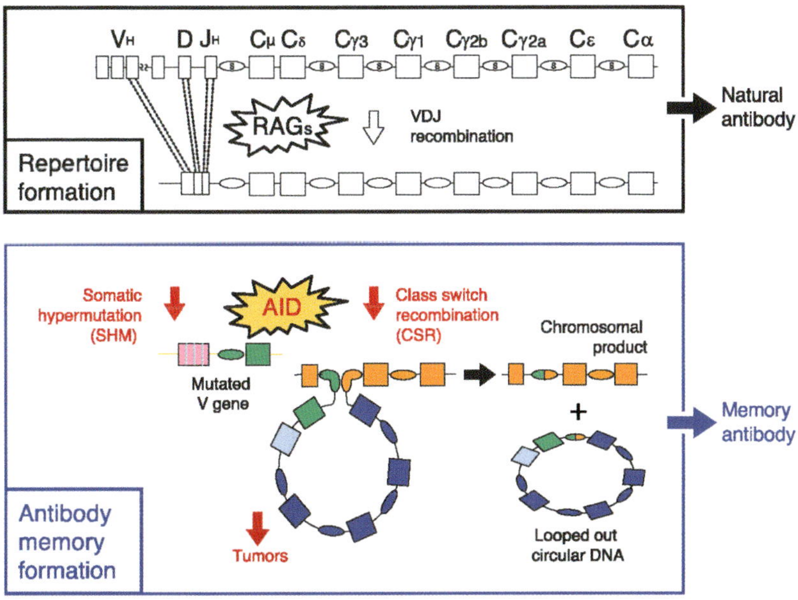

Figure 7. AID engraves Ab memory in the genome for effective vaccination.

believe that AID edits the DNA directly (Meng *et al.*, 2015). Nonetheless, it is now clear that B cell DNA rearrangements occur in two stages. Developing B cells in the bone marrow first generate the V region repertoire by V(D)J recombination. When B cells encounter antigens in the peripheral lymphoid tissues, they are activated to express the AID enzyme that induces SHM and CSR (Fig. 7). Occasionally, however, AID expression can cause aberrant chromosomal translocation between Ig and oncogenes like c-myc, generating leukemias like B cell lymphoma. The idea that immune activation could lead to tumorigenesis was totally unexpected by many immunologists at the time.

DISCOVERY OF PD-1 AND ITS APPLICATION

The great mystery that comes with our immense immune diversity is how to avoid immune activity against oneself. The phenomenon, known as immune tolerance, was theoretically proposed by Sir Frank Macfarlane Burnet and demonstrated by Sir Peter Brian Medawar, for which they shared the 1960 Nobel Prize in Physiology or Medicine. Sir Frank also proposed that tumor cells may express non-self antigens which the immune system would recognize and eliminate by the mechanism called immune surveillance (Burnet, 1957). Tumor cells, however, can grow when they induce the immune system into an inactive state i.e. immune tolerance by reasons unknown even today. Such ideas and observations initiated what would become the field of cancer immunotherapy.

Many scientists tested various approaches to cure cancers; for example, some identified tumor-specific antigens and injected them as vaccines into patients. Others infused patients with their own T cells after expanding the cells in the laboratory with IL-2. Injections with immune-stimulating cytokines such as interferon-γ, IL-2 or IL-12 were also extensively pursued. Unfortunately, these trials did not provide any clear clinical benefits. In retrospect, all these trials were analogous to pressing an accelerator of the immune system under the condition of strong braking, because the immune system of cancer patients was in a state of immune tolerance. Indeed, until the mid-1990s, none of the molecules responsible for this negative immune regulation had been identified.

The new era of the immune regulation came when Pierre Golstein discovered the CTLA-4 molecule and when subsequent studies by Jim Allison, Jeffrey Bluestone, Craig Thomson, Tak Mak and Arlene Sharpe's groups revealed that CTLA-4 is a negative regulator involved in preventing the initial activation of the immune system (Brunet *et al.*, 1987; Krummel *et al.*, 1995; Tivol *et al.*, 1995; Walunas *et al.*, 1994; Waterhouse *et al.*, 1995). Indeed, CTLA-4 regulates the immune activation in an on-off manner in pair with CD28, in a manner equivalent to the parking brake (Fig. 8). Independently, we found another negative regulator called PD-1 that regulates the immune reaction in a rheostatic manner in pair with the accelerator ICOS. PD-1 was fortuitously isolated by Yasumasa Ishida, who had worked intensely on isolating the proteins involved in T cell selection in thymus, in order to answer the major question of how immune cells establish self-tolerance during development. He stimulated T cell lines derived from the thymus followed by cDNA subtraction to compare the

Action	Drive	Stop	Action mode
Parking [Activation]	Ignition [CD28]	Parking brake [CTLA4]	ON/OFF [Drastic]
Driving [Attack]	Accelerator [ICOS]	Brake [PD-1]	~ 100 km/h [Mild]

Figure 8. Brakes and accelerators control immune reactions like those in a car.

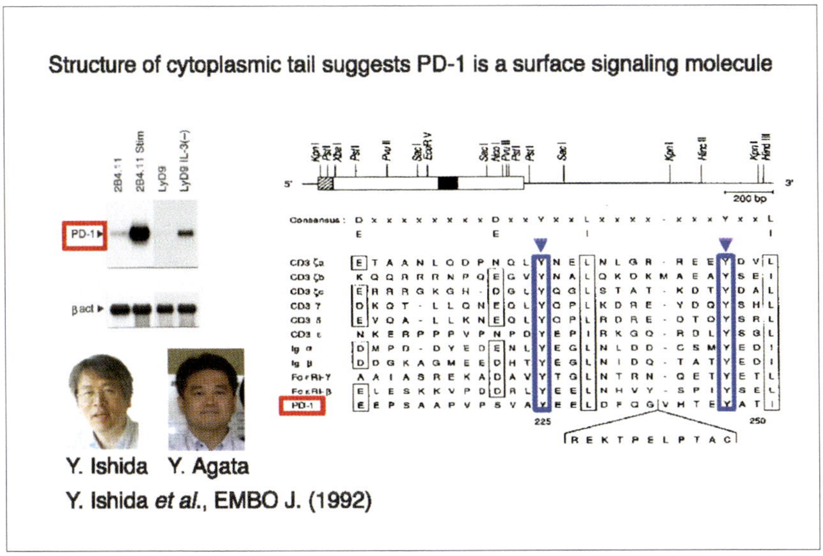

Figure 9. Discovery of PD-1 (programmed death-1) cDNA and its structure.

genes expressed by either growing or dying cells. He isolated four independent clones, which were surprisingly derived from a single mRNA. The cDNA sequence revealed a very interesting molecule we named programmed cell death 1 (PD-1) (Ishida *et al.,* 1992). PD-1 is a cell surface receptor with a cytoplasmic domain containing a pair of tyrosines, which is conserved among then known activation-signal transducing receptors. However, PD-1 is distinct from the rest by the double length between the two tyrosine residues (Fig. 9). We subsequently found that PD-1 expression is strongly induced in T cells and B cells upon stimulation (Agata *et al.*, 1996). We easily excluded the possibility that PD-1 is involved in cell death by the absence of PD-1 expression during the normal cell apoptosis induced by dexamethazone.

Although Hiroyuki Nishimura generated the PD-1 knockout (KO) mice in 1994, it took us a long time to understand the function of PD-1 *in vivo*, as deleting the gene in mice of a mixed genetic background seemed to confer no ill effects. The first hint of the negative regulatory function of PD-1 was obtained when he found that B cells from PD-1 deficient mice had stronger responses to antigen stimulation *in vitro* (Nishimura *et al*, 1998). When crossing the PD-1 KO mice with autoimmune-prone *lpr/lpr* mice, PD-1 deficiency accelerated the development of severe nephritis and arthritis around five months after birth. Eventually, we also found that when using the common laboratory C57BL/6 strain, PD-1 KO mice also developed autoimmune nephritis and arthritis but only by around 14 months (Nishimura *et al.*, 1999). Surprisingly, on the BALB/c background, the PD-1 deficiency induced dilated cardiomyopathy, suggesting that while PD-1 is important for fending off the initiation of autoimmune disease, the specific organs targeted might be affected by other genes (Nishimura *et al.*, 2001). To confirm this idea, we examined consequences of the lack of PD-1 inhibitory function by crossing PD-1 KO mice with other autoimmune-prone mice commonly used by immunologists. Taku Okazaki found that PD-1 deficiency on the NOD background causes exacerbation of diabetes, while on the MRL background the absence of PD-1 causes severe myocarditis (Wang *et al.*, 2005; Wang *et al.*, 2010) (Fig. 10). Regarding the signaling characteristics of PD-1, Taku was able to elucidate the molecular mechanism for its immune brake function. He showed that the engagement of PD-1 induces phosphorylation of the conserved tyrosine residues, which recruits SHP2 phosphatase and dephosphorylates signaling molecules phosphorylated by antigen receptor engagement, resulting in down-modulation of activation signals (Okazaki *et al.*, 2001). All the above results provided strong evidence that PD-1 is the second new negative regulator of the immune system. However, in contrast to CTLA-4 deficiency, which develops severe autoimmunity leading to death by four to five weeks after birth, PD-1 deficiency results in autoimmunity developing more slowly and with milder phenotypes. Such results

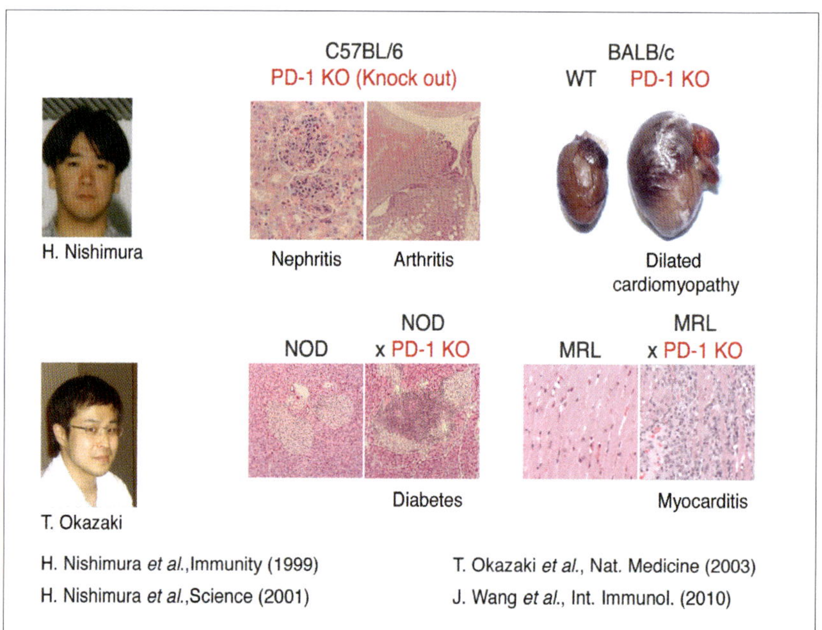

Figure 10. PD-1 is a negative regulator.

convinced us that PD-1 modulation might be a very good approach for the treatment of various immune-related diseases, including cancer.

Meantime, I wanted to isolate the ligand for PD-1 and collaborated with Steve Clark at the Genetic Institute (GI) in Boston. Unexpectedly, Gordon Freeman from the Dana-Farber Cancer Institute chose many cDNAs encoding B7 family member proteins from a database list, and one of the clones, 292, was shown to bind our PD-1-expressing cell line by the GI group. We confirmed their results and further demonstrated that 292 protein binding to the PD-1 receptor inhibits T cell proliferation. The identification of PD-1 ligand, which we named PD-L1, was reported in 2000 (Freeman *et al.* 2000).

Before the dawn of the new millennium, we started to test the effect of PD-1 deficiency on tumor cell growth. Yoshiko Iwai performed experiments comparing the growth of myelomas expressing PD-L1 between mice with or without PD-1 by the middle of 2000. She found that PD-1 deficiency suppressed the growth of PD-L1 expressing tumor cells (Fig. 11). The results convinced me that blocking PD-1 with an antibody would be effective to suppress the tumor growth. We urgently needed a PD-1 blocking Ab to confirm that this strategy could be used for cancer therapy. Our long-term collaborator, Nagahiro Minato, and his colleagues had already begun generating good Ab against PD-1 and PD-L1. Our laboratories embarked on an exciting and fruitful collaborative effort, and Yoshimasa Tanaka from Minato's laboratory showed that the PD-L1 Ab

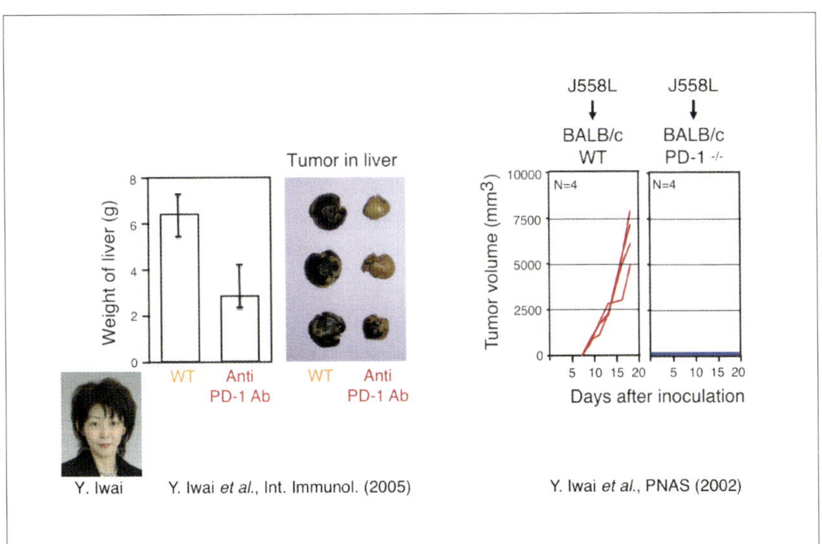

Figure 11. Suppression of B16 melanoma metastasis by PD-1 Ab and inhibition of tumorigenesis of myeloma (J558L) in PD-1-/- mice.

they had generated could efficiently suppress tumor growth in a variety of mouse models (Iwai *et al.*, 2002) (Fig. 12). Yoshiko further confirmed that PD-1 blockade by either genetic manipulation or PD-1 Ab treatment clearly suppressed the metastasis of B16 melanoma cells from spleen to liver (Iwai *et al.*, 2005) (Fig. 11). These findings showed that PD-1 blockade by Ab against either PD-1 or PD-L1 is a powerful tool for treatment of cancers, preventing and delaying the growth and spread of tumor cells (Fig. 13).

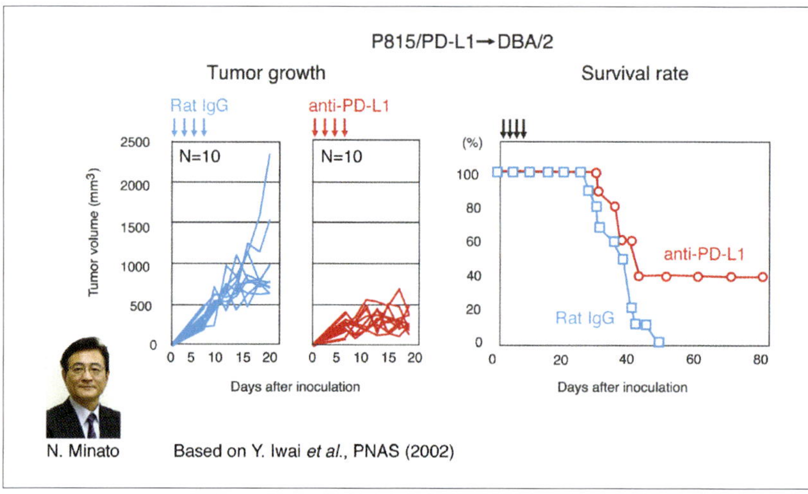

Figure 12. Growth inhibition of PD-L1 expressing mastocytoma (P815/PD-L1) and extension of animal survival by PD-L1 Ab.

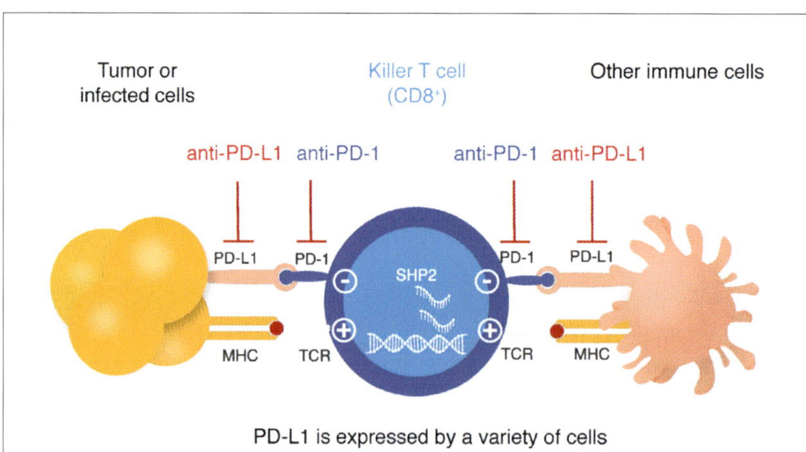

Figure 13. Schematic representation of tumor killing using Ab against either PD-1 or PD-L1.

CANCER IMMUNOTHERAPY BY PD-1 BLOCKADE

In order to apply this treatment to humans, we needed to file a patent application and obtain help from the pharmaceutical industry. After much struggle and many illusions and disillusions – which I describe in my biography – Medarex, a small venture capital already working with Jim Allison on a CTLA-4 monoclonal Ab, began to generate blocking PD-1 Abs in collaboration with Ono Pharmaceutical Co., Ltd. The human anti-PD-1 Ab named Nivolumab was an IgG4 with a mutation at S228P to reduce Ab-dependent cell-mediated cytotoxicity and approved as an investigation new drug by the FDA on August 1, 2006. Clinical trials were initiated immediately in the United States and two years later in Japan. The first Phase I clinical trial was comprehensively summarized by Suzanne Topalian and her colleagues (Topalian *et al.*, 2012). The results of this study surprised the scientific immunology community and many clinicians, because about 20 to 30 percent of terminal-stage patients with very aggressive forms of cancers such as non-small cell lung cancer, melanoma or renal cancer showed complete or partial responses. Perhaps even more surprising was the finding that 20 out of 31 responders of melanoma patients continued to respond for more than one year, some even after stopping the treatment (Fig. 14). Such durable responses had not been experienced with any other chemotherapy regimens. We also started

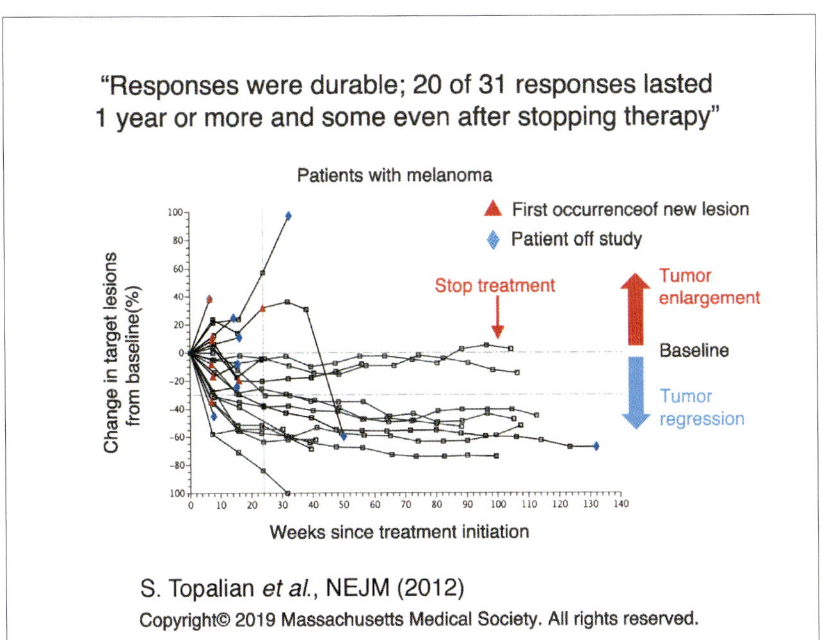

Figure 14. Durable response to PD-1 blockade.

a Phase II trial on platinum-resistant ovarian cancer patients in collaboration with the Gynecology Department of Kyoto University Hospital. Although the study was rather small, the results were promising. We were astonished to find that in about 40 percent of patients, responded or completely stopped the tumor growth (Hamanishi *et al.*, 2015) (Fig. 15). Two patients who showed a complete response after one year of PD-1 blockade are still healthy almost five years after stopping the treatment (Fig. 16).

Dose	total (n)	CR	PR	SD	PD	NE	RR	DCR
1 mg/kg	10	0	1	4	4	1	1/10 (10%)	5/10 (50%)
3 mg/kg	10	2	0	2	6	0	2/10 (20%)	4/10 (40%)
Total	20	2	1	6	10	1	3/20 (15%)	9/20 (45%)

Oct. 21, 2011-Dec.7, 2014

J. Hamanishi

Tumor growth stopped in 40-50% of terminal stage patients

J. Hamanishi *et al.*, J. Clin. Oncol. 33 (2015)

Figure 15. Phase II trial of anti-PD-1 Ab in patients with platinum-resistant ovarian cancer.

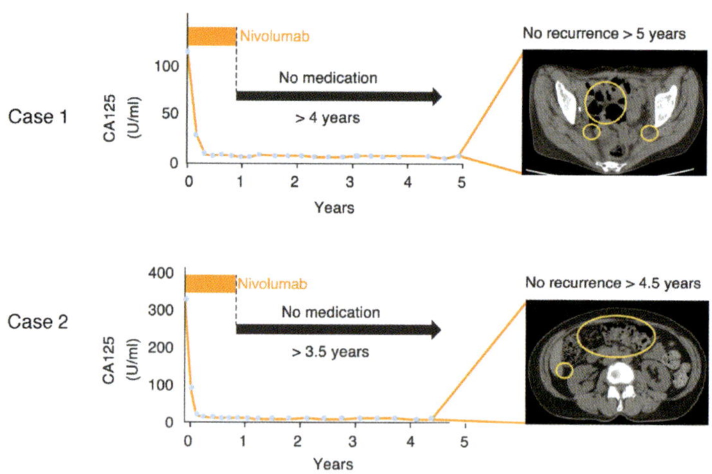

J. Hamanishi. International Federation of Gynecology and Obstetrics FIGO (2018)

Based on: J. Hamanishi *et al.*, J. Clin. Oncol. 33 (2015)

Figure 16. Durable complete responses of ovarian cancer patients to Nivolumab.

Numerous clinical trials carried out around the world during the last six years were recently summarized (Gong *et al.*, 2018). I was most impressed by a randomized study on 418 melanoma patients, on whom the effect of either Nivolumab or chemotherapy (Dacarbazine) was compared. The outcome was quite impressive. Almost 70 percent of the Nivolumab group survived after a year and a half, compared with less than 20 percent of the Dacarbazine group. At this point the study was stopped as it was deemed unethical to continue Dacarbazine treatment in the face of such clear clinical benefits from Nivolumab (Robert *et al.*, 2015) (Fig. 17). This large-scale study was carried out by an international team of clinicians in the European Union, Canada and Australia. Another striking clinical study has been reported for Hodgkin lymphoma cases. Without exception, all of the 23 Hodgkin lymphoma patients with various treatment histories showed either complete responses (four cases), partial responses (16 cases) or stable diseases (three cases) (Ansell *et al.*, 2015). Thanks to the enormous effort of scientists, clinical doctors, patients, bioinformaticians and other researchers in these studies and many more, by 2018 more than a dozen tumor types have been approved for the treatment by PD-1 blockade therapy. Most recently, the FDA approved this treatment to all microsatellite instability-high (MSI-H) cancers, regardless of the origin of their tumors. I view this as a very wise decision by the regulator, as it lifts the financial burden from drug companies to run expensive clinical trials for the many potential tumor types that might benefit from PD-1 treatment, greatly speeding up access to this new drug for many additional patients. Clearly, PD-1 blockade treatment has brought a paradigm shift in cancer therapy due to its efficacy over a wide range of tumors; durability of its effects in those who respond; and importantly, there are few adverse effects compared with previous treatments.

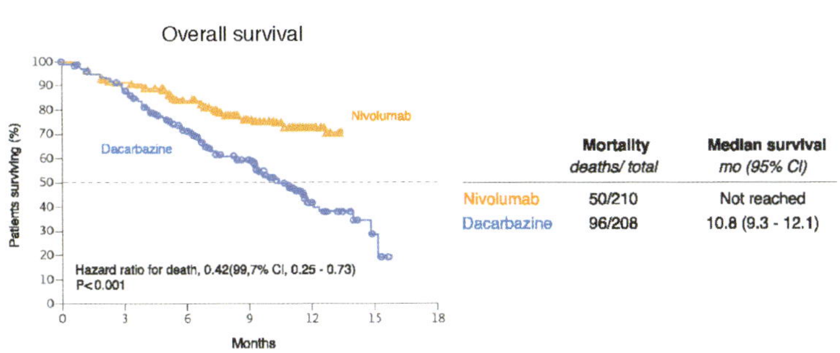

Based on C. Robert *et al.*, NEJM (2015)

Figure 17. Randomized study on untreated melanoma patients with Nivolumab and Dacarbazine (chemotherapy).

But how can we explain the success of checkpoint immunotherapy such as PD-1 Ab treatment? Extensive DNA sequencing of a variety of tumors revealed enormous accumulation of mutations in the DNA of almost all cancer cells. Mutation frequencies in coding regions can reach the levels of one thousand to 10 thousand-fold higher than those of normal cells, which generally have few, if any, mutations (Alexandrov *et al.*, 2013; Martincorena *et al.*, 2015). Melanomas have the highest mutation frequency, followed by lung cancers. In fact, almost all types of tumors have 100 times more mutations compared with normal cells. Summarizing the results of many studies as follows: Firstly, cells become cancerous by accumulating a large number of mutations, changing their appearance from normal cells. These changes can be detected by our immune system's ability to home in on 'foreignness', meaning cancer immunotherapy is potentially effective for all types of cancers; secondly, this large increase in mutations also makes it difficult to identify which among them can be targeted by chemotherapy drugs; thirdly, frequent mutations continue to generate tumor cells that are resistant to chemotherapy, whereas the T cells with enormous immune repertoire can recognize and attack not only original but also mutated tumor cells.

Nevertheless, there are many hurdles ahead before declaring immunotherapy's complete victory over cancer. There are several obvious issues emerging from the clinical trials. First of all, the efficacy of this treatment is still limited. For example, for melanoma, the success rate of PD-1 blockade therapy as the first line of treatment reached about 70 percent at best (Robert *et al.*, 2015). In most other cases, especially when used as the second or the third line of treatment, the efficacy rate was 20–30 percent. Thus, we must urgently address two problems. One is how we can predictively distinguish responders from non-responders to PD-1 blockade. The other is how we can significantly improve the efficacy of the PD-1 immunotherapy.

Predictive markers could be related to tumor characteristics. For example, a high mutation rate of tumor cells is one good indicator for responsiveness to PD-1 blockade therapy as mentioned above. PD-L1 expression by tumors was also considered but unfortunately was not efficiently predictive of responder status (Colwell, 2015). The other factors influencing the response rate of cancer immunotherapy are related to the individual's immune capacity, which is at present very difficult to predict. Since there are hundreds of genes involved in the immune responses, and multiple layers of regulation, it is not easy to predict an individual's immune power, for example, by looking at their genetic polymorphisms. Environmental factors, such as the host microbial communities in the gut, also seem likely to play an important role in modulating immune responses and perhaps tumor features as well. Thus, in the near future, we will strive to understand the best combinatorial factors predictive for a positive outcome of immunotherapy.

Needless to say, improving the immunotherapy response rate would help a great many additional cancer patients, and we are working towards this in two different ways in our current research. Firstly, we are testing ways to increase the number of cytotoxic T cells (CTL) that infiltrate and kill tumors. We recently showed that the first important steps in CTL activation and proliferation occur in tumor-draining lymph nodes. Once they are activated and decorated with responsive receptors, CTLs migrate to the tumor site in response to a gradient of chemokines secreted from the tumor site (Chamoto *et al.*, 2017). We showed that tumor growth could no longer be suppressed by PD-L1 Ab when draining lymph nodes were surgically removed in tumor-bearing mice that were otherwise responsive to PD-1 blockade. This is because activation of CTL in draining lymph nodes is prerequisite for upregulation of CXCR3, a receptor responding to several chemokines, such as CXCL9, CXCL10 and CXCL11. Interestingly, CXCL9 is actually produced by tumor tissues in response to interferon-γ secreted by pre-existing tumor, infiltrating CTLs upon stimulation with PD-1 blocking Ab. Thus, lymph nodes appear to be critical to initiate the activation and to deploy the CTLs to the tumor sites, in turn further facilitating the recruitment of more CTLs from lymph nodes in what appears to be a positive feedback loop. These results suggest that surgeons should avoid removing healthy lymph nodes during dissection of tumors in order to safeguard protective immune responses.

Another approach to strengthening PD-1 blockade which is being pursued by many scientists and pharmaceutical companies around the world is combination therapy. Amazingly, in the United States alone, more than one thousand clinical trials testing PD-1 blockade in combination therapy are currently ongoing (Tang *et al.*, 2018). The most commonly used combination partners for PD-1 blockade are CTLA-4 Ab, followed by chemotherapy, radiotherapy, anti-VEGF agents, and chemoradiotherapy. It is important to remember that neither chemo- nor radiotherapy could show appropriate effects on tumor growth if the immune system is defective in animal models (Apetoh *et al.*, 2007; Takeshima *et al.*, 2010).

Our own laboratory took a totally different approach. We and others found that T cell activation is accompanied by a burst of activity by the cells' mitochondria, the energy-producing organ of the cell (Chamoto *et al.*, 2017; Fox et al., 2005). This energy is probably needed to support the cell division that quickly follows antigenic stimulation. CTLs divide almost every 8 hours, requiring a large amount of building blocks and energy to produce the proteins and nucleic acids required to build a daughter cell several times per day. We therefore searched for chemicals that can boost mitochondrial activity in active T cells. PGC-1α is a key transcriptional cofactor boosting fatty-acid oxidation and oxidative phosphorylation, which are crucial to mitochondrial activation (Chamoto *et al.*, 2017; Ventura-Clapier *et al.*, 2008). In my lab, Kenji Chamoto screened for chemicals which influenced PGC1-α activity in the context of PD-1 blockade therapy. He found that Bezafibrate, already approved as a drug for hyperlipidemia, synergizes with anti-PD-L1 Ab for suppression of tumor growth (Chamoto *et al.*, 2017) (Fig. 18). Subsequently, we found that Bezafibrate has two important activities for increasing the number of tumor-specific CTLs at the tumor site (Chowdhury *et al.*, 2018). By activating PGC-1α and PPAR signaling, Bezafibrate facilitates rapid proliferation of effector CTLs. Also, enhanced PPAR signaling by Bezafibrate appears to prevent cell death, and to facilitate long-term survival of the activated CTLs. We are now carrying out the Phase I clinical trial in Japan, combining PD-1 blockade with Bezafibrate. I hope the large number of trials currently underway will soon identify many other strategies to improve the efficacy of PD-1 blockade.

FUTURE PROSPECTS FOR PD-1 BLOCKADE CANCER THERAPY

In parallel with improvement strategies for successful therapy, I see the need to push forward the basic fundamental understanding of the whole body. In other words, we need to invest more effort into understanding physiology, a challenging task as I have already pointed out. We found that PD-1 deficiency manifests different types of organ-specific autoimmunity in different laboratory mouse strains. As expected, some of the

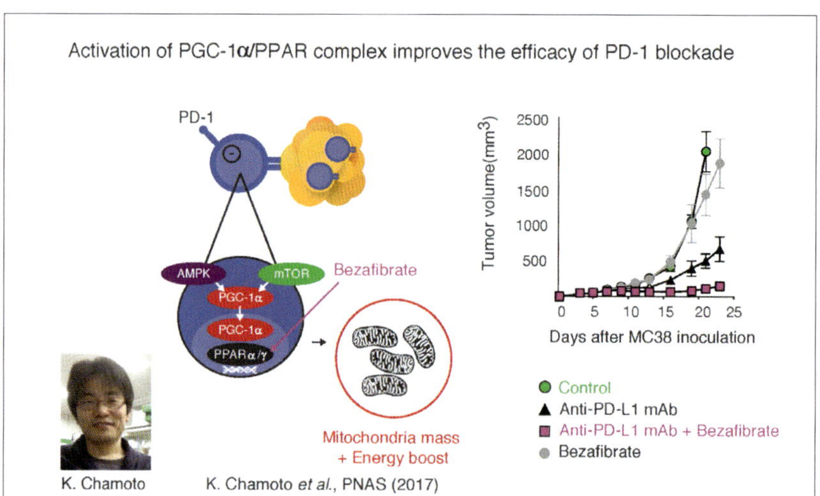

Figure 18. Improvement of PD-1 blockade therapy by Bezafibrate.

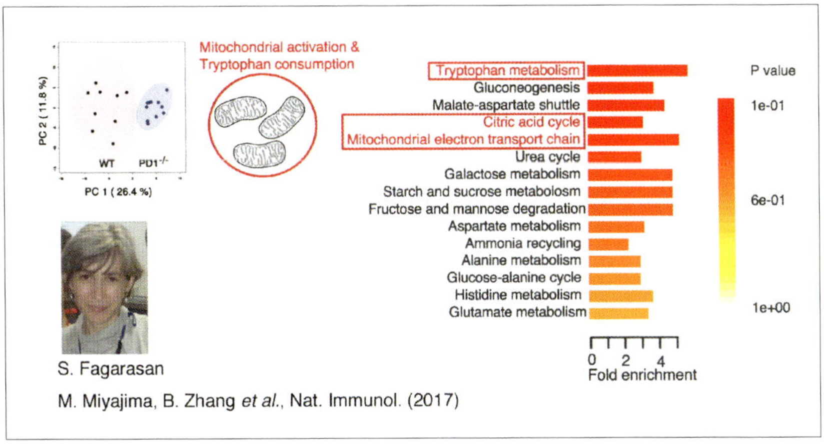

Figure 19. Hyperimmune activity can be read in the blood biochemistry of PD-1 KO mice.

patients undergoing PD-1 blockade therapy develop similar autoimmune features. Some adverse effects of PD-1 blockade are partly due to the fact that T cells proliferate vigorously after stimulation to generate functional progeny, requiring a huge amount of energy and metabolites to build new nucleic acids and proteins. Indeed, Sidonia Fagarasan's group recently showed that activation of the immune system limits the systemic availability of amino acids, hindering the synthesis of neurotransmitters in the brain (Miyajima *et al.*, 2017) (Fig. 19). Serum depletion of tryptophan and tyrosine, for example, reduces the amounts of serotonin and dopamine in the brain and causes behavioral changes like enhanced fear responses and anxiety. Such systemic biochemical changes were observed in PD-1 deficient mice as well as in normal mice under strong antigenic stimulation.

Furthermore, the harmful hyperimmune activation in PD-1 deficient mice is caused not only by the absence of this inhibitory receptor but also by dysregulated microbial communities in the gut (Kawamoto *et al.*, 2012). Indeed, Sidonia's group showed that in addition to AID, PD-1 is an essential molecule required to generate and select proper IgAs responsible for maintenance of a healthy microbial ecosystem in the gut (Fig. 20). The absence of PD-1 causes a skewing of balance between T and B cells in the germinal centers, with expansion of germinal center T cells perturbing the normal process of selection and maturation of IgA repertoire needed to maintain gut symbiosis.

Sidonia's amazing research in this area extended our understanding of PD-1 functions and grew from her own findings on hyperimmune activation caused by microbiota dysbiosis in AID-deficient mice (Fagarasan *et al.*, 2002; Suzuki *et al.*, 2004). She is the first person who showed that skewed microbiota in the gut can activate the immune system throughout

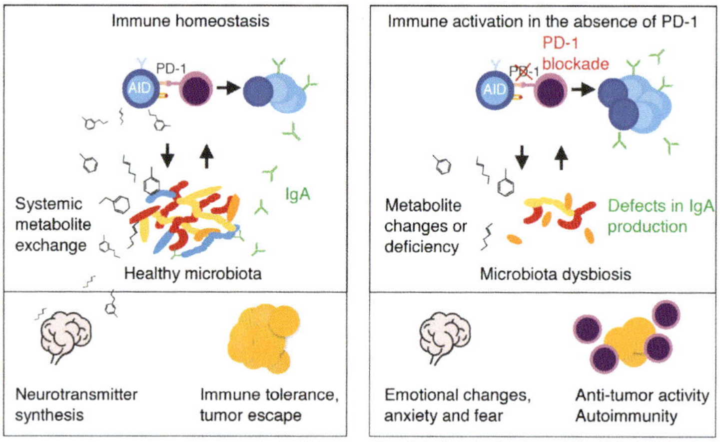

Figure 20. AID and PD-1 cooperate in germinal centers for IgA selection to maintain microbiome and body homeostasis.

the body in the AID-deficiency mouse model. Most strikingly, in the absence of IgA, expansion of certain species of anaerobic bacteria caused enlarged nodular structures along the whole intestinal tract, as well as hypertrophia of all lymphoid tissues. How fortunate I am to have discovered the two critical molecules, AID and PD-1, which by a certain amount of chance became the focus of my research, and collaborate in such a fascinating way to control not only the immune system but also the ecosystem of the gut which is so important for whole-body balance.

Based on these studies, it was perhaps not surprising to find that PD-1 blockade therapy can be influenced by the bacterial communities in the gut (Sivan *et al.*, 2015). We also put into test this by evaluating the anti-tumor activity of AID-deficient mice and found that tumor growth is dramatically inhibited in AID-deficient mice even without PD-1 Ab treatment (M. Akrami and R. Menzies, unpublished data). Sidonia's group further confirmed that the hyperimmune activity against tumor growth in AID-deficient mice depends on microbiota, because AID-deficient mice raised in germ-free conditions lost strong anti-tumor activity (M. Miyajima and S. Fagarasan, unpublished data). Thus, PD-1 is critical for regulation of CTLs against tumors but AID as well as PD-1 is important for boosting the immune potential against tumors through microbiota regulation, strongly indicating that gut microbiota is critical to the anti-tumor activity of individual immunity (Fig. 20).

In summary, I feel that despite enormous recent progress, our knowledge of the immune system and its interactions with other systems is still very limited. We are just at the beginning of an exploration that will lead us to a better understanding of how our bodies function. Even a single molecule, PD-1, has so many unanticipated effects on aspects of health and disease, from tipping the balance of the immune system towards autoimmunity, to regulating bacterial communities, to major changes in the biochemistry of the immune, circulatory and nervous systems. Achieving successful immunotherapy implies learning more not only about improving cancer outcomes, but also understanding all the effects of the treatment on the rest of the body.

In 2016 Andy Coghlan published an article entitled "Closing in on cancer" in *New Scientist magazine.* Regarding the PD-1 blockade therapy, he quoted Dan Chen from Genentech saying "We are at the point where we have discovered the cancer equivalent to penicillin." They considered PD-1 blockade therapy to be comparable to the discovery of penicillin for infectious diseases. "Although penicillin itself couldn't cure all infections, it gave rise to a whole generation of antibiotics that changed medicine forever, consigning most previously fatal infections to history." (Coghlan, 2016). I really hope this dream about the future of cancer and immunotherapy will come true. At this stage in 2018, however, first line treatment by PD-1 blockade or in combination with CTLA-4 blockade is limited to melanoma, lung cancer and renal cancer only. In the near future, I hope many more types of cancer may be treated by immunotherapy as the first-line therapy. I hope that by fine-tuning a therapeutic balance between tumor growth and immune surveillance using PD-1 blockade, this devastating disease may become instead of a death sentence, a chronic and manageable condition. It might sometimes be preferable to merely control or slow tumor growth, especially in more fragile, elderly patients, making cancer a disease we live with, rather than die from. Looking back, vaccination and antibiotic discoveries were two great leaps in human healthcare which greatly reduced infectious diseases. WHO has even declared the human smallpox virus to be completely eradicated. I truly hope that within this century the next leap forward will be controlling cancer by immunotherapy and its improved protocols, which might include microbiome manipulation. We are still at a very early stage in our understanding of the microbiome, metabolome, and all the other possible factors affecting immunity and physiology, including the aging process. However, as we have seen very rapid development of cancer immunotherapy during last 10 years, I believe the future of cancer treatment is promising and really hope that the 21st century will be remembered as the century in which cancer was conquered.

Neither infectious disease control nor cancer immunotherapy would have been possible without acquired immunity, which developed at a very early stage of the vertebrate evolution. Acquired immunity originated in jawless fish, in which AID orthologues with a unique enzymatic ability were responsible for generating the enormously diverse molecules for

Figure 21. Current colleagues and long-term collaborators.

both the T and B lineage equivalents (Guo *et al.*, 2009). Later, in jawed fish, a transposon containing ancestral RAG-1 and RAG-2 'jumped' into a primordial immunoglobulin-like gene. The transposon insertion generated a split gene, an ancestor of the V(D)J segments as well as recombinase enzymes. RAGs replaced the function of AID in the repertoire generation, while AID kept its function for antigen-induced antibody memory including SHM and CSR. All these drastic genetic mechanisms were selected during evolution for the advantage they conferred in the fight against infectious pathogens. As a result, humans now benefit from the sophisticated acquired immune system in the fight against cancer. Cancer cells accumulate mutations and develop into non-self-cells expressing neoantigens that are easily recognized by the acquired immune system. This is obviously an extra bonus from acquired immunity that evolution did not anticipate. In that sense, we humans are very lucky to enjoy this unexpected gift from evolution.

ACKNOWLEDGEMENTS

I would like to thank former collaborators, current lab members and many others in the laboratories of Fumihiko Matsuda and Sidonia Fagarasan who contributed to all these studies (Fig. 21). I thank Sidonia Fagarasan, Alexis Vogelzang, Blake Thompson, Maki Kobayashi and Kenji Chamoto for their help with the design and editing for the Nobel presentation and lecture. I also have an enormous number of outside collaborators, the following of whom collaborated with us on antibody diversity and cancer immunotherapy: E. Severinson [Stockholm Univ.], F. Alt [Harvard Univ.], M. Nussenzweig [Rockefeller Univ.], A. Fischer [Necker Hospital], A. Durandy [Necker Hospital], T. Chiba [Kyoto Univ. Hospital], G. Freeman [Dana Farber Cancer Center], N. Minato [Kyoto Univ.], S. Fujii [Kyoto Univ.], I. Konishi [Kyoto Univ.]. I am very much obliged to my devoted supporting staffs, without whom my achievements would have been impossible. I would also like to acknowledge the many funding agencies that supported my research for many years. Last but not least, I would like to thank my family for their generous support.

REFERENCES

Agata, Y., Kawasaki, A., Nishimura, H., Ishida, Y., Tsubata, T., Yagita, H., and Honjo, T. (1996). Expression of the PD-1 antigen on the surface of stimulated mouse T and B lymphocytes. *International Immunology* **8**, 765–772.

Alexandrov, L. B., Nik-Zainal, S., Wedge, D. C., Aparicio, S. A., Behjati, S., Biankin, A. V., Bignell, G. R., Bolli, N., Borg, A., Borresen-Dale, A. L., et al. (2013). Signatures of mutational processes in human cancer. *Nature* **500**, 415–421.

Ansell, S. M., Lesokhin, A. M., Borrello, I., Halwani, A., Scott, E. C., Gutierrez, M., Schuster, S. J., Millenson, M. M., Cattry, D., Freeman, G. J., et al. (2015). PD-1 blockade with nivolumab in relapsed or refractory Hodgkin's lymphoma. *The New England Journal of Medicine* **372**, 311–319.

Apetoh, L., Ghiringhelli, F., Tesniere, A., Obeid, M., Ortiz, C., Criollo, A., Mignot, G., Maiuri, M. C., Ullrich, E., Saulnier, P., et al. (2007). Toll-like receptor 4-dependent contribution of the immune system to anticancer chemotherapy and radiotherapy. *Nature Medicine* **13**, 1050–1059.

Azuma, C., Tanabe, T., Konishi, M., Kinashi, T., Noma, T., Matsuda, F., Yaoita, Y., Takatsu, K., Hammarström, L., Smith, C. I. E., et al. (1986). Cloning of cDNA for human T-cell replacing factor (interleukin-5) and comparison with the murine homologue. *Nucleic Acids Research* **14**, 9149–9158.

Begum, N. A., Nagaoka, H., Kobayashi, M., and Honjo, T. (2015). Chapter 18 – Molecular Mechanisms of AID Function. In *Molecular Biology of B Cells* (Second Edition), F.W. Alt, T. Honjo, A. Radbruch, and M. Reth, eds. (London: Academic Press), pp. 305–344.

Brack, C., Hirama, M., Lenhard-Schuller, R., and Tonegawa, S. (1978). A complete immunoglobulin gene is created by somatic recombination. *Cell* **15**, 1–14.

Brown, D. D., and Weber, C. S. (1968). Gene linkage by RNA-DNA hybridization: I. Unique DNA sequences homologous to 4 s RNA, 5 s RNA and ribosomal RNA. *Journal of Molecular Biology* **34**, 661–680.

Brunet, J.-F., Denizot, F., Kuciani, M. F., Roux-Dosseto, M., Suzan, M., Mattei, M.-G., and Golstein, P. (1987). A new member of the immunoglobulin superfamily – CTLA-4. *Nature* **328**, 267–270.

Burnet, M. (1957). Cancer: a biological approach. III. Viruses associated with neoplastic conditions. IV. Practical applications. *British Medical Journal* **1**, 841–847.

Chamoto, K., Chowdhury, P. S., Kumar, A., Sonomura, K., Matsuda, F., Fagarasan,

S., and Honjo, T. (2017). Mitochondrial activation chemicals synergize with surface receptor PD-1 blockade for T cell-dependent antitumor activity. *Proceedings of the National Academy of Sciences of the United States of America* **114**, E761–E770.

Chowdhury, P. S., Chamoto, K., and Honjo, T. (2018). Combination therapy strategies for improving PD-1 blockade efficacy: a new era in cancer immunotherapy. *Journal of Internal Medicine* **283**, 110–120.

Coghlan, A. (2016). Closing in on Cancer. New Scientist 229, 34–37.

Colwell, J. (2015). Is PD-L1 Expression a Biomarker of Response? *Cancer Discovery* **5**, 1232.

Cory, S., Webb, E., Gough, J., Adams, J.M. (1981). Recombination events near the immunoglobulin Cmu gene join variable and constant region genes, switch heavy-chain expression, or inactivate the locus. *Biochemistry* **20**, 2662–2671.

Davis, M. M., Kim, S. K., and Hood, L. (1980). Immunoglobulin class switching: Developmentally regulated DNA rearrangements during differentiation. *Cell* **22**, 1–2.

Edelman, G. M. (1959). Dissociation of γ-globulin. *Journal of the American Chemical Society* **81**, 3155–3156.

Fagarasan, S., Muramatsu, M., Suzuki, K., Nagaoka, H., Hiai, H., and Honjo, T. (2002). Critical roles of activation-induced cytidine deaminase in the homeostasis of gut flora. *Science* **298**, 1424–1427.

Fox, C. J., Hammerman, P. S., and Thompson, C. B. (2005). Fuel feeds function: energy metabolism and the T-cell response. *Nature Reviews. Immunology* **5**, 844–852.

Freeman, G. J., Long, A. J., Iwai, Y., Bourque, K., Chernova, T., Nishimura, H., Fitz, L. J., Malenkovich, N., Okazaki, T., Byrne, M. C., Horton, H. F., Fouser, L., Carter, L., Ling, V., Bowman, M. R., Carreno, B. M., Collins, M., Wood, C. R. and Honjo, T. (2000). Engagement of the PD-1 immunoinhibitory receptor by a novel B7 family member leads to negative regulation of lymphocyte activation. *Journal of Experimental Medicine* **192**, 1027–1034.

Gong, J., Chehrazi-Raffle, A., Reddi, S., and Salgia, R. (2018). Development of PD-1 and PD-L1 inhibitors as a form of cancer immunotherapy: a comprehensive review of registration trials and future considerations. *Journal for Immunotherapy of Cancer* **6**, 8.

Guo, P., Hirano, M., Herrin, B.R., Li, J., Yu, C., Sadlonova, A., Cooper, M.D. (2009) Dual nature of the adaptive immune system in lampreys. *Nature* **459**, 796–801.

Hamanishi, J., Mandai, M., Ikeda, T., Minami, M., Kawaguchi, A., Murayama, T., Kanai, M., Mori, Y., Matsumoto, S., Chikuma, S., et al. (2015). Safety and Antitumor Activity of Anti-PD-1 Antibody, Nivolumab, in Patients with Platinum-Resistant Ovarian Cancer. Journal of Clinical Oncology: official journal of the American Society of Clinical Oncology 33, 4015–4022.

Hilschmann, N., and Craig, L. C. (1965). Amino acid sequence studies with Bence-Jones proteins. *Proceedings of the National Academy of Sciences of the United States of America* **53**, 1403–1409.

Honjo, T., Packman, S., Swan, D., Nau, M., and Leder, P. (1974). Organization of immunoglobulin genes: reiteration frequency of the mouse kappa chain constant region gene. *Proceedings of the National Academy of Sciences of the United States of America* **71**, 3659–3663.

Honjo, T., and Kataoka, T. (1978). Organization of immunoglobulin heavy chain genes and allelic deletion model. *Proceedings of the National Academy of Sciences of the United States of America* **75**, 2140–2144.

Ishida, Y., Agata, Y., Shibahara, K., and Honjo, T. (1992). Induced expression of PD-1, a novel member of the immunoglobulin gene superfamily, upon programmed cell death. *The EMBO Journal* **11**, 3887–3895.
Iwai, Y., Ishida, M., Tanaka, Y., Okazaki, T., Honjo, T., and Minato, N. (2002). Involvement of PD-L1 on tumor cells in the escape from host immune system and tumor immunotherapy by PD-L1 blockade. *Proceedings of the National Academy of Sciences of the United States of America* **99**, 12293–12297.
Iwai, Y., Terawaki, S., and Honjo, T. (2005). PD-1 blockade inhibits hematogenous spread of poorly immunogenic tumor cells by enhanced recruitment of effector T cells. *International Immunology* **17**, 133–144.
Iwasato, T., Shimizu, A., Honjo, T., and Yamagishi, H. (1990). Circular DNA is excised by immunoglobulin class switch recombination. *Cell* **62**, 143–149.
Kataoka, T., Kawakami, T., Takahashi, N., and Honjo, T. (1980). Rearrangement of immunoglobulin gamma 1-chain gene and mechanism for heavy-chain class switch. *Proceedings of the National Academy of Sciences of the United States of America* **77**, 919–923.
Kataoka, T., Miyata, T., and Honjo, T. (1981). Repetitive sequences in class-switch recombination regions of immunoglobulin heavy chain genes. *Cell* **23**, 357–368.
Kawamoto, S., Tran, T. H., Maruya, M., Suzuki, K., Doi, Y., Tsutsui, Y., Kato, L. M., and Fagarasan, S. (2012). The inhibitory receptor PD-1 regulates IgA selection and bacterial composition in the gut. *Science* **336**, 485–489.
Krummel, M. F., and Allison, J. P. (1995). CD28 and CTLA-4 have opposing effects on the response of T cells to stimulation. *The Journal of Experimental Medicine* **182**, 459–465.
Maki, R., Traunecker, A., Sakano, H., Roeder, W., and Tonegawa, S. (1980). Exon shuffling generates an immunoglobulin heavy chain gene. *Proceedings of the National Academy of Sciences of the United States of America* **77**, 2138–2142.
Martincorena, I., and Campbell, P. J. (2015). Somatic mutation in cancer and normal cells. *Science* **349**, 1483–1489.
Meng, F.-L., Alt, F. W., and Tian, M. (2015). Chapter 19 – The Mechanism of IgH Class Switch Recombination. In *Molecular Biology of B Cells* (Second Edition), F.W. Alt, T. Honjo, A. Radbruch, and M. Reth, eds. (London: Academic Press), pp. 345–362.
Miyajima, M., Zhang, B., Sugiura, Y., Sonomura, K., Guerrini, M. M., Tsutsui, Y., Maruya, M., Vogelzang, A., Chamoto, K., Honda, K., et al. (2017). Metabolic shift induced by systemic activation of T cells in PD-1-deficient mice perturbs brain monoamines and emotional behavior. *Nature Immunology* **18**, 1342–1352.
Muramatsu, M., Sankaranand, V. S., Anant, S., Sugai, M., Kinoshita, K., Davidson, N. O., and Honjo, T. (1999). Specific Expression of Activation-induced Cytidine Deaminase (AID), a Novel Member of the RNA-editing Deaminase Family in Germinal Center B Cells. *Journal of Biological Chemistry* **274**, 18470–18476.
Muramatsu, M., Kinoshita, K., Fagarasan, S., Yamada, S., Shinkai, Y., and Honjo, T. (2000). Class Switch Recombination and Hypermutation Require Activation-Induced Cytidine Deaminase (AID), a Potential RNA Editing Enzyme. *Cell* **102**, 553–563.
Nakamura, M., Kondo, S., Sugai, M., Nazarea, M., Imamura, S., and Honjo, T. (1996). High frequency class switching of an IgM+ B lymphoma clone CH12F3 to IgA+ cells. *International Immunology* **8**, 193–201.
Nishimura, H., Minato, N., Nakano. T., Honjo, T. (1998). Immunological studies on PD-1 deficient mice: implication of PD-1 as a negative regulator for B cell responses. *International Immunology* **10**, 1563–72.

Nishimura, H., Nose, M., Hiai, H., Minato, N., and Honjo, T. (1999). Development of lupus-like autoimmune diseases by disruption of the PD-1 gene encoding an ITIM motif-carrying immunoreceptor. *Immunity* **11**, 141–151.

Nishimura, H., Okazaki, T., Tanaka, Y., Nakatani, K., Hara, M., Matsumori, A., Sasayama, S., Mizoguchi, A., Hiai, H., Minato, N., and Honjo, T. (2001). Autoimmune dilated cardiomyopathy in PD-1 receptor-deficient mice. *Science* **291**, 319–322.

Noma, Y., Sideras, P., Naito, T., Bergstedt-Lindquist, S., Azuma, C., Severinson, E., Tanabe, T., Kinashi, T., Matsuda, F., Yaoita, Y., and Honjo, T. (1986). Cloning of cDNA encoding the murine IgG1 induction factor by a novel strategy using SP6 promoter. *Nature* **319**, 640.

Okazaki, T., Maeda, A., Nishimura, H., Kurosaki, T., and Honjo, T. (2001). PD-1 immunoreceptor inhibits B cell receptor-mediated signaling by recruiting src homology 2-domain-containing tyrosine phosphatase 2 to phosphotyrosine. *Proceedings of the National Academy of Sciences of the United States of America* **98**, 13866–13871.

Ono, M., Kondo, T., Kawakami, M., and Honjo, T. (1977). Purification of Immunoglobulin Heavy Chain Messenger RNA by Immunoprecipitation from the Mouse Myeloma Tumor, MOPC-31C. *The Journal of Biochemistry* **81**, 949–954.

Porter, R. R. (1959). The hydrolysis of rabbit γ-globulin and antibodies with crystalline papain. *Biochemical Journal* **73**, 119–127.

Putnam, F. W., and Easley, C. W. (1965). Structural Studies of the Immunoglobulins: I. THE TRYPTIC PEPTIDES OF BENCE-JONES PROTEINS *Journal of Biological Chemistry* **240**, 1626–1638.

Rabbitts, T. H., Forster, A., and Milstein, C. P. (1981). Human immunoglobulin heavy chain genes: evolutionary comparisons of Cμ, Cδ and Cγ genes and associated switch sequences. *Nucleic Acids Research* 9, 4509–4524.

Revy, P., Muto, T., Levy, Y., Geissmann, F., Plebani, A., Sanal, O., Catalan, N., Forveille, M., Dufourcq-Lagelouse, R., Gennery, A., *et al.* (2000). Activation-Induced Cytidine Deaminase (AID) Deficiency Causes the Autosomal Recessive Form of the Hyper-IgM Syndrome (HIGM2). *Cell* **102**, 565–575.

Robert, C., Long, G. V., Brady, B., Dutriaux, C., Maio, M., Mortier, L., Hassel, J. C., Rutkowski, P., McNeil, C., Kalinka-Warzocha, E., et al. (2015). Nivolumab in previously untreated melanoma without BRAF mutation. *The New England Journal of Medicine* **372**, 320–330.

Shimizu, A., Takahashi, N., Yaoita, Y., and Honjo, T. (1982). Organization of the constant-region gene family of the mouse immunoglobulin heavy chain. *Cell* **28**, 499–506.

Sivan, A., Corrales, L., Hubert, N., Williams, J. B., Aquino-Michaels, K., Earley, Z. M., Benyamin, F. W., Lei, Y. M., Jabri, B., Alegre, M. L., et al. (2015). Commensal Bifidobacterium promotes antitumor immunity and facilitates anti-PD-L1 efficacy. *Science* **350**, 1084–1089.

Suzuki, K., Meek, B., Doi, Y., Muramatsu, M., Chiba, T., Honjo, T., and Fagarasan, S. (2004). Aberrant expansion of segmented filamentous bacteria in IgA-deficient gut. *Proceedings of the National Academy of Sciences of the United States of America* **101**, 1981–1986.

Takeshima, T., Chamoto, K., Wakita, D., Ohkuri, T., Togashi, Y., Shirato, H., Kitamura, H., and Nishimura, T. (2010). Local radiation therapy inhibits tumor growth through the generation of tumor-specific CTL: its potentiation by combination with Th1 cell therapy. *Cancer Research* **70**, 2697–2706.

Tang, J., Shalabi, A., and Hubbard-Lucey, V. M. (2018). Comprehensive analysis of the clinical immuno-oncology landscape. *Annals of Oncology: official journal of the European Society for Medical Oncology* **29**, 84–91.

Titani, K., Whitley, E. Jr., Avogardo, L., Putnam, F.W. (1965). Immunoglobulin Structure: Partial Amino Acid Sequence of a Bence Jones Protein. *Science* **149**, 1090–1092.

Tivol, E. A., Borriello, F., Schweitzer, A. N., Lynch, W. P., Bluestone, J. A., and Sharpe, A. H. (1995). Loss of CTLA-4 leads to massive lymphoproliferation and fatal multiorgan tissue destruction, revealing a critical negative regulatory role of CTLA-4. *Immunity* **3**, 541–547.

Topalian, S. L., Hodi, F. S., Brahmer, J. R., Gettinger, S. N., Smith, D. C., McDermott, D. F., Powderly, J. D., Carvajal, R. D., Sosman, J. A., Atkins, M. B., et al. (2012). Safety, activity, and immune correlates of anti-PD-1 antibody in cancer. *The New England Journal of Medicine* **366**, 2443–2454.

Ventura-Clapier, R., Garnier, A., and Veksler, V. (2008). Transcriptional control of mitochondrial biogenesis: the central role of PGC-1alpha. *Cardiovascular Research* **79**, 208–217.

Walunas, T. L., Lenschow, D. J., Bakker, C. Y., Linsley, P. S., Freeman, G. J., Green, J. M., Thompson, C. B., and Bluestone, J. A. (1994). CTLA-4 can function as a negative regulator of T cell activation. *Immunity* **1**, 405–413.

Wang, J., Yoshida, T., Nakaki, F., Hiai, H., Okazaki, T., and Honjo, T. (2005). Establishment of NOD-Pdcd1-/- mice as an efficient animal model of type I diabetes. *Proceedings of the National Academy of Sciences of the United States of America* **102**, 11823–11828.

Wang, J., Okazaki, I. M., Yoshida, T., Chikuma, S., Kato, Y., Nakaki, F., Hiai, H., Honjo, T., and Okazaki, T. (2010). PD-1 deficiency results in the development of fatal myocarditis in MRL mice. *International Immunology* **22**, 443–452.

Waterhouse, P., Penninger, J. M., Timms, E., Wakeham, A., Shahinian, A., Lee, K. P., Thompson, C. B., Griesser, H., and Mak, T. W. (1995). Lymphoproliferative disorders with early lethality in mice deficient in Ctla-4. *Science* **270**, 985–988.

Physiology or Medicine 2019

William G. Kaelin, Jr., Sir Peter J. Ratcliffe and Gregg L. Semenza

"for their discoveries of how cells sense and adapt to oxygen availability"

The Nobel Prize in Physiology or Medicine, 2019

Presentation speech by Professor Anna Wedell, Member of the Nobel Assembly at Karolinska Institutet, Adjunct member of the Nobel Committee for Physiology or Medicine, 10 December 2019.

Your Majesties, Your Royal Highnesses, Esteemed Nobel Laureates, Ladies and Gentlemen,

Elixir of Life, was the name used in the 1500s for the gas we now know as oxygen.

Air of Fire was the more concise term chosen by Scheele in the 1700s, for the component of air that is needed for burning.

We are all on fire.

We burn our food within our bodies in order to extract energy.

Without oxygen, no energy, no life.

We all know that without oxygen, we die.

What all may not know is, that oxygen deficiency constantly arises within our bodies, for example in our muscles while they work.

This year's Nobel Prize in Physiology or Medicine concerns how we sense and adapt, in the finest details of our cells, when there is lack of oxygen.

Already at the end of the 1800s we learned that at high altitude, with less oxygen, we adapt by producing more red blood cells.

In the middle of the 1900s we realized that this is mediated by erythropoietin from the kidney. How this regulation works was however not at all understood.

Gregg Semenza and Sir Peter Ratcliffe decided, independently, to find out how the erythropoietin gene can have such an extraordinary ability to react when oxygen levels drop. Semenza discovered an essential DNA element. Ratcliffe was on the same track and they showed that the element is active in all cells. Oxygen sensing thus takes place everywhere in our bodies. Semenza then discovered the critical player that activates our defense genes. It was named HIF. HIF was subjected to an advanced form of control. It is continuously produced, but when oxygen is ample, it disappears. Only when oxygen levels drop, HIF will remain and can mobilise our defense.

William Kaelin studied a different problem, von Hippel-Lindau disease, with inherited increased risk of certain types of cancer. Cancer cells without the gene, VHL, had activated genes normally controlled by HIF. Sir Peter Ratcliffe proved, in a crucial experiment, that VHL is required for HIF to be removed.

But what was the signal to VHL that HIF needs to disappear?

In the early 2000s, Kaelin and Ratcliffe both solved this mystery. The signal was formed by attaching oxygen atoms onto HIF.

Without oxygen, no signal to VHL, HIF is left intact and can activate our defense.

Piece by piece of the puzzle, the Laureates explained a sensitive machinery that compensates when the vital oxygen is not available in exactly the right amount.

Today we know that the machinery affects a vast range of functions.

When oxygen is lacking, oxygen transport is enhanced by generation of new blood vessels and red blood cells. Our cells are also instructed to economize with the oxygen available, by reprogramming their energy metabolism. Oxygen sensing is also involved in many diseases. As a result of the Laureates' discoveries, intense activities are under way to develop treatments against for example anemia and cancer.

Professors Semenza, Ratcliffe and Kaelin:

Your groundbreaking discoveries have shed light on a beautiful mechanism explaining our ability to sense and react to fluctuating oxygen levels. The system you have clarified is of fundamental importance for all aspects of physiology and for many human diseases. Without it, animal life would not be possible on this planet.

On behalf of the Nobel Assembly at Karolinska Institutet, it is my great privilege to convey to you our warmest congratulations. I now ask you to step forward to receive the Nobel Prize from the hands of His Majesty the King.

William G. Kaelin, Jr. © Nobel Prize Outreach AB. Photo: A. Mahmoud

William G. Kaelin, Jr.

Biography

I WAS BORN November 23, 1957 in Jamaica, New York City, to William George and Nancy Priscilla (Horn) Kaelin. My mother's first pregnancy ended in a miscarriage. The obstetrician who delivered me was a friend of my mother's family. He initially thought I was stillborn and lamented that such obstetrical problems seemed to disproportionately affect his friends. He apparently gave me one last whack, whether for his benefit or mine. Regardless, I began to cry. Eventually, my mother would have four more children, each spaced 2 years apart.

My father went to college and law school at Duke University before becoming an international tax and estate lawyer in Manhattan. My mother majored in mathematics at Adelphi College before becoming an actuary at the Metropolitan Life Insurance Company. She became a homemaker after my birth.

We lived in Jamaica, New York until I was two, at which point we moved to Rockville Centre – a middle-class suburb of New York City on Long Island. There I was enrolled in a public elementary school. I recall the fear of polio and remember being given the oral polio vaccine. That was perhaps my first introduction to the miracles that modern medicine could produce.

On the second day of first grade some male classmates came to my house to play in our small backyard. At one point, while we were roughhousing, I became pinned to the ground under the other boys. Among them was a boy who, I later learned, had been held back in first grade twice for behavioral issues. He was bigger and stronger than the rest of us. He twisted my leg like it was a propeller on a toy airplane (perhaps to emulate the antics of the Three Stooges). I screamed in pain until the boys finally dispersed. I crawled on the ground, unable to walk, until I got to our back door and found my mother. She couldn't believe that my leg could be broken, so she asked me to stand. I promptly collapsed onto the ground. She drove me to my pediatrician, who confirmed that my leg was broken

Figure 1. William G. Kaelin, Jr. childhood photos, 4 and 11 years old.

and put me into a plaster cast to my midthigh. I remember looking at my XRAY on our way home (even I could tell my leg was broken).

For the next 8 weeks I was homebound and largely immobilized, although I could motor around the house if I slid on my buttocks. A tutor came to my house each day so that I wouldn't fall behind in my classwork. When I returned to school, I was actually so far ahead that I thought I might somehow get in trouble. I remember sitting at my desk and surreptitiously erasing and then redoing afresh the assignments that my classmates were being given for the first time. I sometimes wonder whether all of the individual attention I received while homeschooled contributed to my intellectual development and academic achievements later on. At the same time, I also wonder whether it contributed to my feeling of not quite belonging. This was exacerbated further when I changed schools the next year to attend a Catholic elementary school (St. Agnes). There, I found myself once again immersed among students who already knew one another.

At St. Agnes I was taught by lay women and nuns. This was an era, sadly, when most bright women were told that their career options were: nurse, teacher, or nun. I am sure that some of my female teachers in elementary school (and later high school) would today be lawyers, doctors, college professors, or CEOs. I did extremely well at school until the fifth grade, when two things happened. First, the coolest kid in the class befriended me. By association, I suddenly became cool. Second, I was taught by a young, attractive, nun rather than a lay sexagenarian (at least that's how old they felt to me). Both of these things led to a lot of acting out on my part, including misbehavior and failure to do my homework. The standard punishments included afterschool detentions and writing out multiple copies of the multiplication table from 1x1 to 12x12. The worse the offense, the more copies you had to write out. I remember once sitting in my backyard writing out 40 copies of the multiplication table. The silver lining was that I was already good in math and now knew multiplication (and hence division) cold.

In those days, I was getting in the low 90s or high 80s (out of 100) in my subjects despite my antics. One day, my fifth grade teacher called my parents and told them that I should be doing *much* better in school given my aptitude (i.e. I was coasting). After that, my study habits improved somewhat, but they were not exemplary.

My father loved to fish, and we spent many hours fishing together. I think there are many analogies between fishing and science. Two keys to success in fishing are technical expertise (e.g., how to bait a hook) and intuition (e.g., knowing where to fish). Both are easier to learn by apprenticeship than from a book. Similar considerations apply in science. But technical expertise and intuition only get you so far; you also need luck. On any given day the biggest fish might be caught by a novice rather than an expert. The same is true of scientific discovery.

In the 1960s, science was celebrated in the United States due to the cold war and the space race. Like many young boys, I had a toy microscope, chemistry sets (with chemicals that were actually dangerous), electric cars and trains, an erector set, and other toys that stimulated an interest in science. I threw tickertape and confetti at the triumphant Apollo astronauts from my father's Wall Street office. I also had ample unstructured time to play, whether it was going to the beach, tossing a ball, collecting postage stamps, running a lemonade stand, or making up games with neighborhood children. I regret, however, spending so much time watching television rather than reading.

By the late 1960s, my father could afford a much larger home. He wanted to live in a town within commuting distance of New York that had strong public schools rather than pay five private school tuitions. We moved to Fairfield, Connecticut in the summer of 1970, and I began the eighth grade that September. Entering adolescence is always challenging. In this case, it was complicated by entering a new school system. This exacerbated my shyness. We now had a beautiful house with a swimming pool, but only a handful of other children lived within walking or bicycling distance. Fortunately, I had a few neighborhood friends with whom to play touch football or ice hockey when a nearby pond froze over in winter.

The other problem was that many of my school subjects were taught in tracks, with track one reserved for the most gifted students. As a newcomer, I was placed into track two. This lack of stimulation likely contributed to my poor study habits until and including my high school junior year. I would typically sit in the back of the room and silently scoff at the kids who sat at the front, raising their hands obsequiously at every opportunity. I did the minimum amount of homework necessary and never reviewed the material covered in class that day. My one saving grace was that I did well in mathematics without really trying, and consistently did well on mathematic aptitude tests. As a result, my report cards usually contained an "A" in mathematics and a mixture of "A"s and (mostly) "B"s in other subjects. During this period, it became even clearer that I liked quantitative/objective subjects and disliked qualitative/subjective subjects. I liked learning concepts and ideas and abhorred rote memorization. I was interested in science, but I found biology pretty boring because it was largely descriptive. Chemistry and physics were more my speed. During this time, I learned that I tended to like the things that I was good at and tended to be good at the things that I liked.

During my junior year, my high school, Roger Ludlowe, got a computer terminal that was connected to a mainframe at Fairfield University. Although I was good at mathematics, I did not see a career for myself in mathematics. Computers seemed like an exciting way to put mathematics to work. And I liked that in computer science, as in mathematics, there were objective ways of knowing whether your solutions were correct or "worked". I began learning various programming languages and would sometimes bicycle to Fairfield University after school to get input from the faculty there.

One day, during my junior year, I was sitting at the Roger Ludlowe computer terminal and saw a lone pamphlet in the waste bin. It described an eight-week National Science Foundation Student Science Training Program in Mathematics and Computer Science held at Florida Atlantic University for 32 "gifted" high school students. I applied and was accepted.

On my flight to Florida, I met another boy on his way to the same program. He had many books he was reading simply for pleasure, including mathematics books and challenging fiction books such as "Gravity's Rainbow". He, like all of the other high school students that summer, were amongst the brightest students I had ever met. I was pleasantly surprised, however, to discover that I could hold my own with them in the college level courses that we took that summer. And I discovered that I did better in school when the material was interesting and challenging and when I was surrounded by exceptionally smart people. For the first time in my academic life, I went to the library to checkout extra books to supplement my assigned reading and would work past midnight solving mathematics problems or writing code.

Figure 2. Participants at 1974 National Science Foundation Student Science Training Program at Florida Atlantic University. William G. Kaelin, Jr. is circled.

My fellow summer students and I were given a Russian Math Olympiad test (or its equivalent). It contained the kind of advanced, abstract mathematics problems that could be solved without having had prior formal instruction if you were naturally gifted in mathematics. It was the most challenging test I had ever seen. I was initially crestfallen when I received a score of ~45, the worst grade I had ever gotten. I then looked around and saw that many other students had done much worse. My ego fleetingly recovered until I saw that one of the other students, however, had scored ~145. I shouldn't have been surprised. He was the one student who was a much better programmer than I was. For example, while I wrote a computer program for playing draw poker as one of my final projects, he wrote a program for playing bridge. I decided that he was the kind of genius who would become an academic mathematician, whereas I better find something more practical. I didn't see the personal computer revolution coming and didn't want to write computer code for industrial or military applications. I thought a career in medicine could be the answer (ironically, "Mr. 145" also became a medical doctor).

Although I was delighted to learn that I wasn't the dumbest of the 32 students that summer, I certainly had the worst high school grades because of my prior poor study habits. I decided I would try an experiment for my senior year. I would sit in the front of the classroom and

Figure 3. William G. Kaelin, Jr., high school yearbook photo.

actively participate, befriend (rather than belittle) the smart kids in class, do all my homework (and extra credit problems) on time, and review the lessons that we had each day. I successfully petitioned to be placed in all track "1" classes and to take precalculus and advanced placement calculus in parallel rather than in series. I put a bulletin board in my bedroom for my hoped for "A"s. And I determined that I would only have fun on the weekends if had done well at school that week.

My plan worked. I got straight "A"s. I also did very well on the scholastic aptitude test (SAT) and achievement tests required for college. I was a "late bloomer".

My father, the son of a laborer, was the first member of his family to attend college. Given his parents' modest means, he sought an up-and-coming school that was still affordable. He applied to Duke, which seemed to fit the bill, and was rejected. He then hitchhiked from his Long Island home to Durham, North Carolina (in the 1950s, this took him several days) to plead his case. The admissions staff, either due to admiration or pity, interviewed him and gave him an aptitude test. His performance earned him admission. While at Duke, my father did a number of odd jobs to pay for his education, including illegally selling sandwiches in the dormitories. He received a scholarship to attend Duke Law School and was allowed to combine his senior year of college with his first year of law school to defray costs. My father was fiercely loyal to Duke and dreamt that I, his eldest son, would also attend Duke.

I applied to Duke and flew to Durham to be interviewed. My father was elated because the associate admissions director, who was one of his former Duke classmates, winked at him as I left the interview to let him know that I would likely be admitted.

I also applied to Harvard and MIT (in those days, "late bloomers" could still get into an elite school. Sadly, this doesn't seem true today). I visited Harvard with "Mr. 145" and one of his classmates. We stayed overnight in a small dormitory. The 12 students it housed had us stand in the middle of their commons area that evening to tell them our name, our hometown, and our SAT scores. When "Mr. 145" and his friend finished both were told "you will get in". When I gave my SAT scores, they said "you might want to apply to Brown". In fairness, my verbal score wasn't as strong as my math score, and they were correct because I was waitlisted.

I was immensely proud, however, when I was accepted to MIT. I walked into my father's den and told him I was leaning toward MIT. He put down his scotch and asked, "Don't you have to be good in math to get into MIT?", to which I of course said "Yes". He then said, "I am going to give you a math problem, if you go to MIT, you'll pay your tuition, but if you go to Duke, I'll pay your tuition". I quickly did the math. Truthfully, however, I did go back and forth as to whether Duke might be a better choice for a future physician. And I thought I might get more attention at Duke than at MIT (where I assumed they had a sea of "145"s).

My Duke freshman class contained many students who clearly had functioned at a high level for years. I felt like I was just getting started, and I was fascinated to see what kind of grades I could get if I really applied myself (I am embarrassed that it was still largely about grades for me back then).

I majored in mathematics and, partly to fulfill premedical requirements, chemistry. There were many premedical students at Duke, and it was widely believed that you needed an "A" in Organic Chemistry to get into medical school. I excelled at freshman Inorganic Chemistry because it was logical, conceptual, and quantitative (plus I had an outstanding high school chemistry teacher, Mr. Ralph Minopoli). Organic Chemistry required much more rote memorization. My first Organic Chemistry exam I got a "C" despite studying very hard. The next two exams I got a "B" and then an "A–." Miraculously, I got an "A+" on the final exam and therefore got an "A" for the term. I had dodged a bullet. I promised myself that I would not dig myself into such a hole the next term, and yet the same test score pattern repeated itself. My assumption is that the rote memorizers couldn't remember the material from the first exam by the time they got to the final, whereas perhaps I assimilated some concepts after all. However, I haven't excluded divine intervention.

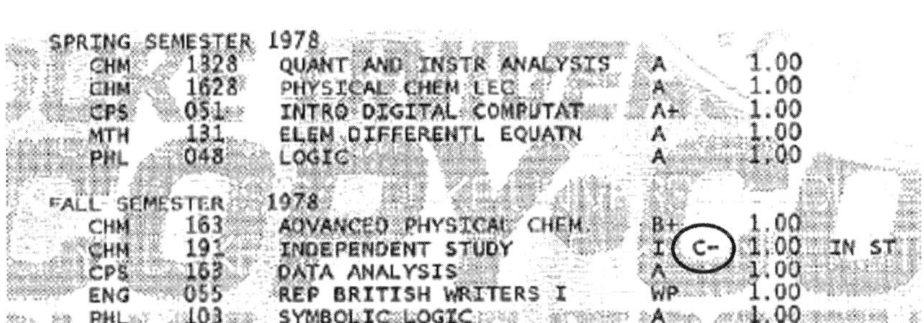

SPRING SEMESTER 1978					
CHM	132S	QUANT AND INSTR ANALYSIS	A	1.00	
CHM	162S	PHYSICAL CHEM LEC	A	1.00	
CPS	051	INTRO DIGITAL COMPUTAT	A+	1.00	
MTH	131	ELEM DIFFERENTL EQUATN	A	1.00	
PHL	048	LOGIC	A	1.00	
FALL SEMESTER 1978					
CHM	163	ADVANCED PHYSICAL CHEM	B+	1.00	
CHM	191	INDEPENDENT STUDY	I C–	1.00	IN ST
CPS	163	DATA ANALYSIS	A	1.00	
ENG	055	REP BRITISH WRITERS I	WP	1.00	
PHL	103	SYMBOLIC LOGIC	A	1.00	

Figure 4. William G. Kaelin, Jr. Duke University Transcript. Note "C–" for laboratory independent study in chemistry.

I earned a higher grade point average at Duke than I did in high school. Ironically, one of my few "Bs" was in Human Physiology, which only reaffirmed my inclination to avoid (largely descriptive) biology courses. I was one of the best mathematics students at Duke, but I worried that I was simply a big fish in a smallish pond. I therefore continued to think about medicine. A friend suggested that I do a laboratory research project to embellish my medical school applications.

My junior year, I met with a physical chemist at Duke about a research project that involved studying the folding of cytochrome C using electron paramagnetic resonance (EPR). I could begin the following summer and continue into the fall semester. He told me that the seven undergraduates who had worked on this project previously had all gone onto medical school. I left his office thinking I had found the golden ticket to medical school.

Now, looking back, I realize I should have asked why my seven predecessors couldn't bring this project to completion. I soon discovered why. I was given a bench to work at and my predecessors' notebooks, but no real instructions from my mentor. In fact, he and I barely interacted. None of the other scientists in his laboratory worked on anything even remotely related to my project. Many of them did exclusively dry bench research, and none of them were protein biochemists. I was supposed to covalently couple a spin label to cytochrome C, separate the modified cytochrome C from the unmodified cytochrome C, and then study the labeled Cytochrome by EPR in response to chemical denaturants. I was lost. I assumed that some people intuitively knew what to do in the laboratory and that others, including me, did not. This experience convinced me that I could assimilate old knowledge but lacked what it took to create new knowledge (at least in the laboratory).

That fall, I attended a lecture by a visiting cytochrome C expert. He was kind enough to briefly discuss my project. His first question was simply to ask me "why?" – specifically, "why are you doing this?" He proceeded to tell me that the conditions proposed to covalently add the spin label were already going to denature my protein, and that the purification method recommended by my mentor was going to separate denatured protein from native protein rather than labeled native protein from unlabeled native protein.

I presented my various concerns about my project to my professor. During my senior year Christmas break, I received a special delivery letter indicating that he had given me a "C–" for my independent study project. He also wrote in the margin that "Mr. Kaelin appears to be a bright young man whose future lies outside the laboratory."

Despite this debacle, I was accepted to several medical schools, including Duke (I was again rejected by Harvard). The medical school basic sci-

Figure 5. William G. Kaelin. Jr. medical school and housestaff badges.

ence curriculum in those days involved a mind-numbing amount of memorization. One thing that saved me was that Duke compressed the standard two years of basic science into one year. The first year was hard for me, but I felt I could withstand anything for one year (although I did wonder if I had made a mistake not pursuing mathematics and computer science). The other saving grace was that the courses were honors/pass/fail, and it was understood that almost no one got honors. You just plowed through.

In the second year, I took the classical clinical rotations of obstetrics/gynecology, internal medicine, psychiatry, surgery, and pediatrics. I loved each of my rotations and could now see the knowledge gained in the first year in action. It was exhilarating and reaffirmed my decision to go to medical school. My internal medicine chief resident was Dr. Paul Klotman. I thought he was the smartest, funniest, coolest person I had ever met. I decided then and there that I would be a chief resident one day.

Duke medical students were encouraged to work in a laboratory full-time for their third year. With some trepidation, I gave laboratory research another try. I worked with Dr. Randy Jirtle, studying tumor blood flow. Randy, in contrast to my first mentor, was very helpful and encouraging. During this time, I read about tumor angiogenesis and the highly vascular tumors linked to von Hippel-Lindau (VHL) disease.

During my third year, Dr. Michael Bishop came to Duke Medical School to give a lecture, which was exhilarating. In it he described the first oncogenes, some of which encoded kinases. I was sure I was seeing the future of oncology and assumed that this would lead to new therapeutics (e.g., kinase inhibitors). Sadly, over the ensuing years, naysayers would tell me why this was naive. It would take people like Alex Levitzki and Alex Matter to prove them wrong.

Figure 6. Johns Hopkins 1983–1984. Internal Medicine housestaff. William G. Kaelin, Jr., then an intern, is circled.

Although my second laboratory experience was much better than my first, it wasn't enough to convince me that I had a future in the laboratory. I did my internship and residency in Internal Medicine at Johns Hopkins. I liked Internal Medicine partly because I liked solving complex clinical puzzles.

My chairman when I was an intern was Dr. Victor McKusick. He taught me the importance of medical history and the power of human genetics. My chairman when I was a resident was Dr. Jack Stobo. He conducted weekly "bench to bedside" chalk talks that introduced me to emerging molecular techniques, such as DNA restriction and Southern blotting. It was pretty clear a revolution was about to happen, even if I didn't think I would be part of it.

A colleague introduced me to Eric Fearon, who was then a medical student with Dr. Bert Vogelstein. Eric had already published several landmark papers in which he applied modern molecular techniques to gain new insights into cancer genetics. He encouraged me to attend one of Bert's seminars. At the time, it was believed that solid tumors, in contrast to hematological malignancies, were too heterogeneous and complex to study with modern molecular techniques. Bert's colon cancer studies shattered that view. This was the second lecture, after Michael Bishop's lecture, where I knew I was seeing the future.

I loved clinical medicine and was honored to be selected to spend an extra year at Johns Hopkins as one of four assistant chiefs of service

Figure 7. William G. Kaelin as Johns Hopkins Internal Medicine Assistant Chief of Service, Circa 1986. Top – With fellow ACSs. Bottom – With his future wife, Carolyn Scerbo, at her medical school graduation in May 1987.

(equivalent to chief resident). I learned even more clinical medicine that year and, more importantly, met a beautiful fourth year medical student (although she never reported to me!). Carolyn Scerbo would later become my wife, and the mother of my two beautiful children, Kathryn Grace and Tripp.

Chief residents typically love to learn about rare eponymous syndromes, such as von Hippel-Lindau (VHL) Disease. That way, if a trainee challenges their authority on rounds, they can silence them with questions about diseases the trainee has never heard of. For the same reason, chief residents like long differential diagnoses for various signs and symptoms. I had memorized, for example, all of the causes of polycythemia. It struck me as odd that the three hallmark tumors of VHL Disease – kidney cancers, hemangioblastomas, and pheochromocytomas – were among them. This, together with their increased vascularity, suggested to me that the VHL gene was involved in oxygen sensing because angiogenesis and erythropoiesis are normally induced by hypoxia (low oxygen level).

I decided to specialize in medical oncology because I liked multisystem disorders and thought cancer biology was fascinating. I also appreciated that the bane of any internist is distinguishing the sick from the not so sick. This is not an issue with cancer patients.

I moved to Boston in 1987 to be a medical oncology fellow at the Dana-Farber Cancer Institute. I was finally arriving at Harvard, which prompted me to wonder whether Harvard (or I!) had made a mistake. My seven fellow fellows were extremely impressive and most of them had more research experience than I did. I was deemed the most likely of us to become a clinical oncologist rather than a laboratory-based researcher.

One path to medical oncology board certification involves working in a laboratory for 2 years. I thought it would be exciting to see what life was like in a top laboratory, even if I didn't expect to succeed myself. I interviewed with Dr. Robert Weinberg at MIT, who had just cloned the retinoblastoma gene *RB1.* He was clearly receptive to having a physician-scientist in the laboratory, having trained Dr. Steven Friend, who cloned *RB1.* Bob then got out a ledger and determined that he would have space for me in 3 years. My heart sank because I needed to start that year.

While researching Bob's laboratory I met Dr. Shelly Bernstein. Shelly, a Weinberg physician-scientist trainee, had just established his own laboratory at the DFCI to study metastasis. I joined Shelly, but I can now appreciate that I was a bit adrift. In my free time, I read a *Science* paper that described PCR-amplification of a K-Ras exon from a paraffin block. Just for fun, I tried to replicate the experiment. To my astonishment it worked! I remember the exhilaration when I turned on the UV box and saw an ethidium-stained band of the correct size. About 4 months after joining Shelly, he told me that he would be closing his laboratory to return to clinical practice. This seemed like another sign that I was not meant to do science! I was an orphan.

My last attending physician during my clinical oncology year was Dr. David Livingston. I knew he ran a laboratory and I went to him for advice. I showed him my PCR result, which perhaps impressed him. Or maybe he saw something else. In any event, he invited me to join his laboratory. It turned out that he, like Bob Weinberg, was studying the *RB1* gene. David's group and Ed Harlow's group had discovered that the SV40 T and the adenovirus E1A proteins, respectively, bound to the *RB1* gene product pRB. I was to map the T/E1A-binding region of pRB, and the PCR technique I had learned proved useful for rapidly making pRB mutants. The minimal T/E1A-binding region, which we called the pRB "pocket", turned out to be a hotspot for cancer-associated *RB1* mutations. We therefore hypothesized that the pRB pocket might bind to one or more cellular proteins. Our initial strategy to find them was to look for coimmunoprecipitating proteins in anti-pRB immunoprecipitates from cells. The problem, however, was that the supply of our workhorse anti-pRB antibody was unreliable.

David decided we needed to make our own anti-pRB antibodies. I was to express pRB fragments in *E. coli* and then inject them into rabbits. The trick was to find a suitable prokaryotic expression vector, since protein expression in *E. coli* was still fickle back then.

One day, a company representative left me a brochure for a new prokaryotic expression vector that encoded glutathione S-transferase fused to a protein of interest. Such GST fusions were frequently more soluble than their unfused counterparts and could be purified using glutathione sepharose. Ordinarily, the captured protein would be eluted with glutathione prior to its intended use. I realized, however, that I could express GST fused to the pRB pocket, capture it with glutathione sepharose, and then use the immobilized fusion protein to capture pRB-binding proteins. I eventually found a series of anonymous 35S-labelled proteins that had the properties I was looking for: they bound to wild-type, but not mutant, pRB, and were displaced by T/E1A-derived peptides.

I was thrilled when my findings were published in Cell and when I was allowed to present it at major meetings. After speaking at the 1990 Cold Spring Harbor Symposium, I walked down to the Robertson House pool, where a lone person was enjoying the afternoon. He told me that my talk was wonderful. I told him that his praise was much appreciated because I had just learned that my National Institutes of Health (NIH) K08 physician-scientist fellowship application to support my work had received a very unfavorable score. Indignant, he asked me which study section evaluated my application. When I told him, he said, "that's outrageous, I chair that study section". Shortly thereafter, I was notified by the NIH that my grant would be funded after all.

I still didn't know the identities of my pRB-binding proteins, or which of them bound directly to pRB. My colleague Myles Brown then told me about the work of Michael Blanar, who had radiolabeled a recombinant protein using radioactive ATP, heart muscle kinase, and a genetically encoded phosphoacceptor site. I redesigned my GST fusion vector so that this phosphoacceptor site was encoded between the GST and pRB moieties. After considerable troubleshooting, I finally produced biologically active, radiolabeled GST-pRB, which I used to probe whole cell extracts in "far western blot" assays. I knew the system worked because I could detect E1A and T in appropriate cell extracts. I also saw a doublet of ~50 kD, which prompted me to team up with one of Dr. Sam Speck's postdoctoral fellows, Erik Flemington. Erik helped me screen a lambda phage expression library with my radiolabeled GST-pRB. David Livingston, Joe Nevins, and Pradid Raychaudhuri had reported that pRB could bind (directly or indirectly) to a DNA-binding activity called "E2F". I was elated when my expression cloning strategy yielded the first E2F cDNA (coined E2F1).

Why was I so successful as a postdoctoral fellow? In stark contrast to my first laboratory experience, I was given a great project to work on, in a great laboratory, and with a great mentor. My expectations of myself were so low that I was exhilarated when even the most prosaic experiment, such as restricting DNA or transforming bacteria, actually worked! Importantly, I was still relatively naïve, which can be a blessing in the laboratory. I ambitiously tried many experiments that a more jaded scientist would dismiss as being too risky. And I didn't know the usual solutions for many problems. This forced me to be resourceful and to look at problems afresh.

In 1992, I started my own laboratory down the hallway from David's laboratory. I initially did a few E2F experiments, but it was obvious that there were many good laboratories ready to work on this protein. I didn't want to look back on my scientific career and realize that everything I had done would have been done with or without me. That would hardly justify leaving clinical medicine, which I loved. Moreover, people wisely advised me to begin differentiating my science from David's science. I briefly attempted to expression clone the *ATM* gene using a functional complementation assay, but without success. In the summer of 1993, however, I read a Science paper describing the cloning of the *VHL* tumor suppressor gene and decided that it would be my focus. I assumed that studying the VHL gene would teach us about the pathogenesis of kidney cancer (which is one of the ten most common cancers), about the control of angiogenesis, and potentially about oxygen sensing. This work is described more fully in my written Nobel Lecture. It was supported by the NIH, the Howard Hughes Medical Institute (which I joined in 1998), the Doris Duke Charitable Foundation, the Murray Foundation, and the late Dr. William Shelton.

As a young faculty member, I served as an inpatient clinical attending physician one month a year and would also "moonlight" several nights a month in a community hospital intensive care unit. I stopped seeing patients, however, by the late 1990s. Firstly, I never wanted people to say I was a good scientist for a *physician*-scientist. I tell young physicians that if their choice is to be an "A" clinician or a "B" scientist, they should choose the former. If I was going to try to be an "A" scientist, I knew I had to do it full-time. Second, I soon realized that most of the cancer patients referred to the DFCI had already been diagnosed. The diagnostic puzzles I was solving often related to the iatrogenic complications of their treatments. There is also a saying on the wards that common things are common. There are many things in medicine that are interesting the first or second time you see them, but much less interesting the 20th time you see them. Solving puzzles in the laboratory began to give me the joy I had once found seeing patients. It was also painfully clear that the therapies available for most of my cancer patients were woefully inadequate. I thought that more effective therapies would require the kind of deep insights into cancer pathogenesis that modern molecular biology offered. Lastly, you can't, by definition, devote yourself fully to two careers. When

I was younger, I believed that my patients were lucky to have me as a doctor and I promised myself that I would stop seeing patients if I could no longer defend that view. I had too much respect for my full-time clinical colleagues and my patients to become a part-time clinician with waning skills.

In 2003, several years after the work that led to my Nobel Prize, my wife Carolyn, herself a beloved breast cancer surgeon, developed breast cancer. She ultimately required a mastectomy and chemotherapy. Even though I had taken care of hundreds of cancer patients and their families, I now had a much more visceral understanding of their anguish and pain. Carolyn survived her breast cancer only to develop an unrelated glioblastoma in 2010. She underwent two major brain operations, each requiring months of rehabilitation in order for her to walk again, as well as radiation therapy and chemotherapy. We assembled a “dream team” of talented clinicians and scientists who helped us. For example, her tumor’s genome was sequenced early on as a guide to specific targeted agents. Ultimately, however, we lost her in 2015. Carolyn and I would often joke about what it would be like if I ever won the Nobel Prize. It was bittersweet to receive the Prize without her at my side.

The von Hippel-Lindau Tumor Suppressor Gene: Insights into Oxygen Sensing and Cancer

Nobel Lecture, December 7, 2019 by
William G. Kaelin, Jr., M.D.
Sidney Farber Professor of Medicine, Dana-Farber Cancer Institute and Brigham and Women's Hospital, Harvard Medical School Investigator, Howard Hughes Medical Institute, USA.

INTRODUCTION

As a third-year medical student at Duke University, I worked in Dr. Randy Jirtle's laboratory studying tumor blood flow using radioactive microspheres (Kaelin *et al.*, 1982, Shrivastav et al., 1983, Kaelin *et al.*, 1984). During that year I read about tumor angiogenesis and became intrigued by the idea, then championed by Dr. Judah Folkman, that blocking angiogenesis would be a way to arrest tumor growth (years later I learned that this idea actually can be found in the 19th century German medical literature). I also learned that some tumors, such as kidney cancers, are particularly rich in new blood vessels and that highly angiogenic tumors, including kidney cancers and hemangioblastomas, are a feature of a rare hereditary cancer syndrome now referred to as von Hippel-Lindau (VHL) disease.

I did my house staff training in internal medicine at Johns Hopkins. As a house office I would often memorize the potential causes, or differential diagnosis, for various symptoms and signs encountered in patients. This, and my knowledge of rare eponymous syndromes such as VHL disease, served me well when I was later selected to be assistant chief of service, equivalent to a chief resident, for the Johns Hopkins internal medicine service (also called the Osler medical service). One such differential diagnosis was for polycythemia, or excess red blood cell production. Included among the causes of polycythemia are certain tumors that can secrete EPO or EPO-like substances, including the three hallmark tumors seen in VHL disease: kidney cancer, hemangioblastoma, and pheochromocytoma (more broadly, paraganglioma) (Golde and Hocking, 1981). This struck me as odd, especially as hemangioblastomas and pheochromocytomas are otherwise rare tumors. The fact that VHL-associated tumors could promote angiogenesis and erythropoiesis suggested to me that these tumors must have dysregulated oxygen sensing, because angiogenesis and erythropoiesis are normal responses to hypoxia. These tumors were acting like they had an exaggerated response to hypoxia, as though their response to oxygen was "stuck" in the hypoxic position. My chairman when I was an intern at Johns Hopkins was Dr. Victor McKusick, who taught me the importance of medical history and the power of human genetics. My chairman when I was a resident was Dr. Jack Stobo, who began introducing me to new molecular techniques that were being used to study human diseases.

THE RB1 TUMOR SUPPRESSOR GENE

I was a medical oncology fellow at the Dana-Farber Cancer Institute. In 1988, shortly after beginning the second year of my fellowship, I joined Dr. David Livingston's laboratory. The retinoblastoma tumor suppressor gene (*RB1*) was cloned about a year earlier by several groups. The predicted primary sequence of the *RB1* gene product, pRB, was relatively featureless and hence did not provide any real clues regarding pRB's likely biochemical functions. Dr. Ed Harlow had shown, however, that the Adenovirus E1A oncoprotein could bind to pRB and David's group, shortly thereafter, showed that SV40 T antigen, via a short peptidic region that was homologous to E1A, could do the same (Whyte *et al.*, 1988, DeCaprio *et al.*, 1988). My first project was to map the minimal region of pRB that could bind to T and E1A. This region, which David coined "the pocket", turned out to be a hotspot for RB1 missense mutations (Kaelin *et al.*, 1990). This suggested that both pocket mutations and the viral oncoproteins caused cancer by preventing pRB from binding to one or more cellular proteins.

It fell upon me to find these putative cellular proteins. Our initial strategy was to look for cellular proteins that coimmunoprecipitated with pRB in anti-pRB immunoprecipitates. Unfortunately, there was really only one anti-pRB antibody available at that time and its continued supply from a competitor laboratory was in doubt. We had an emergency laboratory meeting to make our own anti-pRB antibodies. I was charged with producing fragments of pRB in *E. coli* with which to immunize rabbits and mice. My first task was to find a suitable prokaryotic expression vector. Producing human proteins in *E. coli* was still a bit of an art back then, and success or failure could depend on choosing the right vector.

One day a company representative left a brochure on my desk for a new commercial prokaryotic expression plasmid that encoded glutathione S-transferase (GST) fused to the protein (or protein fragment) of interest. The GST moiety often improved protein solubility (a big issue with bacterially produced proteins) and enabled rapid purification using glutathione sepharose capture followed by elution with reduced glutathione. After looking at the plasmid map it dawned on me that I could use it to express GST-fused to the pRB pocket and employ the fusion protein, rather than an anti-pRB antibody, to look for pRB-associated proteins. Specifically, I could capture the GST-pRB protein on glutathione sepharose and, after washing the sepharose, use the immobilized GST-pRB pocket to capture ^{35}S-labelled cellular proteins. I ultimately identified a series of anonymous ^{35}S-labelled cellular proteins that had the right properties: they bound to GST fused to the wild-type pRB pocket, but not tumor-derived pRB pocket mutants, and their binding to the pRB pocket was prevented by T/E1A-derived peptides (Kaelin *et al.*, 1991).

In parallel, I worked with another Livingston trainee, Xiao Qin, to show that reintroducing the wild-type pRB pocket, but not tumor-derived pRB pocket mutants, into *RB1–/–* tumor cells suppressed their growth (Qin *et al.*, 1992). This strengthened the idea that tumor suppression by pRB was linked to its ability to bind to one or more of my cellular pRB-binding proteins. It also showed me the power of marrying biochemical and functional assays when studying tumor suppressor proteins.

My pulldown experiments did not distinguish direct from indirect binders. Moreover, the identities of my pRB-binding proteins remained unknown, partly because I could not recover them in the amounts required for mass spectrometry sequencing back then. Fortunately, my colleague Myles Brown told me about the work of Michael Blanar, who had radiolabeled recombinant proteins using ^{32}P-ATP, heart muscle kinase, and genetically encoded phosphoacceptor sites (Blanar and Rutter, 1992). I modified my GST-pRB pocket expression vector so that this phosphoacceptor site was inserted between the GST moiety and pRB moiety. In pilot experiments I could readily label my GST-pRB fusion proteins in this way, but they no longer bound to E1A or T. After some troubleshooting, I realized that the brief exposure to 30°C required for the kinase reaction was the culprit. Fortunately, I discovered that the kinase reaction also proceeded at 4°C, and that the protein labeled at this temperature retained the ability to bind to E1A and T.

I proceeded to do "far-western" blot assays using cell extracts that were resolved by SDS-polyacrylamide gel electrophoresis, transferred to nitrocellulose, and then incubated with ^{32}P-labelled GST-pRB pocket in the presence of excess unlabeled GST. After a bit more troubleshooting, I was able to clearly detect direct binding of the GST-pRB pocket to E1A and SV40 T in suitable cell extracts. Notably, I also could see a cellular protein (or proteins) that migrated as a doublet of ~50 kD (Kaelin *et al.*, 1992).

In parallel studies, I worked with another Livingston Laboratory trainee, Thomas Chittenden, who used the immobilized GST-pRB to show that pRB could bind, directly or indirectly, to a DNA-binding activity that Joe Nevins had earlier coined "E2F" (Chittenden *et al.*, 1991). The ability of pRB to bind to this activity was also shown by Drs. Joe Nevins, Nicholas LaThangue, and Pradip Raychaudhuri (Chellappan *et al.*, 1991, Bagchi *et al.*, 1991, Bandara *et al.*, 1991a, Bandara and La Thangue, 1991b).

Encouraged by my far-western blot results, I teamed up with Erik Flemington, who was a postdoctoral fellow in Dr. Sam Speck's laboratory, to screen phage expression libraries with ^{32}P-labelled GST-pRB pocket in the presence of excess unlabeled GST. The dominant clone we recovered represented the first member of the E2F family, now called E2F1 (Kaelin *et al.*, 1992).

THE VON HIPPEL-LINDAU TUMOR SUPPRESSOR GENE

I started my own laboratory at the DFCI in 1992, working down the hallway from David Livingston's laboratory. I did a few additional experiments related to pRB and E2F1 but was appropriately advised that I should probably start to carve out a line of investigation that was clearly distinct from David's. I briefly tried to clone the *ATM* gene using a functional complementation strategy, without success. In 1993 I read a paper in *Science*, from a group led by Drs. Bert Zbar, Michael Lerman, and Marston Linehan at the National Cancer Institute (NCI) and Dr. Eamonn Maher at the University of Birmingham, describing the isolation of a partial cDNA for the gene that, when mutated, caused the von Hippel-Lindau hereditary cancer syndrome (henceforth von Hippel-Lindau disease now that the causative gene is known) (Latif *et al.*, 1993). It was clear that the *VHL* gene, like the *RB1* gene, was a "Knudson 2-hit" tumor suppressor gene. Patients with *VHL* disease have typically inherited a defective maternal or paternal *VHL* allele. Tumor development, such as development of clear cell renal cell carcinoma (ccRCC), is linked to loss of the remaining wild-type allele in a susceptible cell. The authors of the *Science* paper also showed that biallelic VHL inactivation was a common feature of sporadic ccRCCs, which is by far the most common form of kidney cancer in the general population (Latif *et al.*, 1993, Gnarra et al., 1994). In this setting both mutational events, or "hits", occur somatically, in contrast to VHL disease, where the first hit is typically present in the germline. As was true for *RB1*, the predicted *VHL* open reading frame (ORF) did not provide immediate clues as to its likely biochemical functions.

I immediately thought I should work on the *VHL* gene for multiple reasons. In the early 1990s it seemed to me that most of the advances regarding our molecular understanding and treatment of cancer involved cancers that were fascinating, but uncommon. It was clear to me that we needed to make progress on the ten most common cancers (many of which are epithelial cancers) if we wanted to make a big dent in overall cancer mortality. I hoped that studying the *VHL* gene would provide insights into the pathogenesis of kidney cancer, which is an epithelial cancer and one of the ten most common cancers in the developed world.

Judah Folkman was beginning to describe the purification of endogenous angiogenesis inhibitors such as angiostatin (O'Reilly *et al.*, 1994, O'Reilly *et al.*, 1997). His group identified angiostatin in a phenotypic screen for angiogenesis inhibitors, but its mechanism of action was unknown. I had been intrigued by the idea of treating cancers with angiogenesis inhibitors dating back to my third year of medical school, but I thought this would require targeted angiogenesis inhibitors with defined mechanisms of action and that this, in turn, would require a much deeper understanding of the molecular circuits that govern new blood vessel formation. The highly vascular nature of VHL-associated neoplasms suggested to me that studying the *VHL* gene would provide insights into such a circuit.

As a physician I knew that oxygen (or lack thereof) played a role in many diseases, including anemia, myocardial infarction, and stroke, in addition to cancer. The fact that VHL-associated tumors were linked to increased angiogenesis and erythropoiesis suggested to me that studying the *VHL* gene would ultimately provide insights into how human cells and tissues sense and respond to oxygen. This seemed like a great puzzle to work on.

I also had some prosaic reasons for working on the *VHL* gene. There were many great laboratories working on the *RB1* gene and the competition sometimes felt intense. I naively thought it might be more fun to work in relative obscurity on a gene linked to a disease few people had heard of. In addition, the *VHL* cDNA published by the NCI group was thought to be missing about 2 kB of 5' sequence and its predicted ORF was open at the 5' end (Latif *et al.*, 1993). In particular, the *Science* paper reported a *VHL* mRNA size of 6-6.5 kB, based on a Northern Blot, while the cDNA was closer to 4 kB (Latif *et al.*, 1993). I had become very facile at screening phage cDNA libraries with radioactive cDNA probes in the course of my E2F1 cloning work. I had also made an excellent 293T kidney cell cDNA library. I thought we would quickly isolate the 5' end of the cDNA and then be off to the races.

The next postdoctoral fellow to join my laboratory was Othon Iliopoulos, a medical oncology clinical fellow at the Dana-Farber Cancer Institute. He screened multiple cDNA libraries, including the 293T cDNA library described above, with 32P-labelled cDNA probes derived from the published *VHL* cDNA clone (kindly provided by Dr. Michael Lerman). Over the ensuing year he screened tens of millions of cDNAs and isolated scores of new *VHL* cDNA clones. The problem, however, was that none of them extended the published *VHL* cDNA sequence in the 5' direction. In fact, none of them even contained as much 5' sequence as the published VHL cDNA sequence. Othon also tried 5' RACE (Rapid Ampification of cDNA ends) based on the published *VHL* cDNA sequence, without success.

Fortunately for us, however, we had used the partial *VHL* cDNA to express a GST-VHL fusion protein in *E. coli,* which was then used to immunize rabbits. The affinity purified anti-pVHL (pVHL = protein product of *VHL* gene) we made recognized an ~30 kD protein in 293T cells that was absent in human *VHL−/−* ccRCC lines that we had purchased from the ATCC (American Type Culture Collection) (Iliopoulos et al., 1995). We realized that the ORF in the published *VHL* cDNA clone could produce a protein with 213 amino acid residues if a particular upstream methionine was used to initiate translation. I was still working in the laboratory back then and had become fairly proficient at making mammalian expression plasmids and at *in vitro* translation while a postdoctoral fellow in David Livingston's laboratory. I made mammalian expression plasmids that would encode this theoretical protein of 213 amino acids and that could support in *vitro* translation. Othon used one of these plasmids to make the theoretical protein as an ^{35}S-labelled protein by in vitro translation. In parallel, he immunoprecipitated the ~30kD protein with the anti-pVHL antibody using cell extracts from 293T cells labeled with ^{35}S-labelled methionine. The two proteins comigrated when resolved by SDS-PAGE and their partial proteolytic peptide maps were identical. We (and the NCI group) had the complete *VHL* ORF cDNA after all (Iliopoulos *et al.*, 1995). Othon also showed that pVHL resides largely in the cytoplasm under steady-state conditions (Iliopoulos *et al.*, 1995), as determined by immunofluorescence and cellular fractionation assays, although Dr. Richard Klausner would later show that pVHL can dynamically shuttle between the nucleus and the cytoplasm (Lee *et al.*, 1996).

When I was a clinician working in the hospital we used to sometimes say "believe nothing, trust no one, and do it yourself". Knowing that the published "partial" VHL cDNA contained the complete ORF and having been unable to extend the published 5' cDNA sequence, we finally did what we should have done on day 1. Othon did his own Northern Blot and determined that the VHL mRNA is closer to 4.4 kB (Iliopoulos et al., 1995). When I approached the NCI group about their published 5' end that we had never seen in our cDNA clones, they confessed that they were so sure the cDNA was open at the 5' end that they had added some 5' VHL genomic sequence to the VHL cDNA sequence *in silico.*

While this early work was on-going, I travelled to the basement of the Countway Medical Library at Harvard Medical School to find the original papers by Eugen von Hippel and Arvind Lindau (von Hippel, 1904, Lindau, 1927). In the 1970s I was probably among the last students to be told that if you wanted to study medicine, you should first learn German. So, I studied German for 3 years in high school. In medical school I learned the origin of this advice. One of my medical school professors told us that if you thought you had discovered something new in clinical medicine, you should take a sabbatical and read the German medical literature, because there was a good chance your discovery was already reported there. At least one of my medical school professors referred to the late 19th century, when the German medical literature was so dominant, as the great descriptive era in medicine.

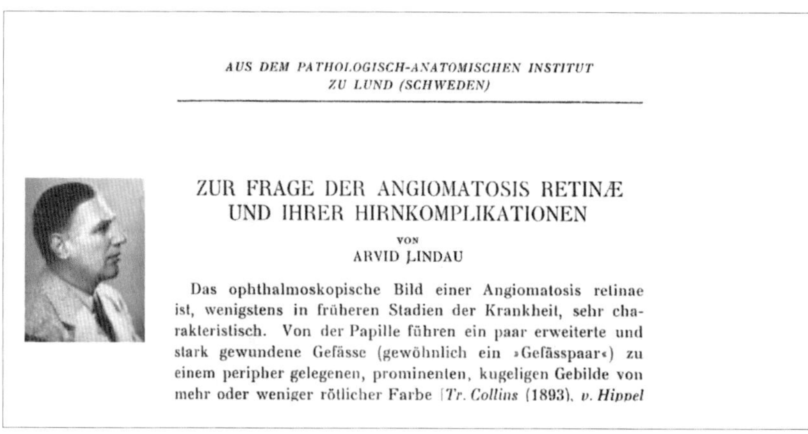

AUS DEM PATHOLOGISCH-ANATOMISCHEN INSTITUT ZU LUND (SCHWEDEN)

ZUR FRAGE DER ANGIOMATOSIS RETINÆ UND IHRER HIRNKOMPLIKATIONEN

VON

ARVID LINDAU

Das ophthalmoskopische Bild einer Angiomatosis retinae ist, wenigstens in früheren Stadien der Krankheit, sehr charakteristisch. Von der Papille führen ein paar erweiterte und stark gewundene Gefässe (gewöhnlich ein »Gefässpaar«) zu einem peripher gelegenen, prominenten, kugeligen Gebilde von mehr oder weniger rötlicher Farbe [*Tr. Collins* (1893), *v. Hippel*

Figure 1. Arvid Lindau's description of von Hippel-Lindau Syndrome. Shown is first page of Arvid Lindau's 1927 paper *Acta Ophthalmologica* (1927) **4**: 193–226. Also included is a picture of Arvid Lindau.

Eugen von Hippel was a German ophthalmologist and Arvid Lindau was a Swedish neuropathologist. Despite my high school German, I could not translate Eugen von Hippel's original paper published in 1904 or Arvid Lindau's seminal paper published in 1927 (Figure 1). I came to learn that Eugen von Hippel's paper described the familial occurrence of retinal hemangioblastomas (Figure 2). Years later my former Chairman, Dr. Victor McKusick, informed me that a similar family had been described in 1894 by Treacher Collins (Figure 3) (Collins, 1894). von Hippel's paper was more widely known, however, because it was published in German (the Collins paper was published in English) and because von Hippel's father was himself an academic physician who would disseminate copies of his son's paper at medical congresses. I came to learn that it was really Lindau, however, who noted that these familial retinal hemangioblastomas were a marker for a systemic disease that also affected the brain, spinal cord, and kidney. Now, approximately 100 years after the paper by Treacher Collins, we had the reagents to begin understanding why *VHL* mutations caused cancer.

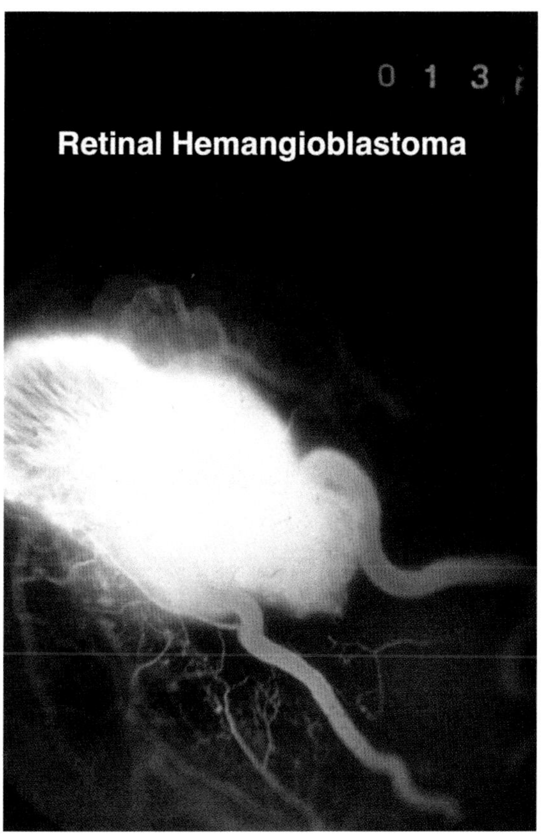

Figure 2. Retinal hemangioblastoma seen on retinal angiography.

My postdoctoral *RB1* studies involved developing functional assays for pRB that could plausibly map to its ability to suppress tumors (e.g. suppression of tumor cell proliferation) and assays for pRB-associated proteins. It seemed reasonable to employ the same strategies here.

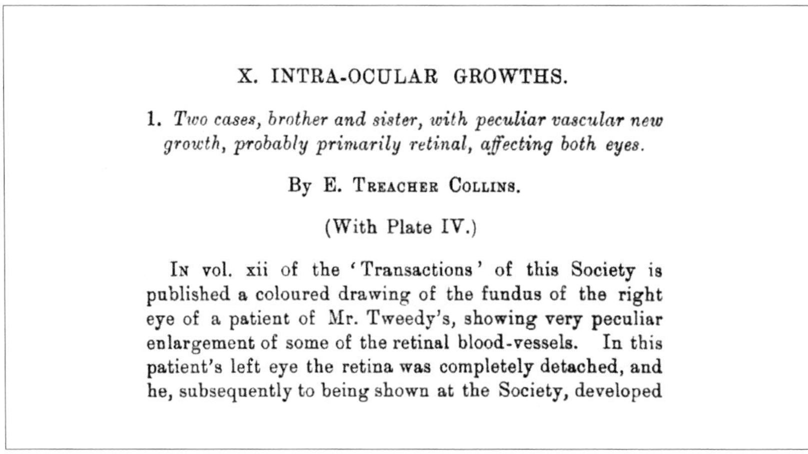
X. INTRA-OCULAR GROWTHS.

1. *Two cases, brother and sister, with peculiar vascular new growth, probably primarily retinal, affecting both eyes.*

By E. Treacher Collins.

(With Plate IV.)

In vol. xii of the 'Transactions' of this Society is published a coloured drawing of the fundus of the right eye of a patient of Mr. Tweedy's, showing very peculiar enlargement of some of the retinal blood-vessels. In this patient's left eye the retina was completely detached, and he, subsequently to being shown at the Society, developed

Figure 3. First report of a family with von Hippel-Lindau syndrome. Shown is first page of Treacher Collins' 1894 paper in *Transactions of the Ophthalmological Society* (1894) **14**: 141–149.

THE *VHL* GENE PRODUCT (PVHL) SUPPRESSES TUMOR GROWTH AS PART OF A MULTIPROTEIN COMPLEX

Othon stably transfected 786-O *VHL* mutant/– (hereafter "–/–" for simplicity) human ccRCC cells to now produce a wild-type version of pVHL ("WT" cells) or a C-terminally truncated version that contained only the first 115 amino acid residues of pVHL ("ARZ" cells). The latter was based on the knowledge that some ccRCC-associated *VHL* mutations caused similar truncations. As an additional control he transfected cells with the empty backbone vector ("pRC" or "EV" cells). He used cloning cylinders to isolate multiple independent clones for each, cognizant that there would likely be some heterogeneity amongst the parental 786-O cells.

The first surprise, which was a bit disappointing, was that it was relatively easy to generate 786-O subclones that now contained wild-type pVHL; they proliferated as well as the control cells under standard cell culture conditions. This was in stark contrast to what would have been seen with other well-studied tumor suppressor proteins, such as pRB and p53.

When Othon implanted the cells subcutaneously in nude mice, however, he saw a dramatic phenotype. The WT cells did not form tumors, in contrast to the ARZ and pRC cells (Iliopoulos *et al.*, 1995). This strengthened our suspicion that tumor suppression by pVHL was linked, at least partly, to suppression of angiogenesis.

In parallel, Adam Kibel, a urology resident at the Brigham and Women's Hospital, joined my laboratory and began looking for pVHL-associated proteins. He labelled 786-O cells that stably expressed hemagglutin (HA)-tagged versions of wild-type pVHL, pVHL (1-115), or a pVHL disease-associated mutant, pVHL (R167W), with ^{35}S methionine and captured potential pVHL-binding partners in immunoprecipitiation assays with either anti-pVHL or anti-HA antibodies. In both cases he saw two proteins with molecular weights of ~18kd (p18) and ~14 kD (p14), respectively(Kibel *et al.*, 1995). The same two proteins coimmunoprecipitated with endogenous pVHL in *VHL*+/+ ccRCC lines but were not detected in anti-pVHL immunoprecipitates prepared from pVHL-defective 786-O ccRCC cells. Adam then mapped the minimal p18/14 binding region of pVHL to pVHL residues 157–172 using GST pulldown assays and peptide competition assays. In the latter, the binding of p18/14 to pVHL was blocked by a peptide in which residues 157–172 corresponded to wild-type pVHL, but not if the peptide contained the R167W mutation. Gratifyingly, we saw that the 157–172 was clearly a hotspot for mutations in VHL families, arguing that the binding to p18 and/or p14 was linked to pVHL's ability to suppress tumors (Kibel *et al.*, 1995).

We purified p18 and p14 and sent them for mass spectrometry sequencing. The first sequences that came back were from p18 and they did not match any known protein. About this time, I attended an Oncogene Meeting in Frederick, MD and presented a poster on our findings. An adjacent poster from Dr. Richard Klausner described a similar line of investigation. Rick was a legendarily good scientist and decided to work on a human cancer syndrome, namely VHL disease, having recently become the Director of the National Cancer Institute. So much for working in a field without competitors! Over the ensuing years, however, I learned that you actually want competitors, as long as they hold themselves and others to a high scientific standard, and they play fairly. Rick, as well as countless other members of the hypoxia-community, are examples of such investigators.

My heart sank when I learned that Rick's group had identified p18 and p14 as elongin B and elongin C, respectively (I don't remember if this was on his poster or revealed in conversation during the meeting). His group had obtained peptide sequences for both p14 and p18, and p14 matched elongin C. They had then reached out to Drs. Joan and Ron Conaway, who had identified elongin C as part of a transcriptional elongation complex (Elongin or SIII), and learned that p18 was another component of the complex, soon to be deposited in databases as elongin B. Moreover, Rick's group and the Conaway group collaborated to show that pVHL could inhibit Elongin/SIII by competing with elongin A for elongin B and C. It looked like, for the first time in my young scientific career, I had been scooped!

I spoke to Rick and learned that he had already made a prepublication inquiry to *Science* about his paper. It would have been within his rights to squash me like a bug at this point. Instead, he graciously offered to have us co-submit with him. As is often the case in science, our independent papers were complimentary of one another (for example, our pVHL-binding studies and mapping studies were very strong) and provided simultaneous corroboration of the key findings. As a postdoctoral fellow I had cloned *E2F1*. At that time David and I decided to orchestrate the co-publication of a competitor paper from Ed Harlow's group (Helin *et al.*, 1992). It seemed I was now to be the beneficiary of such generosity.

Rick's paper emphasized a model in which tumor suppression by pVHL was linked to its ability to inhibit the Elongin/SIII complex (Duan et al., 1995). Our data didn't really speak to this. Instead, we made the more conservative conclusion that tumor suppression by pVHL required that it bind to elongin B and elongin C (Kibel et al., 1995). In truth, we were also concerned that the low abundance of pVHL relative to elongin B and elongin C challenged whether pVHL could truly outcompete elongin A for binding to elongin B and elongin C.

pVHL REGULATES HYPOXIA-INDUCIBLE mRNAs

I remained intrigued by the idea that pVHL suppressed angiogenesis, which would explain both the increased angiogenesis of VHL-associated tumors and potentially explain our finding that pVHL suppressed ccRCC cell growth in nude mice, but not on plastic dishes (at least under standard culture conditions). I reached out to Dr. Judah Folkman and provided him with our isogenic WT, ARZ, and pRC 786-O cells. My idea was for him to test the conditioned medias from these cells for their ability to promote angiogenesis in a battery of *ex vivo* bioassays he had developed over the years. To my dismay, his laboratory found no difference (I still cannot explain this).

Partly for this reason I couldn't convince anyone in my laboratory to measure *VEGF* mRNA levels in these cells by Northern blot analysis. One day, however, I went to a lecture by Dr. Hal Dvorak, who showed angiograms of retinal neoangiogenesis driven by VEGF. The pictures reminded me of retinal hemangioblastomas. I remember thinking, "it has to be VEGF!".

I then reached out to Dr. Mark Goldberg, who was then at the Brigham and Women's Hospital. He was experienced measuring hypoxia-inducible mRNAs, including the *VEGF* mRNA, by Northern blot. Better still, he had an incubator for growing cells at 1% oxygen. He and I designed an experiment to measure hypoxia-inducible mRNAs, including the *VEGF, PDGF B, and GLUT1* mRNAs, in WT, ARZ, and pRC cells grown at 21% or 1% oxygen. Hep3B *VHL+/+* hepatoma cells, a workhorse in the hypoxic gene regulation field, were included as controls. The actual experiment was done by one of his fellows, Andrew Levy.

I will never forget the day they walked into my office with the autoradiogram (Figure 4). As expected, the hypoxia-inducible mRNAs in the Hep3B cells were only detected at high levels if the cells were grown at 1% oxygen. In contrast, the hypoxia-inducible mRNAs were constitutively produced at high levels in the 786-O ccRCC cells, unless they had been engineered to once again produce wild-type pVHL (Iliopoulos *et al.*,

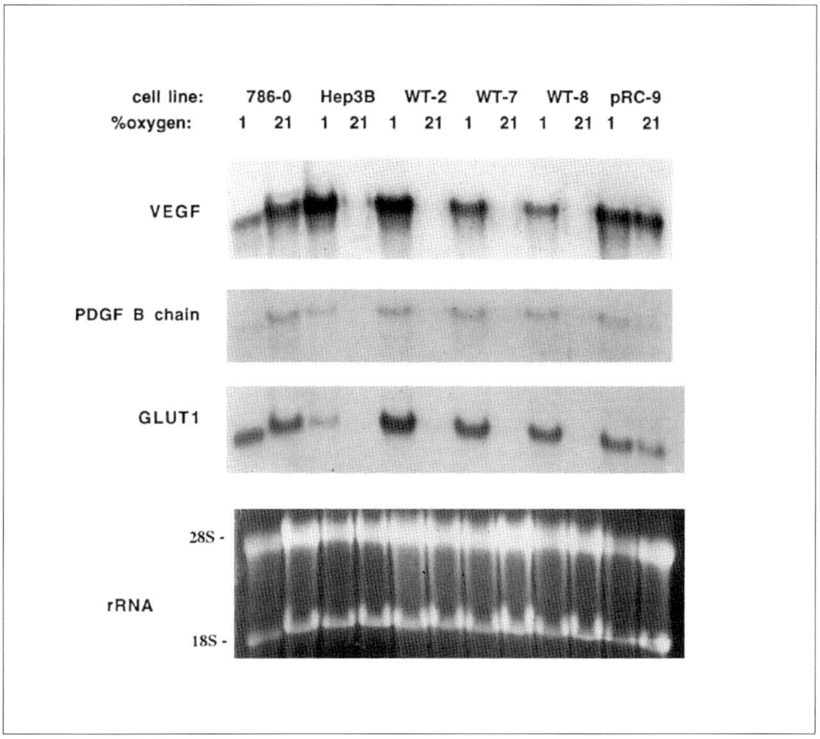

Figure 4. Deregulation of hypoxia-inducible mRNAs in cells lacking wild-type pVHL. Northern blots with the indicated probes using RNA from 786-O *VHL-/-* clear cell renal cell carcinoma cells, ep3B *VHL+/+* hepatoma cells, or independent 786-O subclones stably transfected to produce wild-type pVHL (WT-2, WT-7, and WT-8 subclones) or stably transfected with the backbone vector (pRC-9 subclone). Cells were grown in the presence of 21% oxygen or 1% oxygen, as indicated. Also included is ethidium stained gel prior to transfer to assess loading. From: Iliopoulos, O., *et al.*, Negative Regulation of Hypoxia-Inducible Genes by the von Hippel-Lindau Protein. *Proc. Natl. Acad. Sci.*, 1996. **93**: p. 10595–10599. Copyright (1996) National Academy of Sciences, U.S.A.

1996). Mark told me at the time that this was the first example of a mammalian gene that was strictly required for the transcriptional response to hypoxia. Years later I would hear the expression "the first harpoon in the whale". This was our first real harpoon in the whale. The ability of pVHL to suppress *VEGF* mRNA levels under normoxia was also reported by Marston Linehan's group in collaboration with Rick Klausner (Gnarra *et al.,* 1996).

pVHL FORMS AN E3 UBIQUITIN LIGASE THAT REGULATES THE HIF TRANSCRIPTION FACTOR

Kim Lonergan, a postdoctoral fellow in my laboratory, discovered that Cul2, a member of the cullin family, also bound to pVHL in coimmunoprecipitation assays and did so via elongin C (Lonergan *et al.,* 1998). We collaborated with Joan and Ron Conaway, who showed that Cul2 also copurified with pVHL in native cell extracts. The conclusion that Cul2 bound to pVHL was also reached by the Klausner group working independently (Pause *et al.,* 1997).

Importantly, Kim then tested a series of N-terminal, C-terminal, and missense pVHL mutants for 1) their ability to bind to the elongins and Cul2, and 2) their ability to suppress hypoxia-inducible mRNAs. Binding to the elongins and Cul2 was necessary, but not sufficient, for pVHL to suppress hypoxia-inducible mRNAs (Lonergan *et al.*, 1998). The latter also required sequences N-terminal of the elongin (and hence Cul2)-binding region.

At around this time I had an extremely important conversation with Dr. Steven Elledge, then at Baylor College of Medicine, who was visiting Harvard to give a lecture. He had noticed that the elongin C protein sequence resembled that of the yeast protein Skp1, and that the cullins likewise resembled the yeast protein Cdc53(Bai *et al.*, 1996). He and others had shown that Skp1 and Cdc53, when bound to a so-called F-box protein, generated an ubiquitin ligase complex in which the F-box protein served as the substrate recognition unit (Bai *et al.*, 1996). He therefore predicted that the pVHL, elongin, Cul2 complex was a ubiquitin ligase.

We collaborated with Nikola Pavletich to solve the three-dimensional structure of pVHL bound to elongin B and elongin C. The 3D structure revealed that pVHL actually had two hotspots for mutations linked to VHL disease (Stebbins *et al.*, 1999). The first, which Nikola termed the alpha domain, gratifyingly corresponded to the elongin C-binding domain we had identified biochemically. The second, which Nikola called the beta domain, had the properties of a potential substrate docking site. Finally, the structure confirmed the potential similarity between elongin C and Skp1, and between pVHL and an F-box protein. Collectively, these findings strengthened the case that the pVHL complex was a ubiquitin ligase complex. In further support of this idea, both the Klausner group and Dr. Willy Krek's group showed that anti-pVHL immunoprecipitates contained ubiquitin ligase activity, as though pVHL was, or was associated with, an ubiquitin ligase (Lisztwan *et al.*, 1999, Iwai *et al.*, 1999). Moreover, our collaborators, the Conaways, detected the ubiquitin-conjugating enzyme (E2) Rbx1 bound to the pVHL, elongin, Cul2 complex (Kamura et al., 1999).

The question then became: is the pVHL complex really a ubiquitin ligase and, if so, what are its substrates? Both our group and the Linehan group observed that the suppression of *VEGF* mRNA levels by pVHL involved a modest effect on *VEGF* mRNA transcription and a much more pronounced effect on *VEGF* mRNA stability (this is still not explained given the findings outlined below) (Iliopoulos *et al.*, 1996, Gnarra *et al.*, 1996). For this reason, we hypothesized that pVHL regulated the stability one or more mRNA-binding proteins. Nonetheless, the heterodimeric HIF (Hypoxia-Inducible Factor) was such an attractive target, given its central role in hypoxia-inducible gene regulation and the oxygen-dependent turnover of its alpha subunit, that we did western blots for HIF1α using cell extracts from the WT, ARZ, and pRC cells grown at 1% oxygen or 21%, as well as hypoxic control cells. We did not detect HIF1α in 786-O cells, which furthered our suspicion that pVHL regulated an mRNA-binding protein (and unfortunately quelled any enthusiasm for doing electrophoretic mobility shift assays for HIF, which would have given us the answer).

The VHL Family Alliance is a support network for VHL patients and their family members. They hold annual meetings that bring together scientists, clinicians, and lay stakeholders. The first such meeting I attended, in 1996, was one of the most moving experiences of my life. Up until then, I had never met a patient with VHL disease – they had only existed in textbooks. I became truly inspired by their tenacity, resilience, and courage. In 1998 I attended the VHL Family Alliance Annual Meeting that was held in Paris that year. I met one of Dr. Peter Ratcliffe's postdoctoral fellows, Patrick Maxwell, who kindly told me that he had found that pVHL was strictly required for the degradation of the HIFα subunits in ccRCC under normoxia (Maxwell *et al.*, 1999). The Ratcliffe Laboratory had made a good antibody for HIF2α. Unfortunately for us, our workhorse cell lines at the time, 786-O and A498, produced HIF2α, but not HIF1α (explaining our negative western blot data). In ccRCC cell lines that express both HIF1α and HIF2α, both are upregulated by pVHL loss (Maxwell *et al.*, 1999). Patrick and I agreed that if we both discovered the oxygen-dependent signal that regulated the binding of pVHL to HIFα, we would try to coordinate the publications of our papers.

OXYGEN-DEPENDENT PROLYL HYDROXYLATION OF HIF RECRUITS THE PVHL UBIQUITIN LIGASE

The seminal Maxwell paper appeared in *Nature* (Maxwell *et al.*, 1999), but it did not address whether pVHL directly or indirectly regulated the HIF alpha subunits and it did not describe the mechanism. I had a new post-doctoral fellow in my laboratory, "John", who had "golden hands" at the bench when it came to biochemical experiments. He quickly showed that pVHL, via its beta domain, bound to a region of HIF1α (the Oxygen-Dependent Degradation Domain) that was known to be responsible for its oxygen-dependent turnover. Moreover, with the help of Dr. Vincent Chau (then at Millenium), "John" and Michael Ohh showed that the pVHL complex is indeed the E3 ubiquitin ligase for HIF1α and that this activity is abrogated by tumor-associated *VHL* mutations (Figure 5). We eventually published these findings in *Nature Cell Biology* in 2000 (Ohh *et al.*, 2000). The ability of pVHL to polyubiquitylate HIF1α was subsequently also reported by Ratcliffe, Conaway, and Dr. Lorenz Poellinger laboratories (Kamura *et al.*, 2000, Cockman *et al.*, 2000, Tanimoto et al., 2000).

While our ubiquitinylation paper was out for review, "John" also generated data suggesting that HIFα underwent an oxygen-and iron-dependent posttranslational modification that determined whether it could bind to pVHL. In the summer of 1999, it seemed like we were learning a new

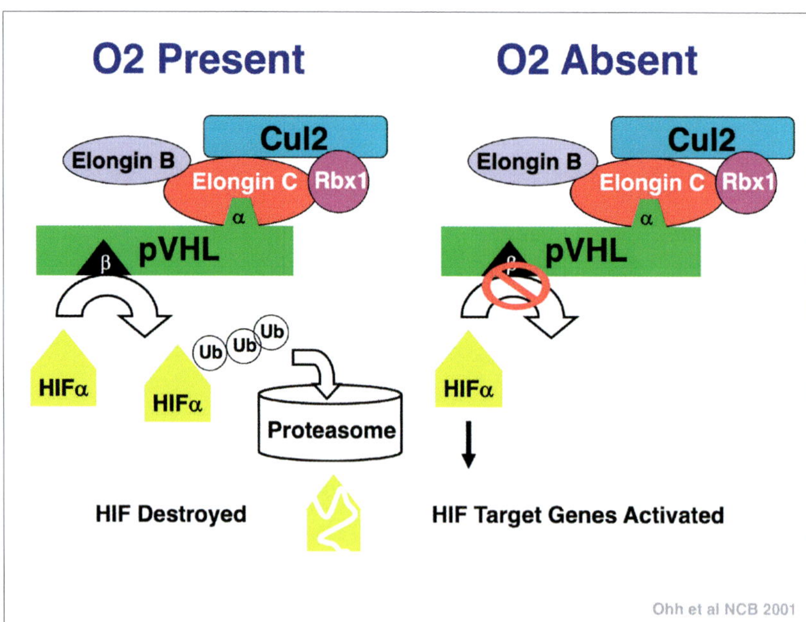

Figure 5. pVHL is the substrate recognition component of a ubiquitin ligase that regulates HIF.

piece of the puzzle almost every week. "John" then told me that he wanted to go to his native country for one week to deal with family matters. One week became two, which became four, which became 7. No word. In the meantime, another postdoctoral fellow in my laboratory confided that "John" had gone home to apply to medical school and to discuss a possible marriage, and that there was a chance he wasn't coming back. He did come back the day I was to leave for my July vacation. I told him that I was disappointed that he had not been more honest with me and that I would have tried to help him with his medical school application if that was what he really wanted. I then left for my vacation. While I was gone, he came to the laboratory one night, took his notebooks, data, and reagents, and returned to his native country.

I returned to the laboratory a week later and was naturally devastated. Some of the most exciting preliminary data I had ever seen had now vanished. I had an emergency meeting of the laboratory and my postdoctoral fellows Michael Ohh and Mircea Ivan stepped in to remake the key reagents and redo the key experiments. Michael did the additional experiments that allowed us to publish the *Nature Cell Biology* paper showing that pVHL was the ubiquitin ligase for pVHL, while Mircea, with help from two other fellows in the laboratory, Haifeng Yang and Billy Kim, pursued the oxygen-dependent posttranslational modification of HIFα. We were still at the leading edge of the field, but this debacle cost us months.

Mircea did pVHL far-western blot assays modeled on my pRB work. Here, however, he used the recombinant pVHL/elongin B/elongin C complex (leftover from the crystallography studies) and detected the bound pVHL with an anti-pVHL antibody (rather than ^{32}P). This taught us that the pVHL bound directly to HIF1α and HIF2α, which was not known previously (Ohh *et al.*, 2000) (Ivan *et al.*, 2001).

Haifeng Yang did a powerful experiment in which TS20 *VHL*+/+ cells, which have a temperature-sensitive mutation in the E1 ubiquitin activating enzyme, were cultured at the restrictive temperature in 1% oxygen or 21% oxygen. As expected, HIF1α accumulated at the restrictive temperature compared to the permissive temperature, as determined by anti-HIF1α immunoblot analysis, consistent with loss of ubiquitylation. In far-western assays, however, pVHL, could only bind to the HIF1α that had accumulated under normoxic conditions (Figure 6) (Ivan *et al.*, 2001).

It was known for years that treating cells with iron chelators or iron antagonists could induce a hypoxic response. Later this was shown to be due to HIF induction. Gratifyingly, Mircea discovered that pVHL could not bind to HIF1α that had been induced by iron chelators or iron antagonists (Ivan *et al.*, 2001). This strongly suggested that HIF1α was indeed undergoing an oxygen-and iron-dependent posttranslational modification that was required for its recognition by pVHL.

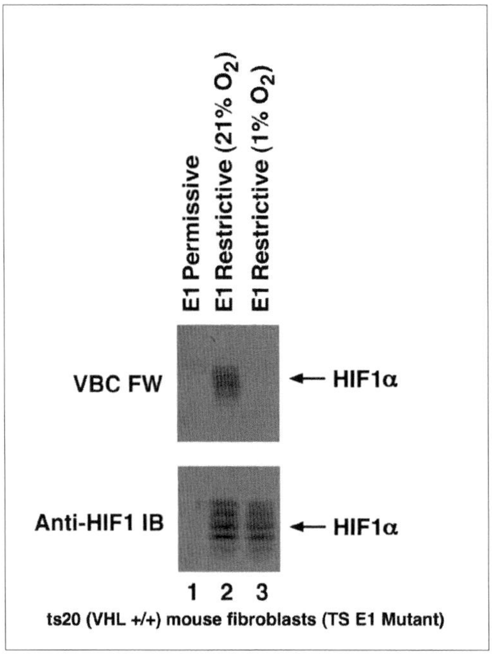

Figure 6. An oxygen-dependent modification of HIFα allows it to bind directly to pVHL. Far-western (FW) Blot with recombinant pVHL/elongin B/elongin C complex and anti- HIFα immunoblot (IB) of TS20 *VHL*+/+ cells grown at the restrictive or permissive temperature. At the restrictive temperature the cells were grown in 21% or 1% oxygen, as indicated. TS20 cells have a temperature-sensitive E1 mutation and do not support ubiquitylation at the restrictive temperature. From: Ivan, M., *et al.*, HIFalpha targeted for VHL-mediated destruction by proline hydroxylation: implications for O_2 sensing. *Science*, 2001. **292**: p. 464–8.

Mircea discovered that recombinant HIF1α fragments produced in *E. coli* (or as synthetic peptides) could only bind to pVHL if first preincubated with a mammalian cell extract (such as unprogrammed rabbit reticulocyte lysate) under conditions permissive for enzymatic activity (presumably to provide the modifying activity) (Ivan *et al.*, 2001). He also found that HIF1α *in vitro* translated in rabbit reticulocyte lysate contained an electrophoretically distinct band that 1) was absent when HIFα was translated using wheat germ extracts and 2) specifically bound to pVHL (Ivan *et al.*, 2001). The assumption was that this band reflected HIFα with the modification we were seeking.

To identify this modification, Mircea mapped the minimal pVHL-binding fragment of HIF1α to a 20mer (corresponding to HIF1α residues 556–575), which contained at its core the sequence MLAPYIPM (Ivan *et al.*, 2001). This 20mer synthetic peptide, immobilized on streptavidin beads via an N-terminal biotin, could bind to 35S-labelled pVHL (made by *in vitro* translation), but again only if the peptide was preincubated with unprogrammed rabbit reticulocyte lysate at 30°C (Figure 7) (Ivan *et al.*, 2001). Fortuitously, Dr. Jaime Caro at Thomas Jefferson had previously done linker scanner mutagenesis of HIF1α and shown that replacing the MLAPYIPM sequence with 8 consecutive alanines rendered HIF1α constitutively stable (Srinivas *et al.*, 1999). Mircea then repeated his binding assays using biotinylated HIF1α peptides in which the entire MLAPYIPM

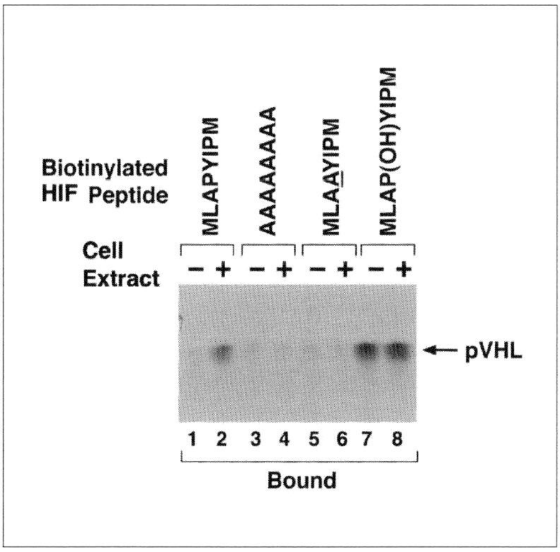

Figure 7. pVHL binds to prolyl hydroxylated HIFα. Binding of ^{35}S-labeled pVHL to biotinylated HIF1α (556–575) peptides with the indicated mutations of the core MLAPYIPM sequence. Peptide was preincubated at 30°C with an unprogrammed reticulocyte lysate where indicated with the "+". From: Ivan, M., *et al.*, HIFalpha targeted for VHL-mediated destruction by proline hydroxylation: implications for O_2 sensing. *Science*, 2001. **292**: p. 464–8.

sequence was replaced with 8 alanines or in which individual residues within the MLAPYIPM were replaced with alanine one at a time (i.e. he did an alanine scan). The two critical residues were the leucine residue and the first proline residue (Figure 7) (Ivan *et al.*, 2001).

I then ran to the computer to look for posttranslational modifications of proline and leucine that might be oxygen-and iron-dependent. Prolyl hydroxylation fit the bill (in fact, I encountered hundreds of papers about collagen prolyl hydroxylation). Mircea also suspected the modification was prolyl hydroxylation, having remembered this posttranslational modification from his college biochemistry course in Romania. He then synthesized his biotinylated HIF1α peptide so that the prolyl residue was already hydroxylated. This was one of those rare "eureka" experiments. The hydroxylated peptide bound to pVHL, and no longer required the reticulocyte lysate preincubation step (presumably because the relevant modification was already present) (Figure 7) (Ivan *et al.*, 2001).

I always tell trainees to think of the least interesting interpretations of their data before jumping to the rosiest interpretations. It was formally possible that the hydroxylated HIF1α peptide was very "sticky" and would have bound non-specifically to any ^{35}S-labelled protein (we just happened to look at pVHL). Billy Kim addressed that by doing peptide pulldown experiments with ^{35}S-labelled cell extracts rather than 35S-labelled pVHL. Here you could see that the binding was exquisitely specific (Figure 8) (Ivan *et al.*, 2001).

Another concern was that the peptide pulldown experiments implied that prolyl hydroxylation was the signal *or* that prolyl hydroxylation mimicked the authentic signal used by cells. The methionines in our 20mer

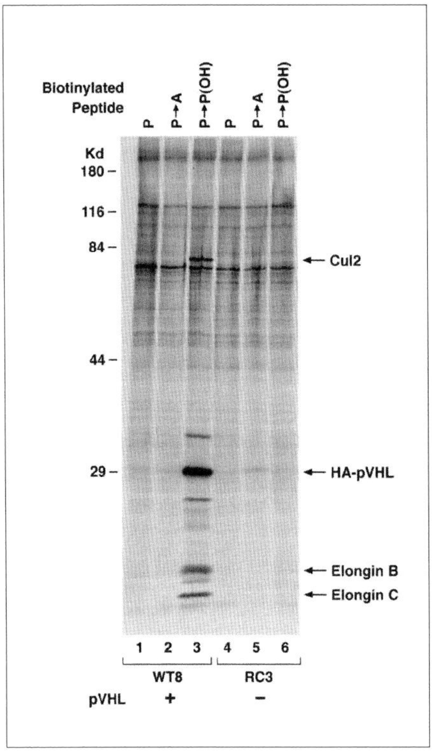

Figure 8. Binding of the pVHL complex to prolyl hydroxylated HIFα is highly specific. 786-O *VHL−/−* cells that were stably transfected to produce wild-type pVHL (WT8) or with the empty vector (RC3) were metabolically labeled with 35S-methionine, lysed, and incubated with biotinylated HIF1α (556–575) peptides with the indicated proline 564 substitutions. Bound proteins after washing were detected by autoradiography. From: Ivan, M., *et al.*, HIFalpha targeted for VHL-mediated destruction by proline hydroxylation: implications for O2 sensing. *Science*, 2001. **292**: p. 464–8.

peptide were not required for binding to pVHL. This was fortunate, because methionines are prone to spontaneous oxidation, which could have complicated our subsequent mass spectrometry analyses. Mass spectrometry of the wild-type HIF1α peptide, but not the HIF1α peptides in which the critical leucine and proline residues were converted to alanine, after incubation with reticulocyte lysate clearly demonstrated prolyl hydroxylation at proline 564 (Ivan *et al.*, 2001). Mircea also showed that HIF1α *in vitro* translate made with rabbit reticulocyte lysate, but not wheat germ extract, in the presence of radioactive proline contained hydroxyproline, as determined by amino acid hydrolysis and thin layer chromatography (Ivan et al., 2001). Finally, Mircea showed that Pro564A mutation stabilized full-length HIF1α in cells (Ivan *et al.*, 2001).

As promised, I reached out to Patrick Maxwell before submitting our findings for publication. He appropriately referred me to his mentor, Peter Ratcliffe. I had never met Peter before and was now doing so by phone. I didn't want to deprive him of the joy of discovering the oxygen sensing mechanism on his own nor, I suppose, was I interested in co-publishing with him if he was hopelessly behind us. We therefore proceeded to speak a bit in code, turning over one card at time so to speak, until it was clear that we had independently arrived at the same conclusion (Ivan *et al.*, 2001, Jaakkola *et al.*, 2001).

Years earlier, when I was just getting started as an independent investigator, an editor at *Science* named Paula Kiberstis met with me at a Keystone Symposium to find out what I was doing in my laboratory. I felt like I had arrived (recently she confessed to me that she thought the work I described that day was pretty boring). I suggested to Peter that we submit our papers to Science and that Paula handle the papers. Peter and I agree that this was a great decision. Firstly, at least in my case, she improved my writing considerably (I still use some of her conventions). More importantly, the reviews were not all positive. One reviewer said we had to be wrong because prolyl hydroxylation only took place in the endoplasmic reticulum. Paula realized this reviewer had tunnel vision based on their own work (presumably on collagen). Another more serious debate amongst the reviewers and editorial board was whether identification of the relevant enzyme(s) should be required for publication. Paula intervened, arguing that our paper was already a major step forward and would open the search for the enzyme(s) to others. Sadly, such editorial interventions seem less common today.

I have learned there are the competitors you know about and the competitors you don't know about. A few months after our two *Science* papers appeared, Frank Lee also reported that HIFα was prolyl hydroxylated (Yu *et al.*, 2001). The Ratcliffe Laboratory would also later report that both HIF1α and HIF2α contained second potential prolyl hydroxylation sites (Masson *et al.*, 2001).

DEVELOPMENT OF PROLYL HYDROXYLASE INHIBITORS FOR THE TREATMENT OF ANEMIA AND ISCHEMIA

Once we knew that HIFα stability was controlled by prolyl hydroxylation, I became intrigued with the idea of developing prolyl hydroxylase inhibitors as a way of treating diseases where HIF might be beneficial, such as anemia, myocardial infarction, and stroke (Ivan *et al.*, 2001). By now I was familiar with the collagen prolyl hydroxylases and I wondered if anyone had developed small molecule inhibitors against them. An internet search led me to a small company in South San Francisco called Fibrogen, Inc., which was making collagen prolyl hydroxylase inhibitors as antifibrotic agents. My hope was that some of their chemical inhibitors might fortuitously also inhibit the prolyl hydroxylation of HIFα and, if so, be good starting points for making more specific inhibitors of the (as yet unidentified) HIF prolyl hydroxylases.

I called Fibrogen and spoke to a scientist named "Max" (I believe this was in early 2001, before the publication of our *Science* paper). He listened very politely as I told him about our findings. He then proceeded to tell me that I had to be wrong because prolyl hydroxylation takes place in the endoplasmic reticulum, whereas HIF is nuclear. He closed the call by saying we must be studying an artefact. Fifteen minutes later he called me back and said, "Please start again from the beginning". Years later I would learn that the second phone call was due to Tom Neff, the late CEO of Fibrogen. Tom had passed "Max" in the hallway immediately after the first call and, seeing that "Max" was amused, asked him what was so funny. "Max" proceeded to tell him about the crazy scientist from Boston who thought HIF was prolyl hydroxylated. Tom stopped in his tracks and said, "Don't you remember that some of our early prolyl hydroxylase inhibitors caused polycythemia in rabbits. This scientist is trying to tell you why. Call him back!". My longstanding collaboration with Fibrogen began with that second phone call. Shortly thereafter I had dinner with Tom in Boston. He said that a new druggable class of enzymes only comes along every 30 or 40 years. He also laid out his vision for developing a new class of anemia drugs and he predicted that I would one day win the Nobel Prize. I told him that if I was ever so lucky it would probably be because Fibrogen eventually succeeded and that, as a result, our legacies were now joined. Sadly, Tom died suddenly shortly before the Nobel announcement.

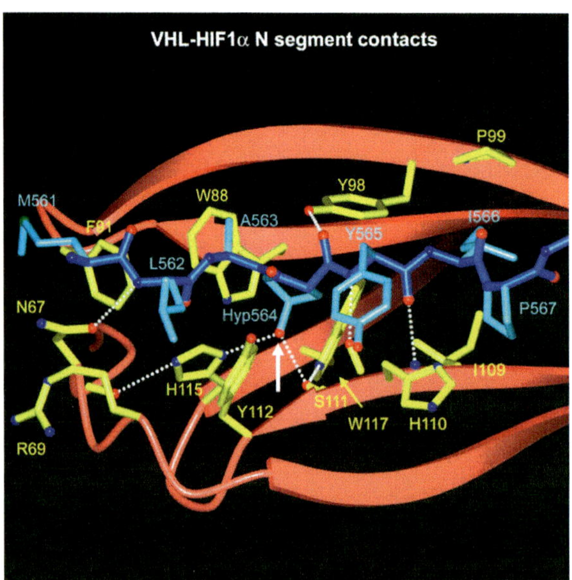

Figure 9. Structure of pVHL bound to prolyl hydroxylated HIFα peptide. The backbone of pVHL is shown in red and its side chains in yellow. The HIF1α backbone is in medium blue and its side chains in light blue. Hydrogen bonds are shown with dotted lines. Note hydrogen bonds between His115 and hydroxylated Pro564 and Ser111 and hydroxylated Pro564. From: Min, J.H., *et al.*, Structure of an HIF-1alpha -pVHL complex: hydroxyproline recognition in signaling. *Science*, 2002. **296** (5574): p. 1886–9.

To understand how pVHL discriminated between hydroxylated and non-hydroxylated HIF1α, we again collaborated with Nikola Pavletich, who then solved the structure of the pVHL/elongin B/elongin C complex bound to our prolyl hydroxylated HIF1α peptide (Figure 9) (Min *et al.*, 2002). It confirmed that the hydroxylated HIF1α peptide bound to the pVHL beta domain. The HIF1α binding site on pVHL consisted largely of hydrophobic residues except for two hydrophilic residues, S112 and H115, that normally interact with water molecules. When the HIF1α peptide interacts with pVHL, this water is displaced, which is energetically unfavorable. However, if the HIF1α is hydroxylated on proline 564, the hydroxyl group can hydrogen bond with S112 and H115, thus compensating for the displaced water molecules. As Nikola would later tell me, if you were trying to design a protein that was specific for hydroxylproline, you couldn't do much better than this. A similar paper appeared from the Ratcliffe group (Hon *et al.*, 2002).

At this point we still didn't know the identity of the enzyme (or enzymes) that hydroxylated HIFα. The collagen prolyl hydroxylases were unlikely candidates because of their subcellular localizations and high oxygen affinities (making them relatively insensitive to oxygen). Moreover, the sequences surrounding the collagen and HIFα prolyl hydroxylation sites were highly dissimilar.

I made two assumptions that proved to be correct. The first was that we could biochemically purify the HIFα prolyl hydroxylase by classical biochemical fractionation, monitoring the ability of fractions to hydroxylate our synthetic, biotinylated, HIF1α peptide (as determined by capture of ^{35}S-labelled pVHL). That way at least the field would have the answer and could move forward. The second assumption, however, was that someone might get their first using a clever genetic approach.

We again collaborated with Joan and Ron Conaway, who had and have far more biochemical expertise than me. They worked closely with my postdoctoral fellow, Mircea Ivan, who spent time in their laboratory. We ultimately identified a biochemical fraction of rabbit reticulocyte lysate that could hydroxylate HIF1α despite containing fewer than 10 proteins (Figure 10). One of those proteins was EglN1 (Figure 10) (Ivan *et al.*, 2002). This was an excellent candidate, as a bioinformatic paper by Drs. Aravind and Koonin at the NCI predicted that EglN1 was actually a 2-oxoglutarate-dependent dioxygenase (Aravind and Koonin, 2001). We then confirmed that recombinant EglN1 could hydroxylate HIF1α (Ivan *et al.*, 2002).

While this work was being completed, I visited UCLA to give a lecture. While there I got a message to call Mircea right away. He informed me that a paper had just appeared from the Ratcliffe group reporting that the three mammalian EglN family members (originally called EglN1, EglN2, and

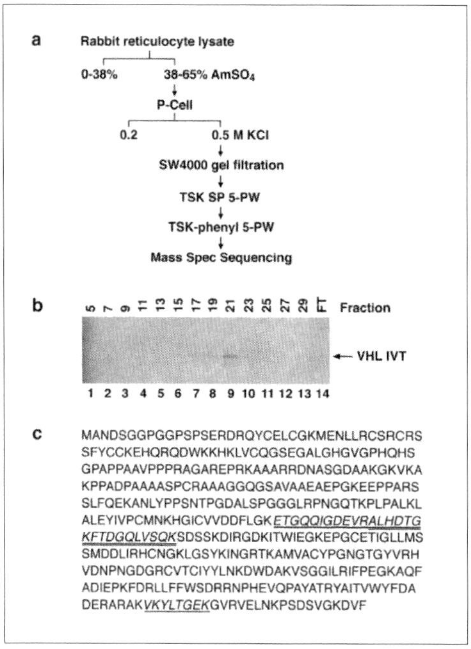

Figure 10. Purification of EglN1 as HIFα prolyl hydroxylase. Purification of EglN1. (a) purification schema. (b) Binding of ^{35}S-labeled pVHL to biotinylated a HIF1α (556–55) peptide. Peptide was preincubated at 30°C with indicated column fractions after TSK SP 5-PW chromatography. FT = flow through. (c) Conceptual open reading frame of EglN1. Peptides identified by mass spectrometry of partially purified HIF prolyl hydroxylase are shown with single or double underline. From: 1. Ivan, M., *et al.*, Biochemical purification and pharmacological inhibition of a mammalian prolyl hydroxylase acting on hypoxia-inducible factor. *Proc Natl Acad Sci U S A*, 2002. **99** (21): p. 13459–64. Copyright (2001) National Academy of Sciences, U.S.A.

EglN3, which they renamed PHD2, PHD1, and PHD3, respectively) were the HIFα prolyl hydroxylases (Epstein *et al.*, 2001). Peter, I would later learn, had a "secret weapon" in his Oxford colleague Chris Schofield, who was an expert regarding 2-oxoglutarate-dependent dioxygenases. He, like Aravind and Koonin, saw that the EglNs were likely to be 2-oxoglutarate-dependent dioxygenases and hence candidate prolyl hydroxylases. Peter's group then showed that eliminating the sole EglN family member in C. *elegans* using RNAi, but not other dioxygenases, induced HIF (Epstein *et al.*, 2001). They went on to confirm that human EglN1, EglN2, and EglN3 could all hydroxylate HIF1α (Epstein *et al.*, 2001). Drs. Rick Bruick and Steven McKnight came to the same conclusion shortly thereafter using a very similar strategy in Drosophila cells (Bruick and McKnight, 2001). Later work by Jacques Pouyssegur indicated that EglN1 is the major enzyme responsible for regulating HIFα turnover in human cells (Berra *et al.*, 2003).

We quickly finished our paper describing the biochemical purification of EglN1 as the HIFα prolyl hydroxylase (Ivan et al., 2002). Partly to enhance the novelty of our biochemical paper, we included data showing that a variety of structurally unrelated prolyl hydroxylase inhibitors from Fibrogen could, as I suspected, inhibit EglN1 in biochemical assays and stabilize HIF in cell culture experiments (Ivan *et al.*, 2002). This then launched medicinal chemistry efforts at Fibrogen and elsewhere to make more specific EglN inhibitors.

The development of EglN inhibitors was enhanced by a powerful mouse model created by a postdoctoral fellow in my laboratory, Michal Safran (Safran *et al.*, 2006). Michal made a reporter consisting of the HIF1α oxygen-dependent degradation domain fused to firefly luciferase, which she then inserted into the murine *ROSA26* locus (to ensure ubiquitous expression). Firefly luciferase expression in mice can be imaged non-invasively by monitoring photon emissions after administration of luciferin. In pilot experiments, Michal observed clear signals over the thymus and kidneys in her reporter mice breathing room air, consistent with the knowledge that these organs are hypoxic at rest (Safran *et al.*, 2006). She also confirmed that the reporter was systemically induced when mice were switched to 8% oxygen (Safran *et al.*, 2006). Using her mouse, she quickly identified which of the Fibrogen compounds we had been given were bioactive *in vivo* after administration by oral gavage, in which organs, and for how long (for example, as in Figure 11). This helped identify promising scaffolds for further medicinal chemistry efforts. Michal did a proof-of-concept study showing that an early Fibrogen EglN inhibitor could induce erythropoietin and red blood cell production in functionally anephric mice, setting the stage for the eventual development of Fibrogen's clinical candidate, Roxadustat (Safran *et al.*, 2006).

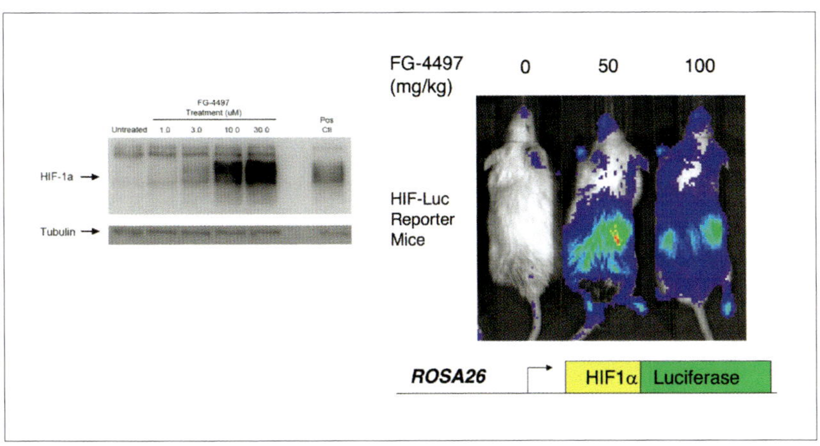

Figure 11. Stabilization of HIFα with orally available EglN inhibitor. Immunoblot of HL-1 Cardiomyocytes (left panel) and bioluminescent imaging of HIF1α-Luciferase mice (right panel) treated with EglN inhibitor FG-4497 in media or by oral gavage, respectively. The right panel also includes a schematic for the reporter, which consists of the HIF1α oxygen-dependent degradation domain fused to firefly luciferase, inserted into the ROSA26 locus (first described in {Safran, 2006 #2948). From: Olenchock, B.A., *et al.*, EGLN1 Inhibition and Rerouting of alpha-Ketoglutarate Suffice for Remote Ischemic Protection. *Cell*, 2016. **164** (5): p. 884–95.

Consistent with EglN1 being the major HIF regulator amongst the 3 EglN paralogs, *EglN1–/–* mouse embryos are not viable, whereas *EglN2–/–* mice and *EglN3* are grossly normal (Takeda *et al.*, 2006). To circumvent the *EglN1–/–* embryonic lethality, a postdoctoral fellow in my laboratory, Andy Minamishima, made mice where he could conditionally inactivate *EglN1* using a floxxed *EglN1* allele and a tamoxifen-inducible Cre recombinase (Minamishima *et al.*, 2008). He showed that inactivation of *EglN1* in adult mice caused massive polycythemia (the same conclusion was reached by Guo-Hua Fong working in parallel (Takeda *et al.*, 2008)) due to increased renal erythropoietin expression (Minamishima *et al.*, 2008).

During fetal life the liver is the major source of erythropoietin. Shortly after birth, however, the hepatic *erythropoietin* locus is silenced, and the kidney becomes the major source of erythropoietin. This has both medical and economic implications. For example, over 20 million Americans have chronic kidney disease and 2–4 million of them will be anemic due, at least partly, to a relative deficiency of erythropoietin. To ask if the hepatic erythropoietin locus could be reactivated in adults by manipulating EglN activity, Andy eliminated *EglN1, EglN2,* and *EglN3* – either singly, in every pairwise combination, or all three – in the livers of adult mice and measured hepatic *erythropoietin* mRNA and serum erythropoietin (Minamishima and Kaelin, 2010). Eliminating all 3 paralogs led to robust and sustained induction of hepatic *erythropoietin* mRNA levels and circulating erythropoietin, thus explaining Michal's earlier results in functionally anephric mice (Figure 12) (Minamishima and Kaelin, 2010).

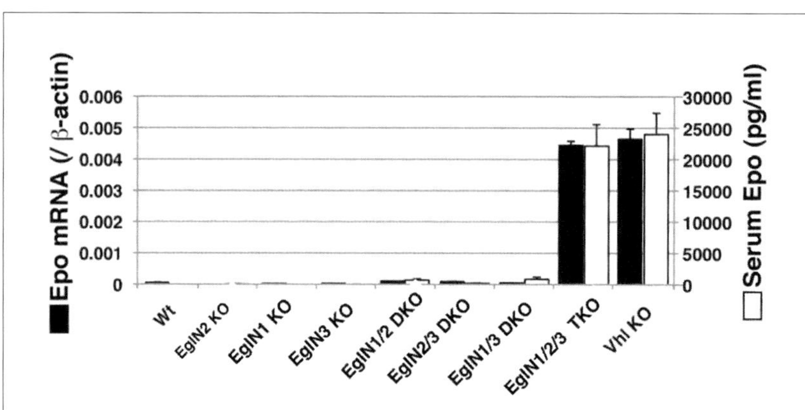

Figure 12. Reactivation of hepatic *EPO* locus after EglN inactivation. Hepatic erythropoietin (Epo) mRNA levels (solid bars), as determined by real-time RT-PCR, and serum Epo (open bars) in mice with the indicated genotypes. N= 3 mice per group. Note that the EglN1 and VHL knockouts were liver-specific. From: Minamishima, Y.A. and W.G. Kaelin, Jr., *Reactivation of hepatic EPO synthesis in mice after PHD loss.* Science, 2010. **329** (5990): p. 407.

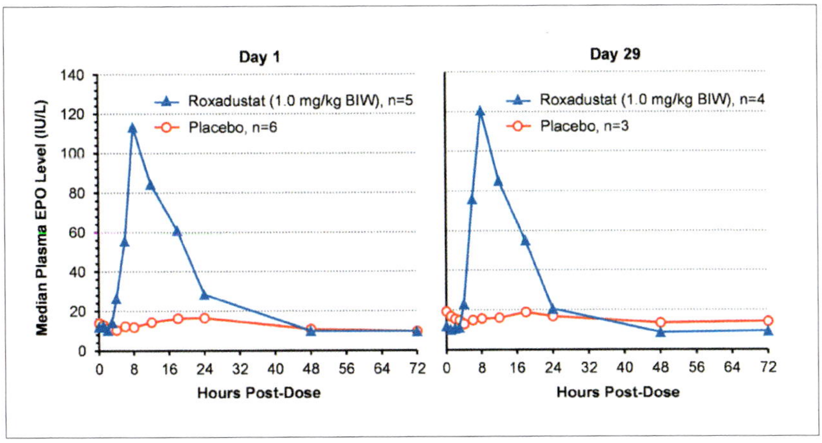

Figure 13. Induction of serum EPO in chronic kidney disease patients taking Roxadustat. From: Besarab, A., *et al., Randomized placebo-controlled dose-ranging and pharmacodynamics study of roxadustat (FG-4592) to treat anemia in nondialysis-dependent chronic kidney disease (NDD-CKD) patients.* Nephrol Dial Transplant, 2015. **30** (10): p. 1665–73.

Early work by us and others showed that EglN3 is induced by HIF and that EglN2 and EglN3 can partially compensate for loss of EglN1 (Minamishima *et al.*, 2009). Andy found that acute inactivation of EglN1 transiently induced hepatic *erythropoietin* mRNA levels, which were then quickly dampened (presumably by EglN3) (Minamishima and Kaelin, 2010). Likewise, we collaborated with Alnylam Pharmaceuticals and showed that nanoparticle-based deliver of *EglN1* siRNAs in mice transiently induced erythropoietin (Querbes *et al.*, 2012).

At least four orally available prolyl hydroxylase inhibitors, including Roxadustat (Fibrogen), Vadadustat (Akebia), Daprodustat (GSK) and Molidustat (Bayer), have advanced to late stage clinical trials for the treatment of anemia in the setting of chronic kidney disease (Gupta and Wish, 2017). Roxadustat, the most advanced of these, has been approved in China and Japan based on positive phase 2 and phase 3 data (Figures 13 and 14). The United States new drug application was submitted in December of 2019 (Besarab *et al.*, 2015, Chen *et al.*, 2019, Kaplan, 2019). We and others have also shown that prolyl hydroxylase inhibitors are tissue protective in preclinical models of regional ischemia (Olenchock *et al.*, 2016).

KIDNEY CANCER TREATMENTS EMERGING FROM STUDIES OF PVHL AND HIF

Inactivation of the *VHL* gene is the initiating genetic event in ccRCCs arising in VHL patients, but is not sufficient to cause ccRCC. Loss of the remaining wild-type *VHL* allele in the kidneys of VHL disease patients

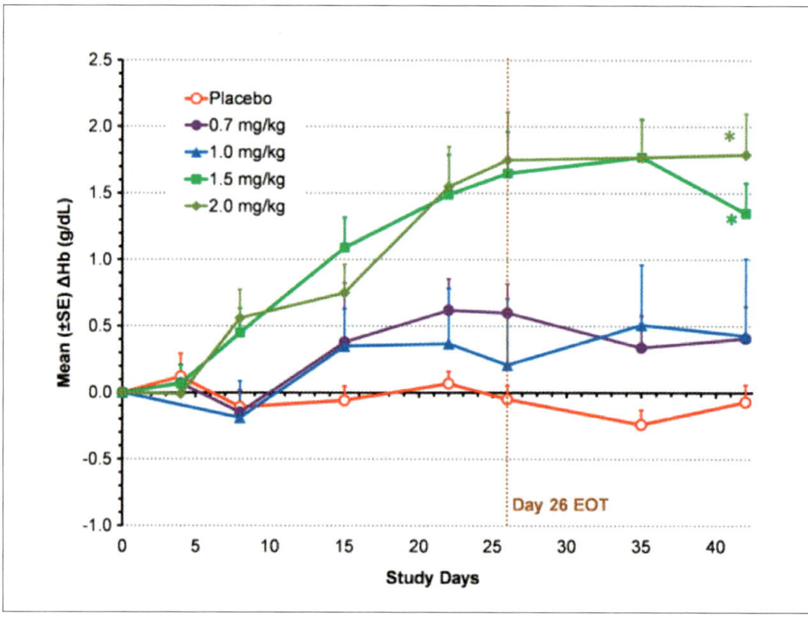

Figure 14. Increased red blood cell production in chronic kidney disease patients taking Roxadustat. From: Besarab, A., *et al.*, *Randomized placebo-controlled dose-ranging and pharmacodynamics study of roxadustat (FG-4592) to treat anemia in nondialysis-dependent chronic kidney disease (NDD-CKD) patients.* Nephrol Dial Transplant, 2015. **30** (10): p. 1665–73.

causes the development of preneoplastic renal cysts. Additional stereotypical cooperating events, such as loss of the *PBRM1* or *BAP1* tumor suppressor genes, are required for progression to ccRCC (Kaelin, 2015). Dr. Charles Swanton and colleagues have done deep sequencing of multiple spatially distinct regions of ccRCCs and used the mutant allele frequencies from those regions from a given patient to infer their evolutionary histories (Gerlinger *et al.*, 2012) (Gerlinger *et al.*, 2014) (Turajlic *et al.*, 2018, Mitchell *et al.*, 2018). Here it again appears that biallelic *VHL* inactivation is the usual initiating or "truncal" mutation in most ccRCCs. Despite (or perhaps because of) this genetic complexity, our reconstitution experiments suggested that at least a subset of ccRCCs require an ongoing loss of pVHL function to grow as tumors in nude mice.

An open question in 2000 was whether HIF dysregulation was causal, or merely correlative, with respective to pVHL-defective tumors. It seemed plausible that it was causal, although HIF biology is very complex. HIF regulates many genes, some of which are likely to promote tumor growth (e.g. VEGF), and others that are likely to constrain tumor growth (e.g. the mTOR inhibitor REDD1 (Reiling and Hafen, 2004, Brugarolas *et al.*, 2004)). We had already shown that restoring pVHL function in *VHL−/−* ccRCC cells suppressed their ability to form tumors in nude mice (Iliopoulos *et al.,* 1995). A postdoctoral fellow in my laboratory, Keiichi Kondo, then showed that such tumor suppression by pVHL could be overridden by cointroducing a HIF2α variant that escapes recognition by pVHL because of proline to alanine substitutions at its two potential prolyl hydroxylation sites (HIF2 dPA mutant) (Kondo *et al.,* 2002). Importantly, this was a specific property of HIF2α. Conversely, Keiichi showed that eliminating HIF2α with shRNAs suppressed tumor growth by two different *VHL−/−* ccRCC cell lines, as did my former postdoctoral fellow, Othon Iliopoulos, in his own laboratory (Kondo *et al.,* 2003) (Zimmer *et al.,* 2004). By then we were already growing leery of potential off-target effects with shRNAs, but Keiichi confirmed that the HIF2 shRNA effects were on-target because he could rescue them with an shRNA-resistant *HIF2α* mRNA (Kondo *et al.,* 2003). Billy Kim showed that transgenic expression of *HIF2α dPA,* but not *HIF1α dPA,* phenocopied VHL inactivation in the liver and skin (Kim *et al.,* 2006). In the former he observed vascular lesions that very loosely resemble hemangioblastomas. And Lianjie Li showed that VHL families with a high risk of developing ccRCC have *VHL* alleles that cause higher levels of HIF than those families with a low risk of developing ccRCC (Li *et al.,* 2007).

Another postdoctoral fellow in the laboratory, Chuan Shen, took this a bit further. She showed that restoring wild-type HIF1α expression in *VHL−/−* ccRCC cell lines that lacked HIF1α suppressed their ability to form tumors in nude mice, while eliminating HIF1α in *VHL−/−* ccRCC cells lines that retained wild-type HIF1α promoted their tumor forming capability (Shen *et al.,* 2011). The differential effects of HIF1α and HIF2α in ccRCC were also noted by Drs. Celeste Simon and Peter Ratcliffe (Gordan *et al.,* 2008, Gordan *et al.,* 2007, Covello *et al.,* 2005, Raval *et al.,* 2005, Mandriota *et al.,* 2002).

Dating back to the mid-1990s we knew that inactivation of pVHL upregulated VEGF and PDGF, providing a potential explanation for the highly angiogenic nature of VHL-associated neoplasms. VEGF is a mitogen and survival factor for endothelial cells, while PDGF stimulates pericytes. The work of Dr. Eli Keshet and coworkers showed that VEGF was particularly important for newly sprouting endothelial cells that lacked pericyte coverage (Benjamin *et al.,* 1998, Benjamin and Keshet, 1997).

I couldn't believe my luck that a number of pharmaceutical companies were, by the late 1990s, developing drugs that could inhibit VEGF itself (e.g. bevacizumab from Genentech) or the VEGF receptor KDR (e.g. SU5416 from Sugen and PTK787 from Novartis/Schering). Better still, some of the KDR inhibitors also fortuitously inhibited the PDGF receptor. It seemed like God was smiling on me. I began arguing that if VEGF inhibitors were going to work against any solid tumor, they would work against VHL-associated tumors such as kidney cancers and hemangioblastomas. In other words, I thought these tumors offered the best chance to demonstrate proof of concept for this class of drugs in cancer. I was particularly optimistic about these drugs in hemangioblastomas because there were reasons to think that hemangioblastomas were genetically simpler that kidney cancers and because there were preclinical data to suggest that local deregulation of VEGF was sufficient (and one hoped necessary) to produce such vascular lesions (Benjamin and Keshet, 1997).

During this time I was consulting for Novartis, which had partnered with Schering to develop PTK787 (Drevs *et al.*, 2000). I advocated for a randomized trial of PTK787 versus placebo in kidney cancer. Unfortunately, the Novartis clinical development team was interested in a commercially more attractive cancer, such as colon cancer, and were leery of kidney cancer because so many prior drugs had failed in kidney cancer trials. Novartis also invoked the conventional wisdom that randomized phase 2 trials lack sufficient statistical power to draw meaningful conclusions. Ultimately, they ran two large randomized trials in colon cancer (which failed), but they did support a single arm trial of PTK787 in kidney cancer. A few responses were seen (de Bazelaire *et al.*, 2008). More remarkably, however, the time to progression data appeared to be remarkably better than one would have expected for a similar cohort of patients. Novartis convened a panel of clinical key opinion leaders (KOLs) who reviewed the data (I would later deduce that almost all of these KOLs were by then working with other VEGF inhibitors). The KOLs said the data were indeed exciting, but that it was formally possible the data were driven by patient selection or chance. This effectively ended PTK787 as a kidney cancer drug.

It became clear to me that small biotechnology companies might be less conservative than large, well-established, drug companies. Through the VHL family alliance I learned of a VHL patient who had lost one eye and was now legally blind because of a retinal hemangioblastoma adjacent to the optic nerve in her remaining eye. I worked with an ophthalmologist at the Joslin Diabetes Center, Dr. Lloyd Aiello, Jr., to treat her with an early KDR inhibitor from Sugen called SU5416 (Aiello *et al.*, 2002). She had dramatic improvement in her vision, likely because of decreased vessel leakiness (VEGF is also known as vascular permeability factor), but her hemangioblastoma did not shrink (Aiello *et al.*, 2002). Later attempts by others to treat VHL-associated hemangioblastomas with KDR inhibitors likewise produced modest results (Girmens *et al.*, 2003, Jonasch *et al.*, 2011, Migliorini *et al.*, 2015).

Sometime in the late 1990s I shared a car to Logan Airport with Rick Klausner, who at that time was still NCI director. I thought the NCI could do the type of clinical experiment that was needed to test whether VEGF inhibitors could alter the natural history of kidney cancer. He enthusiastically agreed. The NCI did a clinical trial of bevacizumab in patients with metastatic kidney cancer, randomizing patients to placebo, low dose, or high dose bevacizumab (I don't actually know if my discussion with Rick was instrumental in this regard) (Yang *et al.*, 2002, Yang *et al.*, 2003). The fact that a placebo arm was ethically justifiable underscores the dearth of treatment options for kidney cancer patients at that time. The tumor shrinkage, or objective response, rates for the bevacizumab arms was very modest. Only 10% of patients on the high dose bevacizumab arm met the criteria for partial responses (Yang *et al.*, 2002, Yang *et al.*, 2003). But there was a clear, dose-dependent, improvement in time to progression in the patients who received bevacizumab (Yang *et al.*, 2002, Yang *et al.*, 2003). Plotting tumor size over time, in so-called "spider plots", you could clearly see that bevacizumab altered the natural history of these tumors (Yang *et al.*, 2002, Yang *et al.*, 2003). I was thrilled.

I loved this trial for two other reasons. One of the placebo patients actually had a partial response (spontaneous tumor regressions, although rare, have been reported in kidney cancer). Had this been a single arm trial of some new drug, this patient might have created false hope. More importantly, one of the early patients on the trial had massive hematuria and was taken to the NCI emergency room. The code was broken, at which point it was learned that this was also a placebo patient. Of course, there was a 2/3 chance the patient would have been getting bevacizumab. It gives me chills to think that if this patient been on bevacizumab the development of VEGF inhibitors might have been significantly delayed for safety concerns.

Another company, Onyx Pharmaceuticals (in collaboration with Bayer), had developed a Raf kinase inhibitor called BAY 43-9006 (sorafenib). In phase 1 trials it appeared that kidney cancers might be particularly sensitive to BAY 43-9006, which did not make sense given what was known about Raf and kidney cancer at that time (Awada *et al.*, 2005). Frank McCormick, a founder at Onyx, was familiar with our work. He suggested they test sorafenib against KDR. Sorafenib proved to be a fairly effective KDR inhibitor and was then developed as a kidney cancer drug (Ratain *et al.*, 2006, Escudier *et al.*, 2007). Sugen produced additional KDR inhibitors, including the drug now called sunitinib (Motzer *et al.*, 2007). Both sorafenib and sunitinib were approved for the treatment of kidney cancer based on positive phase 3 clinical trial data (Escudier *et al.*, 2007, Motzer *et al.*, 2007). Currently there are 7 approved KDR inhibitors for kidney cancer (avastin, sorafenib, sunitinib, axitinib, pazopanib, cabozantinib, levantinib) (Choueiri and Motzer, 2017). The objective response rates of these drugs differ somewhat, possibly due to differences in their ability to inhibit KDR, PDGFR, and other off-targets (e.g. cabozantinib also inhibits MET, which has also been linked to kidney cancer growth and VHL biology (Koochekpour *et al.*, 1999, Bommi-Reddy et al., 2008, Nakaigawa et al., 2006, Pennacchietti *et al.*, 2003)).

Although VEGF inhibitors have become mainstays of kidney cancer treatment, not all patients respond to them and all will eventually become resistant to them. It seemed likely, a priori, that targeting HIF2 would be a more effective ccRCC treatment than targeting any single HIF2-responsive protein (e.g. VEGF). The dogma, however, was that DNA-binding transcription factors such as HIF2, with the exception of the steroid hormone receptors (which have hydrophobic ligand-binding pockets), were undruggable.

In ~2010, however, I was approached to join the scientific advisor board of a small biotechnology start-up company called Peloton Therapeutics in Dallas, Texas. One of the cofounders of this company, Steven McKnight at the University of Texas, Southwestern (UTSW), had originally cloned HIF2α (Tian *et al.*, 1997) and was aware of our work on HIF2α in ccRCC. Two of his UTSW colleagues, Rick Bruick and Kevin Gardner, had a identified a potentially druggable pocket in the HIF2α PAS B domain(Scheuermann *et al.*, 2009). Moreover, they identified chemicals that could bind to this pocket and, in so doing, induce an allosteric change in HIF2α such that it could no longer bind to its partner, ARNT(Rogers *et al.*, 2013, Scheuermann *et al.*, 2013, Scheuermann *et al.*, 2009).

These chemicals were then outlicensed to form Peloton Therapeutics. The Peloton chemists did a brilliant job making these UTSW chemicals more drug-like: increasing their potency, specificity, and bioavailability (Wallace *et al.*, 2016). They provided us with a tool compound, PT2399, which was highly similar to their initial clinical candidate, PT2385.

A major challenge in pharmacology, especially when doing so-called "down assays" (e.g. decreased viability, decreased proliferation, or decreased tumor growth), is to make sure that chemical-induced phenotypes are on-target and not off-target. Hyejin Cho, a postdoctoral fellow in my laboratory, showed that treating *VHL−/−* ccRCC cell lines with PT2399 decreased HIF2-responsive mRNAs, decreased soft agar colony formation, and decreased tumor growth in nude mouse xenograft assays (Cho *et al.*, 2016). Importantly, Hyejin showed that these effects were largely rescued by a PT2399-resistant version of HIF2α that had been created by Bruick and Gardner, thus establishing that the PT2399 effects were on-target (Cho *et al.*, 2016). The activity of this compound in preclinical ccRCC models was also reported by my former postdoctoral fellow, Dr. James Brugarolas, after he started his own laboratory (Chen *et al.*, 2016).

To our surprise, we discovered that some *VHL−/−* ccRCC lines are insensitive to genetic (CRISPR/Cas9-mediated) and pharmacologic (PT2399) inhibition of HIF2 (Cho *et al.*, 2016). These same lines, when tested, are insensitive to restoration of pVHL function (unpublished data). It is possible that some of these lines lost HIF2-dependence over time in culture. Alternatively, some cooperating genetic events in ccRCC (i.e. ensuing mutations after *VHL* mutation) might render ccRCC insensitive to pVHL and loss of HIF.

The first HIF2 inhibitor to be tested in the clinic, PT2385, demonstrated some activity in heavily pretreated ccRCC patients (Courtney *et al.*, 2018). This compound, however, has a metabolic liability that caused unacceptable variability in pharmacokinetics (Courtney *et al.*, 2018). PT2977 is a more potent HIF2 inhibitor with more favorable pharmacokinetic properties. It is about to enter Phase 3 testing in ccRCC based on very promising Phase 2 data in patients who have failed standard of care agents (Figure 15) (Jonasch *et al.*, 2019).

Approximately 50 VHL patients with measurable kidney tumors who had not received prior medical therapy have also been treated with PT2977. These patients were being followed in active surveillance programs in an attempt to delay or avoid multiple surgeries. As expected, many of these patients have other tumors, such as hemangioblastomas.

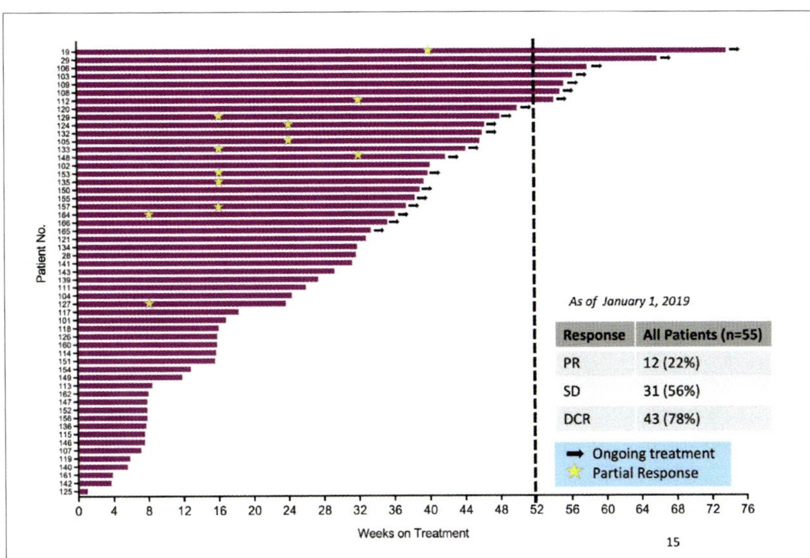

Figure 15. Swimmers plot of patients with metastatic kidney cancer treated with PT2399. Each horizontal bar represents an individual patient with metastatic kidney cancer and how long they remained on therapy once they began treatment with HIF2 inhibitor PT2399. Arrows indicate patients who were still on therapy at the time of this analysis. Yellow stars indicate patients who had a partial response (PR) by RESIST criteria. SD = Stable disease. DCR = disease control rate. From: 1. Jonasch, E., *et al., A first-in-human phase I/II trial of the oral HIF-2a inhibitor PT2977 in patients with advanced RCC.* 2019 European Society for Medical Oncology annual meeting, ESMO 2019, 27 Sept–1 Oct 2019 in Barcelona, Spain (911PD), 2019.

The response data for these patients have not been made public yet (and certainly have not undergone peer review). However, discoverable social media posts from these patients are very encouraging. With luck the natural history of the disease described by Collins, von Hippel, and Lindau will be favorably altered.

REFERENCES

Aiello, L. P., George, D. J., Cahill, M. T., Wong, J. S., Cavallerano, J., Hannah, A. L. & Kaelin, W. G., Jr. 2002. Rapid and durable recovery of visual function in a patient with von hippel-lindau syndrome after systemic therapy with vascular endothelial growth factor receptor inhibitor su5416. *Ophthalmology*, **109**, 1745–51.

Aravind, L. & Koonin, E. V. 2001. The DNA-repair protein AlkB, EGL-9, and leprecan define new families of 2-oxoglutarate- and iron-dependent dioxygenases. *Genome Bio*, **2**, research0007.1–0007.8.

Awada, A., Hendlisz, A., Gil, T., Bartholomeus, S., Mano, M., De Valeriola, D., Strumberg, D., Brendel, E., Haase, C. G., Schwartz, B. & Piccart, M. 2005. Phase I safety and pharmacokinetics of BAY 43-9006 administered for 21 days on/7 days off in patients with advanced, refractory solid tumours. *Br J Cancer*, **92**, 1855–61.

Bagchi, S., Weinmann, R. & Raychaudhuri, P. 1991. The retinoblastoma protein copurifies with E2F-I, an E1A-regulated inhibitor of the transcription factor E2F. *Cell*, **65**, 1063–1072.

Bai, C., Sen, P., Hofmann, K., Ma, L., Goebl, M., Harper, J. W. & Elledge, S. J. 1996. SKP1 connects cell cycle regulators to the ubiquitin proteolysis machinery through a novel motif, the f-box. *Cell*, **86**, 263–274.

Bandara, L. R., Adamczewski, J. P., Hunt, T. & La Thangue, N. B. 1991a. Cyclin A and the retinoblastoma gene product complex with a common transcription factor. *Nature*, **352**, 249–251.

Bandara, L. R. & La Thangue, N. B. 1991b. Adenovirus E1A prevents the retinoblastoma gene product from complexing with a cellular transcription factor. *Nature*, **351**, 494–497.

Benjamin, L. E., Hemo, I. & Keshet, E. 1998. A plasticity window for blood vessel remodelling is defined by pericyte coverage of the preformed endothelial network and is regulated by PDGF-B and VEGF. *Development*, **125**, 1591–8.

Benjamin, L. E. & Keshet, E. 1997. Conditional switching of vascular endothelial growth factor (VEGF) expression in tumors: induction of endothelial cell shedding and regression of hemangioblastoma-like vessels by VEGF withdrawal. *Proc Natl Acad Sci (USA)*, **94**, 8761–6.

Berra, E., Benizri, E., Ginouves, A., Volmat, V., Roux, D. & Pouyssegur, J. 2003. HIF prolyl-hydroxylase 2 is the key oxygen sensor setting low steady-state levels of HIF-1alpha in normoxia. *Embo J*, **22**, 4082–90.

Besarab, A., Provenzano, R., Hertel, J., Zabaneh, R., Klaus, S. J., Lee, T., Leong, R., Hemmerich, S., Yu, K. H. & Neff, T. B. 2015. Randomized placebo-controlled dose-ranging and pharmacodynamics study of roxadustat (FG-4592) to treat anemia in nondialysis-dependent chronic kidney disease (NDD-CKD) patients. *Nephrol Dial Transplant*, **30**, 1665–73.

Blanar, M. A. & Rutter, W. J. 1992. Interaction cloning: isolation of a cDNA encoding FIP, a basic-HLH-zip protein that interacts with cFos. *Science*, In Press.

Bommi-Reddy, A., Almeciga, I., Sawyer, J., Geisen, C., Li, W., Harlow, E., Kaelin, W. G., Jr. & Grueneberg, D. A. 2008. Kinase requirements in human cells: III. Altered kinase requirements in VHL−/− cancer cells detected in a pilot synthetic lethal screen. *Proc Natl Acad Sci U S A*, **105**, 16484–9.

Brugarolas, J., Lei, K., Hurley, R. L., Manning, B. D., Reiling, J. H., Hafen, E., Witters, L. A., Ellisen, L. W. & Kaelin, W. G., Jr. 2004. Regulation of mTOR function in response to hypoxia by REDD1 and the TSC1/TSC2 tumor suppressor complex. *Genes Dev*, **18**, 2893–904.

Bruick, R. & Mcknight, S. 2001. A conserved family of prolyl-4-hydroxylases that modify HIF. *Science*, **294**, 1337–40.

Chellappan, S. P., Hiebert, S., Mudryj, M., Horowitz, J. M. & Nevins, J. R. 1991. The E2F transcription factor is a cellular target for the RB protein. *Cell*, **65**, 1053–1061.

Chen, N., Hao, C., Liu, B. C., Lin, H., Wang, C., Xing, C., Liang, X., Jiang, G., Liu, Z., Li, X., Zuo, L., Luo, L., Wang, J., Zhao, M. H., Liu, Z., Cai, G. Y., Hao, L., Leong, R., Wang, C., Liu, C., Neff, T., Szczech, L. & Yu, K. P. 2019. Roxadustat treatment for anemia in patients undergoing long-term dialysis. *N Engl J Med.*

Chen, W., Hill, H., Christie, A., Kim, M. S., Holloman, E., Pavia-Jimenez, A., Homayoun, F., Ma, Y., Patel, N., Yell, P., Hao, G., Yousuf, Q., Joyce, A., Pedrosa, I., Geiger, H., Zhang, H., Chang, J., Gardner, K. H., Bruick, R. K., Reeves, C., Hwang, T. H., Courtney, K., Frenkel, E., Sun, X., Zojwalla, N., Wong, T., Rizzi, J. P., Wallace, E. M., Josey, J. A., Xie, Y., Xie, X. J., Kapur, P., Mckay, R. M. & Brugarolas, J. 2016. Targeting renal cell carcinoma with a HIF-2 antagonist. *Nature.*

Chittenden, T., Livingston, D. M. & Kaelin, W. G. 1991. The T/E1A-binding domain of the retinoblastom product can interact selectively with a sequence-specific DNA-binding protein. *Cell*, **65**, 1073–1082.

Cho, H., Du, X., Rizzi, J. P., Liberzon, E., Chakraborty, A. A., Gao, W., Carvo, I., Signoretti, S., Bruick, R., Josey, J. A., Wallace, E. M. & Kaelin, W. G., JR. 2016. On-Target Efficacy of a HIF2alpha Antagonist in Preclinical Kidney Cancer Models. *Nature*, **539**, 107–111.

Choueiri, T. K. & Motzer, R. J. 2017. Systemic therapy for metastatic renal-cell carcinoma. *N Engl J Med*, **376**, 354–366.

Cockman, M., Masson, N., Mole, D., Jaakkola, P., Chang, G., Clifford, S., Maher, E., Pugh, C., Ratcliffe, P. & Maxwell, P. 2000. Hypoxia inducible factor-alpha binding and ubiquitylation by the von Hippel-Lindau tumor suppressor protein. J *Biol Chem*, **275**, 25733–41.

Collins, E. T. 1894. Intra-ocular growths (two cases, brother and sister, with peculiar vascular new growth, probably retinal, affecting both eyes). *Trans. Ophthal. Soc. U.K.*, **14**, 141–149.

Courtney, K. D., Infante, J. R., Lam, E. T., Figlin, R. A., Rini, B. I., Brugarolas, J., Zojwalla, N. J., Lowe, A. M., Wang, K., Wallace, E. M., Josey, J. A. & Choueiri, T. K. 2018. Phase I dose-escalation trial of PT2385, a first-in-class hypoxia-inducible factor-2alpha antagonist in patients with previously treated advanced clear cell renal cell carcinoma. *J Clin Oncol*, **36**, 867–874.

Covello, K. L., Simon, M. C. & Keith, B. 2005. Targeted replacement of hypoxia-inducible factor-1alpha by a hypoxia-inducible factor-2alpha knock-in allele promotes tumor growth. *Cancer Res*, **65**, 2277–86.

De Bazelaire, C., Alsop, D. C., George, D., Pedrosa, I., Wang, Y., Michaelson, M. D. & Rofsky, N. M. 2008. Magnetic resonance imaging-measured blood flow change after antiangiogenic therapy with PTK787/ZK 222584 correlates with clinical outcome in metastatic renal cell carcinoma. *Clin Cancer Res,* **14**, 5548–54.

Decaprio, J. A., Ludlow, J. W., Figge, J., Shew, J.-Y., Huang, C.-M., Lee, W.-H., Marsilio, E., Paucha, E. & Livingston, D. M. 1988. SV40 large T antigen forms a specific complex with the product of the retinoblastoma susceptibility gene. *Cell,* **54**, 275–283.

Drevs, J., Hofmann, I., Hugenschmidt, H., Wittig, C., Madjar, H., Muller, M., Wood, J., Martiny-Baron, G., Unger, C. & Marme, D. 2000. Effects of PTK787/ZK 222584, a specific inhibitor of vascular endothelial growth factor receptor tyrosine kinases, on primary tumor, metastasis, vessel density, and blood flow in a murine renal cell carcinoma model. *Cancer Res,* **60**, 4819–24.

Duan, D. R., Pause, A., Burgress, W., Aso, T., Chen, D. Y. T., Garrett, K. P., Conaway, R. C., Conaway, J. W., Linehan, W. M. & Klausner, R. D. 1995. Inhibition of transcriptional elongation by the VHL tumor suppressor protein. *Science,* **269**, 1402–1406.

Epstein, A., Gleadle, J., Mcneill, L., Hewitson, K., O'rourke, J., Mole, D., Mukherji, M., Metzen, E., Wilson, M., Dhanda, A., Tian, Y., Masson, N., Hamilton, D., Jaakkola, P., Barstead, R., Hodgkin, J., Maxwell, P., Pugh, C., Schofield, C. & Ratcliffe, P. 2001. C. elegans EGL-9 and mammalian homologs define a family of dioxygenases that regulate HIF by prolyl hydroxylation. *Cell,* **107**, 43–54.

Escudier, B., Eisen, T., Stadler, W. M., Szczylik, C., Oudard, S., Siebels, M., Negrier, S., Chevreau, C., Solska, E., Desai, A. A., Rolland, F., Demkow, T., Hutson, T. E., Gore, M., Freeman, S., Schwartz, B., Shan, M., Simantov, R. & Bukowski, R. M. 2007. Sorafenib in advanced clear-cell renal-cell carcinoma. *N Engl J Med,* **356**, 125–34.

Gerlinger, M., Horswell, S., Larkin, J., Rowan, A. J., Salm, M. P., Varela, I., Fisher, R., Mcgranahan, N., Matthews, N., Santos, C. R., Martinez, P., Phillimore, B., Begum, S., Rabinowitz, A., Spencer-Dene, B., Gulati, S., Bates, P. A., Stamp, G., Pickering, L., Gore, M., Nicol, D. L., Hazell, S., Futreal, P. A., Stewart, A. & Swanton, C. 2014. Genomic architecture and evolution of clear cell renal cell carcinomas defined by multiregion sequencing. *Nat Genet,* **46**, 225–33.

Gerlinger, M., Rowan, A. J., Horswell, S., Larkin, J., Endesfelder, D., Gronroos, E., Martinez, P., Matthews, N., Stewart, A., Tarpey, P., Varela, I., Phillimore, B., Begum, S., Mcdonald, N. Q., Butler, A., Jones, D., Raine, K., Latimer, C., Santos, C. R., Nohadani, M., Eklund, A. C., Spencer-Dene, B., Clark, G., Pickering, L., Stamp, G., Gore, M., Szallasi, Z., Downward, J., Futreal, P. A. & Swanton, C. 2012. Intratumor heterogeneity and branched evolution revealed by multiregion sequencing. *N Engl J Med,* **366**, 883–92.

Girmens, J. F., Erginay, A., Massin, P., Scigalla, P., Gaudric, A. & Richard, S. 2003. Treatment of von Hippel-Lindau retinal hemangioblastoma by the vascular endothelial growth factor receptor inhibitor SU5416 is more effective for associated macular edema than for hemangioblastomas. *Am J Ophthalmol,* **136**, 194–6.

Gnarra, J. R., Tory, K., Weng, Y., Schmidt, L., Wei, M. H., Li, H., Latif, F., Liu, S., Chen, F., Duh, F.-M., Lubensky, I., Duan, D. R., Florence, C., Pozzatti, R., Walther, M. M., Bander, N. H., Grossman, H. B., Brauch, H., Pomer, S., Brooks, J. D., Isaacs, W. B., Lerman, M. I., Zbar, B. & Linehan, W. M. 1994. Mutations of the VHL tumour suppressor gene in renal carcinoma. *Nature Genetics,* **7**, 85–90.

Gnarra, J. R., Zhou, S., Merrill, M. J., Wagner, J., Krumm, A., Papavassiliou, E., Oldfield, E. H., Klausner, R. D. & Linehan, W. M. 1996. Post-transcriptional regulation of vascular endothelial growth factor mRNA by the VHL tumor suppressor gene product. *Proc Natl Acad Sci,* **93**, 10589–10594.

Golde, D. W. & Hocking, W. G. 1981. Polycythemia: mechanisms and management. *Ann. Intern. Med.*, **95**, 71–87.

Gordan, J. D., Bertout, J. A., Hu, C. J., Diehl, J. A. & Simon, M. C. 2007. HIF-2alpha promotes hypoxic cell proliferation by enhancing c-myc transcriptional activity. *Cancer Cell,* **11**, 335–47.

Gordan, J. D., Lal, P., Dondeti, V. R., Letrero, R., Parekh, K. N., Oquendo, C. E., Greenberg, R. A., Flaherty, K. T., Rathmell, W. K., Keith, B., Simon, M. C. & NatHANSON, K. L. 2008. HIF-alpha effects on c-Myc distinguish two subtypes of sporadic VHL-deficient clear cell renal carcinoma. *Cancer Cell,* **14**, 435–446.

Gupta, N. & Wish, J. B. 2017. Hypoxia-inducible factor prolyl hydroxylase inhibitors: a potential new treatment for anemia in patients with CKD. *Am J Kidney Dis.*

Helin, K., Lees, J. A., Vidal, M., Dyson, N., Harlow, E. & Fattaey, A. 1992. A cDNA encoding a pRB-binding protein with properties of the transcription factor E2F. *Cell,* **70**, 337–350.

Hon, W. C., Wilson, M. I., Harlos, K., Claridge, T. D., Schofield, C. J., Pugh, C. W., Maxwell, P. H., Ratcliffe, P. J., Stuart, D. I. & Jones, E. Y. 2002. Structural basis for the recognition of hydroxyproline in HIF-1alpha by pVHL. *Nature.*

Iliopoulos, O., Jiang, C., Levy, A. P., Kaelin, W. G. & Goldberg, M. A. 1996. Negative regulation of hypoxia-inducible genes by the von Hippel-Lindau protein. *Proc. Natl. Acad. Sci.,* **93**, 10595–10599.

Iliopoulos, O., Kibel, A., Gray, S. & Kaelin, W. G. 1995. Tumor Suppression by the human von Hippel-Lindau gene product. *Nature Medicine,* **1**, 822–826.

Ivan, M., Haberberger, T., Gervasi, D. C., Michelson, K. S., Gunzler, V., Kondo, K., Yang, H., Sorokina, I., Conaway, R. C., Conaway, J. W. & Kaelin, W. G., Jr. 2002. Biochemical purification and pharmacological inhibition of a mammalian prolyl hydroxylase acting on hypoxia-inducible factor. *Proc Natl Acad Sci U S A,* **99**, 13459–64.

Ivan, M., Kondo, K., Yang, H., Kim, W., Valiando, J., Ohh, M., Salic, A., Asara, J., Lane, W. & Kaelin, W. J. 2001. HIFalpha targeted for VHL-mediated destruction by proline hydroxylation: implications for O2 sensing. Science, 292, 464–8.

Iwai, K., Yamanaka, K., Kamura, T., Minato, N., Conaway, R., Conaway, J., Klausner, R. & Pause, A. 1999. Identification of the von Hippel-Lindau tumor-suppressor protein as part of an active E3 ubiquitin ligase complex. *Proc Natl Acad Sci U S A,* **96**, 12436–41.

Jaakkola, P., Mole, D., Tian, Y., Wilson, M., Gielbert, J., Gaskell, S., Kriegsheim, A., Hebestreit, H., Mukherji, M., Schofield, C., Maxwell, P., Pugh, C. & Ratcliffe, P. 2001. Targeting of HIF-alpha to the von Hippel-Lindau ubiquitylation complex by O2-regulated prolyl hydroxylation. Science, 292, 468–72.

Jonasch, E., Mccutcheon, I. E., Waguespack, S. G., Wen, S., Davis, D. W., Smith, L. A., Tannir, N. M., Gombos, D. S., Fuller, G. N. & Matin, S. F. 2011. Pilot trial of sunitinib therapy in patients with von Hippel-Lindau disease. *Ann Oncol,* **22**, 2661–6.

Jonasch, E., Plimack, E., Bauer, T., Merchan, J., Papadopoulos, K., Mcdermott, D., M.D., M., Appleman, L. J., Thamake, S., Zojwalla, N. & Choueiri, T. K. 2019. A first-in-human phase I/II trial of the oral HIF-2a inhibitor PT2977 in patients with advanced RCC. *2019 European Society for Medical Oncology annual meeting, ESMO 2019, 27 Sept–1 Oct 2019 in Barcelona, Spain (911PD).*

Kaelin, W. 2015. Molecular biology of kidney cancer. *Kidney Cancer: Principles and Practice.* 2nd ed.: Springer International Publishing.

Kaelin, W. G., Ewen, M. E. & Livingston, D. M. 1990. Definition of the minimal simian virus 40 large T antigen- and adenovirus E1A-binding domain in the retinoblastoma gene product. *Mol. Cell. Biol.,* **10**, 3761–3769.

Kaelin, W. G., Jr., Shrivastav, S. & Jirtle, R. L. 1984. Blood flow to primary tumors and lymph node metastases in SMT-2A tumor-bearing rats following intravenous flunarizine. *Cancer Res,* **44**, 896–9.

Kaelin, W. G., Jr., Shrivastav, S., Shand, D. G. & Jirtle, R. L. 1982. Effect of verapamil on malignant tissue blood flow in SMT-2A tumor-bearing rats. *Cancer Res,* **42**, 3944–9.

Kaelin, W. G., Krek, W., Sellers, W. R., Decaprio, J. A., Ajchenbaum, F., Fuchs, C. S., Chittenden, T., Li, Y., Farnham, P. J., Blanar, M. A., Livingston, D. M. & Flemington, E. K. 1992. Expression cloning of a cDNA encoding a retinoblastoma-binding protein with E2F-like properties. *Cell,* **70**, 351–364.

Kaelin, W. G., Pallas, D. C., Decaprio, J. A., Kaye, F. J. & Livingston, D. M. 1991. Identification of cellular proteins that can interact specifically with the T/E1A-binding region of the retinoblastoma gene product. *Cell,* **64**, 521–532.

Kamura, T., Koepp, D. M., Conrad, M. N., Skowyra, D., Moreland, R. J., Iliopoulos, O., Lane, W. S., Kaelin, W. G. J., Elledge, S. J., Conaway, R. C., Harper, J. W. & Conaway, J. W. 1999. Rbx1, a component of the VHL tumor suppressor complex and SCF ubiquitin ligase. *Science,* **284**, 657–661.

Kamura, T., Sato, S., Iwain, K., Czyzyk-Krzeska, M., Conaway, R. C. & Conaway, J. W. 2000. Activation of HIF1a Ubiquitination by a Reconstituted von Hippel-Lindau tumor suppressor complex. *Proc. Natl. Acad. Sci. (USA),* **97**, 10430–10435.

Kaplan, J. 2019. Roxadustat and anemia of chronic kidney disease. *N Engl J Med.*

Kibel, A., Iliopoulos, O., Decaprio, J. D. & Kaelin, W. G. 1995. Binding of the von Hippel-Lindau tumor suppressor protein to elongin B and C. *Science,* **269**, 1444–1446.

Kim, W. Y., Safran, M., Buckley, M. R., Ebert, B. L., Glickman, J., Bosenberg, M., Regan, M. & Kaelin, W. G., Jr. 2006. Failure to prolyl hydroxylate hypoxia-inducible factor alpha phenocopies VHL inactivation in vivo. *Embo J,* **25**, 4650–62.

Kondo, K., Kim, W. Y., Lechpammer, M. & Kaelin, W. G., Jr. 2003. Inhibition of HIF2alpha is sufficient to suppress pvhl-defective tumor growth. *PLoS Biol,* **1**, 439–444.

Kondo, K., Klco, J., Nakamura, E., Lechpammer, M. & Kaelin, W. G. 2002. Inhibition of HIF is necessary for tumor suppression by the von Hippel-Lindau protein. *Cancer Cell,* **1**, 237–46.

Koochekpour, S., Jeffers, M., Wang, P., Gong, C., Taylor, G., Roessler, L., Stearman, R., Vasselli, J., Stetler-Stevenson, W., Kaelin, W. J., Linehan, W., Klausner, R., Gnarra, J. & Vande Woude, G. 1999. The von Hippel-Lindau tumor suppressor gene inhibits hepatocyte growth factor/scatter factor-induced invasion and branching morphogenesis in renal carcinoma cells. *Mol Cell Biol 1999,* **19**, 5902–12.

Latif, F., Tory, K., Gnarra, J., Yao, M., Duh, F.-M., Orcutt, M. L., Stackhouse, T., Kuzmin, I., Modi, W., Geil, L., Schmidt, L., Zhou, F., Li, H., Wei, M. H., Chen, F., Glenn, G., Choyke, P., Walther, M. M., Weng, Y., Duan, D.-S. R., Dean, M., Glavac, D., Richards, F. M., Crossey, P. A., Ferguson-Smith, M. A., Pasiler, D. L., Chumakov, I., Cohen, D., Chinault, A. C., Maher, E. R., Linehan, W. M., Zbar, B. & Lerman, M. I. 1993. Identification of the von Hippel-Lindau disease tumor suppressor gene. *Science*, **260**, 1317–1320.

Lee, S., Chen, D. Y. T., Humphrey, J. S., Gnarra, J. R., Linehan, W. M. & Klausner, R. D. 1996. Nuclear/cytoplasmic localization of the von Hippel-Lindau tumor suppressor gene product is determined by cell density. *Proc Natl Acad Sci*, **93**, 1770–1775.

Li, L., Zhang, L., Zhang, X., Yan, Q., Minamishima, Y. A., Olumi, A. F., Mao, M., Bartz, S. & Kaelin, W. G., Jr. 2007. Hypoxia-inducible factor linked to differential kidney cancer risk seen with type 2A and type 2B VHL mutations. *Mol Cell Biol*, **27**, 5381–92.

Lindau, A. 1927. Zur Frage der Angiomatosis Retinae und Ihrer Hirncomplikation. *Acta Opthal*, **4**, 193–226.

Lisztwan, J., Imbert, G., Wirbelauer, C., Gstaiger, M. & Krek, W. 1999. The von Hippel-Lindau tumor suppressor protein is a component of an E3 ubiquitin-protein ligase activity. *Genes Dev*, **13**, 1822–33.

Lonergan, K. M., Iliopoulos, O., Ohh, M., Kamura, T., Conaway, R. C., Conaway, J. W. & Kaelin, W. G. 1998. Regulation of hypoxia-inducible mRNAs by the von Hippel-LIndau protein requires binding to complexes containing elongins B/C and Cul2. *Mol. Cell. Biol.*, **18**, 732–741.

Mandriota, S. J., Turner, K. J., Davies, D. R., Murray, P. G., Morgan, N. V., Sowter, H. M., Wykoff, C. C., Maher, E. R., Harris, A. L., Ratcliffe, P. J. & Maxwell, P. H. 2002. HIF activation identifies early lesions in VHL kidneys: evidence for site-specific tumor suppressor function in the nephron. *Cancer Cell*, **1**, 459–68.

Masson, N., Willam, C., Maxwell, P., Pugh, C. & Ratcliffe, P. 2001. Independent function of two destruction domains in hypoxia-inducible factor-a chains activated by prolyl hydroylation. *EMBO*, **20**, 5197–5206.

Maxwell, P., Weisner, M., Chang, G.-W., Clifford, S., Vaux, E., Pugh, C., Maher, E. & Ratcliffe, P. 1999. The von Hippel-Lindau gene product is necessary for oxgyen-dependent proteolysis of hypoxia-inducible factor a subunits. *Nature*, **399**, 271–5.

Migliorini, D., Haller, S., Merkler, D., Pugliesi-Rinaldi, A., Koka, A., Schaller, K., Leemann, B. & Dietrich, P. Y. 2015. Recurrent multiple CNS hemangioblastomas with VHL disease treated with pazopanib: a case report and literature review. *CNS Oncol*, **4**, 387–92.

Min, J. H., Yang, H., Ivan, M., Gertler, F., Kaelin, W. G., Jr. & Pavletich, N. P. 2002. Structure of an HIF-1alpha -pVHL complex: hydroxyproline recognition in signaling. *Science*, **296**, 1886–9.

Minamishima, Y. A. & Kaelin, W. G., Jr. 2010. Reactivation of hepatic EPO synthesis in mice after PHD loss. *Science*, **329**, 407.

Minamishima, Y. A., Moslehi, J., Bardeesy, N., Cullen, D., Bronson, R. T. & Kaelin, W. G., Jr. 2008. Somatic inactivation of the PHD2 prolyl hydroxylase causes polycythemia and congestive heart failure. *Blood*, **111**, 3236–3244.

Minamishima, Y. A., Moslehi, J., Padera, R. F., Bronson, R. T., Liao, R. & Kaelin, W. G., Jr. 2009. A feedback loop involving the Phd3 prolyl hydroxylase tunes the mammalian hypoxic response in vivo. *Mol Cell Biol*, **29**, 5729–41.

Mitchell, T. J., Turajlic, S., Rowan, A., Nicol, D., Farmery, J. H. R., O'brien, T., Martincorena, I., Tarpey, P., Angelopoulos, N., Yates, L. R., Butler, A. P., Raine,

K., Stewart, G. D., Challacombe, B., Fernando, A., Lopez, J. I., Hazell, S., Chandra, A., Chowdhury, S., Rudman, S., Soultati, A., Stamp, G., Fotiadis, N., Pickering, L., Au, L., Spain, L., Lynch, J., Stares, M., Teague, J., Maura, F., Wedge, D. C., Horswell, S., Chambers, T., Litchfield, K., Xu, H., Stewart, A., Elaidi, R., Oudard, S., Mcgranahan, N., Csabai, I., Gore, M., Futreal, P. A., Larkin, J., Lynch, A. G., Szallasi, Z., Swanton, C., Campbell, P. J. & Consortium, T. R. R. 2018. Timing the landmark events in the evolution of clear cell renal cell cancer: TRACERx Renal. *Cell*, **173**, 611–623 e17.

Motzer, R. J., Hutson, T. E., Tomczak, P., Michaelson, M. D., Bukowski, R. M., Rixe, O., Oudard, S., Negrier, S., Szczylik, C., Kim, S. T., Chen, I., Bycott, P. W., Baum, C. M. & Figlin, R. A. 2007. Sunitinib versus interferon alfa in metastatic renal-cell carcinoma. *N Engl J Med,* **356**, 115–24.

Nakaigawa, N., Yao, M., Baba, M., Kato, S., Kishida, T., Hattori, K., Nagashima, Y. & Kubota, Y. 2006. Inactivation of von Hippel-Lindau gene induces constitutive phosphorylation of MET protein in clear cell renal carcinoma. *Cancer Res,* **66**, 3699–705.

O'Reilly, M. S., Boehm, T., Shing, Y., Fukai, N., Vasios, G., Lane, W. S., Flynn, E., Birkhead, J. R., Olsen, B. R. & Folkman, J. 1997. Endostatin: an endogenous inhibitor of angiogenesis and tumor growth. *Cell,* **88**, 277–85.

O'Reilly, M. S., Holmgren, L., Shing, Y., Chen, C., Rosenthal, R. A., Moses, M., Lane, W. S., Cao, Y., Sage, E. H. & Folkman, J. 1994. Angiostatin: a novel angiogenesis inhibitor that mediates the suppression of metastases by a Lewis lung carcinoma. *Cell,* **79**, 315–28.

Ohh, M., Park, C. W., Ivan, M., Hoffman, M. A., Kim, T.-Y., Huang, L. E., Chau, V. & Kaelin, W. G. 2000. Ubiquitination of HIF requires direct binding to the von Hippel-Lindau protein beta domain. *Nature Cell Biology,* **2**, 423–427.

Olenchock, B. A., Moslehi, J., Baik, A. H., Davidson, S. M., Williams, J., Gibson, W. J., Pierce, K. A., Miller, C. M., Hanse, E. A., Kelekar, A., Sullivan, L. B., Wagers, A. J., Clish, C. B., Vander Heiden, M. G. & Kaelin, W. G., Jr. 2016. Egln1 inhibition and rerouting of alpha-ketoglutarate suffice for remote ischemic protection. *Cell,* **164**, 884–95.

Pause, A., Lee, S., Worrell, R. A., Chen, D. Y. T., Burgess, W. H., Linehan, W. M. & Klausner, R. D. 1997. The von Hippel-Lindau tumor-suppressor gene product forms a stable complex with human CUL-2, a member of the Cdc53 family of Proteins. *Proc. Natl. Acad. Sci. (USA),* **94**, 2156–2161.

Pennacchietti, S., Michieli, P., Galluzzo, M., Mazzone, M., Giordano, S. & Comoglio, P. M. 2003. Hypoxia promotes invasive growth By Transcriptional Activation Of The Met Protooncogene. *Cancer Cell,* **3**, 347–61.

Qin, X.-Q., Chittenden, T., Livingston, D. M. & Kaelin, W. G. 1992. Identification of a growth suppression domain within the retinoblastoma gene product. *Genes and Devel.,* **6**, 953–964.

Querbes, W., Bogorad, R. L., Moslehi, J., Wong, J., Chan, A. Y., Bulgakova, E., Kuchimanchi, S., Akinc, A., Fitzgerald, K., Koteliansky, V. & Kaelin, W. G., Jr. 2012. Treatment of erythropoietin deficiency in mice with systemically administered siRNA. *Blood,* **120**, 1916–22.

Ratain, M. J., Eisen, T., Stadler, W. M., Flaherty, K. T., Kaye, S. B., Rosner, G. L., Gore, M., Desai, A. A., Patnaik, A., Xiong, H. Q., Rowinsky, E., Abbruzzese, J. L., Xia, C., Simantov, R., Schwartz, B. & O'dwyer, P. J. 2006. Phase II placebo-controlled randomized discontinuation trial of sorafenib in patients with metastatic renal cell carcinoma. *J Clin Oncol,* **24**, 2505–12.

Raval, R. R., Lau, K. W., Tran, M. G., Sowter, H. M., Mandriota, S. J., Li, J. L., Pugh, C. W., Maxwell, P. H., Harris, A. L. & Ratcliffe, P. J. 2005. Contrasting properties of hypoxia-inducible factor 1 (HIF-1) and HIF-2 in von Hippel-Lindau-associated renal cell carcinoma. *Mol Cell Biol,* **25**, 5675–86.

Reiling, J. H. & Hafen, E. 2004. The hypoxia-induced paralogs Scylla and Charybdis inhibit growth by down-regulating S6K activity upstream of TSC in Drosophila. *Genes Dev,* **18**, 2879–92.

Rogers, J. L., Bayeh, L., Scheuermann, T. H., Longgood, J., Key, J., Naidoo, J., Melito, L., Shokri, C., Frantz, D. E., Bruick, R. K., Gardner, K. H., Macmillan, J. B. & Tambar, U. K. 2013. Development of inhibitors of the PAS-B domain of the HIF-2alpha transcription factor. *J Med Chem,* **56**, 1739–47.

Safran, M., Kim, W. Y., O'connell, F., Flippin, L., Gunzler, V., Horner, J. W., Depinho, R. A. & Kaelin, W. G., Jr. 2006. Mouse model for noninvasive imaging of HIF prolyl hydroxylase activity: assessment of an oral agent that stimulates erythropoietin production. *Proc Natl Acad Sci U S A,* **103**, 105–10.

Scheuermann, T. H., Li, Q., Ma, H. W., Key, J., Zhang, L., Chen, R., Garcia, J. A., Naidoo, J., Longgood, J., Frantz, D. E., Tambar, U. K., Gardner, K. H. & Bruick, R. K. 2013. Allosteric inhibition of hypoxia inducible factor-2 with small molecules. *Nat Chem Biol,* **9**, 271–6.

Scheuermann, T. H., Tomchick, D. R., Machius, M., Guo, Y., Bruick, R. K. & Gardner, K. H. 2009. Artificial ligand binding within the HIF2alpha PAS-B domain of the HIF2 Transcription Factor. *Proc Natl Acad Sci U S A,* **106**, 450–5.

Shen, C., Beroukhim, R., Schumacher, S. E., Zhou, J., Chang, M., Signoretti, S. & Kaelin, W. G., Jr. 2011. Genetic and functional studies implicate HIF1alpha as a 14q kidney cancer suppressor gene. *Cancer Discov,* **1**, 222–235.

Shrivastav, S., Kaelin, W. G., Jr., Joines, W. T. & Jirtle, R. L. 1983. Microwave hyperthermia and its effect on tumor blood flow in rats. *Cancer Res,* **43**, 4665–9.

Srinivas, V., Zhang, L., Zhu, X. & Caro, J. 1999. Characterization of an oxygen/redox-dependent degradation domain of hypoxia-inducible factor alpha (HIF-alpha) proteins. *Biochem Biophys Res Commun,* **260**, 557–61.

Stebbins, C. E., Kaelin, W. G. & Pavletich, N. P. 1999. Structure of the VHL-ElonginC-elonginB complex: implications for VHL tumor suppressor function. *Science,* **284**, 455–461.

Takeda, K., Aguila, H. L., Parikh, N. S., Li, X., Lamothe, K., Duan, L. J., Takeda, H., Lee, F. S. & Fong, G. H. 2008. Regulation of adult erythropoiesis by prolyl hydroxylase domain proteins. *Blood,* **111**, 3229–3235.

Takeda, K., Ho, V. C., Takeda, H., Duan, L. J., Nagy, A. & Fong, G. H. 2006. Placental but not heart defects are associated with elevated hypoxia-inducible factor alpha levels in mice lacking prolyl hydroxylase domain protein 2. *Mol Cell Biol,* **26**, 8336–46.

Tanimoto, K., Makino, Y., Pereira, T. & Poellinger, L. 2000. Mechanism of regulation of the hypoxia-inducible factor-1alpha by the von Hippel-Lindau tumor suppressor protein. *EMBO J,* **19**, 4298–4309.

Tian, H., Mcknight, S. L. & Russell, D. W. 1997. Endothelial PAS domain protein 1 (EPAS1), a transcription factor selectively expressed in endothelial cells. *Genes Dev,* **11**, 72–82.

Turajlic, S., Xu, H., Litchfield, K., Rowan, A., Horswell, S., Chambers, T., O'brien, T., Lopez, J. I., Watkins, T. B. K., Nicol, D., Stares, M., Challacombe, B., Hazell, S., Chandra, A., Mitchell, T. J., Au, L., Eichler-Jonsson, C., Jabbar, F., Soultati, A., Chowdhury, S., Rudman, S., Lynch, J., Fernando, A., Stamp, G., Nye, E., Stewart, A., Xing, W., Smith, J. C., Escudero, M., Huffman, A., Matthews, N.,

Elgar, G., Phillimore, B., Costa, M., Begum, S., Ward, S., Salm, M., Boeing, S., Fisher, R., Spain, L., Navas, C., Gronroos, E., Hobor, S., Sharma, S., Aurangzeb, I., Lall, S., Polson, A., Varia, M., Horsfield, C., Fotiadis, N., Pickering, L., Schwarz, R. F., Silva, B., Herrero, J., Luscombe, N. M., Jamal-Hanjani, M., Rosenthal, R., Birkbak, N. J., Wilson, G. A., Pipek, O., Ribli, D., Krzystanek, M., Csabai, I., Szallasi, Z., Gore, M., Mcgranahan, N., Van Loo, P., Campbell, P., Larkin, J., Swanton, C. & Consortium, T. R. R. 2018. Deterministic evolutionary trajectories influence primary tumor growth: TRACERx Renal. *Cell*, **173**, 595–10 e11.

Von *Hippel*, E. 1904. Ueber eine sehr seltene Erkrankung der Nethaut. Graefe Arch *Ophthal*, **59**, 83–106.

Wallace, E. M., Rizzi, J. P., Han, G., Wehn, P. M., Cao, Z., Du, X., Cheng, T., Czerwinski, R. M., Dixon, D. D., Goggin, B. S., Grina, J. A., Halfmann, M. M., Maddie, M. A., Olive, S. R., Schlachter, S. T., Tan, H., Wang, B., Wang, K., Xie, S., Xu, R., Yang, H. & Josey, J. A. 2016. A small-molecule antagonist of HIF2alpha is efficacious in preclinical models of renal cell carcinoma. *Cancer Res*, **76**, 5491–500.

Whyte, P., Buchkovich, K. J., Horowitz, J. M., Friend, S. H., Raybuck, M., Weinberg, R. A. & Harlow, E. 1988. Association between an oncogene and an antioncogene: the adenovirus E1A proteins bind to the retinoblastoma gene product. *Nature*, **334**, 124–129.

Yang, J., Haworth, L., Sherry, R., Hwu, P., Schwartzentruber, D., Topalian, S., Steinberg, S., Chen, H. & Rosenberg, S. 2003. A randomized trial of bevacizumab, an anti-vascular endothelial growth factor antibody, for metastatic renal cancer. *N Engl J Med*, **349**, 427–34.

Yang, J., Haworth, S., Steinberg, S., Rosenberg, S. & Novotny, W. 2002. A randomized double-blind placebo-controlled trial of bevacizumab (anti-VEGF antibody) demonstrating a prolongation in time to progression in patients with metastatic renal cancer. *Proc Am Soc Clin Oncol 2002; 21:15.*, **21**, 15.

Yu, F., White, S., Zhao, Q. & Lee, F. 2001. HIF-1alpha binding to VHL is regulated by Stimulus-Sensitive Proline Hydroxylation. *Proc Natl Acad Sci (U S A)*, **98**, 9630–5.

Zimmer, M., Doucette, D., Siddiqui, N. & Iliopoulos, O. 2004. Inhibition of hypoxia-inducible factor is sufficient for growth suppression of VHL−/− tumors. Mol *Cancer Res*, **2**, 89–95.

Sir Peter J. Ratcliffe. © Nobel Prize Outreach AB. Photo: A. Mahmoud

Sir Peter J. Ratcliffe

Biography

I WAS BORN 14 MAY 1954, and my childhood in Carnforth, Lancashire was idyllic. At the time I would have said it was unremarkable, as are most things when not viewed by comparison. The near-total freedom I enjoyed seemed perfectly natural. The town was an unpretentious railway town in North Lancashire, England. My father was the local lawyer; my mother left her work as a telephonist when she married, as was the custom. And I was their only child. They were good enough not be overly protective and I survived as much foolhardy behaviour as most young people. I was neither pushed nor discouraged in schoolwork and again saw that as entirely normal. Appreciation of my parents came to me much later, with the recognition that *other* people saw them as special people in the community. My world was a simple one of building tree houses, lighting fires, mock (and occasionally real) combat with other groups of children. But I was always concerned with improving things: a better catapult, a hotter fire, a bigger explosion. I had an encyclopaedic knowledge of the melting temperature of metals; molten lead was a joy; an ambition to create the 2800OF necessary to melt iron was sadly never fulfilled. I learnt that it was possible to create something akin to a hand grenade by compressing thousands of the tiny 'caps', as supplied for toy guns, between two large bolts threaded onto a single nut, then hurled at a brick wall. The result was a neat hole in a nearby garage door. Fortunately, before the inevitable calamity, life evolved.

Aged 11, I started at Lancaster Royal Grammar School. The school was good and undoubtedly the single most formative experience in my education. But it was not idyllic. There will be others who recall their trials with school meals; the UK was not at the top of the culinary league in the 1960s and the Lancaster Royal Grammar School kitchen was not the *cordon bleu* of the nation. When a good meal cropped up there was no question of equal shares: the older boys, in charge of the table, would take the large majority. Again, without a comparator, I saw this as the way of the world. But that blissful childhood was over.

I was good at most subjects, with the exception of English. Lack of talent was reinforced by lack of effort and sadly this dichotomy in my abilities grew until I eventually came to recognise the importance of communication, a decade or two later. Actually, my ethos was not out of keeping with that of the school; little time was wasted on things other than sport (particularly Rugby Football) and University Entrance (particularly entrance to Oxford or Cambridge). This rather blinkered approach to life did exact its penalty in due course. But I was comfortable at the time. The school had some good teachers; several were truly excellent. I remember in particular Gary Sleightholme, who taught me chemistry. The accent was on finding simplicity in complexity, which is of course the essence of molecular biology. I have heard many scientists describe their first appreciation of this, at the hands of some esteemed mentor in a famous research institution. But for me (and I suspect for many others) I'm pretty sure it was the school. Of the two acceptable school options, excellence in sport or Oxbridge entrance, the pathway for myself was the latter, by default. For various reasons I didn't think I'd enjoy either Oxford or Cambridge, but never had quite the bravery to refuse the entrance examination, which I duly took. Most of the questions were opaque to me, but I remember creating and solving a set of equations relating to a complex chemistry problem. When several variables cancelled out and the answer was simple integer, I felt confident of success. Beyond this I hardly managed a single answer. But rather to my own surprise, and greatly to that of some at the school, a telegram arrived announcing an open scholarship to Gonville and Caius College, Cambridge. This was to study medicine.

I remember that decision, as follows. I had intended to study chemistry; my chemistry teacher was inspiring and, as I was told by my parents, a distant relative had been a successful pharmaceutical chemist. But it wasn't to be. The Headmaster, John Lorraine Spencer MA, a rather ethereal figure, who somewhat incongruously wore a gown as he walked about the rough and tumble school, appeared one day in the chemistry classroom. 'Ratcliffe,' he said, 'may I have a word?' I duly followed him with some trepidation to his study. 'Ratcliffe,' he said, 'I think you should study medicine.' His views were never to be taken lightly. 'Yes sir,' I responded, and the University application form was altered without further exchange. I have never been sure whether he thought I would be a good doctor or a bad a chemist, or really whether he was right or wrong in the end.

But I was pleased with myself. Even though I didn't want to go, I had got the Scholarship to Cambridge a year early and there was a little time to kill. The previous obsession with explosive dynamics resurfaced, in a slightly more dangerous guise. I had always wanted to be good at sport, not chemistry; but though the spirit was strong, the flesh was weak. Now, however improbable, I saw a potential solution; motorsport. I took a job in the analytical laboratory of a local textile firm and earned enough money to buy a racing kart (not a go-kart). Here was a real thrill; the acceleration over a short span was truly mind-blowing. It suited my basic interest in combustion. I learnt to 'mechanic' it (the most important bit) and drive tolerably, though not brilliantly well. It held prestige amongst a set of peers I admired. Though I never possessed a powerful motorcycle I had the gear to tune them for my friends. Mercifully for my poor parents, this period came to an effective end when I finally took up the place at University.

As might be predicted from this preparation, medicine at Cambridge was not an unqualified success. I am not going to go into detail; suffice it to say this was not the fault of the University, or the College, or my tutors. Though I readily saw the potential in a different approach to life, it was simply too far away from two-stroke racing engines and the associated culture. Though I made good friends, I couldn't quite adjust. The saving grace was that the clinical three years of the six-year course were generally undertaken elsewhere, usually in London. This offered a second chance.

The clinical medicine course at St. Bartholomew's Hospital was well organised and I found much of it, particularly the process of medical diagnosis, very interesting. There were other good things. I met my wife Fiona there; Fiona was also a medical student. But, after Cambridge, I wasn't taking any chances with examinations. I wasn't quite sure what I wanted to do, but I wanted the freedom to choose. Still lacking those skills in Rugby Football, the easiest route to first choice in a House Physician job (and hence it seemed, to everything beyond) was a good performance in the final examinations. In these, the majority of marks were awarded through 'multiple-choice' questions. Now, as most of those who have set these examinations will know, there are only certain types of information that are suitable for the exercise. Besides, even when a question is not set properly it is possible to discern what question the examiner is trying to set and answer it. I greatly enjoyed the medical course in a wider sense and flirted with a career in most every speciality covered in the curriculum, but I applied the above logic to the examinations with a brutality that still makes me blush. The result, in the main examination, was substantially more marks than any other student. Perhaps naively, I had not seen the difficulty that this might create. I was surprised by a summons to meet with the sub-dean. Most likely drawing on his experience as a former member of the Hospital Rugby Club, he could think of only one explanation; I had acquired copies of the examination papers in advance of the event. The interview has forever made me sympathetic to the cause of minority groups in police stations. In the end however, he was good enough to recognise that without a confession, the evidence was at best circumstantial and I was duly appointed to the House Physician post of my choice, on a 'firm' specialising in Gastroenterology and Nephrology.

During that time, Larry Baker, the Consultant Nephrologist, kindly suggested that I should become a Nephrologist myself. As with the Headmaster at Lancaster Royal Grammar School, I didn't doubt his wisdom and planned my career accordingly. Others were more cautious: Nephrology is an expensive speciality, and the UK National Health Service was not well resourced. I well remember being told by one of leaders in the speciality that there would only be two consultant positions coming up between then (circa 1980) and year 2000, so I had better distinguish myself, somehow. The first statement turned out to be untrue; but the second was chastening. As a busy trainee Nephrologist there was no real opportunity for laboratory research. So, as the necessary route to distinction, I took to writing medical case reports. In most scientific circles this would not be seen as a useful, let alone a distinguished, training. Later, I became adept at re-arranging my publication list, so this phase of my scientific development was less apparent. Later still, I have come to see it as critical. What is important in embarking on a research career is selecting the question; once the question is clear, answers may follow. For most of us this is the joy of academic research; we are free to pick our questions. I moved from London to Oxford to complete my clinical training in Nephrology. The experience of surveying the patients of London postgraduate and Oxford hospitals for case histories from which something new might be securely deduced, was without doubt a key experience. It didn't directly inform the question I eventually chose, but it taught me how to look for potentially soluble problems amidst a mass of insoluble distractions.

There were a few false starts. I thought it would be possible to understand the physico-chemical properties of myoglobin that led to myoglobinuric kidney failure, using isolated perfused kidneys. I learnt kidney perfusion from Brian Ross, who had developed the technology under Hans Krebs, but myoglobin even at massive concentration had no effect on the preparation. I then thought it would be possible to use 31Phosphorus NMR measurements of cellular energetics to understand why the kidneys are susceptible to injury in shock and joined George Radda's laboratory to study this, but the spatial resolution of the method was not sufficient. I thought it would be possible to understand why the kidneys make erythropoietin (the hormone that stimulates red blood cell production) in response to blood loss, but not to reduction in blood flow. I hadn't properly considered the technical difficulty of measuring intra-renal haemodynamics and oxygen fluxes, under those conditions. So, none of these problems were solved, but they brought me progressively closer to the 'oxygen sensing' question.

Most of my colleagues in Nephrology worked on quite different topics in immunology, genetics or the control of blood pressure. The combination of a slightly awkward, non-compliant nature that led me to avoid these subjects, and the experience of looking for soluble problems, derived from that case-report era, brought me to the question. The sensitivity and precision of regulated erythropoietin production by the kidneys in responses to changes in blood oxygen content must, I felt, reflect an answerable and important question as to the nature of the underlying 'oxygen sensing' process. Not everyone agreed; recombinant erythropoietin was being used to great effect in kidney patients, why worry about its regulation? But I was convinced there would be an answer and that it would be interesting. Besides, the identification of the erythropoietin gene opened a new possibility to trace the transduction pathway from the erythropoietin gene locus 'outwards' to the putative oxygen sensor.

I had finished my clinical training in Nephrology and there was a decision to make; should I move from Oxford to take up one of those rare opportunities to become a Consultant Nephrologist in the National Health Service, one with supposedly protected research time? Or should I stay amongst friends in Oxford? By now I had made, at least in part, a cultural adjustment towards the 'Oxbridge' environment. A combination of kind personal assurances of support from David Weatherall and John Ledingham in the Department of Medicine, miserably heavy rainfall when I travelled to look at a Consultant position at a hospital in Wales, and extreme good fortune in an interview at the Wellcome Trust, conspired to convert what was surely a truly impossible ambition (to solve the problem as a working Consultant Nephrologist) to one which, as a well-funded Wellcome Trust Senior Fellow in Clinical Science, might just be possible.

But there were still problems to overcome. I had no technical knowledge at all of molecular cell biology. At the time, I wasn't greatly concerned, though looking back, the level of ignorance must have raised a few eyebrows amongst my colleagues. This attitude owed a lot to my early experience as a junior doctor. The National Health Service was in perpetual crisis. My years as House Physician coincided with the end of the Callaghan government, and the UK's 'winter of discontent.' Practically everyone was on strike and nothing worked. As far as I could see, in the eyes of the all-powerful senior staff, the solution to each and every problem within the hospital (and on occasion outside it) was that the House Physician or Registrar (the next grade up) would sort it out. Against this background, it had never before struck me that total lack of knowledge was a barrier to engagement with a problem. But now, even I could see that some external means of acquiring skills in molecular cell biology was necessary. I invested in a copy of Benjamin's Lewin book *Genes III* and went to see a friend, John Bell, who I had met as a House Physician at the National Hospital for Nervous Diseases in London. John was generous in giving me a bench place, to work on erythropoietin regulation, in the midst of his crowded HLA immunogenetics laboratory. This was critical and set me up. So, here is my advice to the aspiring clinician scientist, take your time, look around carefully, pick your *own* question, *then* find a friend to help.

In fact, I had a number of friends who helped, most importantly David Weatherall, who, after my initiation in John's laboratory, gave me some laboratory space of my own within his newly commissioned Institute of Molecular Medicine. In the adjacent bay was Richard Jones, to whom I owe a lot. Richard taught me much of the gene regulation technology we used in the early stages of the work. Carole Beaumont (on sabbatical with Richard) taught me tissue culture. Martin Johnson at Cambridge made transgenic mice for me, an attempt to derive the erythropoietin producing cells in kidney by expressing 'T' antigen at the erythropoietin locus. By now, there was growing confidence in the project, at least locally; a series of excellent trainee Nephrologists, Tan Chorh Chuan, Chris Pugh, Patrick Maxwell, John Firth, Jonathan Gleadle came, or were steered my way. Ben Ebert joined the laboratory as a Rhodes Scholar from the US; Masaya Nagao joined us from Japan. All were taking a risk, as I had no track record in the field.

It was a busy time. Fiona had also moved from London to Oxford. We married in 1983 and she was by then a trainee anaesthetist. During that critical period when I was finishing my own Nephrology training, deciding what to do and then setting up the laboratory, she bore our four children: Anna, Alice, Robert and David. I occasionally brought the very young children to the laboratory and entertained them with tricks with the dry ice, another iteration of my own childhood theme of explosives. I interfaced week-end experiments with family walks and trips to the seaside, and generally thought I was managing well. It is only in retrospect that I see the sheer enormity of Fiona's task, managing her own training, the children and the household, almost single-handed. I owe so much to her fortitude.

There had been many unsuccessful attempts to understand the nature of the oxygen sensing process through pharmacological interventions on erythropoietin response to hypoxic stimuli. But identification of the erythropoietin gene gave us a new opportunity. It had been established that the erythropoietin gene was regulated by transcription. So, we argued, working from the gene it should be possible define oxygen-regulated control sequences at the locus and then dissect our way through the transcriptional and signal transduction pathways to the putative oxygen sensor. Nevertheless, despite a lot of talent and enthusiasm in the laboratory, none of us had a biochemical training. And beyond the identification of the oxygen-regulated control sequences by gene transfer, the most obvious approaches to dissecting those pathways were biochemical. As Richard Jones explained, the new molecular approaches based around the gene had opened research questions in cell biology to amateurs like myself, who had come into the laboratory with no training in biochemistry, but there were limitations to this. Even the environment David Weatherall had created with his Institute of Molecular Medicine could not address this deficiency. We spent a lot of time discussing alternative genetic approaches to the problem. One of these led directly, but unexpectedly, to our first breakthrough.

It had long been assumed that the extraordinary sensitivity of erythropoietin to reduction in blood oxygen reflected the function of a high specialised oxygen sensor that was specific to the erythropoietin producing cells themselves. Erythropoietin is produced by cells in the kidneys and to a lesser extent the liver. We had spent a lot of time trying unsuccessfully to culture these cells from the kidneys of the erythropoietin - T antigen transgenic mice. In the end, it was cell lines from the liver, shown by Franklin Bunn and colleagues at Harvard to produce erythropoietin in response to hypoxia, which opened the molecular approach. We and others used these cells as the vehicle to define the oxygen-sensing control sequences at the erythropoietin locus. But then there was my biochemistry problem; how to get to the target, that oxygen sensing mechanism lying further upstream. Another scientist in the Institute, Dave Simmons, was successfully using 'expression cloning' in Cos7 cells to identify surface receptors and cell adhesion molecules. I thought I would use this technology to identify upstream components of the oxygen-sensitive pathway by gene transfer. This would be from Franklin Bunn's oxygen-sensitive hepatoma cells to the Cos7 cells, which I believed would not be intrinsically oxygen sensitive as they do not make erythropoietin. To my great surprise, control experiments – designed to check the absence of oxygen sensitivity prior to gene transfer – clearly showed the same oxygen sensitivity of those control sequences isolated from the erythropoietin locus, in Cos7 cells.

This of course disqualified the intended experiment but changed everything in my scientific life. It was the first evidence of a widespread human oxygen sensing system, manifestly operating beyond erythropoietin. The implications were clear, there must be other targets in non-erythropoietin producing cells, which are also regulated with great sensitivity by oxygen levels. We found the first of these: enzymes encoding specific isoforms glycolytic genes that are also upregulated in cancer, connecting us with the cancer metabolism and oncology communities. Although the work didn't immediately attract so much attention, and there were initial difficulties in publication, from that moment we were confident we were on to something important. Nevertheless, the lag between the tremendous excitement running through the still small group of scientists working on this problem, and interest developing in the general scientific community, was very striking. It still colours my advice to young scientists; when deciding what to do, try very hard to ignore the interests and prejudices of those around you, they will likely be a very long way behind the curve. However, the field grew steadily as more and more people found new pathways that responded to HIF (Hypoxia Inducible Factor, the transcription factor binding those oxygen-regulated sequences, which was discovered by Gregg Semenza). However, I think all of us were surprised by the extent of the HIF transcriptional cascade and the extent to which so many responses to hypoxia (low tissue oxygen levels) had previously been overlooked. Many of the new responses to hypoxia were fascinating in their own right, but our intention had always been to work our way upstream to the oxygen sensing mechanism itself and our attention shifted back.

By now we were an established group in David's Institute of Molecular Medicine and we took advantage of everything it had to offer. Half of an entire floor in the building was given over to the Institute's coffee room. David had clearly been impressed by this much championed facility at the Laboratory of Molecular Biology in Cambridge. Coffee was made in advance, dispensed with impressive efficiency and consumed in armchairs of exactly the correct design and spacing to support conversation. The room was carpeted. For those considering institutional design, every detail of this matters. We sat there with anyone who would listen and speculated (mainly unproductively) about what the mechanism of oxygen sensing *might* be, and (a little more productively) on experimental strategies that we might *use* to define it. With every visitor I considered their work from the perspective of its potential to solve the oxygen sensing problem, my problem. We tried to harness a whole range of gene transfer and expression cloning methodologies, we examined model organisms for conservation of the pathway, hoping to harness genetic methods in flies, nematodes worms, or even yeast. I was impressed by George Stark's use of somatic cell genetics for dissection of interferon response pathways. Morwenna Wood and Emma Vaux, two highly competent trainee Nephrologists, spent vast amounts of time engineering Chinese Hamster Ovary cells to express hypoxia inducible transgenes encoding cell surface markers, and then selecting mutants with defective responses to hypoxia. To their enormous credit, they did isolate valuable mutants, but it proved difficult to identify the defective genes beyond those encoding components such as HIF that we already knew about.

In the end, it was a mixture of genetic and biochemical approaches that brought us step by step towards the solution. Each advance was incremental, to use that favourite word of editors when declining manuscripts. Defining the regulatory domains in HIF that mediated oxygen sensitivity of the complex, demonstrating the physical association of those domains with the von Hippel-Lindau protein (pVHL), showing the function of pVHL as a ubiquitin ligase that degrades HIF, the discovery of prolyl hydroxylation as the mechanism governing oxygen-regulated association of pVHL with HIF, evidence that the enzymes catalysing HIF prolyl hydroxylation belonged to the 2-oxoglutarate dependent dioxygenase family, identification of the actual enzymes and their oxygen sensitivity, were all *incremental* steps. I was one of those fortunate enough to be called to Stockholm for our contributions to this work, but there many others whose work contributed to, and was informed by, the timely publication of all those incremental advances.

The final steps, involving the identification of the actual oxygen sensing 2-oxoglutarate dependent dioxygenases that catalyse the prolyl hydroxylation of HIF, were taken together with my friend and colleague Christopher Schofield, with whom my laboratory continues to enjoy great collaborations. Chris (Professor of Organic Chemistry at Oxford) brought the biochemical perspective that I lacked to the work. But the original collaborative work was not a large-scale biochemical purification, which I had previously envisaged being necessary at some stage in the programme. Rather, based on his earlier structural analyses of 2-oxoglutarate dependent dioxygenases and related enzymes, Chris was able to predict genes that might encode the putative oxygen sensing prolyl hydroxylase. Meantime one of our earlier exploratory ventures, hatched in the Weatherall coffee room, had involved identifying the HIF orthologue in *Caenorhabditis elegans* and, very importantly, raising an antibody against the protein. This enabled us to assess mutants for their impact on the proteolytic regulation of the HIF by hypoxia, but so far none had shown an abnormality. One of Chris's predicted 2-oxoglutarate dependent dioxygenases was represented in the libraries of mutant *C. elegans* that were so beautifully catalogued and efficiently provided by the nematode worm genetics community. Worms bearing mutant alleles of the relevant gene were identified via WormBase and duly ordered.

I have a vivid recall of one morning in March 2001; Andy Epstein, a PhD student in the lab burst into my office, exclaiming, 'here's your gene, *Egl9*'. Three different alleles of the mutant *Egl9* gene, previously characterised by the Nobel Laureate Bob Horvitz on the mechanistically agnostic basis of defective expulsion of their eggs, all showed upregulation of their HIF, irrespective of oxygen levels, as would be predicted for a defective enzyme whose oxygen-dependent catalysis of prolyl hydroxylation was physiologically deployed to signal oxygen levels. We were rapidly able to confirm this and identify three human orthologues. Jonathan Gleadle, one of the trainee Nephrologists who joined the laboratory and did some of early work defining HIF-target genes, had returned to the group after completing his clinical training. Jonathan identified the human orthologue of Egl9 on the basis of a highly conserved catalytic domain and called their products the PHD (prolyl hydroxylase domain) enzymes. Those findings completed that journey I embarked on as a young Nephrologist, from erythropoietin to oxygen. The high points in that journey were exciting, very exciting; the low points I tend to forget. The experience was addictive, and I am always looking for the next of those 'eureka' discovery moments.

Though the work provided an answer to a question, defence of oxygen homeostasis is clearly more complex. Severe hypoxia is fatal within minutes and oxygen homeostasis must be maintained over much shorter and longer timescales than those mediated by the transcriptional pathways we have so far unravelled. For instance, the oxygen sensitive signals by which the carotid body controls breathing, part of the work for which the Nobel Prize for Physiology or Medicine was awarded to Corneille Heymans in 1939, are still not understood at the molecular level. The laboratory works on these and other as yet unsolved problems in the physiology of oxygen homeostasis. Although we were at first surprised to find that the HIF system operated generally in mammalian cells, and was not restricted to erythropoietin producing cells, we were then surprised that it was apparently restricted to animal life, there being no obvious HIF orthologue in non-metazoan species. It is now clear that species in all four eukaryotic kingdoms deploy enzymatic protein oxidations coupled to protein degradation to signal oxygen levels in their cells. However, the oxidations are of different types and coupled in different ways to the signalling systems, raising questions as to the origins of these systems, their inter-relations and whether they also function in human oxygen sensing. I also became interested in cancer, in particular the implications of oncogenic 'switching' of very extensive interconnected physiological pathways, such as occurs when the pVHL ubiquitin ligase is inactivated and the HIF system is unphysiologically activated in kidney cancer. I imagine that this paradigm, in particular the mechanisms by which the developing cancer 'accommodates' adverse components of the oncogenically-activated pathway will be important in understanding the disease. For these reasons, supported by loyal staff who have stayed with me for years, the laboratory remains as active, or at least as hopeful, as ever, and I remain as addicted as ever to the discovery process.

But there have been new experiences. I worked with small and large companies on the development of HIF hydroxylase inhibitors for the treatment of anaemia, and perhaps other ischaemic/hypoxic diseases. This is an interesting experience for an experimental scientist used to near-total control of an experimental research programme. It is very satisfying to see the work progressing towards a medicine, though somewhat alarming not to have that control. But there is the same unpredictability. Given that the field started with the regulation of erythropoietin, it is no surprise that the 2-oxoglutarate analogues being developed as HIF hydroxylase inhibitors will induce erythropoietin production and correct the anaemia of erythropoietin deficiency. But given the extent of the HIF transcriptional cascade that became apparent, I don't think the scientific community would have predicted that this could be done without a major dysregulation of other responses to hypoxia, as appears from current clinical experience to be the case. I learned a certain respect for those taking the risks that are intrinsic to the medicine development process.

In 2003, I was approached about taking on the Headship of the Department of Clinical Medicine. John Bell had succeeded David Weatherall into this chair and was now moving (as David had) to the Regius Chair. I hesitated, knowing very well that University administration was not a natural suit. But the Department was interesting. Under the successful tenure of David and John it had grown well beyond the norms of a University Department, more the size of a small University, with a research budget that would come well up the UK university league. It was and is (in accordance with the name) a Department of *Clinical* Medicine but encompassed an extraordinary range of disciplines ranging across structural biology, large-scale human genetics, clinical trials, epidemiology, molecular cell biology, to tropical medicine and the teaching of clinical medicine within the National Health Service. An expectation was that the senior professorial staff would contribute to the Acute Medicine Rota and teach medical students in that context. With my training in Clinical Nephrology and continuing practice in both Nephrology and Acute Medicine for much of the period I have just described, I could actually do this. My perception was that this was slightly to the surprise of a number of the other medical and nursing staff working in the relevant service; it certainty filled some of my post-docs in the laboratory with horror. They felt that this sort of 'high tension' medicine should only be performed by highly professional, and very good-looking staff in surgical scrubs, as they had seen on TV. I have never been completely sure as to which attributes, they felt to be lacking in myself. However, clinical medicine was only part of the experience. Across all its domains the Department encompassed more than 2000 staff, including many working in tropical South-East Asia and Sub-Saharan Africa. Academic management of this scale is a challenge, by and large I learned not to attempt it. I learned a lot about people, particularly creative people. In the Department, as in the laboratory, the most surprising things (to me) turned out to have their value. I learned tolerance and firmly believe that science is, or should be, one of the great unifying forces in a world of diversity. Of course, a persistent problem was balancing Departmental matters with the running of the laboratory, and I'm grateful to many on both sides of that equation who ensured at least a modicum of success. But I avoided further progression in academic management, and after more than a decade in the position, looked for opportunities that might give me more time in the laboratory and new scientific horizons.

It was therefore a piece of good fortune when, in 2015, Paul Nurse approached me about a position at the shortly-to-be-opened Francis Crick Institute, to develop its interface with clinical practitioners and clinical medicine. Here, I thought my unusual experience of different sides of medicine might be useful. And there were new scientific horizons. The key point about the journey that I have just described is that it is the selecting of the research question that matters. The purpose of bringing clinical practitioners to a major biomedical research institute such as the Crick is that they will ask new questions, not necessarily clinical ones, simply different ones. I combined the Crick position on a 50/50 basis with running my laboratory in Oxford as a member of the Ludwig Institute of Cancer Research and felt I had things under control, at least until one morning in October.

I had heard mention of the Nobel in discussions of the merits or otherwise of the work on oxygen sensing, but the news that morning was a surprise. I had reached that all-too-familiar crisis point in writing a grant application with colleagues from Finland and had been writing most of the weekend. In the frenzy of activity, I had quite forgotten the significance of the first Monday in October. But my personal assistant, Catherine, does not forget things. It was Catherine who pulled me out of the Monday morning laboratory meeting to take the call from Thomas Perlmann and then replaced my strong coffee with a calming mug of tea when I returned to finish the meeting. A great deal of champagne was subsequently consumed in the laboratory throughout the day and, as television viewers observed, standards of safety became lax. Many have recounted the rather surreal experience of being bestowed with omniscience, overnight. Even the children (now grown up) thought I might have a useful opinion on things, for a short period of time. But the experience in Stockholm is worth restating.

The attention to making Nobel Week a celebration, and a happy time in every way possible way, was truly remarkable. Every event was managed to perfection, every microphone tuned precisely, all electronic projection flawless. The King was majestic, the Queen was gracious and Princesses stunning. Breakfast in the Grand Hôtel, overlooking the Stockholms ström, is one of most pleasurable experiences on the planet. The week's hectic schedule was efficiently managed by our attaché Jane Viol. It even seemed my motor-racing aspirations had been considered; our driver Bo Cravell was an ex-rally champion. There was just one hiccup. The family travelled to Stockholm with our new-born granddaughter, her mother (Anna) and father (Rupert) having hastily obtained a passport. At the point of maximum tension, dressing for the ceremony, it emerged that it would not be possible to accommodate the baby at the five-hour banquet, in any way whatsoever. My wife Fiona took the matter seriously. After 37 years of marriage, I did not doubt the outcome. I carried on wrestling with buttons on the formal attire. I think Miss Elizabeth Merryn Everleigh is youngest to attend that magnificent Nobel banquet, safely tucked away under that very grand staircase through which we had entered the Hall. What a day.

Elucidation of Oxygen Sensing Mechanisms in Human and Animal Cells

Nobel Lecture, December 7, 2019 by
Peter J. Ratcliffe
Target Discovery Institute and Ludwig Institute for Cancer Research, University of Oxford and Hypoxia Biology Laboratory, the Francis Crick Institute, London, UK.

INTRODUCTION

Oxygen is absolutely required for most forms of life. It is the terminal electron acceptor in energy-generating electron transport systems and deployed in a wide range of other metabolic, biosynthetic and detoxification processes. All such life requires the defence of oxygen homeostasis against variation in environmental concentrations of oxygen. Delivering oxygen homeostasis is more complicated in large, multi-cellular organisms, whose size precludes effective oxygen transport to the respiring cells by simple diffusion. The problem is most acute in higher animals that are not only large, but also use oxidative metabolism to generate the energy for rapid movement. In humans, the lungs, heart, and circulatory systems have all evolved to meet this challenge and many human diseases are complicated by hypoxia or inadequate tissue oxygen delivery. Thus, the physiology of oxygen homeostasis is of central importance to medicine. In this lecture I will outline how I came to be interested in the field and describe advances in the molecular understanding of the sensing and signalling of hypoxia that were made in my laboratory.

Homeostasis requires the detection and response to an error signal and I will refer to mechanisms of 'oxygen sensing' that underpin this process. The term itself requires some clarification. Since very many biological processes involve chemical reactions with oxygen, sufficiently low oxygen levels will cause the dysregulation of countless processes. All such systems will clearly be 'oxygen sensitive'. To distinguish *bona fide* homoeostatic pathways, I use the term 'oxygen sensor' to define systems whose prime function is in biological control. In theory, such control systems might respond to oxygen itself or to some metabolic consequence of hypoxia. These possibilities are difficult to distinguish in the intact animal and this difficulty appears to have held back understanding of oxygen homeostasis for much of the twentieth century. For instance, early studies of capillary density in different tissue beds recorded both the remarkable association with oxidative capacity, and dynamic responses to changes in oxygen demand or supply, but hypothesised mechanisms generally emphasised products of metabolism as candidates for transducing the observed effects. Thus, given the central importance of oxygen to metabolism, it was difficult to distinguish the direct sensing of oxygen from the sensing of metabolic signals. In fact, it is now clear that systems that sense and respond to different metabolic signals operate in parallel with the direct oxygen sensing mechanisms that I am going to describe.

ERYTHROPOIETIN AS A PARADIGM FOR OXYGEN-REGULATED GENE EXPRESSION

Despite the general emphasis on metabolic consequences of hypoxia, early studies of a few highly dynamic responses to hypoxia, including the control of breathing and regulation of blood red cell levels, suggested the existence of physiological systems that respond directly to oxygen availability. It was the study of one of these, the control of red cell production by oxygen-dependent production of the hormone erythropoietin, that led to the current work. Two aspects of this process supported the existence of a specific and direct 'oxygen sensing' process. First was the extraordinary sensitivity of erythropoietin production to alterations in blood oxygen content. This can be traced back to the pioneering work of Mabel FitzGerald, an unsung female scientist who accompanied J.S. Haldane and colleagues on the 1911 Oxford-Yale expedition to Pike's Peak, Colorado to study altitude acclimatisation. FitzGerald's key contribution to this story was to record changes in the blood haemoglobin levels and blood gas concentrations, in altitude-adapted men and women living at intermediate altitudes in the mining towns around Pike's Peak.[1] This work revealed linear correlations with altitude, with even small increases in altitude (and hence decrease in pO2) being associated with a discernible increase in haemoglobin and decrease in blood carbon dioxide (representing increased breathing) (Figure 1). Later in the 20th century, Allan Elslev's plasma transfer experiments established the existence of a plasma factor (so-called erythropoietin) in the regulation of erythropoiesis, [2] paving the way for sensitive assays for erythropoietin in the blood that confirmed the extra-ordinary sensitivity of erythropoiesis and erythropoietin to very minor changes in blood oxygen availability.

Studies of erythropoiesis and erythropoietin also defined a second characteristic that suggested direct control by oxygen as opposed to secondary metabolic compromise. Early observations on poisoning by cobalt had identified increases in blood haemoglobin as an unusual toxicological phenotype.[3] Subsequent work demonstrated that cobalt induced erythropoietin production.[4] Since cobalt, at erythropoietic doses, had been shown not to be a 'metabolic' poison[5] and substances which are, such as cyanide, do not induce similar changes in erythropoietin and erythropoiesis,[6] it was concluded that this response was being governed by a specific 'oxygen sensing' process.

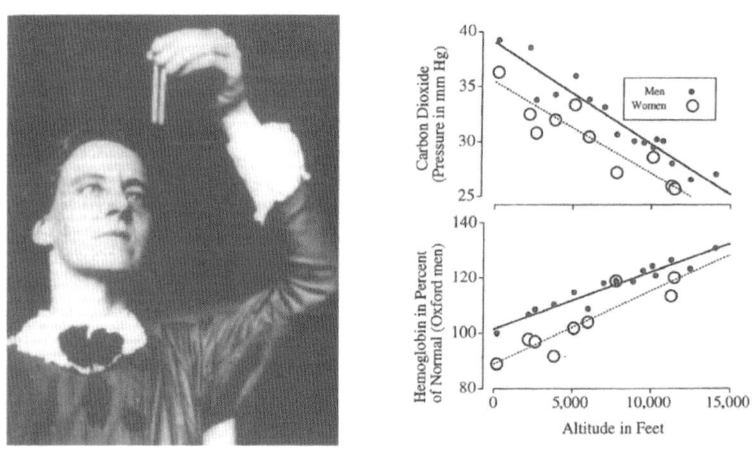

Figure 1. Mabel FitzGerald, the pioneering female scientist who accompanied J.S. Haldane's all male team to study adaptation to altitude on Pike's Peak, Colorado. (A) FitzGerald measuring haemoglobin concentration in the blood, by dilution and colour matching to a standard (photograph from the *Herald Telegraph*, Colorado Springs, 8th July 1911). (B) FitzGerald's measurements of haemoglobin (upper panel) and end expired carbon dioxide concentration (lower panel) in adapted subjects residing at different altitudes. The results demonstrate the effect of even small differences in altitude reflecting the high sensitivity of adaptive responses to hypoxia (adapted from[1]).

Erythropoietin is made in the kidneys (and to a lesser extent in the liver) and, as a recently trained clinical nephrologist, it was this aspect of renal physiology that drew me to the field. Many attempts to understand the process of erythropoietin regulation had been made, but as I completed my clinical training in the 1980s, a new opportunity for study arose. Following purification of erythropoietin from the urine of anaemic subjects,[7] the encoding cDNA and genetic locus had been identified.[8,9] It was established that erythropoietin is not stored at the site of production, but that oxygen-regulated production reflects changes in oxygen-regulated gene transcription. Broadly, our intended strategy was then to identify cis-acting sequences at the erythropoietin locus that mediate oxygen-regulated transcription, then follow the interactions of these sequences to what we believed would be a specific oxygen sensing mechanism lying within specialised cells of the kidney that produce erythropoietin.

ERYTHROPOIETIN-PRODUCING CELLS

The immediate obstacle to this approach was uncertainty as to the identity of the erythropoietin-producing cells in the kidneys and the non-availability in tissue culture of any kidney-derived cell line that produced erythropoietin. A friend at Oxford, John Bell (now Regius Professor) suggested that I tried to overcome this by expressing an oncogene, SV40 'T' antigen, under control of the erythropoietin cis-acting control sequences in transgenic mice. Since at the start of this work there was no knowledge of these sequences, I simply took the largest fragment of genomic DNA that I could isolate at the erythropoietin locus, and inserted 'T' antigen sequences into the erythropoietin 5' untranslated region. Several lines of transgenic mice were produced, and remarkably one of these contained the transgene recombined into the native erythropoietin locus. So we were confident of correct tissue-specific expression and initially optimistic that this would solve the problem. The mice expressed 'T' antigen in a renal interstitial cell population and developed an expansion of that cell population as they aged. Immuno-histochemical studies revealed that this population of cells was a subset of the renal interstitial fibroblasts, expressing 5' ectonucleotidase on the cell surface.[10] To our disappointment, no obvious special features were apparent, nor could we grow a 'T' antigen or erythropoietin expressing population after the explantation of cells into culture. Thus, the experiment failed in its primary intention, although together with work from Kai Eckardt's laboratory,[11] it did identify the erythropoietin producing cells, as a population of interstitial fibroblasts within the kidney cortex. The reasons why the erythropoietin gene (and our transgene) are extinguished in explanted kidney cells are, to my knowledge, still not understood, but likely have a bearing on the mechanism(s) by which erythropoietin production fails in kidney disease.

At the time, the way forward was therefore dependent on two liver-derived hepatoma cell lines that had been demonstrated by Franklin Bunn and colleagues to produce erythropoietin in tissue culture, when ambient oxygen levels were lowered.[12] Not all the physiologists in the field were convinced that this properly represented the physiology of erythropoietin production in the kidneys, but we (my long-standing friend and colleague Chris Pugh and I) set about mapping cis-acting oxygen-regulated control sequences by transient transfection of mouse genomic DNA into these cells, taking advice from Richard Jones in David Weatherall's newly commissioned Institute of Molecular Medicine. This work identified a sequence lying 3' to the mouse erythropoietin gene as a transcriptional enhancer that conferred oxygen regulated expression on a heterologous reporter gene (Figure 2, upper panels).[13] The sequences were highly conserved and lay at exactly the same position with respect to the erythropoietin gene as human sequences identified in parallel work by both Gregg Semenza and Jaime Caro.[14,15]

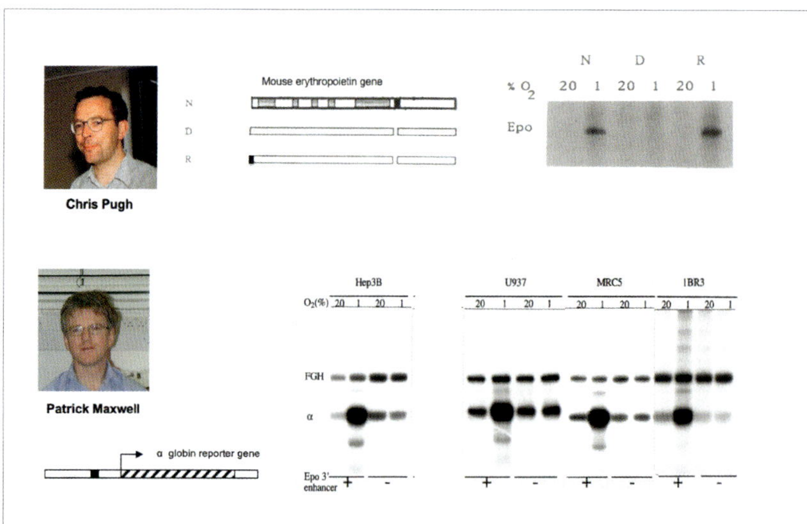

Figure 2. Widespread activity of the oxygen-regulated erythropoietin enhancer in both erythropoietin and non-erythropoietin producing human cells. (A) RNAse protection assays of the human erythropoietin-producing hepatoma cell line Hep3B, transfected with the indicated sequences from the mouse erythropoietin gene; the work identifies a transcriptional enhancer lying 3' to the mouse erythropoietin gene. (B) RNA protection assays demonstrating that the erythropoietin 3' enhancer confers oxygen regulated activity on an α-globin reporter gene (α) not only in erythropoietin-producing Hep3B cells (right panel) but also in a range of non-erythropoietin-producing cells; U937 monocyte/macrophage; MRC5, lung fibroblast; IBR3, skin fibroblast.; α, α-globin; FGH, transfection control without the erythropoietin enhancer sequence. Reproduced from references.[13,16]

DISCOVERY OF A WIDESPREAD SYSTEM OF OXYGEN SENSING IN HUMAN AND ANIMAL CELLS

We immediately recognised that identification of these oxygen-sensitive enhancer sequences would enable another very important question to be addressed. What would be their activity if transfected into cells that did not produce erythropoietin and were not known to be oxygen-sensitive? Initial experiments gave an unclear result; certainly, the sequence did not appear to be as active as in the erythropoietin-producing hepatoma cells. This initially reinforced the prevalent view that the oxygen-sensing mechanism was a specific property of erythropoietin producing cells, and I set about what I thought was an innovative (though at best high risk) experiment attempting to 'expression clone' components of the 'oxygen sensing' pathway by transfer of genes from the 'oxygen sensing' hepatoma cells, to a line that would be predicted to be oxygen insensitive (according to the then existing prejudice, every other cell type). For the latter, I chose Cos 7 cells, for ease of retrieval of plasmid DNA. Cos 7 cells grow rapidly and when I came to check a rather dense culture of these cells transfected with the erythropoietin enhancer alone, to my surprise I observed very clear oxygen-regulated activity. Of course, this disqualified the planned experimental approach, but fundamentally changed our thinking.

Patrick Maxwell joined the laboratory and set about testing other cells. He saw that responses depended on the culture conditions, but that when rapidly dividing, relatively dense cultures were tested, essentially all cultured cells supported oxygen-regulated activity of the erythropoietin enhancer. Though we never measured it directly, the initial variation almost certainly related to dependence of the actual oxygen concentration at the monolayer on the oxygen consumption rate of the cells, and hence on their growth rate and density. Once this was appreciated, the responses across a wide range of cells were very robust (Figure 2, lower panels) and we were very excited indeed by the implications. If essentially all mammalian cells could support oxygen-regulated activity of this erythropoietin enhancer sequence, irrespective of whether they produced erythropoietin, then the oxygen sensing system was clearly widespread and there must be other targets. Moreover, they might also be predicted to respond to hypoxia with high sensitivity, similar to erythropoietin. Though we had no idea at the time of the extent of the field that was to unfold, I had no doubt about the significance of the findings and confidently drove to the London offices of *Nature*, anticipating a dialogue with the editor that would lead to rapid acceptance for publication.

Needless to say, the final response, four months later, was disappointing. Of course, there will be many who have such an experience and the story itself is not particularly remarkable. There are two reasons for telling it. First, as the editor explained, part of the reason for the delay was 'difficulty in finding reviewers.' For any early career readers who have the same experience, this is important evidence that you are entering a new field. Be encouraged by that. Second (from the current vantage point) it is easy to have some sympathy with the editor. Essentially it was only one new fact, and we hadn't then followed through by addressing the obvious next question; what were those additional oxygen-sensitive targets? As a young nephrologist, I had little experience of research practice and it didn't occur to me that deriving follow-on conclusions, or a 'more complete story' might be important before publication. And, actually, I still think this often-sought ambition is a problem. As far as I was concerned, the laboratory was stone cold certain of the actual result and we believed the implications to be profound. Once the findings were published, I argued, we and others could follow up. In the end, with a little help from David Weatherall, we did publish, but in another journal,[16] and there was indeed some follow-up. I met with Gregg, at a meeting organised by Christian Bauer at the Lake Lucerne Palace Hotel. He presented his (then unpublished) work defining the protein binding those control sequences in erythropoietin-producing hepatoma cells (hypoxia inducible factor, HIF-1) and I presented the sadly still unpublished findings described above. We left the meeting with considerable enthusiasm for what lay ahead, meeting regularly at conferences now devoted to the expanding field of hypoxia research (Figure 3).

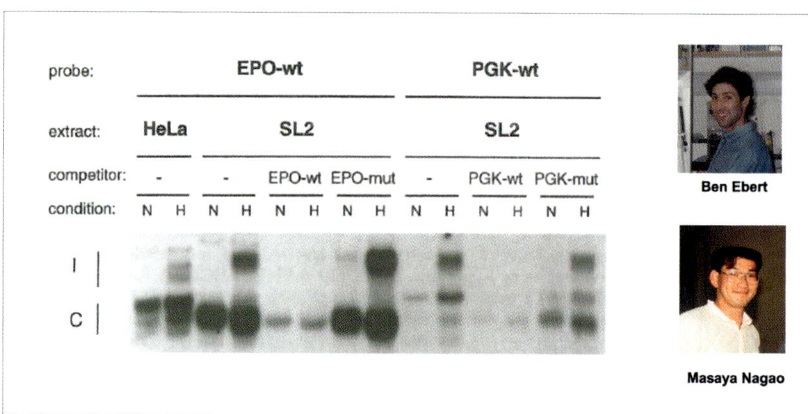

Figure 3. Electrophoretic mobility shift assays demonstrating the presence of hypoxia inducible factor (HIF) in invertebrates. Nuclear extract from normoxic (N) or hypoxic (H) Hela or *Drosophila* (fruit fly) SL2 cells. SL2 cells are shown to contain hypoxia inducible nuclear proteins that bind hypoxia response elements from the human erythropoietin enhancer (EPO-wt) and phosphoglycerate kinase promoter (PGK-wt). Unlabelled oligonucleotides (EPO-wt and PGK-wt), but not mutant version that ablate HIF-binding (EPO-mut and PGK-mut) compete these interactions. Reproduced from[24].

The efforts in my laboratory initially focused on genes encoding glycolytic enzymes, with other clinical trainees, Ben Ebert, John Firth, and Jonathan Gleadle joining the laboratory and contributing much to the work. Dissection of regulatory sequences in the genes encoding the glycolytic enzymes phosphoglycerate kinase and lactate dehydrogenase revealed that their oxygen regulated expression (despite lower amplitude of regulation) was dependent on sequences resembling those in the erythropoietin enhancer.[17] These sequences had a similar core motif 'ACGTG', which bound HIF, and they cross-competed with the erythropoietin enhancer in electrophoretic mobility shift assays of HIF binding.[17,18] These results then clearly confirmed that other genes respond to the same oxygen-regulated signals and we rapidly found many other examples. Angiogenic growth factors such as vascular endothelial growth factor and platelet-derived growth factor,[19] glucose transporters and glycolytic enzymes[18,20] all manifest the classical features associated with erythropoietin regulation; the response to hypoxia was mimicked by cobaltous ions, and (as had more recently been established by Gregg[21]) by iron chelators.[19] All were transcriptional targets of HIF.

Apart from corroborating the evidence for a widespread mechanism of oxygen sensing in mammalian cells, this early work took us in two new directions. I read the work of researchers following up on Warburg's findings on enhanced glycolysis in apparently well oxygenated malignant cells.[22,23] These studies described isoform specificity in the upregulation of genes encoding glycolytic enzymes in cancer cells. It was quickly apparent that the up-regulation of glycolytic enzymes and glucose transporters by HIF manifests a closely similar isoform specific pattern, immediately suggesting a link between HIF activation and cancer.[20] Second, as glycolysis and the genes encoding glycolytic enzymes are more conserved than erythropoiesis and the erythropoietin gene, the new findings raised an important question as to the phylogeny of the pathway, and whether we could apply the powerful genetic technologies available in model organisms to dissect the mechanism of oxygen sensing. Substantial unrewarded effort went into attempting to define an orthologous system in yeast. However, we did obtain clear evidence for the existence of a HIF-like protein that bound to the human erythropoietin control sequences in insect cell extracts derived from *Drosophila melanogaster* SL2 cells (Figure 4).[24] Quite clearly the system ante-dated the appearance of specialised blood vascular systems that have evolved for oxygen delivery in higher animals. Importantly, the findings opened new opportunities for analysis in lower 'model' animal organisms, which we were able to capitalise on later.

Figure 4. The author discussing the future of hypoxia research with Gregg Semenza (Woods Hole, Massachusetts, 1993).

Our own findings on different HIF target genes and transcriptional pathways downstream of HIF were amplified by a great deal work from Gregg and many other groups, with the field undergoing an explosion of interest during the later years of the 1990s. It became clear that the HIF system was involved in many processes beyond those with well-established connections to oxygen homeostasis. In addition to erythropoiesis, angiogenesis and energy metabolism, further studies identified hypoxia and HIF-regulated genes with central roles in iron metabolism, pH regulation, matrix metabolism, cell differentiation, cell motility, immune regulation and stem cell behaviour. Quite quickly, the list (now encompassing hundreds or thousands of genes) extended beyond even our most optimistic perspective.

REGULATORY DOMAINS IN HYPOXIA INDUCIBLE FACTORS

Our major efforts now focussed on defining the oxygen sensing and signalling pathways that lay upstream of HIF. Gregg had identified the HIF DNA binding complex as an α/β heterodimer of basic-helix-loop-helix PAS proteins; HIF-1β being a constitutively expression protein that also dimerises with the aryl hydrocarbon receptor and HIF-1α being a novel protein, specifically transducing the response to hypoxia.[25] Reasoning that definition of isolated sequences from HIF-α sub-units that could confer oxygen sensitivity on heterologous proteins would direct a search for the upstream signal pathways, we set about fusing HIF-α sequences to heterologous transcription factors and assaying for oxygen sensitive responses. Overall, it was possible to define three distinct domains in HIF-α, each of which could convey oxygen sensitivity in heterologous transcriptional reporter systems.[26,27] These experiments demonstrated that two of these sequences operated, at least in part, through oxygen-regulated proteolysis, whereas a third, overlapping the C-terminal activation of HIF-α operated through changes in transcriptional activity that did not reflect altered protein stability. All of them displayed responses to hypoxia, cobaltous ions and iron chelators, consistent with the characteristics of both erythropoietin and HIF regulation. In one way these experiments (together with similar studies from others[28,29]) were very successful in directing the search for upstream oxygen signalling pathways, as each of these discrete domains must interact in some way with those pathways. In other ways, we were disappointed. We had confidently expected to identify a common protein phosphorylation motif that would surely direct us to the signal transduction pathway. But surprisingly, we were initially unable to define a common motif at all, despite the apparently identical regulatory characteristics. Chris Pugh went as far as to mutate every phospho-acceptor amino acid in one of these domains, the more C-terminal of the oxygen dependent degradation domains.[26] None of the mutations made any difference to the oxygen sensitivity. Although initially disappointing, these experiments were critical in excluding what had been believed to be the likely mechanism of signal transduction (a protein phosphorylation cascade linking an 'upstream' oxygen sensor to HIF), and hence were important in directing attention elsewhere.

CONDITIONAL DEGRADATION OF HIF BY THE VON HIPPEL-LINDAU TUMOUR SUPPRESSOR

Our next major step was the recognition that the von Hippel-Lindau tumour suppressor protein was absolutely necessary for the oxygen sensitive proteolysis of HIF-α. We found that pVHL physically bound to HIF-α sequences, that binding was prevented by exposure of cells to the 'hypoxia-mimetics' (cobaltous ions and iron chelators) and that this process governed the stability of HIF-α and the expression of HIF target genes.[30] This work interfaces with contributions from my other co-laureate Bill Kaelin. The interplay between the different contributions illustrates how knowledge gained from different perspectives interacts to build a picture. Bill and others had examined some of the recently-identified hypoxia-inducible transcripts described above (encoding vascular endothelial growth, glucose transporter-1) in renal cancer cells that are defective for the von Hippel-Lindau tumour suppressor.[31,32] They reported constitutive upregulation irrespective of oxygen, but didn't identify the mechanism. We heard about these findings, observed the upregulation of hypoxia inducible transcripts, but initially were also puzzled as to the mechanism as we could not detect HIF-1α at all in the renal cancer cell lines that were being studied.

How then was the matter resolved? In a different line of work, sequencing of the human genome was revealing new opportunities for gene identification; Steve McKnight was hunting for basic-helix-loop-helix PAS proteins. He identified what he proposed to be a HIF-α paralogue strongly expressed in endothelial cells. He termed the gene product EPAS1 (endothelial PAS protein-1).[33] But the regulation of EPAS1 by oxygen was not proven. To study this, it was necessary to measure expression at the protein level. To that end, we made an antibody, PM (Patrick Maxwell) 9, which clearly established that EPAS1 (now more commonly known as HIF-2α) was indeed regulated by oxygen, like HIF-1α.[34] We had, by then, identified more HIF-target transcripts, encoding other glycolytic enzymes and found them also to be constitutively upregulated in VHL deficient cells. So, we went back to HIF, this time also using the new anti-EPAS1 antibody. The result was striking. The renal carcinoma cells that we and others had used (786-0) expressed EPAS1/HIF-2α alone (and not HIF-1α), but the HIF-2α protein was constitutively stabilised and upregulated irrespective of oxygen levels. When we examined other renal carcinoma cell lines, we found some of these expressed HIF-1α as well as HIF-2α; importantly whenever HIF-1α or HIF-2α was expressed, the proteins were constitutively stabilised (Figure 5).[30] We were rapidly able to build on this as described above, with Matthew Cockman and Norma Masson going on to demonstrate directly that pVHL (as predicted by its structure)[35] was a ubiquitin E3 ligase that conditionally targeted HIF-α sub-units for destruction by the ubiquitin-proteasome pathway.[36]

Renal cancer-associated mutations affecting a specific region of pVHL abrogated the interaction with HIF-α, directly linking dysregulation of HIF to the function of a tumour suppressor protein, and hence to can-

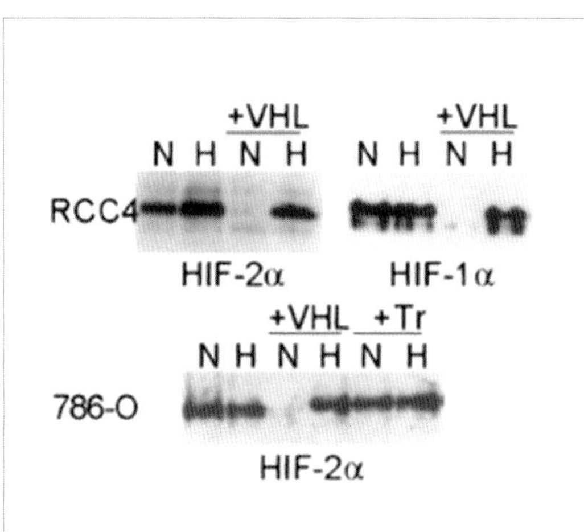

Figure 5. Regulation of HIF by the von Hippel-Lindau protein (VHL). Panels show HIF-1α or HIF-2α immunoblots of extracts from normoxic (N) and hypoxic (H) VHL-defective kidney cancer cell lines, RCC4 and 786-0. HIF-α proteins are stabilised in VHL defective cells irrespective of oxygen levels. Transfection with a wildtype (+VHL) but not a truncated (+Tr) *VHL* gene restored regulation of HIF-α by oxygen. 786-0 cells do not express HIF-1α. Reproduced from[30].

cer.[37] But it was puzzling to find that this unphysiological upregulation of HIF was associated with a very unusual pattern of HIF expression, a very strong bias to HIF-2. By examining kidneys from patients with VHL disease (the familial form of clear cell renal cell carcinoma, ccRCC), we observed that this bias appeared to evolve during progression from morphologically normal renal tubular cells (expressing HIF-1) to dysplastic and overtly malignant cells (dominantly expressing HIF-2).[38] Consistent with this, we (together with Bill and others) subsequently found that, in this setting, HIF-2 appears to be supporting oncogenic drive, with HIF-1 even restraining growth of renal carcinoma cells as experimental tumours.[39] We have since become very interested in the implications of the 'unphysiological switching' of massively interconnected pathways, such as HIF, during oncogenesis, and in the actions of selective pressures that apparently operate during cancer evolution to 'tune' the pathway output (such as by modulating the HIF-1 to HIF-2 bias).[40,41] However, at the time our major quest was to move on to understand the process of oxygen sensing.

A MECHANISM FOR OXYGEN SENSING

Identification of the HIF-α/pVHL connection opened the way to defining the mode of regulation of HIF. We narrowed down sequences in HIF-α that were necessary for interaction with pVHL, initially focusing on the C-terminal of the two regions that showed oxygen dependent proteolysis, the one in which we had mutated every phospho-acceptor amino acid without impairing its function.[26] We found that peptides representing this sequence required treatment with cell extract to promote interaction with pVHL, implying some form of modification (that was presumably not protein phosphorylation). But one result was very puzzling. By whatever means we used to assess the pVHL-HIF association, we always found that the association was blocked when extracts were made from cells treated with hypoxia mimetics (cobaltous ions and iron chelators), but apparently not when they were made from hypoxic cells. Could it be that the oxygen dependent reaction was continuing in the cell extracts post-lysis? This was shown to be the case by Panu Jaakkola and David Mole, who worked with a large oxygen-controlled work-station, a chance purchase from Baker-Ruskinn that I had made with an (unusual) surplus on my grants that year. This enabled extracts to be made, and immuno-precipitations performed, entirely within a hypoxic atmosphere, and using buffers pre-equilibrated to the same level of hypoxia. These and other experiments clearly revealed that the process was dependent on oxygen, but not ATP, and that it was inhibited by cobaltous ions and iron chela-

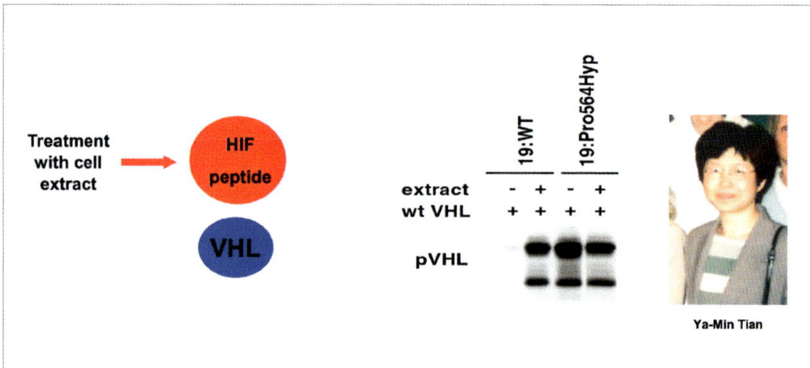

Figure 6. Definition of prolyl hydroxylation as the regulatory modification of HIF-α proteins governing binding to the von Hippel-Lindau protein (VHL). Binding of VHL to synthetic peptides derived from HIF-1α sequences (556–574) was assayed by pull-down. The unmodified HIF peptide (19:WT) requires treatment with cell extract to promote binding of VHL. In contrast a HIF peptide bearing a trans-4-hydroxyprolyl substitution at residue 564 (19:Pro564Hyp) binds VHL without such treatment. Adapted from[42].

tion. Other experiments narrowed the target HIF-α sequences, mutational analyses implicating a specific proline residue (Pro 564 in human HIF-1α). Mass spectrometry revealed oxidation compatible with hydroxylation of that proline, but the eureka moment came with the synthesis of a HIF-1α peptide with a substituted residue (*trans*-4-hydroxyproline) at the suspected site, Pro 564. It was one of those rare moments where things come together. I remember well the moment Ya-Min Tian phoned excitedly to tell me the result; untreated hydroxyproline HIF peptide bound VHL very tightly, just like the extract-treated HIF peptides containing proline (Figure 6).

All findings were therefore consistent with the action of a prolyl hydroxylase in the cell extract. Nevertheless, the known human prolyl hydroxylases, the collagen modifying enzymes, did not seem good candidates for an oxygen sensor. Moreover, the target sequence did not fit with the collagen consensus and indeed Johanna Myllyharju in Oulu, Finland, tested the HIF-α peptides with recombinant pro-collagen prolyl hydroxylases and found no activity. Nevertheless, the procollagen prolyl hydroxylases belong to the family of 2-oxoglutarate dependent dioxygenases and we found that a 2-oxoglutarate analogue, dimethyloxalylglycine, caused striking up-regulation of HIF when applied to cells, strongly suggesting the operation of a similar type of enzyme.

In the catalytic cycle of 2-oxoglutarate dependent dioxygenases, molecular oxygen is split and oxidation of the prime substrate (in this case the target prolyl residue) is coupled to the oxidative decarboxylation of 2-oxoglutarate, yielding succinate and carbon dioxide. The co-ordination of the catalytic Fe (II) ions by the 2-histidine-1-carboxylate motif is sufficiently labile to allow inhibition by iron chelating agents or substitution by cobaltous ions or other transition metals that cannot support or are very ineffective in supporting catalysis. Thus their properties fitted with the regulatory characteristics of the HIF system. Furthermore, the absolute dependence of catalysis on molecular oxygen clearly implied a mechanism of 'oxygen sensing'. In hypoxia, the reaction would slow, enabling HIF to escape destruction and assemble into a transcriptional complex.

The work therefore conceptualised the mechanism of sensing and was published simultaneously with independent work from Bill Kaelin.[42,43] That story may itself be of some interest. Shortly after we first identified the HIF-α/pVHL connection, I had suggested that Patrick Maxwell attend the international VHL meeting to find out current thinking. He reasonably asked 'what should I say we are doing ourselves, given that the work is unpublished.' The answer, we agreed, was straightforward. Since we had been working for some years on HIF, and now attending a meeting on VHL, then it might reasonably be deduced we were working on HIF and VHL and that's what should be said. Patrick returned and indicated that he had indeed revealed our position in conversation with Bill; it was agreed (since we were clearly competing) to keep in touch regarding future publication intentions. Bill was as good as his word, and in due course, shortly after we had done the work I've just described identifiying prolyl hydroxylation as the regulatory modification of HIF, there was a call from Boston saying they 'had something very interesting' and were ready to publish. There was a rapid set of exchanges when it was ascertained that we did indeed have very similar findings. Then things proceeded very quickly to publication, perhaps aided by the fact that two groups were reporting the same key findings, derived by somewhat different experimental approaches.

Further work soon completed other parts of the picture. As I indicated earlier, we had defined not one, but two, distinct sequences that could confer oxygen dependent degradation on heterologous proteins. Although sequence similarities between the two domains are not striking, Norma Masson, working in my laboratory, showed that the more N-terminal of the two sites in HIF-α also contained a site of prolyl hydroxylation (Pro 402 in human HIF-1α), which promotes binding to pVHL, and hence proteolytic destruction in the presence of oxygen.[44] Though the two domains can function independently, they ordinarily operate in a co-ordinated manner; C-terminal hydroxylation precedes N-terminal hydroxylation, hydroxylation of both sites co-operating to promote efficient destruction of the HIF-α protein. Again, binding experiments using synthetic peptides revealed remarkable discrimination of the VHL complex for the *trans*-4-hydroxylated versions. In subsequent work, both my laboratory

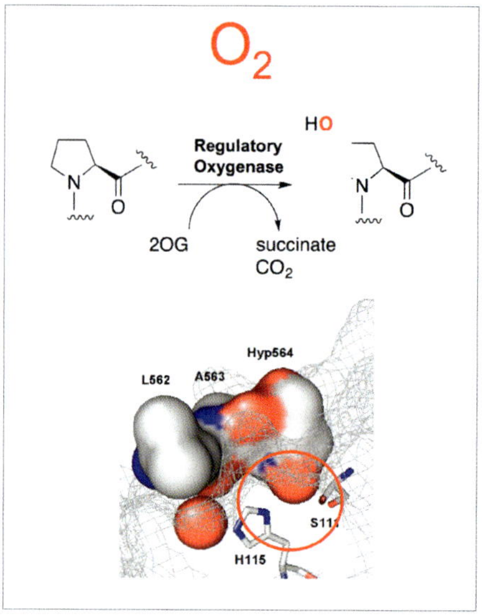

Figure 7. Diagram depicting the molecular 'switch' at the core of the oxygen sensing mechanism. The regulatory prolyl hydroxylase a 2-oxoglutarate dependent dioxygenase, splits molecular oxygen, incorporating one oxygen atom into its prime substrate (the HIF-α polypetide) and uses the other in the coupled oxidative decarboxylation of 2-oxoglutarate yielding succinate and carbon dioxide. The alcohol group of the trans-4-hydroxyprolyl residue in the HIF-α product is able to form an optimised hydrogen bonding network with residues serine 111 and histidine 115 in the floor of a hydroxyprolyl- binding pocket in VHL, increasing affinity *circa* 1000-fold. Adapted from[45].

working with Yvonne Jones and David Stuart[45] and Bill's laboratory, working with Nicola Pavletich,[46] independently showed that the exquisite discrimination (>1000 fold increase in affinity) of the VHL-E3 ligase complex for *trans*-4-hydroxylated over non-hydroxylated HIF was dependent on optimised hydrogen-bonding between the oxygen atom of the alcohol group in *trans*-4-hydroxyproline and each of two specific residues (Ser 111 and His 115) in the floor of the hydroxyproline binding pocket in human pVHL (Figure 7).

IDENTIFICATION OF THE HIF PROLYL HYDROXYLASES AS ENZYMATIC OXYGEN SENSORS

Whilst the demonstration of oxygen sensitive enzymatic prolyl hydroxylation governing the HIF-α/pVHL association, and hence degradation of HIF, clearly suggested a mechanism of oxygen sensing, it left open a key question: the identity of the actual enzyme(s). By this stage we had linked with a colleague who was to become an excellent and much valued collaborator for many years, Christopher Schofield, Professor of Organic Chemistry at Oxford. The story well exemplifies one of the most important aspects of scientific discovery, one often overlooked in the quest for findings of immediate value. Knowledge builds on knowledge in unexpected ways.

Chris had worked on the structure and enzymology of prokaryotic enzymes related to the 2-oxoglutarate dioxygenases, including an enzyme catalysing the key step in penicillin synthesis, isopenicillin *N* synthase.

This work revealed a core jelly-roll structure at the catalytic domain. Together with sequence alignments for a number of 2-oxo-acid-dependent dioxygenases, this led Chris and colleagues to propose that many such enzymes use a 2-histidine-1-carboxylate triad of residues, typically positioned on the 2nd and 7th stands of the jelly-roll, to co-ordinate the catalytic iron atom.[47] As genome sequencing programmes progressed, this work enabled the informatic prediction of other 2-oxoglutarate dioxygenases, including many members encoded by the human genome, in addition to the known procollagen modifying enzymes.

In a wholly different line of work, Bob Horvitz (Nobel Laureate, Physiology or Medicine 2002) and colleagues had been identifying mutations in the nematode worm *Caenorhabditis elegans*, carefully identifying large numbers of distinct mutants whose phenotypes corresponded to broad categories of developmental abnormality. Many different mutants were characterised that were associated with the retention of eggs within a swollen body, 'egg-laying defective', the mutant genes being termed *EGL*.[48]

In my laboratory we had been following up on our earlier work defining a HIF-like DNA binding activity in *Drosophila* SL2 cells. We identified the HIF-α orthologue in *Drosophila*, again using a fusion gene strategy to assay candidate *Drosophila* proteins that could confer oxygen regulated activity on the transcription factor Gal4, when transfected into human cells.[49] We used this information to predict a HIF-α orthologue in *C. elegans*. John O'Rourke, who was doing this work, then made an antibody to this protein. Slightly to our surprise (since we had no idea what level of ambient air would induce the putative HIF system in nematode worms) when we exposed worms to 1% oxygen using a simple bell jar, we saw a dramatic upregulation of the predicted HIF-α protein. We used this system to test mutants affecting genes encoding molecules that had been previously proposed as candidate oxygen sensors, including a range of mitochondrial mutants, but saw no clear alteration in the regulation of HIF.

But the system now enabled us to test candidate prolyl hydroxylases as predicted by Chris Schofield, in particular one group that was highly conserved across human and invertebrate animal genomes and represented in worms by the mutant *egl-9*. The result was striking: I have a vivid memory of Andrew Epstein, who had taken on the laboratory's *C. elegans* work, bursting through the door with the result. Worms bearing different *egl-9* mutations all showed clear upregulation of HIF irrespective of oxygen (Figure 8).[50] We rapidly proceeded to demonstrate that the EGL-9 gene product had HIF prolyl hydroxylase activity. Jonathan Gleadle identified the three human orthologues on the basis of their highly conserved catalytic domains, naming them PHD (prolyl hydroxylase domain) 1, 2 and 3.[50] Each of the PHDs catalysed the prolyl hydroxylation of peptides

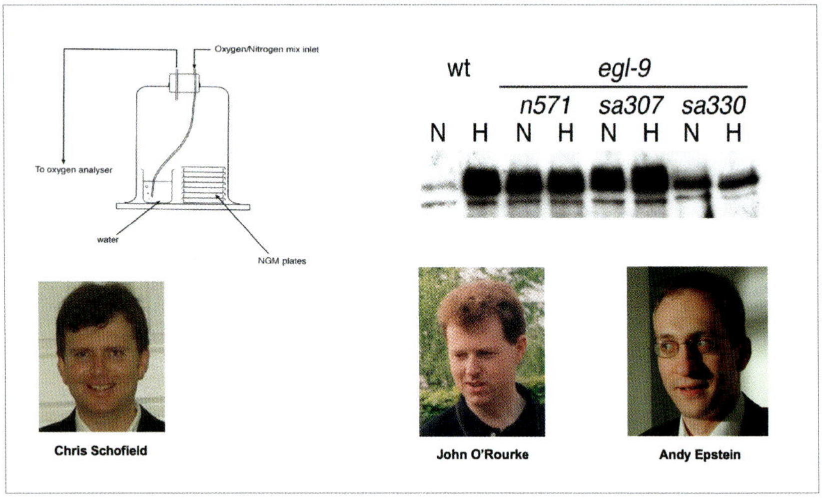

Figure 8. Identification of the 'oxygen sensing' HIF prolyl hydroxylase in the nematode *worm C.* elegans. (A) simple bell jar apparatus that was used to expose worms to a hypoxic atmosphere (B) Immunoblot of *C. elegans* HIF-α protein using an antibody produced 'in house' to test responses of wild-type and mutant worms to hypoxia; whereas HIF-α is strongly induced by hypoxia in wild-type worms (wt), HIF-α protein is stabilised and upregulated irrespective of oxygen in worms bearing different mutant alleles of the gene encoding the HIF prolyl hydrolylase *egl-9*. Adapted from[50].

and polypeptides corresponding to sites of prolyl hydroxylation identified within the oxygen dependent degradation domains of HIF-α proteins, though some differences in site selectivity were observed.[50] We were rapidly able to demonstrate the striking sensitivity of the PHDs to even modest reductions in the ambient oxygen concentration using capture of pVHL by HIF-α peptides as a measure of HIF hydroxylation.[50] More formal kinetic measurements by Johanna Myllyharju and colleagues confirmed these findings, revealing high Michaelis-Menton K_mO_2 values in the range 200–250 μM.[51]

Elucidation of the HIF-α/PHD/pVHL pathway clearly explained the function of two of the three domains in HIF-α that we had identified as being able to independently confer regulation by oxygen on heterologous proteins. But it left open a question as to the mode of regulation of the third oxygen-regulated domain at the C-terminal of HIF-α, which encompasses the C-terminal activation domain. This isolated domain was stable and did not appear to be regulated by proteolysis. Otherwise, it manifested closely similar behaviour to the other two regulatory domains, being activated by cobaltous ions and iron chelators as well as hypoxia.[26] Somewhat to our surprise, despite an apparent consensus (LxxLAP), revealed by identification of the first two prolyl hydroxylated domains, we could see no similar sequence in the C-terminal domain. As we were puzzling over this, two papers appeared; one from Gregg identifying a molecule he termed FIH (factor inhibiting HIF) as a protein associating with HIF-1α,[52] and one from Murray Whitelaw's laboratory describing a new modification, hydroxylation of an asparaginyl residue, within the third domain of HIF-α.[53] Based on a predicted jelly roll motif suggestive of a 2-oxoglurarate dioxygenase, Chris Schofield suggested that FIH might be the implied HIF asparaginyl hydroxylase. Working together we were rapidly able to confirm that and show that hydroxylation prevented association of the C-terminus of HIF-α with the p300 transcriptional co-activator,54 thus explaining, at least in part, the activity of this third regulatory domain.

Thus, HIF is subjected to dual regulation through stability (prolyl hydroxylation) and activity (asparaginyl hydroxylation), both the oxygen sensitive signals being generated by the catalytic activity of 2-OG dioxygenases. Studies of the genes encoding HIF and these HIF hydroxylases across multiple animal genomes identified a basic HIF-α/PHD/pVHL triad in all genomes examined. FIH is also represented in most animal species, generally as a single gene. However, multiple *PHD* and *HIF-a* genes appear to have arisen by gene duplication events at the base of vertebrate evolution.[55] This raised a question as to whether these additional isoforms have specialised roles in the regulation of specialised oxygen delivery systems observed in higher animals.

When expressed together in cells, HIF isoforms have partial overlapping functions.56 However, some key homeostatic responses observed in higher animals appear largely dependent on the cell-specific expression of one isoform. Thus, the induction of erythropoietin in the renal interstitial cells is exclusively driven by HIF-2.[57] In more recent work, Tammie Bishop and others in my laboratory have observed that the increase in respiratory sensitivity to hypoxia observed after acclimatisation is again specifically dependent on HIF-2.[58,59] In keeping with these findings, both the interstitial fibroblasts that make erythropoietin,[60] and the type 1 cells of the carotid body, which mediate hypoxic ventilatory control, specifically express very high levels of the HIF-2α isoform.[59] Thus, remarkably, both the classical features of acclimatisation, as studied by FitzGerald, Haldane and colleagues more than 100 years ago, appear critically dependent on a specific component of this system; HIF-2α.

To summarise at this point, the work achieved my original goal as a young nephrologist, of connecting erythropoietin to oxygen. But so much more was found along the way: a panorama of entirely unforeseen responses to hypoxia, an unanticipated mode of cell signalling that might be used in other types of biological circuitry, and new opportunities for the therapeutic manipulation of the human oxygen sensing system. In the final part of the lecture I will briefly consider some of these extensions.

SIGNALLING OXYGEN BY ENZYMATIC PROTEIN OXIDATION COUPLED TO PROTEOSTASIS IS A COMMON PARADIGM

We were initially surprised that the HIF system operated in all animal cells (and was not restricted to erythropoietin producing cells). However, once it became clear that HIF also regulated highly conserved cellular functions, such as glycolysis, we were *then* surprised not to be able to trace it further across non-animal organisms. Nevertheless, following elucidation of the mode of HIF regulation by oxygenase-catalysed post-translational hydroxylation, it has now become clear that many other life forms use oxygenase-catalysed protein oxidation coupled to proteostasis to signal oxygen levels in their cells, but deploy different types of enzyme-substrate couple.

Work by Christopher West on the slime-mould *Dictyostelium discoideum* has characterised a similar prolyl-4-hydroxylase that catalyses hydroxylation of Skp1, a sub-unit of $E3^{SCF}$ ubiquitin ligases.61 In this system, post-translational prolyl hydroxylation of Skp-1 is followed by glycosylation of the hydroxyproline residue to signal oxygen levels to pathways involved in culmination. In the fission yeast *Schizosaccharomyces pombe*, another prolyl hydroxylase, Ofd1, has been demonstrated by Peter Espenshade to transduce the oxygen regulated stability of the sterol regulatory element-binding protein Sre1.[62] In this case it is not the transcription factor itself that is hydroxylated; rather, the prolyl hydroxylation reaction regulates stability of Sre1 indirectly. Ofd1 is represented in human cells by the enzyme OGFOD1. Unexpectedly we found that OGFOD1 and its orthologues are ribosomal hydroxylases that catalyse hydroxylation of a prolyl residue in the small subunit protein, Rps23.[63] In fission yeast it is proposed that Rps23 hydroxylation imparts oxygen sensitivity on Ofd1's nuclear functions through competitive association;[64] whether an analogous process occurs in human cells is not yet clear. In plants, a group of plant transcription factors belonging to the ERF (ethylene response factor) VII family are subject to control by oxygen dependent instability via the N-degron pathway.[65,66] In this system an N-terminal cysteine is generated by the action of methionine aminopeptidases. Oxidation of this N-cysteine residue in turn generates a substrate for arginyl transferases; addition of arginine then creating an N-terminal destabilising residue that is targeted by ubiquitin E3 ligases.[67] In plants, Francesco Licausi showed that this N-cysteine oxidation is enzymatically catalysed by a set of closely related plant cysteine oxidases and that this is the step conferring oxygen sensitivity on the degradation pathway.[68] Like the animal PHDs, the plant cysteine oxidases are dioxygenases (i.e. they incorporate oxygen atoms derived from molecular oxygen directly into their substrates). Thus, this work identified a system signalling oxygen levels in plants using a quite different type of enzyme, but the same general process of oxygenase-catalysed protein oxidation coupled to proteostasis.

A chance meeting with Francesco led us into a collaborative project, which resulted in the discovery of a human orthologue of the plant cysteine oxidases. To our surprise, introduction into human cells of a fusion protein reporter in which sequences from the *Arabidopsis thaliana* ERF-VII transcription factor RAP2.12 were fused to green fluorescent protein, showed proteolytic regulation by oxygen even without co-transfection of genes encoding the plant cysteine oxidases. We identified the human enzyme orthologue of the plant enzymes, which was implied by this work, as ADO (2-aminoethane thiol dioxygenase, otherwise known as cysteamine dioxygenase) and showed that this enzyme catalyses the dioxygenation of N-cysteine residues on a range of human polypeptides.[69] Like the PHDs, ADO manifested an unusually low affinity (high KmO2) for oxygen. We found that ADO catalysed what had previously been proposed as non-enzymatic oxidation of an N-terminal cysteine,[70] on the R4 family of RGS (regulator of G-protein signalling) proteins to target these proteins for oxygen-dependent degradation, thus modulating G-protein signalling.[69] Potentially, direct proteolytic regulation of these signalling molecules in human cells would allow for more rapid responses to oxygen than those transduced via transcriptional regulation. Overall, what is now clear is that all four eukaryotic kingdoms use enzymatic protein oxidation coupled to proteostasis to signal oxygen levels, and that at least some of these systems contribute, alongside the HIF hydroxylase pathway, to oxygen homeostasis in human cells.

PHYSIOLOGY AND THERAPEUTICS

Before finishing, I'd like to discuss briefly a general question regarding the precision with which a reductionist biochemical understanding, such as has been gained by the work I have described, can be used to predict integrated physiological responses and therapeutic outputs. This question is common to many fields of biomedical research, but here brought into sharp focus by the precision demanded in physiological oxygen homeostasis, the enormous complexity of the pathways defined to date, and the development of drugs with the potential to modulate these pathways therapeutically.

The widespread operation of the HIF hydroxylase oxygen signalling pathways across animal cells and the existence in those cells of thousands of transcriptional targets of HIF, clearly qualifies this simple molecular pathway for a homeostatic function on countless cellular processes. These must operate with great precision across highly heterogeneous tissue oxygen concentrations. To exemplify from my nephrology perspective, within the mammalian kidney, oxygen tension varies by an order of magnitude between the renal cortex and the physiologically hypoxic renal papilla.71 Yet at each site, levels of HIF proteins are low under physiological conditions, but raised by a systemic hypoxic challenge.[60] Presumably this reflects an ability of cells in their native environment to respond to changes in intra-renal tissue pO2 over entirely different ranges. This 'range finding' problem is just one example of many in which the outputs specified by different types of homeostatic response appears to require an adjustment to the operation of the basic oxygen signalling pathway.

Multiple feedback systems have been described by which activity in the HIF-α/PHD/pVHL system is modulated.[72] Such systems might alter the *in cellulo* oxygen kinetics by altering substrate to enzyme ratios. Equally co-factor and co-substrate availability may tune the activity of the PHD enzymes to adjust their regulatory range with respect to oxygen. These and other molecular interactions *might, in principle*, contribute the tuning necessary for organismal homeostasis. The outstanding question is: *do they?*

This physiological challenge of oxygen homeostasis is compounded in therapeutics, where the aim is usually to modulate one component of a complex matrix but not another, again with considerable precision. Conceivably, massive computational modelling might ultimately assist our ability to understand or even predict such outputs. But at present there are limitations in our ability to study in *cellulo* biochemistry at the required resolution, and the complex interconnected signalling networks in the necessary detail, to accurately predict integrated physiological outputs. The nineteenth and early twentieth century physiologists who first considered the problem of oxygen homeostasis were woefully ill-equipped to link their beautifully precise observations on integrated physiology to molecular mechanisms. That came about decades later, based on progress in countless areas of biochemistry and biology that underpinned the studies described in this lecture. But excitement in our power to deliver reductionist biology should not make us over-confident in our ability to go back and precisely predict actions in those integrated systems.

Nevertheless, studies of several classes of drug that modulate the output of the HIF pathway are showing promising results: two in the clinical setting. One class of agent aims to inhibit HIF itself, the most advanced of these being agents that bind a large pocket in the PAS domain of HIF-2α to block dimerisation and activity.[73] Such HIF-2 antagonists are showing efficacy in the treatment of clear cell renal carcinoma,[74–76] which manifests the unusual bias to HIF-2α expression that I described earlier; Bill Kaelin will be speaking more about the efficacy of these agents in cancer. But the data raises a number of interesting questions. Structural work has also revealed pockets in the PAS domains and dimerisation interfaces of other HIF-α iso-forms;[77] although they are smaller, might these also be targeted? Will it be possible to manipulate the HIF-2 transcriptional responses productively in non-malignant disease? As predicted by genetic studies on Hif-2α in the mouse,[58] our recent work reveals effects of pharmacological antagonism of HIF-2 on the control of respiration,[78] indicating that the benefits and risks of systemic inhibition of HIF-2 will need to be considered as clinical development progresses.

The development of PHD inhibitors that activate HIF offers a different strategy for a different set of conditions. These drugs are 2-oxoglutarate analogues that (like hypoxia) reduce the rate of prolyl hydroxylation. Following the principle that augmentation of an endogenous physiological response might be predicted to be an effective therapeutic strategy, a large number of pre-clinical studies support the potential of these molecules to ameliorate ischaemic and hypoxic diseases.[79] So far however, clinical studies have largely been confined to correcting the erythropoietin deficient anaemia that complicates chronic kidney disease. Here, the ability to mimic the natural erythropoietic response to hypoxia, likely entrains other actions (such as direct effects on iron metabolism and on the bone marrow itself), which complement the erythropoietin-stimulating action. As might be predicted from investigations of the erythropoietin gene at the inception of this field, PHD inhibitors have been shown to be very effective in inducing erythropoietin and erythropoiesis. What is less clear is whether effects on the many HIF targets unrelated to erythropoietin and erythropoiesis can be largely avoided by careful adjustment of dose, whether such effects are generally beneficial, or whether they may be harmful in some circumstances. As outlined above, with current (and foreseeable) analytical methods we cannot eliminate these uncertainties. The complexity and inter-connectivity that Darwinian evolution has built, runs counter to precise predictions of therapeutic action, however desirable that might be. Drug development entrains a substantial element of empiricism.

But I will conclude with the prescient words of Sir Michael Jagger and Mr Keith Richards; *'you can't always have what you want, but if you try, sometimes you get what you need'*. As we enjoy this special time in Stockholm, the results of clinical use of PHD inhibitors and HIF-2 antagonists are looking promising.[76,80–83] And as clinician scientists, Gregg, Bill and I look forward to further exploring the therapeutic potential of these agents.

REFERENCES

1 FitzGerald, M. P. The changes in the breathing and the blood at various high altitudes. *Philosophical Transactions of the Royal Society of London. Series B.* **203**, 351–371 (1913).
2 Erslev, A. J. Humoral regulation of red cell production. *Blood* **8**, 349–357 (1953).
3 Waltner, K. & Waltner, K. Kobalt und Blut. *Klinische Wochenschrift* **8**, 313 (1929).
4 Goldwasser, E., Jacobson, L. O., Fried, W. & Plazk, L. F. Studies on erythropoiesis V: The effect of cobalt on the production of erythropoietin. *Blood* **13**, 55–60 (1958).
5 Warren, C. O., Schubmehl, Q. D. & Wood, I. R. Studies on the mechanism of cobalt polycythemia. *American Journal of Physiology* **142**, 173–178 (1944).
6 Necas, E. & Thorling, E. B. Unresponsiveness of erythropoietin-producing cells to cyanide. *American Journal of Physiology* **222**, 1187–1190 (1972).
7 Miyake, T., Kung, K. H. & Goldwasser, E. Purification of human erythropoietin. *Journal of Biological Chemistry* **252**, 5558–5564 (1977).
8 Lin, F.-K., Suggs, S., Lin, C.-H., Browne, J. K., Smalling, R., Egrie, J. C., Chen, K. K., Fox, G. M., Martin, F., Stabinsky, Z., Badrawi, S. M., Lai, P.-H. & Goldwasser, E. Cloning and expression of the human erythropoietin gene. *Proceedings of the National Academy of Sciences,* U.S.A. **82**, 7580–7584 (1985).
9 Jacobs, K., Shoemaker, C., Rudersdorf, R., Neill, S. D., Kaufman, R. J., Mufson, A., Seehra, J., Jones, S. S., Hewick, R., Fritsch, E. F., Kawakita, M., Shimizu, T. & Miyake, T. Isolation and characterization of genomic and cDNA clones of human erythropoietin. *Nature* **313**, 806–810 (1985).
10 Maxwell, P. H., Osmond, M. K., Pugh, C. W., Heryet, A., Nicholls, L. G., Tan, C. C., Doe, B. G., Ferguson, D. J. P., Johnson, M. H. & Ratcliffe, P. J. Identification of the renal erythropoietin-producing cells using transgenic mice. *Kidney International* **44**, 1149–1162 (1993).
11 Bachmann, S., Le Hir, M. & Eckardt, K.-U. Co-localization of erythropoietin messenger RNA and ecto-5'-nucleotidase immunoreactivity in peritubular cells of rat renal cortex indicates that fibroblasts produce erythropoietin. *Journal of Histochemistry & Cytochemistry* **41**, 335–341 (1993).
12 Goldberg, M. A., Glass, G. A., Cunningham, J. M. & Bunn, H. F. The regulated expression of erythropoietin by two human hepatoma cell lines. *Proceedings of the National Academy of Sciences,* U.S.A. **84**, 7972–7976 (1987).
13 Pugh, C. W., Tan, C. C., Jones, R. W. & Ratcliffe, P. J. Functional analysis of an oxygen-regulated transcriptional enhancer lying 3' to the mouse erythropoietin gene. *Proceedings of the National Academy of Sciences, U.S.A.* **88**, 10553–10557 (1991).
14 Beck, I., Ramirez, S., Weinmann, R. & Caro, J. Enhancer element at the 3'-flanking region controls transcriptional response to hypoxia in the human erythropoietin gene. *Journal of Biological Chemistry* **266**, 15563–15566 (1991).
15 Semenza, G. L., Nejfelt, M. K., Chi, S. M. & Antonarakis, S. E. Hypoxia-inducible nuclear factors bind to an enhancer element located 3' to the human erythropoietin gene. *Proceedings of the National Academy of Sciences, U.S.A.* **88**, 5680–5684 (1991).
16 Maxwell, P. H., Pugh, C. W. & Ratcliffe, P. J. Inducible operation of the erythropoietin 3' enhancer in multiple cell lines: evidence for a widespread oxygen sensing mechanism. *Proceedings of the National Academy of Sciences, U.S.A.* **90**, 2423–2427 (1993).

17 Firth, J. D., Ebert, B. L., Pugh, C. W. & Ratcliffe, P. J. Oxygen-regulated control elements in the phosphoglycerate kinase 1 and lactate dehydrogenase A genes: similarities with the erythropoeitin 3' enhancer. *Proceedings of the National Academy of Sciences, USA* **91**, 6496–6500 (1994).
18 Ebert, B. L., Firth, J. D. & Ratcliffe, P. J. Hypoxia and mitochondrial inhibitors regulate expression of glucose transporter-1 via distinct *cis*-acting sequences. *Journal of Biological Chemistry* **270**, 29083–29089 (1995).
19 Gleadle, J. M., Ebert, B. L., Firth, J. D. & Ratcliffe, P. J. Regulation of angiogenic growth factor expression by hypoxia, transition metals, and chelating agents. *American Journal of Physiology* **268**, C1362–C1368 (1995).
20 Ebert, B. L., Gleadle, J. M., O'Rourke, J. F., Bartlett, S. M., Poulton, J. & Ratcliffe, P. J. Isoenzyme specific regulation of genes involved in energy metabolism by hypoxia, cobalt and desferrioxamine: similarities with the regulation of erythropoietin. *Biochemical Journal* **313**, 809–814 (1995).
21 Wang, G. L. & Semenza, G. L. Desferrioxamine induces erythropoietin gene expression and hypoxia-inducible factor 1 DNA-binding activity: implications for models of hypoxia signal transduction. *Blood* **82**, 3610–3615 (1993).
22 Weinhouse, S. Glycolysis, respiration, and enzyme alterations in rat liver neoplasia. *Science* **158**, 537 (1967).
23 Weinhouse, S. Glycolysis, respiration, and anomalous gene expression in experimental hepatomas: G.H.A. Clowes Memorial Lecture. *Cancer Research* **32**, 2007–2016 (1972).
24 Nagao, M., Ebert, B. L., Ratcliffe, P. J. & Pugh, C. W. *Drosophila melanogaster* SL2 cells contain a hypoxically inducible DNA binding complex which recognises mammalian HIF-1 binding sites. *FEBS Letters* **387**, 161–166 (1996).
25 Wang, G. L., Jiang, B.-H., Rue, E. A. & Semenza, G. L. Hypoxia-inducible factor 1 is a basic-helix-loop-helix-PAS heterodimer regulated by cellular O_2 tension. *Proceedings of the National Academy of Sciences, U.S.A.* **92**, 5510–5514 (1995).
26 Pugh, C. W., O'Rourke, J. F., Nagao, M., Gleadle, J. M. & Ratcliffe, P. J. Activation of hypoxia inducible factor-1; definition of regulatory domains within the a subunit. *Journal of Biological Chemistry* **272**, 11205–11214 (1997).
27 O'Rourke, J. F., Tian, Y.-M., Ratcliffe, P. J. & Pugh, C. W. Oxygen-regulated and transactivating domains in endothelial PAS protein 1: comparison with hypoxia inducible factor-1a. *Journal of Biological Chemistry* **274**, 2060–2071 (1999).
28 Jiang, B.-H., Zheng, J. Z., Leung, S. W., Roe, R. & Semenza, G. L. Transactivation and inhibitory domains of hypoxia-inducible factor 1a. Modulation of transcriptional activity by oxygen tension. *Journal of Biological Chemistry* **272**, 19253–19260 (1997).
29 Huang, L. E., Gu, J., Schau, M. & Bunn, H. F. Regulation of hypoxia-inducible factor 1a is mediated by an oxygen-dependent domain via the ubiquitin-proteasome pathway. *Proceedings of the National Academy of Sciences, U.S.A.* **95**, 7987–7992 (1998).
30 Maxwell, P. H., Wiesener, M. S., Chang, G.-W., Clifford, S. C., Vaux, E. C., Cockman, M. E., Wykoff, C. C., Pugh, C. W., Maher, E. R. & Ratcliffe, P. J. The tumour suppressor protein VHL targets hypoxia-inducible factors for oxygen-dependent proteolysis. *Nature* **399**, 271–275 (1999).
31 Iliopoulos, O., Levy, A. P., Jiang, C., Kaelin, W. G., Jr. & Goldberg, M. A. Negative regulation of hypoxia-inducible genes by the von Hippel-Lindau protein. *Proceedings of the National Academy of Sciences, U.S.A.* **93**, 10595–10599 (1996).
32 Gnarra, J. R., Zhou, S., Merrill, M. J., Wagner, J. R., Krumm, A., Papavassiliou, E., Oldfield, E. H., Klausner, R. D. & Linehan, W. M. Post-transcriptional regula-

tion of vascular endothelial growth factor mRNA by the product of the *VHL* tumor suppressor gene. *Proceedings of the National Academy of Sciences, U.S.A.* **93**, 10589–10594 (1996).
33 Tian, H., McKnight, S. L. & Russell, D. W. Endothelial PAS domain protein 1 (EPAS1), a transcription factor selectively expressed in endothelial cells. *Genes & Development* **11**, 72–82 (1997).
34 Wiesener, M. S., Turley, H., Allen, W. E., William, C., Eckardt, K.-U., Talks, K. L., Wood, S. M., Gatter, K. C., Harris, A. L., Pugh, C. W., Ratcliffe, P. J. & Maxwell, P. H. Induction of endothelial PAS domain protein-1 by hypoxia: characterization and comparison with hypoxia-inducible factor-1a. *Blood* **92**, 2260–2268 (1998).
35 Stebbins, C. E., Kaelin, W. G., Jr. & Pavletich, N. P. Structure of the VHL-ElonginC-ElonginB complex: implications for VHL tumor suppressor function. *Science* **284**, 455–461 (1999).
36 Cockman, M. E., Masson, N., Mole, D. R., Jaakkola, P., Chang, G. W., Clifford, S. C., Maher, E. R., Pugh, C. W., Ratcliffe, P. J. & Maxwell, P. H. Hypoxia inducible factor-a binding and ubiquitylation by the von Hippel-Lindau tumor suppressor protein. *Journal of Biological Chemistry* **275**, 25733–25741 (2000).
37 Clifford, S. C., Cockman, M. E., Smallwood, A. C., Mole, D. R., Woodward, E. R., Maxwell, P. H., Ratcliffe, P. J. & Maher, E. R. Contrasting effects on HIF-1a regulation by disease-causing pVHL mutations correlate with patterns of tumourigenesis in von Hippel-Lindau disease. *Human Molecular Genetics* **10**, 1029–1038 (2001).
38 Mandriota, S. J., Turner, K. J., Davies, D. R., Murray, P. G., Morgan, N. V., Sowter, H. M., Wykoff, C. C., Maher, E. R., Harris, A. L., Ratcliffe, P. J. & Maxwell, P. H. HIF activation identifies early lesions in VHL kidneys: evidence for site-specific tumor suppressor funtion in the nephron. *Cancer Cell* **1**, 459–468 (2002).
39 Raval, R. R., Lau, K. W., Tran, M. G., Sowter, H. M., Mandriota, S. J., Li, J. L., Pugh, C. W., Maxwell, P. H., Harris, A. L. & Ratcliffe, P. J. Contrasting properties of hypoxia-inducible Factor 1 (HIF-1) and HIF-2 in von Hippel-Lindau-associated renal cell carcinoma. *Molecular and Cellular Biology* **25**, 5675–5686 (2005).
40 Ratcliffe, P. J. Oxygen sensing and hypoxia signalling pathways in animals: the implications of physiology for cancer. *Journal of Physiology* **591**, 2027–2042 (2013).
41 Schmid, V., Lafleur, V. N., Lombardi, O., Li, R., Salama, R., Colli, L., Choudhry, H., Chanock, S., Ratcliffe, P. J. & Mole, D. R. Co-incidence of RCC-susceptibility polymorphisms with HIF cis-acting sequences supports a pathway tuning model of cancer. *Scientific reports* **9**, 18768 (2019).
42 Jaakkola, P., Mole, D. R., Tian, Y.-M., Wilson, M. I., Gielbert, J., Gaskell, S. J., Kriegsheim, A. v., Hebestreit, H. F., Mukherji, M., Schofield, C. J., Maxwell, P. H., Pugh, C. W. & Ratcliffe, P. J. Targeting of HIF-a to the von Hippel-Lindau ubiquitylation complex by O_2-regulated prolyl hydroxylation. *Science* **292**, 468–472 (2001).
43 Ivan, M., Kondo, K., Yang, H., Kim, W., Valiando, J., Ohh, M., Salic, A., Asara, J. M., Lane, W. S. & Kaelin, W. G. J. HIF*a* targeted for VHL-mediated destruction by proline hydroxylation: implications for O_2 sensing. *Science* **292**, 464–468 (2001).
44 Masson, N., Willam, C., Maxwell, P. H., Pugh, C. W. & Ratcliffe, P. J. Independent function of two destruction domains in hypoxia-inducible factor-a chains activated by prolyl hydroxylation. *EMBO Journal* **20**, 5197–5206 (2001).
45 Hon, W. C., Wilson, M. I., Harlos, K., Claridge, T. D., Schofield, C. J., Pugh, C. W., Maxwell, P. H., Ratcliffe, P. J., Stuart, D. I. & Jones, E. Y. Structural basis for

the recognition of hydroxyproline in HIF-1a by pVHL. *Nature* **417**, 975–978 (2002).

46 Min, J.-H., Yang, H., Ivan, M., Gertler, F., Kaelin, W. G. J. & Pavletich, N. P. Structure of an HIF-1a-pVHL complex: hydroxyproline recognition in signaling. *Science* **296**, 1886–1889 (2002).

47 Roach, P. L., Clifton, I. J., Fulop, V., Harlos, K., Barton, G. J., Hajdu, J., Andersson, I., Schofield, C. J. & Baldwin, J. E. Crystal-structure of isopenicillin N-synthase is the first from a new structural family of enzymes. *Nature* **375**, 700–704 (1995).

48 Trent, C., Tsung, N. & Horvitz, H. R. Egg-laying defective mutants of the nematode *Caenorhabditis elegans*. *Genetics* **104**, 619–647 (1983).

49 Bacon, N. C., Wappner, P., O'Rourke, J. F., Bartlett, S. M., Shilo, B., Pugh, C. W. & Ratcliffe, P. J. Regulation of the Drosophila Basic helix-loop-helix PAS protein Sima by hypoxia: functional evidence for homology with mammalian HIF-1 alpha. *Biochemical and Biophysical Research Communications* **249**, 811–816 (1998).

50 Epstein, A. C. R., Gleadle, J. M., McNeill, L. A., Hewitson, K. S., O'Rourke, J., Mole, D. R., Mukherji, M., Metzen, E., Wilson, M. I., Dhanda, A., Tian, Y.-M., Masson, N., Hamilton, D. L., Jaakkola, P., Barstead, R., Hodgkin, J., Maxwell, P. H., Pugh, C. W., Schofield, C. J. & Ratcliffe, P. J. C. *elegans* EGL-9 and mammalian homologues define a family of dioxygenases that regulate HIF by prolyl hydroxylation. *Cell* **107**, 43–54 (2001).

51 Hirsila, M., Koivunen, P., Gunzler, V., Kivirikko, K. I. & Myllyharju, J. Characterization of the human prolyl 4-hydroxylases that modify the hypoxia-inducible factor HIF. *Journal of Biological Chemistry* **278**, 30772–30780 (2003).

52 Mahon, P. C., Hirota, K. & Semenza, G. L. FIH-1: a novel protein that interacts with HIF-1a and VHL to mediate repression of HIF-1 transcriptional activity. *Genes & Development* **15**, 2675–2686 (2001).

53 Lando, D., Peet, D. J., Whelan, D. A., Gorman, J. J. & Whitelaw, M. L. Asparagine hydroxylation of the HIF transactivation domain: a hypoxic switch. *Science* **295**, 858–861 (2002).

54 Hewitson, K. S., McNeill, L. A., M.V., R., Tian, Y.-M., Bullock, A. N., Welford, R. W., Elkins, J. M., Oldham, N. J., Bhattacharya, S., Gleadle, J. M., Ratcliffe, P. J., Pugh, C. W. & Schofield, C. J. Hypoxia inducible factor (HIF) asparagine hydroxylase is identical to Factor Inhibiting HIF (FIH) and is related to the cupin structural family. *Journal of Biological Chemistry* **277**, 26351–26355 (2002).

55 Loenarz, C., Coleman, M. L., Boleininger, A., Schierwater, B., Holland, P. W., Ratcliffe, P. J. & Schofield, C. J. The hypoxia-inducible transcription factor pathway regulates oxygen sensing in the simplest animal, Trichoplax adhaerens. *EMBO Reports* **12**, 63–70 (2011).

56 Smythies, J. A., Sun, M., Masson, N., Salama, R., Simpson, P. D., Murray, E., Neumann, V., Cockman, M. E., Choudhry, H., Ratcliffe, P. J. & Mole, D. R. Inherent DNA-binding specificities of the HIF-1alpha and HIF-2alpha transcription factors in chromatin. *EMBO Reports* **20**, e46401 (2019).

57 Paliege, A., Rosenberger, C., Bondke, A., Sciesielski, L., Shina, A., Heyman, S. N., Flippin, L. A., Arend, M., Klaus, S. J. & Bachmann, S. Hypoxia-inducible factor-2alpha-expressing interstitial fibroblasts are the only renal cells that express erythropoietin under hypoxia-inducible factor stabilization. *Kidney International* **77**, 312–318 (2010).

58 Hodson, E. J., Nicholls, L. G., Turner, P. J., Llyr, R., Fielding, J. W., Douglas, G., Ratnayaka, I., Robbins, P. A., Pugh, C. W., Buckler, K. J., Ratcliffe, P. J. & Bishop, T. Regulation of ventilatory sensitivity and carotid body proliferation in hypoxia by the PHD2/HIF-2 pathway. *Journal of Physiology* **594**, 1179–1195 (2016).
59 Fielding, J. W., Hodson, E. J., Cheng, X., Ferguson, D. J. P., Eckardt, L., Adam, J., Lip, P., Maton-Howarth, M., Ratnayaka, I., Pugh, C. W., Buckler, K. J., Ratcliffe, P. J. & Bishop, T. PHD2 inactivation in Type I cells drives HIF-2alpha-dependent multilineage hyperplasia and the formation of paraganglioma-like carotid bodies. *Journal of Physiology* **596**, 4393–4412 (2018).
60 Rosenberger, C., Mandriota, S. J., Jurgensen, J. S., Wiesener, M. S., Horstrup, J. H., Frei, U., Ratcliffe, P. J., Maxwell, P. H., Bachmann, S. & Eckardt, K. U. Expression of hypoxia-inducible factor-1a and -2a in hypoxic and ischemic rat kidneys. *Journal of the American Society of Nephrology* **13**, 1721–1732 (2002).
61 West, C. M., van der Wel, H. & Wang, Z. A. Prolyl 4-hydroxylase-1 mediates O2 signaling during development of Dictyostelium. *Development* **134**, 3349–3358 (2007).
62 Hughes, B. T. & Espenshade, P. J. Oxygen-regulated degradation of fission yeast SREBP by Ofd1, a prolyl hydroxylase family member. *EMBO Journal* **27**, 1491–1501 (2008).
63 Singleton, R. S., Liu-Yi, P., Formenti, F., Ge, W., Sekirnik, R., Fischer, R., Adam, J., Pollard, P. J., Wolf, A., Thalhammer, A., Loenarz, C., Flashman, E., Yamamoto, A., Coleman, M. L., Kessler, B. M., Wappner, P., Schofield, C. J., Ratcliffe, P. J. & Cockman, M. E. OGFOD1 catalyzes prolyl hydroxylation of RPS23 and is involved in translation control and stress granule formation. *Proceedings of the National Academy of Sciences, U.S.A.* **111**, 4031–4036 (2014).
64 Clasen, S. J., Shao, W., Gu, H. & Espenshade, P. J. Prolyl dihydroxylation of unassembled uS12/Rps23 regulates fungal hypoxic adaptation. *eLife* **6**, e28563 (2017).
65 Licausi, F., Kosmacz, M., Weits, D. A., Giuntoli, B., Giorgi, F. M., Voesenek, L. A., Perata, P. & van Dongen, J. T. Oxygen sensing in plants is mediated by an N-end rule pathway for protein destabilization. Nature 479, 419–422 (2011).
66 Gibbs, D. J., Lee, S. C., Isa, N. M., Gramuglia, S., Fukao, T., Bassel, G. W., Correia, C. S., Corbineau, F., Theodoulou, F. L., Bailey-Serres, J. & Holdsworth, M. J. Homeostatic response to hypoxia is regulated by the N-end rule pathway in plants. *Nature* **479**, 415–418 (2011).
67 Varshavsky, A. N-degron and C-degron pathways of protein degradation. *Proceedings of the National Academy of Sciences, U.S.A.* **116**, 358–366 (2019).
68 Weits, D. A., Giuntoli, B., Kosmacz, M., Parlanti, S., Hubberten, H. M., Riegler, H., Hoefgen, R., Perata, P., van Dongen, J. T. & Licausi, F. Plant cysteine oxidases control the oxygen-dependent branch of the N-end-rule pathway. *Nature communications* **5**, 3425 (2014).
69 Masson, N., Keeley, T. P., Giuntoli, B., White, M. D., Puerta, M. L., Perata, P., Hopkinson, R. J., Flashman, E., Licausi, F. & Ratcliffe, P. J. Conserved N-terminal cysteine dioxygenases transduce responses to hypoxia in animals and plants. *Science* **365**, 65–69 (2019).
70 Hu, R. G., Sheng, J., Qi, X., Xu, Z., Takahashi, T. T. & Varshavsky, A. The N-end rule pathway as a nitric oxide sensor controlling the levels of multiple regulators. *Nature* **437**, 981–986 (2005).
71 Leichtweiss, H.-P., Lübbers, D. W., Weiss, C., Baumgartl, N. & Reschke, W. The oxygen supply of the rat kidney: measurements of intrarenal pO_2. *Pflugers Archives* **309**, 328–349 (1969).

72 Ivan, M. & Kaelin, W. G., Jr. The EGLN-HIF O2-Sensing System: Multiple Inputs and Feedbacks. *Molecular Cell* **66**, 772–779 (2017).
73 Scheuermann, T. H., Li, Q., Ma, H. W., Key, J., Zhang, L., Chen, R., Garcia, J. A., Naidoo, J., Longgood, J., Frantz, D. E., Tambar, U. K., Gardner, K. H. & Bruick, R. K. Allosteric inhibition of hypoxia inducible factor-2 with small molecules. *Nature Chemical Biology* **9**, 271–276 (2013).
74 Chen, W., Hill, H., Christie, A., Kim, M. S., Holloman, E., Pavia-Jimenez, A., Homayoun, F., Ma, Y., Patel, N., Yell, P., Hao, G., Yousuf, Q., Joyce, A., Pedrosa, I., Geiger, H., Zhang, H., Chang, J., Gardner, K. H., Bruick, R. K., Reeves, C., Hwang, T. H., Courtney, K., Frenkel, E., Sun, X., Zojwalla, N., Wong, T., Rizzi, J. P., Wallace, E. M., Josey, J. A., Xie, Y., Xie, X. J., Kapur, P., McKay, R. M. & Brugarolas, J. Targeting renal cell carcinoma with a HIF-2 antagonist. *Nature* **539**, 112–117 (2016).
75 Cho, H., Du, X., Rizzi, J. P., Liberzon, E., Chakraborty, A. A., Gao, W., Carvo, I., Signoretti, S., Bruick, R. K., Josey, J. A., Wallace, E. M. & Kaelin, W. G. On-target efficacy of a HIF-2alpha antagonist in preclinical kidney cancer models. *Nature* **539**, 107–111 (2016).
76 Courtney, K. D., Infante, J. R., Lam, E. T., Figlin, R. A., Rini, B. I., Brugarolas, J., Zojwalla, N. J., Lowe, A. M., Wang, K., Wallace, E. M., Josey, J. A. & Choueiri, T. K. Phase I dose-escalation trial of PT2385, a first-in-class hypoxia-inducible Factor-2alpha antagonist in patients with previously treated advanced clear cell renal cell carcinoma. *Journal of Clinical Oncology* **36**, 867–874 (2018).
77 Wu, D., Rastinejad, F. Structural characteristics of mammalian bHLH-PAS transcription factors. *Current Opinion in Structural Biology* **43**, 1–9 (2017).
78 Cheng, X., Prange-Barczynska, M., Fielding, J. W., Zhang, M., Burrell, A. L., Lima, J. D., Eckardt, L., Argles, I., Pugh, C. W., Buckler, K. J., Robbins, P. A., Hodson, E. J., Bruick, R. K., Collinson, L. M., Rastinejad, F., Bishop, T. & Ratcliffe, P. J. Marked and rapid effects of pharmacological HIF-2alpha antagonism on hypoxic ventilatory control. *Journal of Clinical Investigation* **130**, 2237–2251 (2020).
79 Bishop, T. & Ratcliffe, P. J. HIF hydroxylase pathways in cardiovascular physiology and medicine. *Circulation Research* **117**, 65–79 (2015).
80 Meadowcroft, A. M., Cizman, B., Holdstock, L., Biswas, N., Johnson, B. M., Jones, D., Nossuli, A. K., Lepore, J. J., Aarup, M. & Cobitz, A. R. Daprodustat for anemia: a 24-week, open-label, randomized controlled trial in participants on hemodialysis. *Clinical Kidney Journal* **12**, 139–148 (2019).
81 Haase, V. H., Chertow, G. M., Block, G. A., Pergola, P. E., deGoma, E. M., Khawaja, Z., Sharma, A., Maroni, B. J. & McCullough, P. A. Effects of vadadustat on hemoglobin concentrations in patients receiving hemodialysis previously treated with erythropoiesis-stimulating agents. N*ephrology, Dialysis, Transplantation* **34**, 90–99 (2019).
82 Akizawa, T., Macdougall, I. C., Berns, J. S., Yamamoto, H., Taguchi, M., Iekushi, K. & Bernhardt, T. Iron regulation by molidustat, a daily oral hypoxia-inducible factor prolyl hydroxylase inhibitor, in patients with chronic kidney disease. *Nephron* **143**, 243–254 (2019).
83 Kaplan, J. Roxadustat and anemia of chronic kidney disease. *New England Journal of Medicine* **381**, 1070–1072 (2019).

Gregg L. Semenza. © Nobel Prize Outreach AB. Photo: A. Mahmoud

Gregg L. Semenza

Biography

CHILDHOOD

I was born on July 12, 1956 in Flushing, which is a neighborhood in the Borough of Queens, in New York City. My mother's family was of German-English-Irish descent and lived in Queens, whereas my father's family was of Italian descent and lived in the Bronx. I was the first born of five children, followed by Laurie, Beth, Matt and Paul. Our family lived in Old Bridge, New Jersey for five years, not far from Bruce Springsteen's childhood home. When I was nine years old, we moved to the village of Tarrytown in New York. Tarrytown is basically one long uphill climb, extending from the Hudson River to Main Street and Broadway (the center of town) and further up the hill to the aptly named Altamont Avenue, where my family resided.

I attended *Sleepy Hollow High School,* so named because Washington Irving had set his famous story The Legend of Sleepy Hollow in Tarrytown. In the story, a Headless Horseman chases a superstitious schoolmaster named Ichabod Crane, who is never seen again. The mascot of the high school's sports teams was the Headless Horseman and the players were referred to as the Horsemen. My freshman biology teacher, Rose S. Nelson (Figure 1), was the inspiration for my career in biomedical research. Dr. Nelson had earned a Ph.D. in Endocrinology and then worked as a postdoc at Woods Hole. Because of her research training, she did not recite facts but instead described discoveries in vivid detail and told us about the scientists who had made those discoveries. She transmitted to us the thrill of performing biomedical research. I also took Advanced Placement (AP) Biology with Dr. Nelson in my junior year and she recommended that I apply to attend a National Science Foundation-sponsored summer program at the Boyce-Thompson Institute for Plant Research in Yonkers, New York. I was accepted into the program, which was my first hands-on laboratory research experience and, despite the fact that my pipetting was clumsy, I loved it. During Dr. Nelson's AP Biology course, I became particularly intrigued by genetics.

Figure 1. Mentors and collaborators.

COLLEGE LIFE

In the fall of 1973, I made a road trip with my mother to visit Cornell University in Ithaca, New York. I had recently visited Harvard, where a former Sleepy Hollow High student who had graduated the previous year was enrolled. I had purchased a Harvard t-shirt as a souvenir of my visit and was wearing it when we stopped in Albany for the night. My mother's cousin looked at me and said with disdain: "Why are you wearing that shirt? You will never get in there."

My first year at Harvard, I lived in Wigglesworth Hall, which had several unique attributes that made it a memorable college residence. We occupied the first-floor suite, under which was located the terminus of the Red Line. When the subway train rumbled into Harvard Square station, the floor of my bedroom rumbled with it. Fortunately, the last train came into the station around midnight during the week and around 1 a.m. on the weekends. Wigglesworth also served as a sound barrier between the noise of Harvard Square and the tranquility of Harvard Yard. Public transit buses queued outside my bedroom window and the hustle and bustle of the Square was just a few feet away. On Saturday mornings in the fall, when the football team was playing at home, I would wake up to the sound of the Harvard Band marching up Massachusetts Avenue performing the fight song "Ten Thousand Men of Harvard Want Victory Today."

The first two years as a Harvard undergraduate were challenging and at times unpleasant for me, as they were filled with math and chemistry courses about which I possessed neither interest nor ability, which I attributed, at least in part, to my prior experiences. I took chemistry as a high school sophomore. The class met right after lunch and the teacher often had a strong odor of liquor on her breath. She did not inspire me. I took AP Calculus in my senior year of high school and had great difficulty with it. I decided that I would need to take Calculus over again in college, so there was no reason to take the AP exam. My teacher was quite displeased to learn about my plans. He informed me that if I did not take the AP exam, I would have to take his exam. I told him that because I was a senior with an 85 grade-average in the class, I was exempt from final exams. He replied that I did not have an 85 average, because he was giving me a 65 for the final quarter. My guidance counselor, Bill Burnette, who was a man of infinite wisdom and patience, brokered a compromise, in which I received a grade of 65 for the final quarter but did not have to take either exam. My indifferent attendance also led to a 65 in physical education, so my high school career ended not with a bang, but a Bronx cheer.

Starting in the spring of my junior year at Harvard, I worked in the laboratory of Dr. Park Gerald on the Genetics Unit of Children's Hospital in Boston. My introduction to the lab involved helping to prepare karyotypes from blood samples. Karyotypes identify chromosome abnormalities, such as the extra copy of chromosome 21 that is present in the cells of children with Down syndrome. After several months spent "paying my dues" in the karyotype lab, I was invited to move down the hall to the research lab. The goal was to map a human gene encoding glutathione peroxidase (GPX) using somatic cell hybrids. Increased GPX enzyme activity had been reported in cells from patients with Down syndrome, which suggested that a gene coding for GPX might be located on chromosome 21. We now know that there are eight genes encoding GPX in the human genome. Sadly, none of them are located on chromosome 21. Nevertheless, as I was working out the conditions for assaying GPX activity, which took several months of trial and error, Dr. Gerald provided me with valuable advice regarding the nature of the scientific enterprise. He told me: "Search and re-search."

PENN, CHOP, AND MSTP

Although I had planned to go to graduate school to study genetics, the human gene-mapping project increased my interest in medical genetics, and I decided to apply to MD-PhD programs with the intention of pursuing both a clinical and a research career in human genetics. I was accepted into the Medical Scientist Training Program (MSTP) at the University of Pennsylvania School of Medicine. The National Institutes of Health had funded the MSTP at two-dozen medical schools in order to encourage training of clinician investigators. The MSTP covered my medical and graduate school tuition and provided a small monthly stipend, all of which I would be required to repay only if I chose a career other than biomedical research. During my first semester at Penn, I spent pretty much all of my savings on Gray's Anatomy and other expensive medical school textbooks. I had a meal plan that covered Monday through Saturday. On Sundays, when the dining hall was closed, I would venture to a greasy spoon called Troy's, which featured a 99-cent breakfast that would tide me over until Monday morning.

My initial efforts to earn a PhD in genetics did not go well. I started my thesis research in a lab as a junior faculty member's first graduate student. After a year attempting two projects, neither of which was generating any encouraging data, I sought advice from the graduate program director and then, with his blessing, met with Elias Schwartz, who was the director of the hematology division at Children's Hospital of Philadelphia (CHOP), and Saul Surrey, who was in charge of Eli's research laboratory. They generously provided me with the opportunity to join their lab team.

The goal of my thesis project at CHOP was to perform a molecular analysis of the β-globin gene in a family with a variant form of β-thalassemia known as the silent carrier. β-thalassemia is a caused by defects in the expression of the β-globin polypeptide, which associates with the α-globin polypeptide and a heme ring to form hemoglobin, the protein in red blood cells that carries O_2. In patients with β-thalassemia, β-globin production is defective, leading to the precipitation of α-globin and subsequent destruction of red blood cells, causing anemia. Affected individuals carry two mutant copies of the β-globin gene, one inherited from the mother and one inherited from the father. Because the parents have one mutated and one wild-type copy of the β-globin gene, they do not develop the severe anemia associated with β-thalassemia, but they usually have smaller red blood cells that contain less hemoglobin than normal and can thus be identified as carriers of a mutant β-globin gene.

In the family that I studied, the son and daughter had β-thalassemia, indicating that both parents were carriers. The mother's red blood cells showed typical findings of a carrier, whereas the father's red cells appeared completely normal and thus he was described as a 'silent' carrier. My goal was to determine whether the silent carrier allele, like all other β-thalassemia alleles that had been studied so far, contained a mutation in the β-globin gene. We hoped that the mutation would have some unique feature at the molecular level that was responsible for its unique clinical presentation.

I drew blood samples from the parents and children, isolated genomic DNA, constructed a library of DNA fragments in a bacteriophage vector, and then isolated and sequenced a 4.4-kilobase DNA fragment containing the β-globin gene that the affected daughter had inherited from her father. At that time, obtaining the sequence of 4 kilobases of DNA on both strands of the double helix was a manual process that took about a year to complete. To my horror, after a year's work, I identified the same mutation that had been previously discovered by someone else in the lab studying an unrelated patient. Further analysis indicated that the mutation was not present in the daughter's DNA but was apparently the result of contamination with another DNA sample somewhere along the process of the gene cloning and sequencing. The source of the contamination will forever remain a mystery.

At this point, I was quite depressed and went to see a psychiatrist, who happened to be the father of two of my medical school classmates. I told him my tale of woe regarding the prospect of starting my thesis research over yet again. He looked at me and said "Gregg, it certainly seems that you have good reason to be depressed!" After that validation, I felt much better and went back to the lab with renewed enthusiasm. I found that the β-thalassemia allele that the daughter had inherited from her mother contained a mutation in the β-globin gene at the junction between exon 1 and intron 1 that prevented proper processing of β-globin mRNA, leading to a failure to translate the β-globin mRNA into protein. In contrast, the β-globin gene that the daughter had inherited from her father had a completely normal DNA sequence. Moreover, the analysis of DNA polymorphisms spanning the entire β-globin gene cluster, which includes the genes for embryonic and fetal as well as adult β-globin, revealed that the two affected children did not even inherit the same β-globin gene cluster from their father, indicating that the silent carrier allele was due to a mutation somewhere else in the genome, a truly novel finding.

After my thesis presentation, I sat in the hallway for an hour while my committee members debated whether the bar was lower for MD-PhD students, who needed to apply for internship and residency well in advance of their thesis defense, thus exerting "pressure" on thesis committees to approve the student's work in time for graduation and commencement of clinical training. The implication was that the work I had done was in some way not up to the lofty standards associated with a PhD from Penn. Fortunately, the editors of the *Journal of Biological Chemistry* and *Cell* did not have such misgivings when they accepted for publication the manuscripts I had written reporting the results of my thesis research.

DUKE

Following graduation from Penn, I spent two years as a pediatrics intern and resident at Duke University Medical Center. When I drove a rented U-Haul truck into Durham, North Carolina in June of 1984, I was greeted by the pungent aroma of tobacco, which was still being made into cigarettes in a downtown factory. Mr. Duke had made his fortune in the tobacco trade and cigarettes were still offered for sale from a vending machine in the hospital cafeteria. My residency training there preceded the major national reforms that placed limits on the amount of time that house officers were permitted to spend on the wards, both in terms of consecutive hours and hours per week. Our schedule involved working more than a hundred hours per week, and being "on call" every third night, which meant that all or part of that night would be spent awake taking care of acutely ill children (except for a month on the Pediatric Surgery service, when the call was every other night).

The most difficult rotation was the neonatal intensive care unit (the "NICU"), where extremely premature infants with severely immature lungs were kept alive on ventilators. Two residents would split the night's work when possible, thereby allowing each to have a few hours of sleep, provided that no emergencies arose that required the simultaneous efforts of both physicians. During my first rotation as an intern in the NICU, I had managed a brief sleep and then reported for duty; the senior resident observed my half-awake state and said "Gregg, go back and splash some water on your face." Working a full day, then a full night, then a full day again before finally going home was a grueling experience, especially when repeated over many weeks.

Another feature of the Duke pediatrics residency at the time was that the delivery room was located in the old South hospital, whereas the NICU and pediatric wards were in the new North hospital. There was an electric tram that ferried patients, staff, and visitors at a leisurely pace between the two buildings. However, when a newborn baby required resuscitation at night, the on call resident literally had to sprint from North to South and hope to arrive in time to save the neonate's life.

When I reminisce about my time at Duke, I usually mention that by some miracle, I only fell asleep at the wheel driving from the hospital to my apartment once. Fortunately, the car did not drift across the centerline, but rather towards the side of the road, which was covered with gravel, and the noise of the tires on the rough surface woke me before an accident occurred.

THE FAST TRACK TO BALTIMORE

Sue Church, who was one of the senior pediatric residents at Duke, noticed that I did not appear to be particularly happy in my position as a house officer. She suggested that I forego my final year of residency and fast track to a medical genetics fellowship. The pediatric genetics unit at Johns Hopkins was well known to me because Haig Kazazian and

Stylianos Antonarakis (Figure 1), in collaboration with Stuart Orkin at Harvard, had led the effort to identify and characterize the functional consequences of mutations in the β-globin gene that caused β-thalassemia. I had read their papers with the intensity and devotion reserved by others for the Bible. In addition to the fantastic research environment, the Johns Hopkins fellowship provided the opportunity to learn Medical Genetics at the foot of the master, Victor McKusick (Figure 1). I interviewed at several genetics programs and when Haig Kazazian offered me a position at Hopkins, I immediately accepted. In order to fulfill the requirements for board certification in Pediatrics, I worked in the pediatric emergency department at Johns Hopkins Hospital on those evenings when I was not already on call as a Medical Genetics fellow.

During this first year in Baltimore, I also contemplated the possibilities for a research project. I was interested in studying the regulation of gene expression in vivo using transgenic mice. The Kazazian and Antonarakis labs had moved on from identifying mutations in the β-globin gene that cause β-thalassemia to identification of mutations in the gene encoding factor VIII that cause hemophilia A. I contacted Chuck Shoemaker at Genetics Institute, who had isolated the factor VIII gene, to ask whether he could provide me with cloned genomic DNA. After I told him what I had in mind, he said that he could provide factor VIII gene sequences, but that there was this other gene that I might want to consider studying called *EPO*, which he had cloned before starting work on factor VIII.

My education about EPO was facilitated by the happy coincidence that one of the world's experts on the role of EPO in erythropoiesis was Jerry Spivak (Figure 1), a Hopkins hematologist, whose office was just down the hall from mine in Johns Hopkins Hospital. Based on Chuck's suggestion and Jerry's encouragement, I decided to focus on *EPO* gene regulation and was fortunate that John Gearhart (Figure 1), who was a faculty member in the physiology department at Hopkins, agreed to collaborate with me to generate *EPO* transgenic mice. Stylianos Antonarakis served as my mentor on a daily basis in the lab and Haig Kazazian provided advice about the direction of the project and funding opportunities.

At this point I should emphasize how uncommon – if not inappropriate – it was for me, as a starting postdoc, to: (a) enter the Kazazian-Antonarakis lab, in which the investigators were establishing a paradigm for molecular dissection of human genetic disease, which subsequently served as the template for hundreds of investigators studying dozens of inherited disorders, and blithely assume that I would initiate a completely unrelated (and unfunded) research project; and (b) enlist the participation of a senior faculty member in another department as well. Such was the incredible generosity of my mentors.

My good fortune grew larger when John Gearhart called me up one day and asked me if I had seen the RFA on EPO. I replied, "What's an RFA?" John told me that the National Institutes of Health (NIH) had released a Request for Proposals (a.k.a. RFA) to study the molecular mechanisms of *EPO* gene regulation. It seemed to be a tailor-made source of funding for my project. The only catch was that the RFA funded R01 grants, which are awarded to faculty members, not postdoctoral trainees. Stylianos agreed to serve as the principal investigator, with John as a co-investigator. I composed the application on a Macintosh SE computer in my apartment during a midwinter blizzard. The application scored in the fifth percentile and was funded for three years.

Around the same time, Haig Kazazian was on a flight back to Baltimore from California with Tom Pollard, a faculty colleague in Cell Biology who was an expert on the workings of actin and myosin in driving cell motility. Tom told Haig about his involvement with the Lucille P. Markey Charitable Trust, which was at that moment soliciting applications to fund research programs for young investigators as a bridge from postdoctoral to faculty positions. Mrs. Markey had been married to, and outlived, two wealthy men, including the owner of Calumet Farms in Lexington, Kentucky, a famous stable that had produced eight Kentucky Derby champions. She had several very enlightened advisors, who suggested that she use her considerable wealth to establish a trust to foster biomedical research. At its peak, the annual funding expenditures by the Markey Charitable Trust exceeded those of the Howard Hughes Medical Institute.

Although this grant application seemed like a great opportunity, there was a catch: the application was due the following week. I worked nonstop that weekend to prepare the application, secured letters of recommendation from Haig and Victor McKusick, and sent the application off to the Trust headquarters in Lexington. Incredibly, the application was funded for six years (one extra year as a postdoc and five years as an assistant professor). Although Tom Pollard was on the selection committee, he was no doubt recused when my application was considered, which was probably a good thing: several years later, when a faculty member in Tom's department told him that she had discovered the *Drosophila* gene responsible for a developmental mutant, and that it encoded a transcription factor, he said to her, "Oh, I'm so sorry!"

The studies of *EPO* gene regulation in transgenic mice, and later in human cells, that were funded by these two grants led directly to the discovery of hypoxia-inducible factor 1 (HIF-1), as described in my Nobel Lecture. One of the great benefits of receiving funding from the Lucille P.

Markey Charitable Trust was that the grantees met annually to present their research progress to the members of the selection committee, which was a Who's Who of Science that included Mike Brown, Joe Goldstein, George Palade, and Torsten Wiesel. At the annual meeting in 1993 or 1994, Joe Goldstein, who was one of my personal scientific heroes, gave an after dinner speech that made a deep impression on me. He spoke of what he called "technical courage," by which he meant a determination to follow the science wherever it led, rather than being limited by the particular experimental techniques that one was most comfortable using.

This advice was soon put to the test. We had attempted to identify cDNA sequences encoding HIF-1 using a bacteriophage expression library cloning approach without any success. In mulling over my options, I came up with three choices: (1) We could continue to use the same screening approach and hope that our luck would somehow change after screening millions of bacteriophages; this seemed very close to Einstein's definition of insanity. (2) We could give up and wait for someone else to do the job; that wasn't an appealing option. (3) We could take an entirely different experimental approach, which was to perform a biochemical purification of the protein. This seemed improbable, given that I was trained in molecular biology and we did not even possess a fraction collector at the time. Fortunately, Tom Kelly (Figure 1), who was then the director of the department of molecular biology and genetics at Hopkins, was one of the pioneers of DNA affinity chromatography. Tom's lab provided advice and assistance that was essential to our eventual success in purifying HIF-1 and isolating DNA sequences encoding the HIF-1α and HIF-1β subunits. This work was reported in papers published in 1995 in the *Journal of Biological Chemistry and the Proceedings of the National Academy of Sciences.* The latter manuscript, which was communicated by Victor McKusick, was submitted to and returned without review by the editors of *Cell, Science,* and *Nature*. This PNAS paper has now been cited over 6,000 times.

It is satisfying that the first translation of our discovery of hypoxia-inducible factors to the clinic has been the development of drugs that increase HIF activity as a means of stimulating red blood cell production in patients with chronic kidney disease. Unlike recombinant human EPO, which must be injected, these new HIF stabilizers are small molecules that can be taken by mouth. Four different drugs are currently in phase III clinical trials that have enrolled over 20,000 subjects and it seems likely that one or more of these will be approved by the FDA sometime in 2020, exactly 25 years after the publication of our papers reporting the purification, cloning and sequencing of HIF-1 of the HIF-1α and HIF-1β subunits.

FRIENDS, COLLABORATORS, AND TRAINEES

By now, the reader is fully aware that our discovery of HIF-1 could not have been achieved without the mentoring that I received from Haig Kazazian, Stylianos Antonarakis, Victor McKusick, Jerry Spivak, John Gearhart, and Tom Kelly at Johns Hopkins. I have also been extremely fortunate to have an abundance of wonderful collaborators at Johns Hopkins and at other universities with whom I have investigated the role of HIF-1 in development, physiology, medicine, and evolution as described in my Nobel Lecture. Among the longest and most valued friendships, forged over decades of collaboration, have been those with Nanduri Prabhakar (University of Chicago) and Joe Prchal (University of Utah), and my Hopkins colleagues Chi Dang, Larissa Shimoda and Akrit Sodhi (Figure 1). I am honored to serve as the C. Michael Armstrong Professor at Johns Hopkins University. Mike Armstrong (Figure 1) has been a tremendous supporter of our work and with Mike's help we hope to one day reach our goal of developing drugs that block HIF activity for the treatment of cancer.

I have had the distinct pleasure of training an extremely talented and hardworking group of 55 postdoctoral fellows, 30 graduate students, and 40 undergraduates. Since 1996, when I decided that I would not be able to adequately mentor the members of my lab unless I gave up my spot at the bench, all of our experimental data has been produced by these trainees and collaborators. Some, but by no means all of them, are mentioned in my Nobel Lecture.

Among my fondest memories of the whirlwind that occurred between the Nobel Prize announcement on October 10 and the Nobel Lectures on December 10, 2019, two stand out. The first of these was a lab reunion at our home in November, which was attended by over fifty present and former lab members from as far away as Hong Kong. The second was a private dinner, which Laura and I hosted for thirty of our closest family members and friends at the Gondolen restaurant, that lasted late into a snowy December night in Stockholm.

SAVING THE BEST FOR LAST

My fulfilling scientific career has been made possible by the love, understanding, and support of my wife Laura and my children Allie, Evan, and Gabe (Figure 2). They have transformed my pedestrian life into a fantastic journey and give meaning to every breath I take.

Figure 2. Family.

Hypoxia-Inducible Factors in Physiology and Medicine

Nobel Lecture, December 7, 2019 by
Gregg L. Semenza
C. Michael Armstrong Professor of Genetic Medicine, Pediatrics, Medicine, Oncology, Radiation Oncology, and Biological Chemistry; and Director, Vascular Program, Institute for Cell Engineering, Johns Hopkins University School of Medicine, Baltimore, MD, USA.

OXYGEN HOMEOSTASIS: A BALANCING ACT

In 1865, the French physiologist Claude Bernard famously stated: "*La fixité du milieu intérieur est la condition de la vie libre* (the constancy of the internal environment is necessary for independent life)." In 1932, the American physiologist Walter B. Cannon published his influential book *The Wisdom of the Body*, in which he captured Bernard's concept of a stable equilibrium in a single word: homeostasis, a modern Latin term that is derived from the ancient Greek words *homoios* 'similar' and *histemi* 'standing still'. Although this term has been applied to virtually every physiological system, there is no steady state that is more fundamental, more essential to survival of the organism, than oxygen homeostasis. Indeed, each of the hundred trillion cells in the adult human body requires a constant supply of O_2, principally for mitochondrial respiration. If the delivery of O_2 is disrupted for mere minutes, cells will die, as is most dramatically demonstrated by the catastrophic pathological processes of myocardial infarction (heart attack) and cerebrovascular accident (stroke).

Before considering the mechanisms by which O_2 homeostasis is maintained, let us first consider how this concept also serves an organizing principle for understanding the evolution of life on Earth. We will focus on two extraordinary biological milestones. The first of these occurred approximately 2.5 billion years ago, with the evolution of primitive single-celled organisms capable of transducing solar energy into the chemical energy of carbon bonds through the process of photosynthesis, which utilizes light, carbon dioxide and water as substrates to generate glucose and, as a side product, O_2. These primitive organisms were capable of breaking one of the carbon bonds in glucose, through the process of glycolysis, to generate a small amount of adenosine triphosphate (ATP), which was used to power other biochemical reactions. Initially, the O_2 released from these cells was consumed by reaction with the abundant iron in the Earth's crust, but after much of the iron had been oxidized, the concentration of O_2 in the atmosphere began to rise.

The second extraordinary milestone occurred about 1.5 billion years ago when (slightly less) primitive single-celled organisms evolved the ability to catalyze a complex series of chemical reactions in which glucose and O_2 were utilized to generate ATP, with carbon dioxide and water as side products. Compared to the fermentation of glucose to lactate, the complete oxidation of glucose yielded approximately 18-fold more ATP per mol of glucose. Next, some of these primitive respiring cells established a symbiotic relationship within other cells and became the mitochondria of present day eukaryotic organisms.

Just as the evolution of photosynthetic organisms and the resulting increase in atmospheric O_2 levels was a necessary prerequisite for the evolution of organisms capable of respiration, so the evolution of multicellular organisms was dependent upon the efficient generation of energy, which was needed in much greater amounts to generate and maintain complex collections of different cell types acting in a coordinated manner over space and time. Finally, because small animals tend to be eaten by large animals, metazoan evolution over a period of several hundred million years made a quantum leap with the increased body mass of vertebrates, which in turn required more complex systems for O_2 delivery. Thus, the roundworm *Caenorhabditis elegans* consists of a thousand cells, all of which obtain O_2 by direct diffusion from the atmosphere. By contrast, the fruit fly *Drosophila melanogaster* consists of hundreds of thousands of cells and developed tracheal tubes to conduct air to the interior of the organism, whereas *Homo sapiens* has evolved complex respiratory and circulatory systems to capture and deliver O_2 to trillions of cells.

O_2 homeostasis is a simple matter of supply and demand – matching O_2 consumption with O_2 delivery – but achieving this goal is not so simple when multiplied by trillions of cells integrated over the millions of minutes of a human life. Worse yet, there is little room for error at the cellular level, as either too little or too much O_2 for too long can be lethal to cells. The basis for this conundrum is the mechanism by which ATP is synthesized in the mitochondria. During the oxidation of glucose, reducing equivalents are captured in the molecules NADH and $FADH_2$ and passed to complex I and II, respectively, of the electron transport chain, which in turn transfer the electrons to complex III and then to complex IV. This process of electron transport generates a proton gradient that is used to drive the synthesis of ATP. In the final step, complex IV (cytochrome c oxidase) catalyzes the transfer of electrons to O_2 to form H_2O. However, the process of electron transfer is not completely efficient, and some electrons escape prior to complex IV and react with O_2 to form superoxide anion, which is converted to hydrogen peroxide. These reactive oxygen species (ROS) oxidize cellular macromolecules and, if present in large quantities for prolonged periods of time, result in cell dysfunction and death. The electron transport chain is optimized to function most efficiently at physiological O_2 levels, such that any decrease or increase in O_2 availability results in less efficient electron transport and greater production of ROS. Hence, the need to precisely maintain O_2 levels within a narrow physiological range. How is this formidable task accomplished?

ERYTHROPOIETIN AND THE CONTROL OF RED BLOOD CELL PRODUCTION

I arrived at Johns Hopkins University School of Medicine to start my postdoctoral fellowship in Medical Genetics in 1986. I had earned an M.D. degree and a Ph.D. in genetics at the University of Pennsylvania in 1984 by performing thesis research in the laboratory of Elias Schwartz and Saul Surrey in the Hematology Division at Children's Hospital of Philadelphia. We studied a family with a variant form of β-thalassemia, which was the first disease in which many different mutations affecting the expression of a gene (in this case, the gene encoding β-globin) were identified in a comprehensive manner. The leading research team in the field was a collaboration involving Stuart Orkin at Harvard and Haig Kazazian and Stylianos Antonarakis at Johns Hopkins. After completing internship and residency training in pediatrics at Duke University Medical Center, I accepted a postdoctoral fellowship at Hopkins because of my familiarity with the genetics research being performed there and the opportunity to train in clinical genetics with the renowned Victor McKusick.

As a pediatric geneticist, my interest at the time was in studying the regulation of gene expression during development. I thought that this could best be accomplished by utilizing a newly developed technique in which cloned DNA fragments were microinjected into fertilized mouse eggs to create transgenic mice. At the time, Haig Kazazian and Stylianos Antonarakis had transitioned from identifying mutations in the β-globin gene that caused β-thalassemia to analyzing mutations in the factor VIII gene that caused hemophilia A. Haig advised me to contact Chuck Shoemaker at Genetics Institute to ask whether he could provide me with cloned genomic DNA from the factor VIII gene. Chuck said that he could, but that there was another gene I might want to consider studying called erythropoietin (EPO). Chuck and his colleagues had previously cloned the cDNA and genomic DNA encoding erythropoietin, which is the hormone that controls red blood cell production.

I learned that EPO was produced in the liver during fetal life and in the kidney after birth. EPO is secreted into the bloodstream and binds to receptors on the surface of erythroid progenitor cells in the bone marrow, stimulating their survival, proliferation, and differentiation. In addition to this interesting developmental regulation, expression of the gene was induced by hypoxia; i.e. EPO producing cells were able to sense a decrease in local O_2 levels and respond with increased *EPO* gene transcription, leading to increased red blood cell production and increased blood O_2-carrying capacity, thereby correcting the tissue hypoxia – a beautiful homeostatic system.

However, patients with chronic kidney disease (CKD) have a progressive loss of EPO production as their disease progresses, resulting in anemia of progressively greater severity. The molecular cloning of EPO was a landmark event in the biotech industry, as it enabled commercial production of recombinant human EPO for the treatment of CKD patients with anemia, which had previously required transfusions and the accompanying risk of transfusion-associated diseases, such as hepatitis and AIDS.

I was fortunate that John Gearhart's lab in the Department of Physiology was accomplished at making transgenic mice. I went to John and asked him whether he would be willing to inject the human *EPO* gene into fertilized mouse eggs. He said yes.

Like the renowned physiologist Walter B. Cannon, I had obtained my undergraduate education at Harvard College. During my junior and senior years, I had worked in Park Gerald's laboratory on the Pediatric Genetics Unit at Children's Hospital in Boston for my honors thesis, using somatic cell hybrids for human gene mapping. I had also spent six years at the University of Pennsylvania and Children's Hospital of Philadelphia for my M.D.-Ph.D. training, so I had already worked in several major biomedical research institutions. When I arrived at Johns Hopkins, I was impressed by the culture of collegiality that is interwoven into the fabric of the institution. I learned that whatever scientific technique I wanted to apply to my research, there was someone who knew how to do it – and was willing to help me.

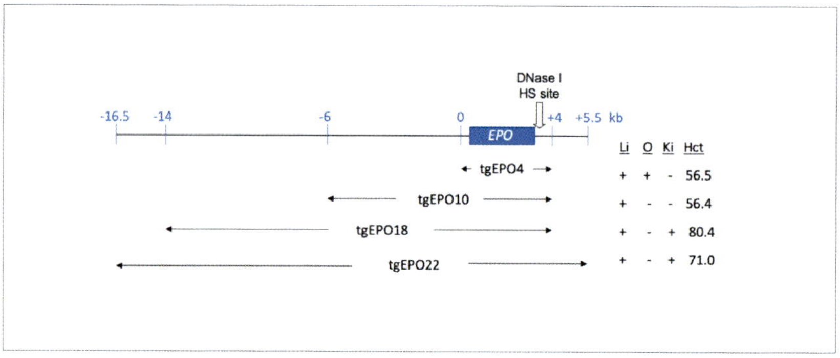

Figure 1. Human EPO transgenic mice. Nucleotide coordinates are given relative to the transcription start site (drawing is not to scale). Expression in liver (Li), kidney (Ki), or other tissues (O) and mean hematocrit (%) are indicated.

The Gearhart lab injected into mouse eggs a 4-kilobase (kb) DNA fragment containing the human *EPO* gene that I had purified from a plasmid vector after amplification in *E. coli.* We obtained mice in which the DNA had stably incorporated into the mouse genome such that the *EPO* transgene was passed on to half of each transgenic mouse's offspring at each mating. These transgenic mice were polycythemic – their hematocrit (i.e. the proportion of blood volume that is occupied by red blood cells), which is normally around 45%, was increased to over 55% (Semenza *et al.*, 1989). When adult mice were subjected to bleeding or were treated with phenylhydrazine to induce hemolysis, the expression of the human EPO transgene was dramatically induced in the liver, but not in the kidney. In addition, the transgene was expressed at high levels in organs where the endogenous gene was not expressed.

We concluded that the 4-kb *EPO* transgene: (i) contained regulatory sequences that mediated inducible human EPO expression in the liver; (ii) lacked a positive regulatory element required for inducible expression in the kidney; and (iii) lacked a negative regulatory element required to silence the gene in other tissues. To test this hypothesis, we next injected a 10-kb transgene that contained an additional 6 kb of 5'-flanking DNA beyond the 0.4-kb present in the first transgene (Figure 1). Mice transgenic for the 10-kb transgene were also polycythemic, with a mean hematocrit of 56%, and expressed human EPO in an inducible manner in the liver, but not in the kidney, and did not express human EPO mRNA in other tissues (Semenza *et al.*, 1990). We concluded that the putative negative regulatory element was present within the additional 6-kb of 5-flanking DNA.

Next we generated mice with an additional 8 kb of 5'-flanking DNA. These mice had a mean hematocrit of 80%, expressed the 18-kb human *EPO* transgene in an inducible manner in both the liver and kidney, and renal expression was localized by *in situ* hybridization (in collaboration with Stephen Koury at Vanderbilt) to the same rare interstitial/peritubular fibroblast-like cell type that expressed the endogenous mouse *Epo* gene (Semenza *et al.*, 1991a). Mice with a 22-kb transgene with additional 5' and 3' flanking sequences showed similar expression, suggesting that there no additional regulatory elements in proximity to the gene. Thus, we had successfully obtained regulated, tissue-specific expression of the human *EPO* gene in mice. I subsequently performed a deoxyribonuclease I (DNase) assay, using nuclei purified from the livers of mice carrying the 10-kb *EPO* transgene, and this experiment suggested the presence of an important transcriptional regulatory element in the 0.7-kb of 3'-flanking DNA, which was present in all three of the transgenes (Semenza *et al.*, 1991b).

Our next task was to establish the basis for hypoxia-induced *EPO* gene transcription. EPO was discovered by Alan Erslev, a hematologist at Thomas Jefferson University in Philadelphia (Erslev, 1957). Dr. Erslev's colleague Jaime Caro first reported that mice subjected to anemia or hypoxia responded with increased renal expression of EPO mRNA (Schuster et al., 1987). Exposure of the Hep3B human hepatoma cell line to hypoxia was shown to induce increased EPO expression (Goldberg *et al.,* 1987) and the Caro lab demonstrated that this was due to increased transcription (Costa-Giomi et al., 1990). I decided that we would utilize Hep3B cells to localize the *EPO* gene sequences required for hypoxia-inducible transcription and Jaime Caro generously sent us a flask of Hep3B cells to start our work.

My understanding of transcriptional regulation was greatly influenced by the work of Harold Weintraub, who had also received his M.D. and Ph.D. degrees from the University of Pennsylvania School of Medicine. He pioneered use of the DNase sensitivity assay as a means of identifying transcriptional regulatory elements (Weisbrod and Weintraub, 1979). In order for transcription factors to bind to DNA, the chromatin must be unwound from its nucleosomal structure, in which the DNA is tightly associated with histone proteins. Such unwound DNA, because it is not protected by histones, is exposed and accessible for binding of transcription factors, but also for degradation by DNase. Our finding of nuclease sensitivity of the *EPO* gene 3'-flanking region (Semenza *et al.*, 1991b) pointed to this area of the gene as the potential location of a hypoxia response element (HRE).

To extend this analysis to single nucleotide resolution, I performed a DNase footprint assay, in which a purified 256-bp DNA fragment from the *EPO* 3'-flanking region was incubated with nuclear lysate prepared from the livers of *EPO* transgenic mice that were either untreated or subjected to phlebotomy to cause anemia. The assay revealed a region of the 256-bp fragment that when incubated with nuclear extract from livers of control mice was hypersensitive to DNase I digestion, indicating absence of protein binding, and when incubated with liver nuclear extract from anemic mice was protected from DNase I digestion, suggesting that a transcription factor had been induced to bind to the DNA in response to anemia (Semenza *et al.*, 1991b). Furthermore, we inserted the 256-bp fragment into a reporter gene, in which expression of bacterial chloramphenicol acetyltransferase (CAT) coding sequences was driven by a basal SV40 promoter, and then transfected the reporter into Hep3B cells, which were incubated for 24 hours at 20% O_2 (standard tissue culture conditions of 95% air and 5% CO_2) or in a chamber flushed with a hypoxic gas mixture consisting of 1% O_2, 5% CO_2 and 94% N2. Insertion of the 256-bp fragment led to 2-fold increased CAT expression in cells subjected to hypoxia. Taken together, these experiments suggested that we had localized a cis-acting HRE to the 3'-flanking region of the *EPO* gene and there appeared to be a nuclear factor capable of binding to this sequence that was present specifically in the nuclei of liver cells from anemic mice. The requirement for 3'-flanking region sequences for hypoxia-inducible human *EPO* gene expression was also reported by the Caro lab (Beck *et al.*, 1991) and Peter Ratcliffe's lab at Oxford identified a candidate HRE in the 3'-flanking region of the mouse *Epo* gene (Pugh *et al.*, 1991).

We tested various fragments of the 256-bp region for their ability to function as an HRE and identified a 50-bp fragment that mediated a 7-fold increased reporter gene expression in hypoxic Hep3B cells (Semenza and Wang, 1992). We focused on the 5' end of this fragment (Figure 2), which had the sequence 5'-GCC.CTA.CGT.GCT.GTC.TCA-3'. Scanning mutagenesis performed in six 3-bp intervals revealed that the mutation CTA→TAT, CGT→AAA, or GCT →ATG (underscored in the sequence above) resulted in a complete loss of HRE activity (Semenza and Wang, 1992).

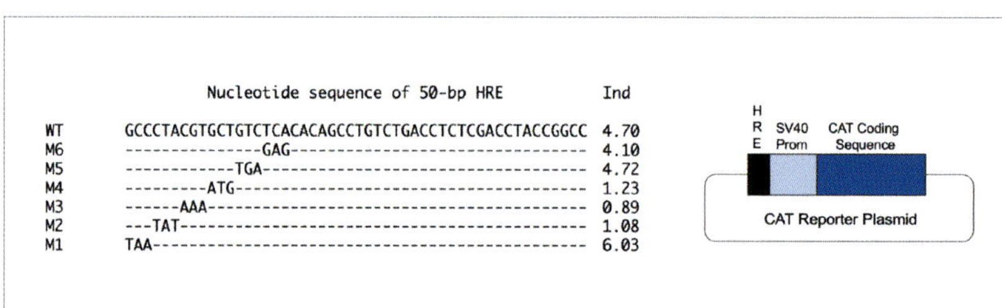

Figure 2. Functional analysis of human *EPO* gene hypoxia response element (HRE). Scanning mutagenesis of the first 18 bp of the HRE was performed. Wild type (WT) and mutant (M) nucleotide sequences were inserted into the chloramphenicol acetyltransferase (CAT) reporter gene and transfected into Hep3B cells, which were exposed to 20% or 1% O_2 for 24 hours. The fold hypoxic induction (Ind; ratio of CAT activity at 1%: 20% O_2) was determined.

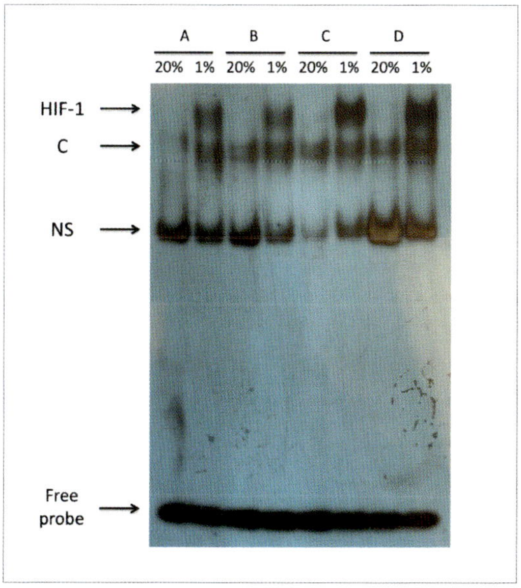

Figure 3. Identification of HIF-1 by electrophoretic mobility shift assay. A ^{32}P-labeled 18-bp oligonucleotide probe W18 was incubated (in four different binding buffers labeled A–D) with aliquots of nuclear extract prepared from Hep3B cells that were exposed to 20% or 1% O_2 for 4 hours. After incubation, polyacrylamide gel electrophoresis was performed to separate the free probe from probe that was bound by nuclear proteins. One of these bound proteins (HIF-1) is only present in nuclear extracts from hypoxic cells. The binding of a protein that is constitutively expressed (C) and a binding activity that is non-specific with respect to the nucleotide sequence of the probe (NS) are also indicated. The identification of HIF-1 as a DNA binding activity in 1992 led to its purification and the isolation of DNA encoding the HIF-1α and HIF-1β subunits in 1995, which opened up the field of oxygen biology to molecular analysis. The autoradiograph shown in the figure was donated to the Nobel Prize Museum in Stockholm, where it is currently on display.

An electrophoretic mobility shift assay (EMSA) was performed using a ^{32}P-labelled double-stranded oligonucleotide probe containing the 18-bp sequence shown above. This probe, designated W18, was incubated with nuclear extracts, which were prepared from Hep3B cells that were exposed to either 20% or 1% O2 for 4 hours (Figure 3). The binding of a nuclear factor that was only present in hypoxic cells was identified. Furthermore, the factor did not bind to an oligonucleotide, designated M18, which contained the same CGT→AAA mutation that abrogated hypoxia-inducible reporter gene expression, thereby linking the hypoxia-induced binding of the factor to hypoxia-induced transcriptional activation (Semenza and Wang, 1992). We designated this DNA-binding activity hypoxia-inducible factor 1 (HIF-1).

My postdoctoral fellow, Guang Wang, performed methylation interference assays, which revealed that HIF-1 directly contacted four guanine residues in the W18 oligonucleotide, two each on the sense and antisense strands in the sequence 5'-CGTGC-3' (Wang and Semenza, 1993a), which was consistent with the mutagenesis studies demonstrating that these residues were required for HRE activity and for HIF-1 DNA-binding activity (Semenza and Wang, 1992), as described above. Remarkably, the induction of HIF-1 DNA-binding activity was detected within 30 minutes and was maximal after 4 hours of hypoxic exposure; when hypoxic cells, were reoxygenated, HIF-1 DNA-binding activity was markedly reduced within 5 minutes (Wang and Semenza, 1993a). These were quite impressive kinetics because this was essentially the amount of time required for O_2 to diffuse out of or into the tissue culture media. Furthermore, we found that both *EPO* gene expression and HIF-1 DNA-binding activity were induced by exposure of Hep3B cells to cobalt chloride (Wang and Semenza, 1993c) or the iron chelator desferrioxamine (Wang and Semenza, 1993b). These findings tightly linked HIF-1 activity to *EPO* gene transcription.

OXYGEN SENSING BEYOND EPO

Hypoxia-induced *EPO* gene expression was demonstrated in Hep3B and HepG2 cells (Goldberg et al., 1989) but was not a general property of mammalian cell lines, which was consistent with its restricted expression *in vivo*. However, we found that HIF-1 DNA-binding activity was induced by hypoxia in all mammalian cell lines that we tested (Wang and Semenza, 1993c). These results were consistent with studies from the Ratcliffe lab indicating that the expression of a reporter gene containing the *Epo* gene HRE was induced by hypoxia after transfection into several different mammalian cell lines (Maxwell et al., 1993). These results indicated that the ability to sense decreased O_2 availability and to respond by induction of HIF-1 activity was a general property of mammalian cells.

Based on our finding that HIF-1 was ubiquitously induced by hypoxia, we hypothesized that HIF-1 regulated the expression of genes in addition to *EPO* and that these genes were likely to also contain HREs. To test this hypothesis, we focused on genes encoding glycolytic enzymes, since the switch from oxidative to glycolytic metabolism is a fundamental and perhaps universal response of animal cells to hypoxia, as described above, and Keith Webster had reported induction of glycolytic gene expression in response to hypoxia (Webster, 1987). We identified the core HIF-1 binding-site sequence 5'-ACGTG-3' in the human genes encoding aldolase A (*ALDOA*), enolase 1 (*ENO*1), and phosphoglycerate kinase 1 (PGK1) and in the mouse genes encoding lactate dehydrogenase A (*Ldha*) and phosphofructokinase L (*Pfkl*) (Semenza *et al.*, 1994). Using Hep3B cells, which express EPO, and HeLa cells, which do not, we demonstrated hypoxia-induced binding of HIF-1 to these glycolytic gene sequences by EMSA; HRE activity of DNA fragments encompassing these HIF-1 binding sites; and induction of endogenous glycolytic gene expression by exposure to hypoxia, cobalt chloride or desferrioxamine (Semenza *et al.*, 1994). The Ratcliffe lab reported similar results for PGK1 and *Ldha* (Firth *et al.*, 1994).

PURIFICATION OF HIF-1

To thoroughly investigate the molecular basis for hypoxia-induced gene expression, we needed to isolate the DNA sequences encoding HIF-1. We performed an experiment in which we fractionated Hep3B nuclear proteins by polyacrylamide gel electrophoresis (PAGE), transferred the proteins to a nitrocellulose membrane, and incubated it with the W18 probe. We reasoned that if HIF-1 was a monomer and was successfully renatured after transfer it might bind to the probe, whereas if HIF-1 consisted of multiple subunits, the binding activity would not be reconstituted. We observed a single band on the autoradiograph, suggesting that the protein was monomeric. Unfortunately, excitement trumped scientific rigor and we did not attempt to replicate this result, but instead employed an expression cloning strategy that Steven McKnight's lab at the Carnegie Institution (which like Johns Hopkins is located in Baltimore) had used to identify cDNA encoding the transcription factor C/EBP (Vinson *et al.*, 1988). We generated a recombinant bacteriophage library expressing cDNA sequences prepared from mRNA that we had isolated from hypoxic Hep3B cells. Several million bacteriophage plaques were screened using the W18 probe with no positive results.

I went to Haig Kazazian for advice. He suggested that I speak with Tom Kelly, the chairman of the Department of Genetics and Molecular Biology, whose lab was one of the first to purify a protein by DNA affinity chromatography (Rosenfeld and Kelly, 1986). Just as when I went to see John Gearhart in 1987, my entire research plan was dependent on obtaining essential assistance from a senior Hopkins faculty member. I met with Tom Kelly, told him what I wanted to do, and asked if he could help. Tom said yes. Two postdoctoral fellows in the Kelly lab, Dan Herendeen and George Brush, provided invaluable advice and instruction to my postdoctoral fellow Guang Wang, who performed the purification.

The purification strategy that I designed had two important components. First, it was based on our critical observation that the 3-bp mutation CGT→AAA abrogated the ability of the 50-bp HRE to mediate hypoxia-inducible transcription and abrogated the ability of HIF-1 to bind to the oligonucleotide probe. Second, it relied on a pilot analytical purification, which involved thirty 15-cm tissue-culture dishes of Hep3B cells that were exposed to 1% O_2 for 4 hours, and a large-scale purification, which involved

120 liters of HeLa S3 cells grown in suspension culture in the presence of 125 μM $CoCl_2$ for 4 hours. [HeLa, the first and still the most commonly-used human cell line, was established by Dr. George Gey from a cervical cancer biopsy obtained from Mrs. Henrietta Lacks at the Johns Hopkins Hospital in 1951 (Scherer *et al.*, 1953).] The HIF-1 DNA-binding activity present in hypoxic Hep3B cells was our gold standard, but it would be very difficult to isolate sufficient HIF-1 from Hep3B cells to perform even limited amino acid sequence analysis, which would require at least 50 picomoles of purified protein. The use of HeLa cells in spinner cultures provided a means for the necessary scale-up, but we needed to be certain that the HIF-1 DNA-binding activity that was cobalt-induced in HeLa cells was in all respects identical to the hypoxia-induced HIF-1 from Hep3B cells, so we analyzed the two preparations in parallel.

Guang first performed anion exchange chromatography using DEAE-Sepharose, and tested aliquots from each fraction by EMSA using the W18 probe. The positive fractions were pooled and DNA affinity chromatography was performed, using the W18 oligonucleotide coupled to Sepharose beads; the bound protein was eluted in high salt and applied to a column containing the M18 oligonucleotide coupled to Sepharose beads; the flow-through from this column was then applied to another W18 column and the bound protein was eluted and fractions were tested by EMSA using the W18 probe. In the purification from HeLa cells, we started with 3,040 mg of nuclear extract containing 608,000 units of HIF-1 DNA-binding activity (we defined one unit as equivalent to the DNA binding activity present in 5 μg of crude nuclear extract) with a specific activity of 0.2 units per μg, and we recovered 0.06 mg of purified protein with a specific activity of 2,250 units per μg, which represented a purification of 11,250-fold and a yield of 22% (Wang and Semenza, 1995).

HIF-1 purified from HeLa and Hep3B cells was fractionated by SDS-PAGE and the gel was silver-stained to visualize the protein, which revealed a polypeptide with an apparent molecular mass of 120 kDa, which was designated HIF-1α, and a series of polypeptides with an apparent molecular mass of 91–94 kDa, which were designated HIF-1β (Wang and Semenza, 1995). Glycerol gradient sedimentation, UV crosslinking to the W18 oligonucleotide, and methylation interference studies suggested that HIF-1 was a heterodimer of HIF-1α and HIF-1β, and that both subunits contacted DNA in the major groove (Wang and Semenza, 1995).

HIF-1 purified from HeLa cells was fractionated by SDS-PAGE and the polypeptides were transferred to a membrane, individually digested with trypsin, and fractionated by reverse-phase HPLC. Protein microsequence analysis was performed in the laboratory of David Speicher at the Wistar Institute in Philadelphia. The sequences of four peptides from HIF-1α were obtained and degenerate oligonucleotides were synthesized, based on the sequences of two of the peptides, and used as primers for PCR with cDNA prepared from mRNA purified from hypoxic Hep3B cells. The PCR product was sequenced and shown to encode the predicted amino acids, demonstrating that the two polypeptides were contiguous. I received advice regarding primer design and PCR strategy from my faculty colleague (and future Nobel laureate) Peter Agre, who had used a similar approach to isolate cDNA that encoded aquaporin 1 several years earlier (Preston and Agre, 1991).

My postdoctoral fellow Bing-Hua Jiang isolated a 3.4-kb cDNA from the Hep3B library that encoded an 826-amino-acid HIF-1α protein, which contained a basic helix-loop-helix (bHLH) domain and a PAS homology domain in its amino-terminal half (Wang *et al.*, 1995). The bHLH domain is a DNA-binding domain that defines a large superfamily of transcription factors. The PAS domain was first identified in the PER, ARNT, and SIM proteins (Hoffman et al., 1991). Analysis of two tryptic peptides from the 94-kDa HIF-1β polypeptide yielded partial amino acid sequences that were used to design degenerate oligonucleotide primers that amplified a PCR product of the expected size from Hep3B cDNA. Nucleotide and amino acid sequence analysis revealed that the HIF-1β sequence was identical to the aryl hydrocarbon nuclear translocator (ARNT), which is a bHLH-PAS protein that dimerizes with the aryl hydrocarbon receptor, another bHLH-PAS protein (Reyes et al., 1992). Two isoforms of ARNT that differ by the presence or absence of a 15-amino-acid sequence encoded by an alternative exon had been described (Hoffman et al., 1991), providing a basis for the purification of 91- and 93/94-kDa HIF-1β polypeptides. Further studies revealed that HIF-1β/ARNT was the common subunit for multiple bHLH-PAS proteins (Wu and Rastinejad, 2017).

We raised antibodies against HIF-1α and HIF-1β and showed that these antibodies supershifted the HIF-1: W18 complex in the EMSA. We analyzed the time course of induction of HIF-1α and HIF-1β in nuclear extracts of Hep3B cells exposed to 1% O_2. Immunoblot assays revealed that the levels of both proteins were increased after 30 min and peak levels were reached after 4 hours (Wang *et al.*, 1995), very similar to the induction of HIF-1 DNA-binding activity as detected by EMSA (Wang and Semenza, 1993a). When hypoxic cells were reoxygenated, HIF-1α and HIF-β protein levels in nuclear extracts were decreased by more than 50% within 5 min and completely lost by 15 min (Wang *et al.*, 1995). Our paper reporting the cloning of HIF-1 concluded with the following statement: "The cDNA and antisera described here will permit a detailed analysis of the molecular mechanisms by which HIF-1 activity is regulated and the role of HIF-1 in mammalian O_2 homeostasis" (Wang *et al.*, 1995). We shared these reagents with all of the other investigators in the field as soon as the work was published.

IDENTIFICATION OF HIF-2α AND HIF-3α

Once we had published the nucleotide and amino acid sequence of HIF-1α, it was possible to perform database queries for related sequences, which led to the discovery by four labs of a protein that they named endothelial PAS domain protein 1 (Tian *et al.*, 1997), HIF-like factor (Ema et al., 1997), HIF-related factor (Flamme *et al.*, 1997) and member of PAS domain family 2 (Hogenesch *et al.*, 1997), respectively, but which is now known as HIF-2α. Another ortholog, HIF-3α, was subsequently identified (Gu et al., 1998). Each of these proteins heterodimerizes with HIF-1β under hypoxic conditions and is subject to O_2-dependent degradation. Unlike HIF-1α and HIF-1β, which are found in almost all metazoan species, HIF-2α and HIF-3α are only found in vertebrates (Duan, 2016; Graham, 2017).

CHARACTERIZATION OF HIF-1α PROTEIN STRUCTURE AND FUNCTION

We demonstrated that efficient dimerization with HIF-1β and binding to the *EPO* gene HRE required the amino terminal half of HIF-1α (residues 1-390), which encompassed the bHLH and PAS domains (Figure 4), whereas transcriptional activation of the HRE-reporter gene required the carboxyl terminal half of HIF-1α (residues 531–826) (Jiang *et al.*, 1996a). When these C-terminal residues were fused to the GAL4 DNA-binding domain, they mediated hypoxia-inducible transactivation of a reporter gene in which firefly luciferase coding sequences were downstream of a basal promoter containing five GAL4 binding sites. Remarkably, expression of the GAL4-HIF-1α fusion protein was not regulated by hypoxia, indicating that hypoxia directly regulated the transactivation domain (TAD) function (Jiang *et al.*, 1997). TAD function was also inducible by treatment of cells with cobalt chloride or desferrioxamine, similar to the induction of endogenous HIF-1α protein levels by these agents (Wang et al., 1993b), but these agents did not affect the levels of the fusion protein, only its TAD function. These results suggested that similar mechanisms might underlie the O_2-dependent regulation of HIF-1α protein stability and TAD function.

A detailed analysis revealed two different regions that functioned as TADs when fused to the GAL4 DNA-binding domain: residues 531–575 and 786–826 (Jiang *et al.*, 1997). The residues between these two TADs, amino acids 576–785, had a negative effect on TAD function, and were designated the inhibitory domain (Jiang *et al.*, 1997). In order to identify proteins interacting with HIF-1α to regulate TAD function, we performed a yeast two-hybrid assay and postdoc Connor Mahon identified a protein

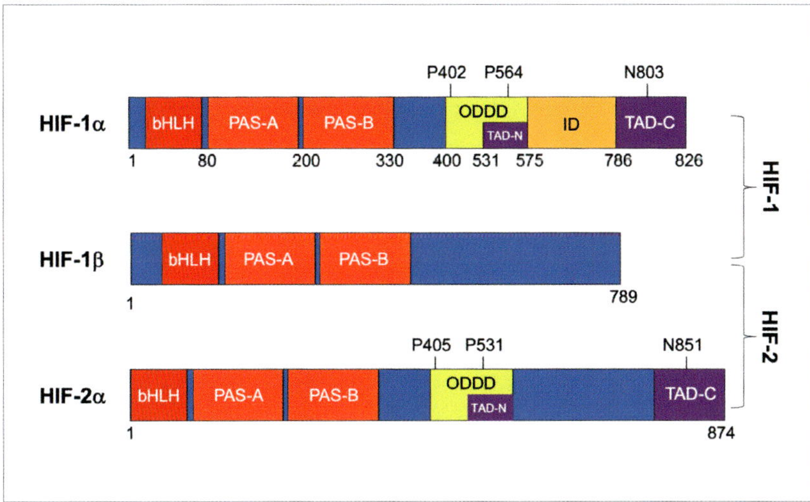

Figure 4. Structure of HIF-1 and HIF-2. The well-defined structural/functional domains of HIF-1a are the basic helix-loop-helix (bHLH) domain, PER-ARNT-SIM homology (PAS) domains, oxygen-dependent degradation domain (ODDD), amino-terminal transactivation domain (TAD-N), inhibitory domain (ID), and carboxy-terminal transactivation domain (TAD-C). Sites of O_2-dependent prolyl (P402, P564) and asparaginyl (N803) hydroxylation are shown. Each of the subunits has multiple isoforms as a result of alternative transcription initiation and splicing; the longest isoform is shown (drawing is not to scale).

that bound to residues 757–826, but not to 786–826, and negatively regulated TAD function; we named this protein factor inhibiting HIF-1 (FIH-1) (Mahon *et al.,* 2001). Remarkably, residues 757–826 functioned as a hypoxia-inducible TAD, whereas residues 786–826 functioned as a constitutively active TAD (Jiang *et al.*, 1997).

O2-DEPENDENT REGULATION OF HIF-1α PROTEIN STABILITY AND TAD FUNCTION

In order to investigate the expression of HIF-1α in cells exposed to a range of O_2 concentrations, we collaborated with Christian Bauer's lab at the Institute of Physiology in Zurich. They exposed HeLa cells to precisely defined O2 levels, ranging from 20% to 0%, for 4 hours in an instrument known as a tonometer. The cells were flash-frozen and sent to my lab in Baltimore where Bing-Hua Jiang prepared nuclear extracts and performed immunoblot assays to quantify the levels of HIF-1α and HIF-1β. He found that HIF-1α levels increased gradually as O_2 levels decreased from 20% to 6% O_2, then rose rapidly at lower O_2 levels, with a half-maximal response at 1.5–2% and maximal response at 0.5% O_2 (Jiang *et al.,* 1996b). It was known that there was considerable variation in the O_2 levels in various organs and even in different areas of the same organ. Other than in the lungs, most cells are exposed to O_2 levels between 3% and 6% O_2, which correspond to a PO_2 of 20 and 40 mm Hg, respectively. This meant that if the induction of HIF-1α *in vivo* was similar to that demonstrated in HeLa cells then, for most cells in the body, any decrease in O_2 levels would occur on the steep portion of the stimulus-response curve, allowing a graded transcriptional response to hypoxia: the greater the decrease in O_2 levels, the greater the increase in HIF-1α protein levels and in the expression of HIF target genes. We subsequently showed a similar graded response in vivo, using an isolated and perfused lung preparation (Yu *et al.*, 1998).

Our initial studies had demonstrated the remarkably rapid degradation of HIF-1α in post-hypoxic cells (Wang *et al.*, 1995) and the Caro lab demonstrated that this degradation was dependent upon ubiquitination and proteasomal degradation (Salceda and Caro, 1997). HIF-1α residues 401–603 were shown to be necessary and sufficient for O2-dependent protein degradation (Huang *et al.*, 1998). The Ratcliffe lab reported that in renal carcinoma cell lines that lacked expression of the wild-type von Hippel-Lindau tumor suppressor protein (VHL), HIF-1α and HIF-2α protein were constitutively expressed regardless of O2 availability (Maxwell *et al.*, 1999). When an expression vector encoding wild-type VHL was transfected into VHL-deficient cells, O_2-dependent degradation of HIF-1α and HIF-2α was restored. VHL directly bound to the HIF-α subunits and this binding was disrupted when cells were treated with cobalt chloride or iron chelator (Maxwell *et al.*, 1999). Structural data indicated that VHL, in a complex with Elongin B and C, likely functioned as a ubiquitin-protein ligase (Stebbins et al., 1999). The trigger for VHL binding and subsequent HIF-1α ubiquitination and proteasomal degradation was the hydroxylation of proline residue 402 and/or 564 in a reaction, which utilized O_2 and α-ketoglutarate as substrates and Fe^{2+} as a co-factor, catalyzed by prolyl hydroxylase domain (PHD) proteins (Epstein *et al.*, 2001; Jaakkola *et al.*, 2001; Ivan *et al.*, 2001; Masson *et al.*, 2001; Yu *et al.*, 2001). Three genes encoding PHD1-3, respectively, were identified in mammalian species (Bruick and McKnight, 2001; Epstein *et al.*, 2001).

Whereas O_2-dependent hydroxylation of Pro-402 and/or Pro-564 was required for binding of VHL, O_2-dependent hydroxylation of asparagine residue 803 abrogated the binding of the coactivators p300 and CBP to the carboxyl-terminal TAD of HIF-α subunits in a reaction that was catalyzed by FIH-1 (Lando *et al.*, 2002a,b). Taken together, these studies indicated that both the half-life and activity of HIF-α subunits were regulated by O_2-dependent post-translational modifications that transduce changes in O_2 availability to the nucleus as changes in HIF–mediated gene transcription. Thus, the fundamental transcriptional mechanism that underlies oxygen homeostasis, composed of cis-acting HREs and trans-acting HIFs regulated by PHDs, VHL, and FIH-1 was delineated by a small number of laboratories over the course of a dozen years from 1991 to 2002.

VASCULAR ENDOTHELIAL GROWTH FACTOR AND THE CONTROL OF ANGIOGENESIS

Systemic hypoxia due to anemia or ascent to high altitude induces a systemic response, including the induction of EPO expression to induce red blood cell production and thereby increase the blood O_2 carrying capacity. In contrast, localized hypoxia due to reduced tissue perfusion in the context of vascular disease results in a local response, the production of new blood vessels, which is known as angiogenesis. In 1992, Eli Keshet's laboratory reported that exposure of cells to hypoxia induced the expression of mRNA encoding vascular endothelial growth factor (VEGF) (Shweiki *et al.*, 1992). Analysis of glioblastoma tumor sections revealed VEGF expression in cells surrounding areas of necrosis – which represent tumor cells that have died because they are too far from a blood vessel to receive adequate O_2 for survival, suggesting that intratumoral hypoxia was a key stimulus for tumor vascularization (Shweiki *et al.*, 1992). Werner Risau's lab also reported a role for VEGF in tumor angiogenesis (Plate *et al.*, 1992).

We identified a 47-bp sequence located approximately 1 kb 5' to the transcription start site of the *VEGF* gene, which mediated hypoxia-inducible reporter gene expression and contained a HIF-1 binding site; furthermore, the same 3-bp mutation (TACGTG to TAAAAG) that inactivated the *EPO* gene HRE and abrogated HIF-1 binding also inactivated the *VEGF* HRE and abrogated HIF-1 binding to an oligonucleotide from the *VEGF* gene (Forsythe *et al.,* 1996). We obtained from Oliver Hankinson's lab wild-type Hepa1 cells; an ARNT/HIF-1β-deficient subclone that they had identified and designated c4; and a subclone of c4 cells (designated VT2) that had been transfected with an ARNT expression vector (Hoffman *et al.*, 1991). We found that in the absence of ARNT/HIF-1β, there was very little induction of HIF-1α protein and no HIF-1 DNA-binding activity under hypoxic conditions in c4 cells, whereas in VT2 cells HIF-1β protein expression and hypoxia-induced HIF-1α protein expression as well as HIF-1 DNA-binding were rescued. Analysis of VEGF mRNA expression revealed hypoxic induction in wild type and VT2 cells but not in c4 cells (Forsythe et al., 1996). In addition, co-transfection of the VEGF HRE reporter gene with expression vectors encoding HIF-1α and HIF-1β dramatically increased reporter gene expression under both hypoxic and non-hypoxic conditions (Forsythe *et al.*, 1996).

Taken together, these studies demonstrated that HIF-1 was necessary and sufficient for hypoxia-induced *VEGF* transcription. Judah Folkman had drawn attention to the critical role of angiogenesis in tumor progression (Folkman, 1971) and had characterized a "tumor angiogenesis factor" (Folkman *et al.*, 1971). Napoleone Ferrara subsequently purified the protein, cloned the gene, and named it VEGF (Leung *et al.*, 1989). The discovery of hypoxia-induced and HIF-mediated *VEGF* gene transcription provided a molecular mechanism underlying the observation of increased VEGF expression in human tumors.

GENERATION OF HIF-1α DEFICIENT EMBRYONIC STEM CELLS AND KNOCKOUT MICE

In order to further investigate the role of HIF-1 in mammalian development, physiology, and disease pathogenesis, my postdoctoral fellow Narayan Iyer engineered (in collaboration with John Gearhart's lab) mouse embryonic stem (ES) cells in which the HIF-1α coding sequence was disrupted by homologous recombination (Iyer *et al.*, 1998). Loss of HIF-1α expression was associated with decreased expression of mRNAs encoding the glucose transporters GLUT1 and GLUT3 and the glycolytic enzymes hexokinase (HK1, HK2), glucose phosphate isomerase (GPI), PFKL, ALDOA, ALDOC, triose phosphate isomerase (TPI), PGK1, ENO1, pyruvate kinase (PKM), and LDHA. Thus, HIF-1 coordinately regulated the entire glycolytic pathway in hypoxic ES cells (Iyer *et al.*, 1998).

Hypoxia-induced VEGF mRNA expression was completely eliminated in the HIF-1α-deficient ES cells. The ES cells were injected into mouse blastocysts and *Hif1a*$^{+/-}$ mice, which were heterozygous for the HIF-1α knockout allele, were obtained. However, when these mice were interbred, the *Hif1a*$^{-/-}$ embryos arrested in their development at embryonic day 8.5 and died by day 10.5. I was fortunate to recruit as a postdoctoral fellow Lori Kotch, who had considerable expertise in mouse embryology. She discovered that complete HIF-1α deficiency resulted in impaired erythropoiesis, cardiac malformations, and vascular defects, demonstrating that all three components of the circulatory system are dependent on HIF-1α expression for their proper development (Iyer *et al.*, 1998).

ADAPTATION OF CELLULAR METABOLISM TO HYPOXIA: A PARADIGM SHIFT

We established mouse embryo fibroblast (MEF) lines from wild type and *Hif1a–/–* embryos. Remarkably, exposure of the HIF-1α-null MEFs to 1% O2 resulted in death of virtually all of the cells over 72–96 hours (Kim et al., 2007; Zhang *et al.*, 2008). The conventional wisdom held that cells switch from oxidative to glycolytic metabolism to maintain ATP production. Yet ATP levels in HIF-1α-null MEFs cultured at 1% O_2 were higher than in wild-type MEFs cultured at 20% O_2. The cells were not dying from lack of ATP, but rather due to excess reactive oxygen species (ROS) production (Kim *et al.*, 2007; Zhang *et al.*, 2008). The reason that both ATP and ROS levels were high in HIF-1α-null MEFs was that they could not switch from oxidative to glycolytic metabolism without HIF-1α. In collaboration with Jay Kim and Chi Dang we found several HIF-1 target genes that were required to prevent excess ROS production under hypoxic conditions. The first of these was pyruvate dehydrogenase kinase 1 (PDK1), which phosphorylates and inactivates the catalytic subunit of pyruvate dehydrogenase, the enzyme that converts pyruvate to acetyl coenzyme A (Acetyl CoA) for entry into the tricarboxylic cycle. Transfection of HIF-1α-null MEFs with an expression vector encoding PDK1 prevented excess ROS production and cell death under conditions of chronic hypoxia (Kim et al., 2006).

Acetyl CoA can also be generated from fatty acid oxidation. The second critical HIF-1 target gene required to prevent excess ROS production under hypoxic conditions was BNIP3, which my postdoc Huafeng Zhang showed mediated mitochondrial-selective autophagy: after 48 hours at 1% O_2 wild-type MEFs reduced their mitochondrial mass by 50%. Forced expression of BNIP3 was sufficient to reduce mitochondrial mass and significantly reduced hypoxia-induced death of *Hif1a*$^{-/-}$ MEFs (Zhang *et al.*, 2008). Thus, by reducing mitochondrial mass, wild-type MEFs reduced respiration and thereby reduced ROS generation.

A third mechanism of adaptation was more subtle and involved cytochrome c oxidase (complex IV of the electron transport chain), which catalyzes the terminal transfer of electrons to O_2 to form H_2O. In response to hypoxia, HIF-1 induces a subunit switch in complex IV by inducing expression of COX4-2 and LON, which is a mitochondrial protease that is required for the degradation of COX4-1 (Fukuda *et al.*, 2007). This COX4 subunit switch optimizes the efficiency of respiration under hypoxic conditions and may represent an initial response (perhaps under conditions of mild hypoxia) that allows cells to maintain oxidative metabolism with its much more efficient energy production. When hypoxia is severe and/or prolonged, the more draconian measures mediated by PDK1 and BNIP3 may be required. Thus, cells switch from oxidative to glycolytic metabolism, long before O_2 is limiting for ATP production, in order to prevent ROS-mediated cell death.

Hif1a$^{+/-}$ MICE ARE PROTECTED FROM HYPOXIA-INDUCED PULMONARY HYPERTENSION

The embryonic lethality of *Hif1a*$^{-/-}$ mice precluded the possibility of studying the effect of complete HIF-1α deficiency on physiological responses to hypoxia in adult mice. Fortunately, the *Hif1a*$^{+/-}$ heterozygous knockout mice developed normally but had impaired responses to hypoxia. It was known that unlike arterioles in the systemic circulation, which dilate in response to hypoxia in order to increase blood flow (and O_2 delivery) to the affected tissue, pulmonary arterioles constrict in response to hypoxia so that blood flow is shunted away from areas of the lung that are not ventilated. This is an adaptive response in the context of a lobar pneumonia; however, during ascent to high altitude, the entire lung is exposed to hypoxia and the generalized pulmonary hypertension (PH) that develops in sensitive individuals can be a life-threatening component of acute mountain sickness. Individuals with chronic obstructive pulmonary disease are also at risk of PH, which leads to increased right ventricular pressure (RVP), right ventricular hypertrophy, and progressive heart failure.

In collaboration with my colleague Larissa Shimoda of the Pulmonary division, wild-type (*Hif1a*$^{+/+}$) and heterozygous knockout (*Hif1a*$^{+/-}$) mice were placed in a chamber in which the O_2 concentration was maintained at 10%, which is comparable to the summit of Mt. Kilimanjaro (19,000 feet), for three weeks. In wild-type mice, the mean RVP increased from 7.3 to 18 mm Hg, whereas in the heterozygous mice, RVP only increased from 6.8 to 12 mm Hg, demonstrating a role for HIF-1α in the pathogenesis of PH (Yu *et al.*, 1999). The Shimoda lab showed that treatment of mice with either of two different HIF inhibitors (digoxin and acriflavine), which we originally identified in order to investigate the role of HIFs in cancer progession (as described below), blocked the development of PH induced by chronic hypoxia and prevented further progression in animals in which PH had already developed (Abud *et al.*, 2012).

ESTABLISHING THE ROLE OF THE HIF PATHWAY IN REGULATING ERYTHROPOIESIS IN HUMANS

George Dover, who was then the director of Pediatric Hematology at Johns Hopkins, referred to me a family affected by a condition known as congenital polycythemia or hereditary erythrocytosis, in which the body makes too many red blood cells – the same phenotype displayed by our *EPO* transgenic mice. However, the cause of the disease in humans had not been established. I contacted the family and explained that I was a geneticist interested in identifying the molecular basis for the condition, which affected a man, his mother, and his two children. This immediately told us that the disease was inherited in an autosomal dominant manner, meaning that affected individuals had one wild type and one mutant allele, and that unaffected family members did not have the mutant allele in their DNA, which we extracted from routine blood samples obtained from family members.

I contacted Josef Prchal, a hematologist who was an expert on the diagnosis of congenital polycythemia, and we began a scientific collaboration, which is now in its third decade. A frameshift mutation, the insertion of an additional guanine residue at codon 430, was identified in one copy of the EPOR gene, which encodes the EPO receptor, in affected family members (Sokol et al., 1995). *EPOR* is expressed on erythroid progenitor cells in the bone marrow and, when activated by EPO binding, stimulates their survival, proliferation and differentiation. The mutation resulted in the synthesis of a receptor that lacked the last 64 amino acids of the protein. This mutant EPOR had increased sensitivity to EPO, providing a molecular basis for the disease in this family (Sokol *et al.,* 1995). Albert de la Chapelle's group studied another family with polycythemia and identified a nonsense mutation in EPOR that led to synthesis of a protein lacking the last 70 amino acids (de la Chapelle *et al.,* 1993). This mutant protein also endowed erythroid progenitors with increased sensitivity to EPO (Sokol *et al.*, 1995).

Joe Prchal became involved in a collaboration to study congenital polycythemia that was endemic in the Chuvash region of the Russian Federation (Sergeyeva *et al.,* 1997). Many affected individuals experienced headaches, which were thought to be due to increased blood viscosity, leading to decreased blood flow in cerebral capillaries. Some patients received relief after partial exchange transfusion, in which blood is removed and only the plasma is returned to the patient's circulation in order to reduce the hematocrit. Joe observed that reduction of the hematocrit from 66% to 48% (upper normal range) was associated with an increase in serum EPO levels from 51 to 121 U/ml (Sergeyeva *et al.*, 1997). In other words, the patients were behaving as if their "normal range" had been shifted upward. Based on these results, Joe hypothesized that these patients had a disorder of oxygen sensing. He queried me regarding candidate genes that he should analyze for the presence of a mutation.

At the top of my list of candidates were the genes encoding HIF-1α, HIF-2α, PHD2, and VHL. Unlike the family from Baltimore, the Chuvash polycythemia pedigrees indicated autosomal recessive inheritance. Affected individuals were found to be homozygous for a missense mutation in the VHL gene that changed the amino acid at position 200 from arginine to tryptophan, which resulted in a mutant VHL protein that bound to hydroxylated HIF-1α or HIF-2α with decreased avidity, resulting in impaired degradation of the HIF-α subunits (Ang *et al.,* 2002). Thus, at any given O2 concentration, there were higher levels of HIF-α proteins and higher expression of HIF target genes, including EPO. Chuvash polycythemia was indeed, as Joe had predicted, a disorder of O_2 sensing.

The *VHL* gene was mutated in every cell of affected individuals and the HIFs regulated many genes in addition to *EPO,* which raised the question as to whether other physiological responses to hypoxia were altered.

Compared to control subjects, affected individuals were shown by Peter Robbins at Oxford to have greater increases in heart rate, respiratory rate, and pulmonary artery pressure in response to a brief decrease in inspired O2 levels (Smith *et al.*, 2006). Other missense mutations in the *VHL* gene were subsequently identified in patients with congenital polycythemia; in one case, severe pulmonary hypertension resulted in death of the patient at 2 years of age (Sarangi *et al.*, 2014).

Subsequent studies by Frank Lee's lab identified heterozygous missense mutations in the EGLN1 gene, which encodes PHD2, in several families with polycythemia; a proline-to-arginine change at residue 317 impaired binding of PHD2 to HIF-1α, whereas a histidine-to-arginine change at residue 374 affected Fe^{2+} binding and catalytic activity, which may also be affected by an arginine-to-histidine substitution at residue 371 (Percy et al., 2006, 2007; Ladroue *et al.*, 2008). Finally, heterozygous missense mutations in the *EPAS1* gene, which encodes HIF-2α, also cause congenital polycythemia (Percy *et al.*, 2008, 2012). Thus, mutations that cause congenital polycythemia have been identified at each level of the HIF-PHD-VHL pathway, demonstrating the key role of these proteins in the regulation of erythropoiesis in humans.

THE PROTECTIVE ROLE OF HIF-1 IN ISCHEMIC CARDIOVASCULAR DISEASE

Atherosclerotic narrowing (stenosis) of major arteries in the brain, heart and leg result in cerebral, myocardial and limb ischemia, respectively. These disorders have a progressive (i.e., age-related) course that leads to stroke, heart attack, or gangrene requiring limb amputation, respectively. Limb ischemia has been modeled in mice by femoral artery ligation (FAL); however, the vast majority of studies were performed in young, healthy mice whereas patients with critical limb ischemia (CLI), meaning that perfusion is not sufficient to maintain tissue viability, are usually of advanced age and suffer from chronic diseases, most notably, diabetes.

To investigate the role of HIF-1 in limb ischemia, my postdoc Marta Bosch-Marcé performed unilateral FAL on $Hif1a^{+/-}$ and $Hif1a^{+/+}$ littermates and used laser Doppler perfusion imaging for serial and non-invasive analysis of blood flow. Mice are sexually mature (adult) by 2 months and have a lifespan of approximately 2 years. In young (2-month-old) wild-type ($Hif1a^{+/+}$) mice, blood flow (compared to the contralateral limb) was reduced to 20% immediately following surgery and gradually recovered to > 70% by 5 weeks after ligation; by contrast, middle-aged (8-month-old) mice recovered < 55% and old (20-month-old) mice recovered only 30% of blood flow (Bosch-Marcé *et al.*, 2007). At each age, $Hif1a^{+/-}$ mice recovered less well than their wild-type littermates.

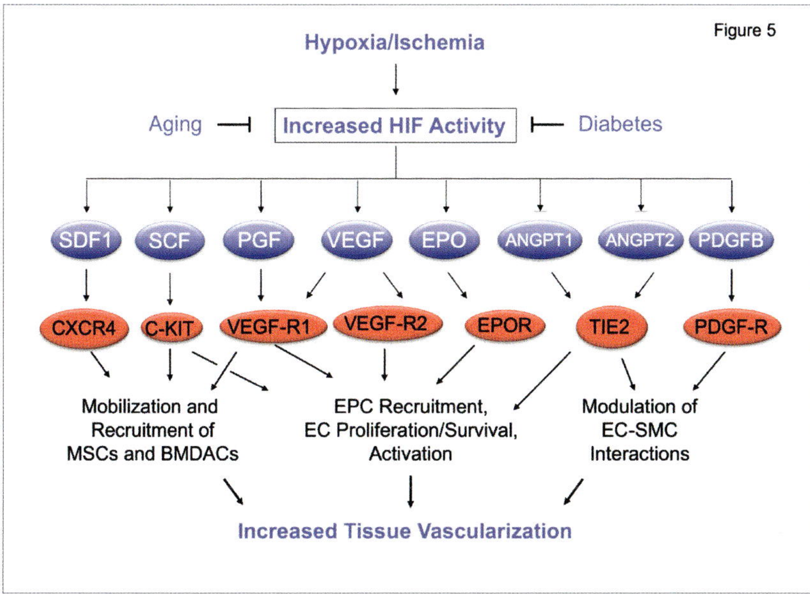

Figure 5. Regulation of tissue vascularization by hypoxia-inducible factors. HIFs activate transcription of genes encoding the angiogenic factors stromal-derived factor 1 (SDF1), stem cell factor (SCF; also known as kit ligand), placental growth factor (PGF), vascular endothelial growth factor (VEGF), erythropoietin (EPO), angiopoietins 1 and 2 (ANGPT1, ANGPT2), and platelet-derived growth factor B (PDGFB), which bind to cognate receptors (CXCR4, CKIT, VEGFR1, VEGFR2, EPOR, TIE2, PDGFR) to recruit and activate vascular endothelial cells, vascular pericytes/smooth muscle cells, and various bone marrow-derived angiogenic cells (BMDACs) to promote angiogenesis and vascular remodeling to increase tissue perfusion. Aging and diabetes impair the induction of HIFs in response to hypoxia and thereby impair ischemia-induced vascular remodeling.

These differences in the recovery of blood flow were correlated with the clinical outcome: all of the young wild-type mice recovered without any permanent tissue damage (limb salvage), whereas 40% of middle-aged mice developed soft tissue necrosis or spontaneous amputation of one or more toes and 15% of old mice lost their entire foot because the recovery of blood flow was insufficient to maintain tissue viability. Again, at every age, the clinical phenotype of *Hif1a*$^{+/-}$ mice was more severe than their wild-type littermates (Bosch-Marcé *et al.*, 2007).

Next, postdoc Ye Liu analyzed the expression of HIF-1α protein in the limb three days after FAL. In young wild-type mice, strong induction of HIF-1α protein expression was observed in skeletal muscle from the ipsilateral compared to contralateral limb. In middle-aged mice, the induction of HIF-1α protein was less robust and in old mice very little HIF-1α protein was detected in the ischemic limb (Bosch-Marcé *et al.*, 2007). Analysis of mRNA expression revealed induction of HIF-1α and mRNAs encoding multiple angiogenic growth factors, including VEGF, stromal-derived factor 1 (SDF1), stem cell factor, angiopoietin 1, angiopoietin 2, placental growth factor, and platelet-derived growth factor B (Bosch-Marcé *et al.*, 2007) (Figure 5). Ischemia-induced expression of all of these mRNAs was impaired by aging or HIF-1α deficiency. These mRNAs encode angiogenic factors that are all secreted proteins, which bind to receptors on vascular endothelial and smooth muscle cells as well as mesenchymal stem cells and other bone marrow-derived angiogenic cells (BMDACs) that are recruited to sites of ischemia and promote vascular remodeling to increase tissue perfusion. Indeed, analysis of circulating $CD34^+/VEGFR^{2+}$ or $Sca1^+/CXCR4^+$ BMDACs, which express a stem cell marker (CD34 or Sca1) and the cognate receptor for VEGF or SDF1, respectively, revealed a two-fold increase in the peripheral blood of wild type mice, but not in *Hif1a*$^{+/-}$ mice, on day 3 after FAL (Bosch-Marcé *et al.*, 2007).

We hypothesized that because (i) HIF-1α was required for expression of multiple angiogenic factors that controlled the homing and activation of multiple cell types contributing to vascular remodeling; and (ii) aging was associated with impaired induction of HIF-1α and all downstream events, that HIF-1α gene therapy might represent a novel therapeutic approach to CLI. Kiichi Hirota, a postdoc in the lab, constructed a replication-defective recombinant adenovirus (designated AdCA5) encoding a modified form of HIF-1α containing a deletion and several missense mutations that rendered the protein resistant to O_2-dependent degradation and, thus, constitutively active. Infection of AdCA5 into various cell types was sufficient to induce cell-type-specific changes in angiogenic gene expression that were identical to those induced by hypoxia (Kelly *et al.*, 2003).

Marta Bosch-Marcé showed that the injection of AdCA5 into the limb of a healthy mouse was sufficient to increase the number of circulating BMDACs in peripheral blood by three-fold within three days, which was similar to the response elicited by FAL (Bosch-Marcé *et al.*, 2007). When injected into the limb muscle of young wild-type mice immediately following FAL, AdCA5 significantly increased the recovery of blood flow to > 90% at three weeks after ligation (Bosch-Marcé *et al.*, 2007). However, young wild-type mice recovered without any therapy and were not a model of CLI. When middle-aged mice were treated with AdCA5 injected into the ischemic limb after FAL, their recovery was comparable to control young mice. Thus, HIF-1α gene therapy could overcome the age-dependent impairment of vascular remodeling. Postdoc Kakali Sarkar showed that injection of AdCA5 in the ischemic limb of *db/db* diabetic mice also dramatically improved recovery of perfusion and limb salvage (Sarkar *et al.*, 2009).

Remarkably, when postdoc Sergio Rey injected AdCA5 into the ischemic limb immediately after FAL of 13-month-old mice, the beneficial effect on the recovery of blood flow observed in 8-month-old mice was completely absent (Rey *et al.*, 2009). We reasoned that in addition to the failure to induce HIF-1 and HIF-dependent angiogenic factors, there was an additional impairment in old mice, which was a failure of BMDACs to respond to angiogenic signals induced by AdCA5, and the failure to respond was due to a failure to induce HIF-1 in BMDACs. To test this hypothesis, Sergio harvested bone marrow mononuclear cells from a donor mouse and cultured these cells for 4 days in the presence of angiogenic growth factors and dimethyloxalylglycine (DMOG), an α-ketoglutarate analog that the Ratcliffe lab had shown inhibits PHD activity, leading to induction of HIF-α protein expression even under aerobic conditions (Epstein *et al.*, 2001). Sergio performed FAL and injected AdCA5 into the ischemic limb muscle and then waited 24 hours before injecting the DMOG-treated bone marrow cells intravenously. We waited in order to allow AdCA5 to induce production of angiogenic factors, which would serve as a homing signal to attract BMDACs to the ischemic tissue, and we performed intravenous injection so that BMDACs bearing cognate receptors for the angiogenic factors induced by AdCA5 would be selectively recruited to the ischemic tissue. This two-stage, combined gene-and-cell therapy resulted in complete limb salvage: old mice had no permanent tissue damage or motor defects after FAL (Rey *et al.*, 2009). Remarkably, like AdCA5 alone, DMOG-treated bone marrow cells alone had no beneficial effect in old mice, because there was no homing signal to attract BMDACs to the ischemic tissue.

In the experiments described above, the recipient of the cell therapy was an old mouse, but the bone marrow donor was a young mouse. Sergio next repeated the experiments using 17-month-old mice as both donors and recipients, in order to model autologous bone marrow donation, which would be used if this approach were to be translated to a clinical trial involving elderly patients with CLI. Again, a dramatic recovery of perfusion was observed over a four-week period after FAL, with a marked reduction in tissue damage and motor impairment (Rey *et al.*, 2011). These results provide a potential novel strategy for treatment of CLI. Remarkably, a Phase I clinical trial, involving injection of a recombinant adenovirus encoding a HIF-1α/VP16 fusion protein into the limbs of elderly CLI patients did not improve perfusion or clinical measures (Rajagopalan *et al.*, 2007), as predicted by the mouse model.

HIF-1 PROMOTES WOUND VASCULARIZATION AND HEALING

Patients with CLI develop non-healing foot ulcers, which precede 85% of limb amputations. As in the case of peripheral arterial disease, wound healing is impaired by both aging and diabetes, and the pathogenic effects of these two conditions are synergistic (Brem *et al.*, 2007). In collaboration with the lab of surgeon John Harmon, we demonstrated that correction of the age-dependent impairment of HIF-1α expression in diabetic mice by gene therapy increased wound healing, angiogenesis, and recruitment of BMDACs (Liu *et al.*, 2008). Subsequent studies revealed important roles for HIF-1 in the vascularization and healing of burn wounds as well (Zhang *et al.*, 2010, 2011; Sarkar *et al.*, 2012).

HIF-1 PLAYS A PROTECTIVE ROLE IN LUNG TRANSPLANTATION

Lung transplant patients have a 5-year survival of 50%, which is the poorest outcome of all solid organ transplants. The lung is the only allograft for which the major blood supply (the bronchial arteries) is not re-established after transplantation. Autopsy data indicate that chronic rejection is preceded by a loss of airway microvasculature. Mark Nicolls, the director of the Pulmonary division at Stanford University developed a mouse model of chronic rejection using orthotopic tracheal transplantation into MHC-mismatched mice. Mark's postdoc Xinguo Jiang showed HIF-1 dependent recruitment of recipient $Tie2^+$ BMDACs into the donor airway was required for repair of the airway microvasculature and maintenance of graft survival (Jiang *et al.*, 2011). Treatment of the allografts with AdCA5 prior to transplantation increased graft perfusion and survival. Mark and Xinguo demonstrated that treatment of allografts with desferrioxamine nanoparticles also improved perfusion and survival in the mouse model of chronic rejection (Jiang *et al.*, 2011). Most recently, they have demonstrated that conditional knockout of HIF-2α in endothelial cells of the allograft accelerates, whereas HIF-2α overexpression protects against, microvascular loss after airway transplantation (Jiang *et al.*, 2019). Taken together, these elegant studies suggest a potential role for HIF inducers in the prevention or treatment of chronic rejection of lung allografts.

HIFS PLAY KEY ROLES IN THE PATHOGENESIS OF SLEEP APNEA-ASSOCIATED HYPERTENSION

Hypoxemia (reduced blood O_2 content) is sensed by glomus cells in the carotid bodies (CBs), which depolarize, triggering adaptive reflex arcs in the sympathetic nervous system that lead to increased breathing, heart rate, and blood pressure, thereby increasing O_2 capture and delivery. These homeostatic responses of the CB are subverted in the context of obstructive sleep apnea (OSA), which is characterized by airway obstruction leading to 15–30 seconds of hypoxemia that eventually awakens the patient, who clears his airway, leading to reoxygenation; this process is repeated dozens of times per night. The major sequela of OSA is hypertension, which places the patient at increased risk for heart failure, heart attack and stroke. Approximately 1 in 5 men and 1 in 15 women in the U.S. have OSA, which is the major cause of treatment-resistant hypertension. Although OSA causes both chronic intermittent hypoxemia (CIH) and chronic intermittent hypercarbia, only exposure of rodents to CIH leads to hypertension and only if the animal's CBs are intact (Fletcher *et al.*, 1992).

Nanduri Prabhakar's lab at the University of Chicago has utilized an OSA model in which rodents are placed in a chamber and the O_2 concentration is rapidly decreased from 21% to 5% O_2 for 15 seconds followed by rapid reoxygenation, which results in CIH similar to what is observed in OSA patients. These cycles of hypoxia and reoxygenation are repeated for 8 hours per day during the sleep cycle, leading to hypertension within 10 days. Nanduri's lab showed that CIH leads to increased ROS in CB glomus cells, which increases HIF-1α levels and transcriptional activity as a result of increased intracellular Ca^{2+} levels, which activate protein kinase C and Ca^{2+}/calmodulin-dependent kinase (Yuan *et al.*, 2008). CBs isolated from *Hif1a*$^{+/-}$ mice have dramatically impaired depolarization in response to hypoxia (Kline *et al.*, 2002) and are completely protected from CIH-induced ROS in the CB and CIH-induced hypertension (Peng *et al.*, 2006). Expression of NADPH oxidase 2 (NOX2) in the CB is induced by CIH in a HIF-1α-dependent manner (Yuan *et al.*, 2011). Remarkably, whereas CIH induces HIF-1α synthesis in the CB, it induces HIF-2α degradation by calpains, which are Ca^{2+}-dependent proteases. Whereas HIF-1α is required for expression of the pro-oxidant enzyme NOX2 and superoxide production in CBs, HIF-2α is required for expression of superoxide dismutase 2 (SOD2), the antioxidant enzyme that converts superoxide to hydrogen peroxide, which can then be converted to water by the action of glutathione peroxidase or catalase. Whereas *Hif1a*$^{+/-}$ mice are completely protected from CIH-induced ROS production in the CB and CIH-induced hypertension, *Hif2a*$^{+/-}$ mice have: increased ROS levels in their CBs due to decreased SOD2 and increased NOX2 expression; augmented depolarization of isolated CBs in response to hypoxia; and hypertension under normoxic conditions (Nanduri *et al.*, 2009; Peng *et al.*, 2011). These studies indicated that the balance between HIF-1α and HIF-2α levels determines the redox state of glomus cells in the CB, which in turn determines the set point of the sympathetic nervous system: high HIF-1α:HIF-2α ratio → high ROS → high sympathetic activity. The dramatic confirmation of this hypothesis came from the analysis of *Hif1a*$^{+/-}$ *Hif2a*$^{+/-}$ double heterozygous mice, which have normal ROS levels in their CBs, a normal CB response to hypoxia, and normal blood pressure (Yuan *et al.*, 2013). Most recently, Gabriela Pavlinkova's lab at the Czech Academy of Sciences has shown that HIF-1α is also required for development of the sympathetic nervous system (Bohuslavova *et al.*, 2019).

HIFS PROMOTE PATHOLOGICAL OCULAR NEOVASCULARIZATION

We collaborated with Peter Campochiaro at the Johns Hopkins Wilmer Eye Institute to study the role of HIF-1 in ocular neovascularization, which occurs in the wet type of age-related macular degeneration (AMD) and in proliferative diabetic retinopathy (PDR). We showed that the induction of HIF-1α protein expression was temporally and spatially correlated with VEGF expression in a mouse model of ischemic retinopathy (Ozaki *et al.*, 1999) and that retinal and choroidal neovascularization could be blocked by administration of a HIF inhibitor (digoxin or acriflavine) by intraocular or intraperitoneal injection (Yoshida *et al.*, 2010; Zeng *et al.*, 2017).

In collaboration with Akrit Sodhi, another retinal surgeon at Hopkins, we found that the two major angiogenic factors induced in the mouse model of ischemic retinopathy were VEGF and angiopoietin-like 4 (ANGPTL4). In diabetic patients, Akrit found that intraocular ANGPTL4 levels were significantly correlated with the presence of PDR, whereas VEGF levels were not (Babapoor-Farokhran *et al.*, 2015). Drugs targeting VEGF are effective in approximately 40% of patients with vision loss due to ocular neovascularization. The Sodhi lab demonstrated that vitreous fluid from the eyes of many patients treated with anti-VEGF therapy still had angiogenic activity, which could be blocked by an anti-ANGPTL4 antibody (Babapoor-Farokhran *et al.*, 2015). In the mouse model of ischemic retinopathy, administration of the HIF inhibitor digoxin blocked the induction of both VEGF and ANGPTL4 mRNA expression (Xin *et al.*, 2013), suggesting that HIF inhibitors may be beneficial in patients who do not respond to anti-VEGF therapy.

HIFS PLAY CRITICAL ROLES IN PROMOTING CANCER PROGRESSION

Most advanced human cancers contain areas of hypoxia as a result of increased O_2 consumption, due to dysregulated cell proliferation, and inadequate O_2 delivery, due to the formation of blood vessels that are structurally and functionally abnormal. In collaboration with oncologist Jonathan Simons, we found high levels of HIF-1α in human primary and metastatic cancers compared to normal tissue by immunohistochemistry (Zhong *et al.*, 1999). Furthermore, high levels of HIF-1α in the diagnostic tumor biopsy was associated with failure to achieve a complete response to radiation therapy in oropharyngeal cancer (Aebersold *et al.*, 2001) and with patient mortality in lymph node negative breast cancer (Bos *et al.*, 2003). Similar results have been reported in many other types of cancer (Semenza *et al.*, 2010).

In collaboration with Professor Jun Liu in the Department of Pharmacology, my postdoctoral fellows David Qian, Huafeng Zhang, and Kang-Ae Lee performed a cell-based screen for inhibitors of hypoxia-induced HIF reporter activity in Hep3B cells using a library of 3000 drugs that were FDA approved or had been evaluated in a phase I trial. Huafeng discovered that digoxin blocked the accumulation of HIF-1α protein in hypoxic cells, whereas Kang-Ae showed that acriflavine bound to HIF-1α and HIF-2α and blocked their dimerization with HIF-1β; both drugs potently inhibited tumor xenograft growth in mice (Zhang *et al.*, 2008; Lee *et al.*, 2009).

In a mouse model of triple-negative breast cancer (TNBC), in which human MDA-MB-231 cells were implanted in the mammary fat pad of immunodeficient mice, postdocs Huafeng Zhang and Carmen Wong, and graduate student Luana Schito showed that treatment of mice with digoxin or acriflavine blocked both blood vessel and lymphatic metastasis (Wong *et al.*, 2012; Zhang *et al.*, 2012; Schito *et al.*, 2012). Through the work of Huafeng, Carmen and postdocs Daniele Gilkes, Pallavi Chaturvedi, and Ting Wang, we subsequently deconvoluted the process of blood vessel metastasis from the mammary gland to the lungs of mice into a series of discrete steps and identified specific HIF target genes required for each step (Semenza, 2016; Table 1).

As part of the metastatic process, we focused on the specification of breast cancer stem cells, because although many cancer cells may successfully metastasize to the lungs, only a cancer stem cell is capable of giving rise to a secondary (metastatic or recurrent) tumor. Graduate students Lisha Xiang, Chuanzhao Zhang, and Jie Lan identified multiple

Table 1. HIF target genes required for metastasis of MDA-MB-231 human triple-negative breast cancer cells from the mammary fat pad to lungs of immunodeficient mice.

METASTATIC STEP	HIF TARGET GENES REQUIRED
MSC* and macrophage cooptation	*CCR5, CSF1, CXCL16, CXCR3, PGF*
Migration and invasion	*P4HA1, P4HA2, PLOD2, RHOA, ROCK1*
Microvesicle formation	*RAB22A*
Margination and extravasation	*ANGPTL4, L1CAM*
Immune evasion	*CD47, CD73, PDL1*
Premetastatic niche formation	*LOX, LOXL2, LOXL4*
Cancer stem cell specification and maintenance	*ALKBH5, CD47, GCLM, GSTO1, IL6, IL8, MDR1 PHGDH, SIAH1, SLC7A11, WWTR1, ZNF217*

*MSC, mesenchymal stromal cell

genes that were induced by hypoxia in a HIF-dependent manner and increased the specification of breast cancer stem cells by increasing the expression of the pluripotency factor Nanog (Xiang and Semenza, 2019; Table 2). Remarkably, postdocs Debangshu Samanta and Haiquan Lu demonstrated that exposure of TNBC cells to cytotoxic chemotherapy (such as paclitaxel, gemcitabine, doxorubicin, or carboplatin), either in vitro or in vivo, induced HIF-dependent expression of genes that increase Nanog expression and breast cancer stem cell specification. Remarkably, whereas chemotherapy treatment markedly increased the percentage of cancer stem cells among the surviving breast cancer cells, this counter-therapeutic effect could be blocked by co-administration of the HIF inhibitor digoxin.

Another critical requirement for cancer cells to form primary and metastatic cancers is that they evade killing by the immune system. Postdoc Debangshu Samanta found that the expression of three genes involved in immune evasion was induced by exposure of breast cancer cells to hypoxia or cytotoxic chemotherapy (Samanta *et al.*, 2018). CD47 encodes a cell surface protein that interacts with a receptor on macrophages called SIRPα, which blocks the ability of macrophages to engulf cancer cells, thereby earning CD47 the nickname of the "don't eat me signal." CD73, also known as ecto-5'-nucleotidase, is a cell surface protein that converts AMP to adenosine, which binds to receptors on T lymphocytes and induces anergy or apoptosis, enabling cancer cells to evade killing. Finally, PDL1 encodes a cell surface protein that interacts with PD1, a cell surface receptor expressed by T lymphocytes, and thereby induces anergy or apoptosis. Remarkably, using flow cytometry we demonstrated coordinate expression of all three genes in individual cancer cells exposed to hypoxia or chemotherapy, thereby rendering the cells resistant to killing by both innate and adaptive immunity (Samanta *et al.*, 2018).

Table 2. HIF target genes that increase Nanog expression and breast cancer stem cell specification in response to hypoxia or cytotoxic chemotherapy.

GENE	STIMULUS TESTED	EFFECT ON NANOG EXPRESSION
A2BR	Hypoxia	Increased Nanog mRNA synthesis mediated by STAT3
ALKBH5	Hypoxia	Increased Nanog mRNA stability
CD73	Hypoxia	Increased Nanog mRNA synthesis mediated by STAT3
DUSP9	Chemotherapy	Increased Nanog mRNA synthesis mediated by FOXO3
GCLM	Chemotherapy	Increased Nanog mRNA synthesis mediated by FOXO3
GSTO1	Chemotherapy	Increased Nanog mRNA synthesis mediated by STAT3
REST	Chemotherapy	Increased Nanog mRNA stability
SLC7A11	Chemotherapy	Increased Nanog mRNA synthesis mediated by FOXO3
ZNF217	Hypoxia	Increased Nanog mRNA stability

Debangshu injected mouse 4T1 mammary carcinoma cells into the mammary fat pad of syngeneic, immunocompetent female BALB/c mice and treated the mice with vehicle, paclitaxel or paclitaxel plus the HIF inhibitor acriflavine. Compared to tumors from vehicle-treated mice, tumors from paclitaxel-treated mice showed a 3-fold increase in the percentage of $PDL1^+/CD47^+/CD73^+$ cancer cells, a 3-fold decrease in $CD8^+/CD44^+/CD69^+$ activated effector T cells, and a 2-fold increase in $CD4^+/CD25^+/FoxP3^+$ regulatory T cells, which promote T cell anergy (Samanta *et al.*, 2018). Thus, treatment with cytotoxic chemotherapy resulted in a profound change in the immune cell microenvironment in the residual breast tumor that favored immune evasion. Remarkably, all of the counter-therapeutic effects of paclitaxel on the immune microenvironment were blocked by co-administration of the HIF inhibitor acriflavine. Similar results were obtained using carboplatin, another cytotoxic chemotherapeutic agent (Samanta *et al.*, 2018).

Taken together, our studies indicate that TNBC cells that survive cytotoxic chemotherapy are more likely that untreated cells to have cancer stem cell and immune evasive phenotypes, i.e., the lethal cancer phenotype, providing a molecular basis for the recurrent and metastatic disease that is responsible for the high mortality of TNBC compared to all other breast cancer subtypes. Our results also suggest that combining a HIF inhibitor with chemotherapy might result in a major improvement in the survival of women with TNBC. Furthermore, HIF inhibitors may improve the response to immune checkpoint blockers such as anti-PD1 or anti-PDL1 antibodies. Although digoxin and acriflavine have been useful HIF inhibitors for animal models, they are not suitable for this use in patients. We are currently devoting considerable efforts towards the identification of novel HIF inhibitors that are safe and effective in cancer patients.

CONCLUSION

Over the last 30 years, our lab with the help of many essential collaborators at Johns Hopkins and other institutions has had the good fortune to discover and characterize HIF-1 at the molecular level as well as to investigate the involvement of HIF-1 in embryonic development, cellular and systemic physiology, and disease pathophysiology. It has been particularly rewarding to witness the translation of our discovery to the clinic with the development of HIF inducers and inhibitors that are now in advanced clinical trials for the treatment of anemia and cancer, respectively. Based on work in animal models described above, we can look forward to many other applications for these two new classes of drugs in the years to come.

ACKNOWLEDGMENTS

I have endeavored to present the work of as many of my lab members and collaborators as possible, and I regret that I was unable to include many others. Likewise, many laboratories have made important contributions to the field and I apologize to those individuals who may be offended by this one-sided accounting of my lab's activities.

REFERENCES

Abud EM, Maylor J, Undem C, Punjabi A, Zaiman AL, Myers AC, Sylvester JT, Semenza GL, Shimoda LA. Digoxin inhibits development of hypoxic pulmonary hypertension in mice. *Proc Natl Acad Sci U S A.* 2012; **109**(4): 1239–44.

Aebersold DM, Burri P, Beer KT, Laissue J, Djonov V, Greiner RH, Semenza GL. Expression of hypoxia-inducible factor-1a: a novel predictive and prognostic parameter in the radiotherapy of oropharyngeal cancer. *Cancer Res.* 2001; **61**(7): 2911–6.

Ang SO, Chen H, Hirota K, Gordeuk VR, Jelinek J, Guan Y, Liu E, Sergueeva AI, Miasnikova GY, Mole D, Maxwell PH, Stockton DW, Semenza GL, Prchal JT. Disruption of oxygen homeostasis underlies congenital Chuvash polycythemia. *Nat Genet.* 2002; **32**: 614–21.

Babapoor-Farrokhran S, Jee K, Puchner B, Hassan SJ, Xin X, Rodrigues M, Kashiwabuchi F, Ma T, Hu K, Deshpande M, Daoud Y, Solomon S, Wenick A, Lutty GA, Semenza GL, Montaner S, Sodhi A. Angiopoietin-like 4 is a potent angiogenic factor and a novel therapeutic target for patients with proliferative diabetic retinopathy. *Proc Natl Acad Sci U S A.* 2015; **112**(23): E3030–9.

Beck I, Ramirez S, Weinmann R, Caro J. Enhancer element at the 3'-flanking region controls transcriptional response to hypoxia in the human erythropoietin gene. *J Biol Chem.* 1991; **266**(24): 15563–6.

Bohuslavova R, Cerychova R, Papousek F, Olejnickova V, Bartos M, Görlach A, Kolar F, Sedmera D, Semenza GL, Pavlinkova G. HIF-1α is required for development of the sympathetic nervous system. *Proc Natl Acad Sci U S A.* 2019; **116**(27): 13414–23.

Bos R, van der Groep P, Greijer AE, Shvarts A, Meijer S, Pinedo HM, Semenza GL, van Diest PJ, van der Wall E. Levels of hypoxia-inducible factor-1alpha independently predict prognosis in patients with lymph node negative breast carcinoma. *Cancer.* 2003; **97**(6): 1573–81.

Bosch-Marcé M, Okuyama H, Wesley JB, Sarkar K, Kimura H, Liu YV, Zhang H, Strazza M, Rey S, Savino L, Zhou YF, McDonald KR, Na Y, Vandiver S, Rabi A, Shaked Y, Kerbel R, LaVallee T, Semenza GL. Effects of aging and HIF-1 activity on angiogenic cell mobilization and recovery of perfusion following limb ischemia. *Circ Res.* 2007; **101**: 1310–18.

Brem H, Tomic-Canic M, Entero H, Hanflik AM, Wang VM, Fallon JT, Ehrlich HP. The synergism of age and db/db genotype impairs wound healing. *Exp Gerontol.* 2007; **42**(6): 523–31.

Bruick RK, McKnight SL. A conserved family of prolyl-4-hydroxylases that modify HIF. *Science.* 2001; **294**(5545): 1337–40.

de la Chapelle A, Träskelin AL, Juvonen E. Truncated erythropoietin receptor causes dominantly inherited benign human erythrocytosis. *Proc Natl Acad Sci U S A*. 1993; **90**(10): 4495–9.

Costa-Giomi P, Caro J, Weinmann R. Enhancement by hypoxia of human erythropoietin gene transcription in vitro. *J Biol Chem*. 1990; **265**(18): 10185–8.

Duan C. Hypoxia-inducible factor 3 biology: complexities and emerging themes. *J Physiol Cell Physiol*. 2016; **310**(4): C260–9.

Ema M, Taya S, Yokotani N, Sogawa K, Matsuda Y, Fujii-Kuriyama Y. A novel bHLH-PAS factor with close sequence similarity to hypoxia-inducible factor 1a regulates the VEGF expression and is potentially involved in lung and vascular development. Proc Natl Acad Sci U S A. 1997; **94**(9): 4273–8.

Epstein AC, Gleadle JM, McNeill LA, Hewitson KS, O'Rourke J, Mole DR, Mukherji M, Metzen E, Wilson MI, Dhanda A, Tian YM, Masson N, Hamilton DL, Jaakkola P, Barstead R, Hodgkin J, Maxwell PH, Pugh CW, Schofield CJ, Ratcliffe PJ. C. elegans EGL-9 and mammalian homologs define a family of dioxygenases that regulate HIF by prolyl hydroxylation. *Cell*. 2001; **107**(1): 43–54.

Erslev AJ. Observations on the nature of the erythropoietic serum factor. II. Erythropoietic activity of serum and bone marrow after time limited exposure to anemic and hypoxic hypoxia. *J Lab Clin Med*. 1957; **50**(4): 543–9.

Firth JD, Ebert BL, Pugh CW, Ratcliffe PJ. Oxygen-regulated control elements in the phosphoglycerate kinase 1 and lactate dehydrogenase A genes: similarities with the erythropoietin 3' enhancer. *Proc Natl Acad Sci U S A*. 1994; **91**(14): 6496–500.

Flamme I, Fröhlich T, von Reutern M, Kappel A, Damert A, Risau W. HRF, a putative basic helix–loop-helix-PAS-domain transcription factor is closely related to hypoxia-inducible factor-1a and developmentally expressed in blood vessels. *Mech Dev*. 1997; **63**(1): 51–60.

Fletcher EC, Lesske J, Behm R, Miller CC 3rd, Stauss H, Unger T. Carotid chemoreceptors, systemic blood pressure, and chronic episodic hypoxia mimicking sleep apnea. *J Appl Physiol*. 1992; **72**(5): 1978–84.

Folkman J. Tumor angiogenesis: therapeutic implications. *N Engl J Med*. 1971; **285**(21): 1182–6.

Folkman J, Merler E, Abernathy C, Williams G. Isolation of a tumor factor responsible for angiogenesis. *J Exp Med*. 1971; **133**(2): 275–88.

Forsythe JA, Jiang BH, Iyer NV, Agani F, Leung SW, Koos RD, Semenza GL. Activation of vascular endothelial growth factor gene transcription by hypoxia-inducible factor 1. *Mol Cell Biol*. 1996; **16**(9): 4604–13.

Fukuda R, Zhang H, Kim JW, Shimoda L, Dang CV, Semenza GL. HIF-1 regulates cytochrome oxidase subunits to optimize efficiency of respiration in hypoxic cells. *Cell*. 2007; **129**(1): 111–22.

Goldberg MA, Glass GA, Cunningham JM, Bunn HF. The regulated expression of erythropoietin by two human hepatoma cell lines. *Proc Natl Acad Sci U S A*. 1987; **84**(22): 7972–6.

Graham AM, Presnell JS. Hypoxia-inducible factor (HIF) transcription factor family expansion, diversification, divergence, and selection in eukaryotes. *PLoS One*. 2017; **12**(6): e0179545.

Gu YZ, Moran SM, Hogenesch JB, Wartman L, Bradfield CA. Molecular characterization and chromosomal localization of a third a-class hypoxia inducible factor subunit, HIF3a. *Gene Expr*. 1998; **7**(3): 205–13.

Hoffman EC, Reyes H, Chu FF, Sander F, Conley LH, Brooks BA, Hankinson O. Cloning of a factor required for activity of the Ah (dioxin) receptor. *Science*. 1991; **252**(5008): 954–8.

Hogenesch JB, Chan WK, Jackiw VH, Brown RC, Gu YZ, Pray-Grant M, Perdew GH, Bradfield CA. Characterization of a subset of the basic-helix-loop-helix-PAS superfamily that interacts with components of the dioxin signaling pathway. *J Biol Chem.* 1997; **272**(13): 8581–93.

Huang LE, Gu J, Schau M, Bunn HF. Regulation of hypoxia-inducible factor 1a is mediated by an O2-dependent degradation domain via the ubiquitin-proteasome pathway. *Proc Natl Acad Sci U S A.* 1998; **95**(14): 7987–92.

Ivan M, Kondo K, Yang H, Kim W, Valiando J, Ohh M, Salic A, Asara JM, Lane WS, Kaelin WG Jr. HIFa targeted for VHL-mediated destruction by proline hydroxylation: implications for O2 sensing. *Science.* 2001; **292**(5516): 464–8.

Iyer NV, Kotch LE, Agani F, Leung SW, Laughner E, Wenger RH, Gassmann M, Gearhart JD, Lawler AM, Yu AY, Semenza GL. Cellular and developmental control of O2 homeostasis by hypoxia-inducible factor 1a. *Genes Dev.* 1998; **12**(2): 149–62.

Jaakkola P, Mole DR, Tian YM, Wilson MI, Gielbert J, Gaskell SJ, von Kriegsheim A, Hebestreit HF, Mukherji M, Schofield CJ, Maxwell PH, Pugh CW, Ratcliffe PJ. Targeting of HIF-a to the von Hippel-Lindau ubiquitylation complex by O2-regulated prolyl hydroxylation. *Science.* 2001; **292**(5516): 468–72.

Jiang BH, Rue E, Wang GL, Roe R, Semenza GL. Dimerization, DNA binding, and transactivation properties of hypoxia-inducible factor 1. *J Biol Chem.* 1996; **271**(30): 17771–8.

Jiang BH, Semenza GL, Bauer C, Marti HH. Hypoxia-inducible factor 1 levels vary exponentially over a physiologically relevant range of O2 tension. *Am J Physiol.* 1996b; **271**(4 Pt 1): C1172–80.

Jiang BH, Zheng JZ, Leung SW, Roe R, Semenza GL. Transactivation and inhibitory domains of hypoxia-inducible factor 1a. Modulation of transcriptional activity by oxygen tension. *J Biol Chem.* 1997; **272**(31): 19253–60.

Jiang X, Khan MA, Tian W, Beilke J, Natarajan R, Kosek J, Yoder MC, Semenza GL, Nicolls MR. Adenovirus-mediated HIF-1α gene transfer promotes repair of mouse airway allograft microvasculature and attenuates chronic rejection. *J Clin Invest.* 2011; **121**(6): 2336–49.

Jiang X, Tian W, Tu AB, Pasupneti S, Shuffle E, Dahms P, Zhang P, Cai H, Dinh TT, Liu B, Cain C, Giaccia AJ, Butcher EC, Simon MC, Semenza GL, Nicolls MR. Endothelial Hypoxia-inducible factor-2α is required for the maintenance of airway microvasculature. *Circulation.* 2019; **139**(4): 502–17.

Kelly BD, Hackett SF, Hirota K, Oshima Y, Cai Z, Berg-Dixon S, Rowan A, Yan Z, Campochiaro PA, Semenza GL. Cell-type-specific regulation of angiogenic growth factor gene expression and induction of angiogenesis in non-ischemic tissue by a constitutively active form of hypoxia-inducible factor 1. *Circ Res.* 2003; **93**: 1074–81.

Kim JW, Tchernyshyov I, Semenza GL, Dang CV. HIF-1-mediated expression of pyruvate dehydrogenase kinase: a metabolic switch required for cellular adaptation to hypoxia. *Cell Metab.* 2006; **3**(3): 177–85.

Kline DD, Peng YJ, Manalo DJ, Semenza GL, Prabhakar NR. Defective carotid body function and impaired ventilatory responses to chronic hypoxia in mice partially deficient for hypoxia-inducible factor 1a. *Proc Natl Acad Sci U S A.* 2002; **99**(2): 821–6.

Ladroue C, Carcenac R, Leporrier M, Gad S, Le Hello C, Galateau-Salle F, Feunteun J, Pouysségur J, Richard S, Gardie B. PHD2 mutation and congenital erythrocytosis with paraganglioma. *N Engl J Med.* 2008; **359**: 2685–92.

Lando D, Peet DJ, Gorman JJ, Whelan DA, Whitelaw ML, Bruick RK. FIH-1 is an asparaginyl hydroxylase enzyme that regulates the transcriptional activity of hypoxia-inducible factor. *Genes Dev.* 2002; **16**(12): 1466–71.

Lando D, Peet DJ, Whelan DA, Gorman JJ, Whitelaw ML. Asparagine hydroxylation of the HIF transactivation domain a hypoxic switch. *Science.* 2002; **295**(5556): 858–61.

Lee K, Zhang H, Qian DZ, Rey S, Liu JO, Semenza GL. Acriflavine inhibits HIF-1 dimerization, tumor growth, and vascularization. *Proc Natl Acad Sci U S A.* 2009; **106**(42): 17910–5.

Leung DW, Cachianes G, Kuang WJ, Goeddel DV, Ferrara N. Vascular endothelial growth factor is a secreted angiogenic mitogen. *Science.* 1989; **246**(4935): 1306–9.

Liu L, Marti GP, Wei X, Zhang X, Zhang H, Liu YV, Nastai M, Semenza GL, Harmon JW. Age-dependent impairment of HIF-1alpha expression in diabetic mice: Correction with electroporation-facilitated gene therapy increases wound healing, angiogenesis, and circulating angiogenic cells. *J Cell Physiol.* 2008; **217**(2): 319–27.

Mahon PC, Hirota K, Semenza GL. FIH-1: a novel protein that interacts with HIF-1a and VHL to mediate repression of HIF-1 transcriptional activity. *Genes Dev.* 2001; **15**(20): 2675–86.

Masson N, Willam C, Maxwell PH, Pugh CW, Ratcliffe PJ. Independent function of two destruction domains in hypoxia-inducible factor-a chains activated by prolyl hydroxylation. *EMBO J.* 2001; **20**(18): 5197–206.

Maxwell PH, Pugh CW, Ratcliffe PJ. Inducible operation of the erythropoietin 3' enhancer in multiple cell lines: evidence for a widespread oxygen-sensing mechanism. *Proc Natl Acad Sci U S A.* 1993; **90**(6): 2423–7.

Maxwell PH, Wiesener MS, Chang GW, Clifford SC, Vaux EC, Cockman ME, Wykoff CC, Pugh CW, Maher ER, Ratcliffe PJ. The tumor suppressor VHL targets hypoxia-inducible factors for oxygen-dependent proteolysis. *Nature.* 1999; **399**(6733): 271–5.

Nanduri J, Wang N, Yuan G, Khan SA, Souvannakitti D, Peng YJ, Kumar GK, Garcia JA, Prabhakar NR. Intermittent hypoxia degrades HIF-2a via calpains resulting in oxidative stress: implications for recurrent apnea-induced morbidities. *Proc Natl Acad Sci U S A.* 2009; **106**(4): 1199–204.

Ozaki H, Yu AY, Della N, Ozaki K, Luna JD, Yamada H, Hackett SF, Okamoto N, Zack DJ, Semenza GL, Campochiaro PA. Hypoxia inducible factor-1a is increased in ischemic retina: temporal and spatial correlation with VEGF expression. *Invest Ophthalmol Vis Sci.* 1999; **40**(1): 182–9.

Peng YJ, Nanduri J, Khan SA, Yuan G, Wang N, Kinsman B, Vaddi DR, Kumar GK, Garcia JA, Semenza GL, Prabhakar NR. Hypoxia-inducible factor 2α (HIF-2α) heterozygous-null mice exhibit exaggerated carotid body sensitivity to hypoxia, breathing instability, and hypertension. *Proc Natl Acad Sci U S A.* 2011; **108**(7): 3065–70.

Peng YJ, Yuan G, Ramakrishnan D, Sharma SD, Bosch-Marce M, Kumar GK, Semenza GL, Prabhakar NR. Heterozygous HIF-1a deficiency impairs carotid body-mediated systemic responses and reactive oxygen species generation in mice exposed to intermittent hypoxia. *J Physiol.* 2006; **577**(Pt 2): 705–16.

Percy MJ, Chung YJ, Harrison C, Mercieca J, Hoffbrand AV, Dinardo CL, Santos PC, Fonseca GH, Gualandro SF, Pereira AC, Lappin TR, McMullin MF, Lee FS. Two new mutations in the HIF2A gene associated with erythrocytosis. *Am J Hematol.* 2012; **87**: 439–42.

Percy MJ, Furlow PW, Beer PA, Lappin TR, McMullin MF, Lee FS. A novel erythrocytosis-associated PHD2 mutation suggests the location of a HIF binding groove. *Blood*. 2007; **110**: 2193–6.

Percy MJ, Furlow PW, Lucas GS, Li X, Lappin TR, McMullin MF, Lee FS. A gain-of-function mutation in the HIF2A gene in familial erythrocytosis. *N Engl J Med*. 2008; **358**: 162–8.

Percy MJ, Zhao Q, Flores A, Harrison C, Lappin TR, Maxwell PH, McMullin MF, Lee FS. A family with erythrocytosis establishes a role for prolyl hydroxylase domain protein 2 in oxygen homeostasis. *Proc Natl Acad Sci U S A*. 2006; **103**: 654–9.

Plate KH, Breier G, Weich HA, Risau W. Vascular endothelial growth factor is a potential tumor angiogenesis factor in human gliomas in vivo. *Nature*. 1992; **359**(6398): 845–8.

Preston GM, Agre P. Isolation of the cDNA for erythrocyte integral membrane protein of 28 kilodaltons: member of an ancient channel family. *Proc Natl Acad Sci U S A*. 1991; **88**(24): 11110–4.

Pugh CW, Tan CC, Jones RW, Ratcliffe PJ. Functional analysis of an oxygen-regulated transcriptional enhancer lying 3' to the mouse erythropoietin gene. *Proc Natl Acad Sci U S A*. 1991; **88**(23): 10553–7.

Rajagopalan S, Olin J, Deitcher S, Pieczek A, Laird J, Grossman PM, Goldman CK, McEllin K, Kelly R, Chronos N. Use of a constitutively active hypoxia-inducible factor-1α transgene as a therapeutic strategy in no option critical limb ischemia patients: phase I dose-escalation experience. *Circulation*. 2007; **115**(10): 1234–43.

Rey S, Lee K, Wang CJ, Gupta K, Chen S, McMillan A, Bhise N, Levchenko A, Semenza GL. Synergistic effect of HIF-1α gene therapy and HIF-1-activated bone marrow-derived angiogenic cells in a mouse model of limb ischemia. *Proc Natl Acad Sci U S A*. 2009; **106**: 20399–404.

Rey S, Luo W, Shimoda LA, Semenza GL. Metabolic reprogramming by HIF-1 promotes the survival of bone marrow-derived angiogenic cells in ischemic tissue. *Blood*. 2011; **117**(18): 4988–98.

Reyes H, Reisz-Porszasz S, Hankinson O. Identification of the Ah receptor nuclear translocator protein (Arnt) as a component of the DNA binding form of the Ah receptor. *Science*. 1992; **256**(5060): 1193–5.

Rosenfeld PJ, Kelly TJ. Purification of nuclear factor I by DNA recognition site affinity chromatography. *J Biol Chem*. 1986; **261**(3): 1398–408.

Salceda S, Caro J. Hypoxia-inducible factor 1α protein is rapidly degraded by ubiquitin-proteasome system under normoxic conditions: its stabilization by hypoxia depends on redox-induced changes. *J Biol Chem*. 1997; **272**(36): 22642–7.

Samanta D, Park Y, Ni X, Li H, Zahnow CA, Gabrielson E, Pan F, Semenza GL. Chemotherapy induces enrichment of CD47+/CD73+/PDL1+ immune evasive triple-negative breast cancer cells. *Proc Natl Acad Sci U S A*. 2018; **115**(6): E1239–48.

Sarangi S, Lanikova L, Kapralova K, Acharya S, Swierczek S, Lipton JM, Wolfe L, Prchal JT. The homozygous VHLD126N missense mutation is associated with dramatically elevated erythropoietin levels with consequent polycythemia and early onset severe pulmonary hypertension. *Pediatr Blood Cancer*. 2014; **61**: 2104–6.

Sarkar K, Fox-Talbot K, Steenbergen C, Bosch-Marcé M, Semenza GL. Adenoviral transfer of HIF-1a enhances vascular responses to critical limb ischemia in diabetic mice. *Proc Natl Acad Sci U S A*. 2009; **106**(44): 18769–74.

Sarkar K, Rey S, Zhang X, Sebastian R, Marti GP, Fox-Talbot K, Cardona AV, Du J, Tan YS, Liu L, Lay F, Gonzalez FJ, Harmon JW, Semenza GL. Tie2-dependent knockout of HIF-1 impairs burn wound vascularization and homing of bone marrow-derived angiogenic cells. *Cardiovasc Res.* 2012; **93**(1): 162–9.

Scherer WF, Syverton JT, Gey GO. Studies on the propagation in vitro of poliomyelitis viruses. IV. Viral multiplication in a stable strain of human malignant epithelial cells (strain HeLa) derived from an epidermoid carcinoma of the cervix. *J Exp Med* 1953; **97**(5): 695–710.

Schito L, Rey S, Tafani M, Zhang H, Wong CC, Russo A, Russo MA, Semenza GL. Hypoxia-inducible factor 1-dependent expression of platelet-derived growth factor B promotes lymphatic metastasis of hypoxic breast cancer cells. *Proc Natl Acad Sci U S A.* 2012; **109**(40): E2707–16.

Schuster SJ, Wilson JH, Erslev AJ, Caro J. Physiologic regulation and tissue localization of renal erythropoietin messenger RNA. *Blood.* 1987; **70**(1): 316–8.

Semenza GL. Defining the role of hypoxia-inducible factor 1 in cancer biology and therapeutics. *Oncogene.* 2010; **29**(5): 625–34.

Semenza GL. The hypoxic tumor microenvironment: A driving force for breast cancer progression. *Biochim Biophys Acta.* 2016; **1863**(3): 382–91.

Semenza GL, Dureza RC, Traystman MD, Gearhart JD, Antonarakis SE. Human erythropoietin gene expression in transgenic mice: multiple transcription initiation sites and cis-acting regulatory elements. *Mol Cell Biol.* 1990; **10**(3): 930–8.

Semenza GL, Koury ST, Nejfelt MK, Gearhart JD, Antonarakis SE. Cell-type-specific and hypoxia-inducible expression of the human erythropoietin gene in transgenic mice. *Proc Natl Acad Sci U S A.* 1991a; **88**(19): 8725–9.

Semenza GL, Nejfelt MK, Chi SM, Antonarakis SE. Hypoxia-inducible nuclear factors bind to an enhancer element located 3' to the human erythropoietin gene. *Proc Natl Acad Sci U S A.* 1991b; **88**(13): 5680–4.

Semenza GL, Roth PH, Fang HM, Wang GL. Transcriptional regulation of genes encoding glycolytic enzymes by hypoxia-inducible factor 1. *J Biol Chem.* 1994; **269**(38): 23757–63.

Semenza GL, Traystman MD, Gearhart JD, Antonarakis SE. Polycythemia in transgenic mice expressing the human erythropoietin gene. *Proc Natl Acad Sci U S A.* 1989; **86**(7): 2301–5.

Semenza GL, Wang GL. A nuclear factor induced by hypoxia via de novo protein synthesis binds to the human erythropoietin gene enhancer at a site required for transcriptional activation. *Mol Cell Biol.* 1992; **12**(12): 5447–54.

Sergeyeva A, Gordeuk VR, Tokarev YN, Sokol L, Prchal JF, Prchal JT. Congenital polycythemia in Chuvashia. *Blood.* 1997; **89**: 2148–54.

Shweiki D, Itin A, Soffer D, Keshet E. Vascular endothelial growth factor induced by hypoxia may mediate hypoxia-initiated angiogenesis. *Nature.* 1992; 359(6398): 843–5.

Smith TG, Brooks JT, Balanos GM, Lappin TR, Layton DM, Leedham DL, Liu C, Maxwell PH, McMullin MF, McNamara CJ, Percy MJ, Pugh CW, Ratcliffe PJ, Talbot NP, Treacy M, Robbins PA. Mutation of von Hippel-Lindau tumor suppressor and human cardiopulmonary physiology. *PLoS Med.* 2006; **3**: e290.

Sokol L, Luhovy M, Guan Y, Prchal JF, Semenza GL, Prchal JT. Primary familial polycythemia: a frameshift mutation in the erythropoietin receptor gene and increased sensitivity of erythroid progenitors to erythropoietin. *Blood.* 1995; **86**(1): 15–22.

Stebbins CE, Kaelin WG Jr, Pavletich NP. Structure of the VHL-Elongin C-Elongin B complex: implications for VHL tumor suppressor function. *Science.* 1999; **284**(5413): 455–61.

Tian H, McKnight SL, Russell DW. Endothelial PAS domain protein 1 (EPAS1), a transcription factor selectively expressed in endothelial cells. *Genes Dev.* 1997; **11**(1): 72–82.

Vinson CR, LaMarco KL, Johnson PF, Landschulz WH, McKnight SL. In situ detection of sequence-specific DNA binding activity specified by a recombinant bacteriophage. *Genes Dev.* 1988; **2**(7): 801–6.

Wang GL, Jiang BH, Rue EA, Semenza GL. Hypoxia-inducible factor 1 is a basic-helix-loop-helix-PAS heterodimer regulated by cellular O2 tension. *Proc Natl Acad Sci U S A.* 1995; **92**(12): 5510–4.

Wang GL, Semenza GL. Characterization of hypoxia-inducible factor 1 and regulation of DNA binding activity by hypoxia. *J Biol Chem.* 1993a; **268**(29): 21513–8.

Wang GL, Semenza GL. Desferrioxamine induces erythropoietin gene expression and hypoxia-inducible factor 1 DNA-binding activity: implications for models of hypoxia signal transduction. *Blood.* 1993b; **82**(12): 3610–5.

Wang GL, Semenza GL. General involvement of hypoxia-inducible factor 1 in transcriptional response to hypoxia. *Proc Natl Acad Sci U S A.* 1993c; **90**(9): 4304–8.

Wang GL, Semenza GL. Purification and characterization of hypoxia-inducible factor 1. *J Biol Chem.* 1995; **270**(3): 1230–7.

Webster KA. Regulation of glycolytic enzyme RNA transcriptional rates by oxygen availability in skeletal muscle cells. *Mol Cell Biochem.* 1987; **77**(1): 19–28.

Weisbrod S, Weintraub H. Isolation of a subclass of nuclear proteins responsible for conferring a DNase I-sensitive structure on globin chromatin. *Proc Natl Acad Sci U S A.* 1979; 76(2): 630–4.

Wong CC, Zhang H, Gilkes DM, Chen J, Wei H, Chaturvedi P, Hubbi ME, Semenza GL. Inhibitors of hypoxia-inducible factor 1 block breast cancer metastatic niche formation and lung metastasis. *J Mol Med.* 2012; **90**(7): 803–15.

Wu D, Rastinejad F. Structural characterization of mammalian bHLH-PAS transcription factors. *Curr Opin Struct Biol.* 2017; **43**: 1–9.

Xiang L, Semenza GL Hypoxia-inducible factors promote breast cancer stem cell specification and maintenance in response to hypoxia or cytotoxic chemotherapy. *Adv Cancer Res.* 2019; **141**: 175–212.

Xin X, Rodrigues M, Umapathi M, Kashiwabuchi F, Ma T, Babapoor-Farrokhran S, Wang S, Hu J, Bhutto I, Welsbie DS, Duh EJ, Handa JT, Eberhart CG, Lutty G, Semenza GL, Montaner S, Sodhi A. Hypoxic retinal Muller cells promote vascular permeability by HIF-1-dependent up-regulation of angiopoietin-like 4. *Proc Natl Acad Sci U S A.* 2013; **110**(36): E3425–34.

Yoshida T, Zhang H, Iwase T, Shen J, Semenza GL, Campochiaro PA. Digoxin inhibits retinal ischemia-induced HIF-1a expression and ocular neovascularization. *FASEB J.* 2010; **24**(6): 1759–67.

Yu AY, Frid MG, Shimoda LA, Wiener CM, Stenmark K, Semenza GL. Temporal, spatial, and oxygen-regulated expression of hypoxia-inducible factor 1 in the lung. *Am J Physiol.* 1998; **275**: L818–26.

Yu AY, Shimoda LA, Iyer NV, Huso DL, Sun X, McWilliams R, Beaty T, Sham JS, Wiener CM, Sylvester JT, Semenza GL. Impaired physiological responses to chronic hypoxia in mice partially deficient for hypoxia-inducible factor 1a. *J Clin Invest.* 1999; **103**(5): 691–6.

Yu F, White SB, Zhao Q, Lee FS. HIF-1a binding to VHL is regulated by stimulus-sensitive proline hydroxylation. *Proc Natl Acad Sci U S A.* 2001; **98**(17): 9630–5.

Yuan G, Khan SA, Luo W, Nanduri J, Semenza GL, Prabhakar NR. Hypoxia-inducible factor 1 mediates increased expression of NADPH oxidase-2 in response to intermittent hypoxia. *J Cell Physiol.* 2011; **226**(11): 2925–33.

Yuan G, Nanduri J, Khan S, Semenza GL, Prabhakar NR. Induction of HIF-1α expression by intermittent hypoxia: involvement of NADPH oxidase, Ca2+ signaling, prolyl hydroxylases, and mTOR. *J Cell Physiol.* 2008; **217**(3): 674–85.

Yuan G, Peng YJ, Reddy VD, Makarenko VV, Nanduri J, Khan SA, Garcia JA, Kumar GK, Semenza GL, Prabhakar NR. Mutual antagonism between hypoxia-inducible factors 1α and 2α regulates oxygen sensing and cardio-respiratory homeostasis. *Proc Natl Acad Sci U S A.* 2013; **110**(19): E1788–96.

Zeng M, Shen J, Liu Y, Lu LY, Ding K, Fortmann SD, Khan M, Wang J, Hackett SF, Semenza GL, Campochiaro PA. The HIF-1 antagonist acriflavine: visualization in retina and suppression of ocular neovascularization. *J Mol Med.* 2017; **95**(4): 417–29.

Zhang H, Bosch-Marce M, Shimoda LA, Tan YS, Baek JH, Wesley JB, Gonzalez FJ, Semenza GL. Mitochondrial autophagy is an HIF-1-dependent adaptive metabolic response to hypoxia. *J Biol Chem.* 2008; **283**(16): 10892–903.

Zhang H, Qian DZ, Tan YS, Lee K, Gao P, Ren YR, Rey S, Hammers H, Chang D, Pili R, Dang CV, Liu JO, Semenza GL. Digoxin and other cardiac glycosides inhibit HIF-1a synthesis and block tumor growth. *Proc Natl Acad Sci U S A.* 2008; **105**(50): 19579–86.

Zhang H, Wong CC, Wei H, Gilkes DM, Korangath P, Chaturvedi P, Schito L, Chen J, Krishnamachary B, Winnard PT Jr, Raman V, Zhen L, Mitzner WA, Sukumar S, Semenza GL. HIF-1-dependent expression of angiopoietin-like 4 and L1CAM mediates vascular metastasis of hypoxic breast cancer cells to the lungs. *Oncogene.* 2012; **31**(14): 1757–70.

Zhang X, Liu L, Wei X, Tan YS, Tong L, Chang R, Ghanamah MS, Reinblatt M, Marti GP, Harmon JW, Semenza GL. Impaired angiogenesis and mobilization of circulating angiogenic cells in HIF-1a heterozygous-null mice after burn wounding. *Wound Repair Regen.* 2010; **18**(2): 193–201.

Zhang X, Wei X, Liu L, Marti GP, Ghanamah MS, Arshad MJ, Strom L, Spence R, Jeng J, Milner S, Harmon JW, Semenza GL. Association of increasing burn severity in mice with delayed mobilization of circulating angiogenic cells. *Arch Surg.* 2010; **145**(3): 259–66.

Zhong H, De Marzo AM, Laughner E, Lim M, Hilton DA, Zagzag D, Buechler P, Isaacs WB, Semenza GL, Simons JW. Overexpression of hypoxia-inducible factor 1a in common human cancers and their metastases. *Cancer Res.* 1999; **59**(22): 5830–5.

Physiology or Medicine 2020

Harvey J. Alter, Michael Houghton and Charles M. Rice

"for the discovery of Hepatitis C virus"

The Nobel Prize in Physiology or Medicine, 2020

Presentation speech by Professor Gunilla Karlsson Hedestam of the Nobel Assembly at Karolinska Institutet.

Your Majesties, your Royal Highnesses, Esteemed Laureates, Ladies and Gentlemen,

Viruses have always been part of our existence. Through evolution, we have acquired versatile functions to handle invading pathogens. This has equipped us with a finely tuned and, in most cases, a highly effective immune system that protects us against daily threats. Different kinds of life forms, bacteria, plants and animals, and different species of these, have their own unique set of viral intruders that circulate within the given population. But new threats continuously arise. Viruses often cross species barriers to increase their host range and spread more effectively, which is their sole reason to exist.

While we cannot see viruses with the naked eye, we know that they come in many different shapes and forms, and they frequently take us by surprise. This year, the world is facing a viral pandemic that affects almost every aspect of our society. The expeditious identification of the causative virus, SARS CoV-2, was unprecedented and our understanding of the virus and the disease it causes is increasing by the day. Progress towards the goal of containing the pandemic is now being made.

Hosting this year's Nobel Prize ceremony in the absence of the Laureates is one of many consequences of the pandemic, but one which also highlights the significance of this year's Nobel Prize in Physiology or Medicine: the discovery of the Hepatitis C virus, a major human pathogen and the causative agent of a much more invisible pandemic.

The work performed by the Laureates spans several decades and illustrates the importance of scientific collaboration and systematic experimentation, using both traditional and emerging technologies. The starting point was the observation that an increasing number of individuals who had received blood transfusions developed chronic liver inflammation years later. The clinical course of the disease was different from that seen after infection with the known hepatitis viruses. This was a silent condition, which rarely gave symptoms in the initial phase, but resulted in progressive damage of the liver over time in most infected individuals.

Dr. Harvey Alter set out to investigate the new disease. Through carefully designed studies of samples from blood donors and individuals who developed this form hepatitis, he laid the puzzle. In 1978, Alter formally demonstrated that the disease was transmissible, and the race to identify the causative agent could start.

Around this time, molecular biology-based techniques had become increasingly available, but they were nowhere near as efficient as they are today. Dr. Michael Houghton and his team set out to clone the unknown virus using a novel screening technique. Almost against all odds, given the limitations of the techniques at the time, the screen was successful and resulted in the isolation of the Hepatitis C virus.

Dr. Charlie Rice immediately embarked on studies to define how the virus copies itself in our cells to make new infectious particles. By defining the details of this process, he could generate a virus clone that alone induced productive infection and liver disease. This provided the final evidence that the Hepatitis C virus caused the new form of blood-borne hepatitis.

The discovery of the Hepatitis C virus by this year's Laureates laid the foundation for our current understanding about how the virus survives in its niche during the long chronic phase of the infection, and how liver disease develops. And importantly, it led to the development of highly effective antiviral medicines that now cure the infection in almost all treated persons.

This year's Nobel Prize in Physiology or Medicine puts viruses at center stage. Viruses will continue to surprise us. They exploit different routes of transmission and are masters at adaptation. To be prepared for new viral threats, we need research such as that performed by this year's Laureates, and we need a deep appreciation of fundamental aspects of different types of viruses and their interactions with the host.

On behalf of the Nobel Assembly at Karolinska Institutet, it is my great privilege to convey to you our warmest congratulations.

Harvey J. Alter. © Nobel Prize Outreach AB. Photo: Joy Asico

Harvey J. Alter

Biography

I never had no Nobel dreams

I was born in 1935, a pivotal year in American history, although it is possible that my birth had nothing to do with these historical events. In 1935, the US was emerging from the Great Depression and the election of Franklin Roosevelt in 1932 instilled hope for a new deal. Hope was on the rise and a newborn such as myself had the opportunity to live in a freer, more progressive and more prosperous world. But for those with foresight, 1935's optimism was tempered by a growing fear of the rising tide of fascism in Europe. The Big Lie was to counter The Big Hope and by the time I was six, the US was at war with a Nazi regime that had already overrun most of continental Europe. Only England and Winston Churchill stood between my childhood and a direct attack on US soil.

I do not have many direct memories of my birth year since my brain cells were as immature as the rest of me. Indeed, I could not talk for that entire first year nor could I play a competitive game of Scrabble. However, at some subliminal level, it is possible that these opposing forces of hope and despair were imprinted on my developing psyche. In retrospect, I see 1935 as a harbinger of the future vicissitudes of my existence, the yin and yang of Harvey Alter.

Most of my conscious early memories were strongly influenced by World War II. After the Japanese surprise attack on Pearl Harbor, our isolationist, conservative Congress declared war on Germany and Japan barely in time to support an embattled Britain and its courageous people. However, we did not enter in time to prevent the annihilation of 6 million Jews. My people were decimated by a monstrous force beyond imagination in the 20th century. Not until I later saw films of piled skeletons did I realize the ravages that occurred as I grew up in pristine and protected Ridgewood Queens.

What I do remember of the war years was that my life was barely affected by it. I played ball in the streets and went to school just as before.

I was reminded of the war only by newsreels in the movie theaters and a rash of exciting war films where the Americans always won the critical battles. There was no TV or social media in those days so the news came only from radio and from newspapers that I did not read. Because of my young age and because there was some censorship of bad news, I didn't realize the truth and horror of war and was always certain America and the allies would ultimately win. This certainty abated fear. To me, the hardship of war was that I had to wait on line for bubble gum and was restricted to only two pieces, that we had to use black-out shades, that my favorite pink Spalding balls were in short supply because the rubber was being used by the military, that my father would leave the house at night to roam the streets as an air-raid warden and that we were encouraged to plant victory gardens which was difficult because the closest earth to our apartment was a mile away. One of the most exciting events of my war experience was that the FBI captured a German spy who was living in the apartment house next to ours, The FBI had picked up signals from his radio-transmitter. Who knows what the spy relayed to Germany about myself and my gang of friends? A second war memory was that I returned home from grammar school one day to find my mother crying. President Roosevelt had died and for my family, it was a time of profound mourning. In contrast, when I returned to the street in my predominantly German neighborhood, there were people cheering because Roosevelt was perceived as anti-German and most critically, a "Jew lover." My neighborhood was not a paean for democracy or emblem of liberalism, but the latent anti-semitism was not generally manifest.

As alluded to, I measured out my youth in pink Spalding balls, thrown on traffic laden city streets, hurled against walls, stoops and curbstones, hit with bats, racquets and shortened broomsticks, and constantly lost to gaping sewer drains or thrust on commercial rooftops lined in jagged glass to thwart retrieval. These balls were my lifeblood and I was seldom without one. A lost ball required immediate replacement which was done at Olsen's candy store where I would get a free chocolate egg-cream using Frequent Spalding Mile credits and where I would also buy a comic book. The ecstasy of buying a new ball and a new comic book was the equivalent of buying a new Lexus today. Harold Olsen was one of my two best friends and after school he had to run his father's candy store until it closed at 10:00 pm. I spent a lot of time there and was always amazed at his business skill even as a teenager, and his hard work and sacrifices bore fruit because 30 years later he was made president of the mid-west division of Macy's; a true Horatio Alger story.

I spent my first 17 years on one street of apartment houses and stoop-fronted duplexes. These were modest edifices in a lower middle-class neighborhood. My family, though not wealthy, were better off financially than most of our neighbors and sufficiently solvent to escape the sweltering city in the summer when I was young and then to send me to summer camp in my early teens. I was privileged in a modest way. Our apartment was on the second floor of a 5-story building. Directly below our apartment was a muscle building studio and a barbershop. Given the proximity of these establishments one would have thought that I would have grown up stronger and with more hair. Putnam Avenue, the street I lived on for 17 years, was nestled between two commercial avenues. Within 3 blocks of my house were 3 movie theaters one or the other of which I attended twice a week, a German delicatessen where I would get delicious ham sandwiches unbeknownst to my kosher mother, a kosher delicatessen where I loved hot pastrami sandwiches even better than ham, innumerable candy and magazine stores like Olsen's, a very good Chinese restaurant, a hobby shop where I spent endless hours selecting the next wood model to put together, my pediatrician, a synagogue and the rabbi who lived in the next apartment house and whom we had to assiduously avoid when driving on the sabbath. There were all manner of clothing stores, small food markets, butcher shops and even a live chicken market. When going to the butcher shop with my mother, I was intrigued by the butcher knives and the way they methodically sliced the fat off the choice cuts that my mother ordered from Abe the Butcher. For a brief period in my childhood I wanted to be a butcher, and I am now surprised that I was not drawn to being a surgeon. In any case, Ridgewood Queens was a thriving and bustling neighborhood of which I knew every inch. Hence, the family did not move for 25 years because there was nothing we were missing, unless, of course, you count luxury. I might still be there, if college did not intervene.

My parents were first generation Americans, their parents emigrating from Western Russia and Poland. Recently, when I had a gene profile, I was found to be 99.5% Ashkenazy Jew. We are still looking for the other 0.5% who disrupted the purity of my family tree. My parents were both good and kind people who showered me with love. My sister, Jackie, who was 5 years older, also treated me well though she often joked that she was an only child. Although my parents tried to be even-handed, I think that I got favored treatment because my sister challenged them more, but Jackie never seemed resentful of me. She died about 3 years ago with dementia and smoking-induced obstructive lung disease. By the time she reached age 75, our parents, her husband and her two children had all died and I became her sole caretaker, albeit with the substantial help of an excellent nursing home. She died in hospice at age 85, my mother died at age 79 and my father at 81. I somehow always expected to die at the same age as my father, but here I am writing this autobiography at age 86 so I am, as they say in the gambling world, living on house money. I am intrigued by Woody Allen's idea of achieving immortality, not by great accomplishments, but by not dying. For me, so far so good.

Despite their good natures, my parents had one major flaw... they didn't get along. Their personalities were diametrically opposed. My father, one of 9 children from the lower east side of Manhattan, was college educated, active in the business world, president of every organization he belonged to, expansive in his thinking, handsome, a fashion plate, interested in the arts, particularly opera, adventurous and flamboyant. My mother came from meager beginnings. Her father died early in her life and she was raised by my favorite grandmother who earned a living by working a push-cart in the Bronx and in old age was supported by my father. Whether because of this hard beginning or other psychological factors, my mother grew up as a fearful person, afraid to take risks, to try new things, to drive, to spend money, to fully engage in life. Hence, my parents often argued, one wanting to move forward, the other holding back. The arguments affected me deeply because to my young mind they portended divorce. I often wondered which parent I would be left with and how I would bear the shame because I had no other friends whose parents had separated, divorce being rare in those days. My fears were not justified, as they remained married for 60 years and between arguments had a reasonably good life together. Nonetheless, I was stuck in their polarization, trying not to take sides while admiring my father more but sympathizing with my mother because she played the role of the eternally injured party. Their dichotomy influenced my personality because I too became dichotomous, on the one side fearful and holding back and on the other trying to break out and be more expansive. While my parents did not separate, their dipole exists in me.

Because my parents were overly protective of me, they allowed my elementary school to place me in a specialized health class, the concept of which was that an extra carton of milk and a rest period after lunch would instill vigor and growth. This decision, over which I had no control, separated me from my usual classmates and curriculum and labelled me as different when all I wanted to be was the same. The Health class was for kids who were too thin, too fat, too lazy, too active or too anything but the school's definition of "normal." I was ashamed to be segregated in this way, but nonetheless adapted, as is my tendency. I made some nice friends among the more normal of the "abnormal" and developed a crush on Roxanne who became perhaps my first sort-of girlfriend and bestowed my first real kiss. Ms. Gunther, the only teacher in the class, deemed me the brightest in the class and had me skip an entire year of elementary school. This magnanimous gesture, however, had its downside as I graduated at age 13 and entered high-school as the smallest in the class. I could have used another year of extra milk.

Although I was likeable and had a "good personality" and was even funny, because I was small and shy and socially inept, despite Roxanne, high school was not a great time for me. I did pretty well scholastically, but was not an outstanding student, wrote for the school paper and magazine, won some minor honors and a national essay contest and most importantly, grew to a reasonable size. They were able to take my face off the milk cartons and send a good progress report to Ms. Gunther. Overall, my high school days were nothing like one sees in the movies; Ferris Bueller or Tom Cruise, I was not. So, I will give short shrift to my high-school years as they were less than memorable.

After visiting six small coed colleges in the Northeast, I selected the University of Rochester because it had a renowned medical school that I ultimately attended. I did not apply to Harvard or Yale, but thought they would call me anyway. Somehow, they did not. If you are listening, Harvard, it is still not too late to call. I got off to a rough start academically getting a D on my first essay, despite the fact that writing was my strong point. My first mid-term GPA was C+ and I saw my medical career careening into an abyss of uncertainty. High school had been too easy and my study habits were poorly founded. I then taught myself to study hard and my grades responded accordingly. I graduated cum laude, but not magna or summa. I was sort of a "cum plus laude." My subsequent achievements may have surpassed my innate intelligence through extra effort and perseverance. I remember a meeting with my college advisor at the end of my junior year where he told me that I had done much better than they had expected. Exceeding expectations may be the keynote of my life.

College represented my social transformation. I grew up fast emotionally in that first year at the University of Rochester. I did not join a fraternity as most freshmen did. Being an independent proved to be a great advantage, giving me some stature among those who were not accepted into or declined to join the fraternity system. I became a small big-man on campus, was managing editor of the school paper, was elected into the junior and senior honorary societies, made great friends and generally thrived. My social life burgeoned, my social skills accelerated and my shyness with women dissipated. I loved those four years, possibly the best in my life, and if I had one wish, it would be to relive them.

Toward the end of my senior year I received a letter that was the fulfillment of my life's dream, an acceptance letter to Rochester Medical School. Only at the birth of my children did I ever experience equal elation. My path was set, though it was yet to take several twists and turns. If there was one week in my life that stands out, it would be the week after finals of my senior year when I awaited graduation, had absolutely no responsibilities and had my medical school acceptance letter in my back pocket. I have been trying to recreate that pressure-less feeling for the past 60 years and fear I am running out of time.

Medical school was everything I had hoped for. I found almost every course interesting, even anatomy, and each succeeding year more interesting as the program became more clinical. We had only 70 students in a class and grades were not given out to decrease competitiveness. Only if you were doing poorly were grades revealed to you. Fortunately, I never knew. The medical curriculum was highly integrated with the psychiatric service headed by John Romano and George Engel, two giants in the field. Great emphasis was given to medical history-taking and to understanding the patient as a person as well as a disease. This approach has stayed with me and has enhanced my patient interactions. I have found that what patients want most from their doctor, in addition to competence, is his or her time and empathy. NIH has allowed me to spend quality time with patients and that has greatly enriched my clinical experience. I found it incredibly rewarding to see patients in the context of a study that had a beginning and an end that could be summarized in a publication. That was the path I ultimately chose but it was a very circuitous path.

In med school I was fascinated by many subjects, the first being pathology which I learned from Lowell Orbison, who wrote a classic text and later became Dean. I wanted to take a year out studying under Orbison but he told me that I would also have to do autopsies and that put a cadaverous chill on the deal. In my clinical rotations, I was intrigued by ophthalmology and how many medical conditions had diagnostic ocular manifestations. By my next rotation, I was on to something else, but I am now back to the eye as I have developed dry macular degeneration, but so far without visual consequences. I am now also fascinated by studies that are restoring vision through stem cell implants. Pediatrics was another attraction, but I realized that my forte was verbal communication and pediatrics required other skills; also, I could not emotionally handle the death of infants and children. Thus, I navigated to internal medicine and particularly hematology. I liked the process of differential diagnosis that internal medicine demanded and still love to read diagnostic dilemmas in the New England Journal and the Health Sections of the New York Times and Washington Post. My big diagnostic breakthrough came as an intern when I had a patient with unexplained acute renal failure. No one could discern the cause until an eavesdropping patient in the same room showed me an article from Readers Digest that described a case of renal failure due to carbon tetrachloride (CCL4) toxicity. It was well known that CCL4 caused hepatotoxicity when ingested, but this lay article noted that it caused renal disease when inhaled. On further questioning of the patient, who was a long-haul truck driver, I found that he slept in the cab of his truck which housed a fire extinguisher in close proximity to his bed.

I obtained the keys to his truck that was parked about a mile from the hospital. I then embarked on my first epidemiologic investigation. I found the truck and the sleeping cabin and the extinguisher just above the bed. Reading the metal tag, I was excited to find that the fire extinguisher was fueled by CCL4 and importantly that it was empty, the fumes already residing in my patient. Case closed! I was a medical hero for a brief time and was asked to give Grand Rounds where I was both the case presenter and the discussant, a rare event at the intern level. I have subsequently given Grand Rounds many times in my life, but this one was the most scary, most exciting and most memorable. The patient recovered fully and was prescribed a new fire extinguisher. The patient who clued me in to the Reader's Digest article shared in my fleeting glory.

I seemed to have segued into my internship year at Strong Memorial Hospital in Rochester, NY. This was another pivotal year in my medical growth curve. I learned very rapidly how little I knew of real medicine coming out of medical school. Learning the Henderson-Hasselbalch Equation in physiology was quite different than managing a patient with renal failure and electrolyte imbalance. Memorizing the Krebs cycle in biochemistry was quite remote from managing diabetic ketoacidosis. Internship engendered a steep learning curve. It was a sink or swim environment. Once I caught on, I felt empowered and enjoyed that year more than any other in my training despite the long hours and intense pressure. Fortunately, Strong Memorial had a humane night-float system so that after 10:00 pm one could pass on new admissions to the night float. This did not mean that my day ended at 10 pm, because one still had to clean up the day residue and then write up cases admitted before 10. Generally, I would be writing notes to 2 or 3:00 am, often cat-napping as I did so. Many years later I returned to these wards and while flipping charts, by chance, found one of my old notes which was characterized by a descending line on the paper where I had fallen asleep mid-note. Unmarried interns lived in the staff house, which was a cross between a dormitory and a bordello, as various nurses slipped in and out of the rooms of the interns they were dating. Some things were even more important than sleep. Having free room and board at that time was essential because the annual intern's salary was $600. This amount seemed to suffice because there was no time to spend it on entertainment or fine meals and no need for clothes beyond the white pants and white jacket that were provided. Only those who smoked were short on cash. My main source of entertainment, beyond that alluded to above, was to watch Robert Stack in the Untouchables, a weekly episode that I prioritized above all else. I hope that no patients were lost as a consequence of my addiction to this FBI drama. My internship year was lightened by a wonderful relationship with a woman whom I will call Sue, because that was her name. Sue was beautiful, sweet and uncomplicated and a person I might have married, if I had not been brainwashed since childhood that the worst thing I could do to my parents was to marry out of my religion. It was a Hebraic curse placed on Jewish children and it was powerful. As you will see, I later broke through this guilt-driven curse, but that was too late for Sue and myself. It is a chapter in my life whose end I will never know.

I continued my tenure at Strong Memorial by entering their Internal Medicine residency. It was easier to be a resident because the program was intern-centric, but I enjoyed being the hands-on intern more than the advisory resident. Near the beginning of my residency, I received a letter that was to dictate the future course of my life ... my career path, my wife-to-be and consequently the children I would have. It was 1961 and the letter, which I still have, was from my draft board and began with the infamous word, "Greetings." I had been drafted into the army and was to report to Fort Dix, New Jersey by Nov. 30 of that year. They even attached a subway token to help me get there, though the subway from upstate New York to New Jersey, encompassing about 400 miles, had not yet been built. I'm still holding that token should it ever come to pass. I can't overstate the pivotal nature of that invitation from the government. Had I gone into the army, I probably would have been sent to Germany for the Berlin crisis that was the basis for calling up previously deferred physicians at that time. After 2 years in the army, I would definitely have gone into clinical practice since that had been my long-term goal and since I would have had no research experience to direct me toward academia. The alternative was that I could spend my military time at NIH where I had already been accepted, but not yet commissioned in the Public Health Service (PHS). This critical fork in the road had my future teetering on divergent paths. Either might have been fine, but in retrospect, the path I ultimately took to NIH had amazing outcomes I could never have anticipated. Through multiple conversations with the PHS, I learned that if I could secure a position at NIH and report there before my draft date, the PHS had authority over the draft. With the help of Scott Swisher, the head of Rochester hematology, I found a position at the Division of Biologic Standards (DBS), the forerunner of the FDA and the division responsible for the Clinical Center Blood Bank. I reported to DBS on Nov. 27, 1961, 3 days before my scheduled visit to Fort Dix. Fortunately, DBS did not require that I do push-ups every morning. DBS had an active research program as well as regulatory responsibilities and there I learned about plasma fractionation and how to do column separations. Within 6 months, the blood bank was transferred to the Clinical Center Pathology Department and I was transferred with it. The blood bank was a miniaturization of the one that now exists and had a staff of only about 10. I, as a first-year fellow, was a de facto assistant Chief and the second year fellow was Chief. When it was my turn to become Chief, they realized the weakness of this hierarchy and hired Paul Schmidt as permanent Chief. Paul began to grow the department and to gain its independence from the Pathology Department. In my long tenure in the blood bank, I have had only three bosses, Paul Schmidt, Paul Holland and Harvey Klein. Each was instrumental in advancing my career and each expanded the Department of Transfusion Medicine to its current stature as perhaps the best and most innovative in the world with a current staff of about 150.

As a first-year blood bank fellow, I had to come up with a research project. I had no research mentor and for reasons I cannot now remember, I decided to study whether blood recipients might have transfusion reactions because they harbored an antibody against a serum protein distinct from their own. I had read about polymorphisms in human albumin and haptoglobin, and this gave the study some scientific underpinnings. To test for these immune reactions, I cut circular wells into agar plates by the double immunodiffusion method of Ouchterlony and allowed serum from a multiply transfused patient to diffuse into the agar toward normal donor serum in surrounding wells. An immune reaction would be indicated by a curved precipitin line. Although agar plates continued to pile up, publications did not. One day, Richard Aster, a soon-to-be famous platelet immunologist, told me that he heard an interesting lecture by a geneticist named Baruch Blumberg who was doing experiments very similar to my own. I went to see Blumberg soon thereafter and found him to be interesting, quite chatty and very collegial. We immediately established a collaboration to, so to speak, join wells. Blumberg had already discovered polymorphisms in human beta-lipoproteins and these precipitin lines characteristically stained blue for their lipid content. One day, I observed a precipitin line that did not stain blue, but stained red when counterstained with azocarmine for protein. The antigen detected became known as the Australia antigen and was later shown by Blumberg to be the surface coating of the hepatitis B virus as described in my current Nobel lecture and in the 1976 Nobel lecture by Blumberg. This serendipitous finding of the Australia antigen was my first introduction to the hepatitis field and served as my conduit to non-A, non-B hepatitis, now known as hepatitis C. As yet another example of the serendipity of my life, had Richard Aster not heard the Blumberg lecture and told me about it, I probably would not have found and published about the Australia antigen and perhaps not wended my way back to NIH to complete my scientific history. The rest of my long NIH career is chronicled in my accompanying lecture. I can summarize briefly to say that none of what has brought me to this Nobel Prize would have happened had I not been at NIH and its nurturing, supportive, intellectually stimulating and highly collaborative environment. It was, and is, a research Valhalla and I owe everything I have achieved to this tower of hope and innovation.

Full bent on entering clinical practice, after completing my NIH fellowship, I did a second- year medical residency at the University of Washington Hospitals in Seattle and then did a hematology fellowship at Georgetown University under a wonderful mentor and life-long friend, Charlie Rath. Rath taught me as much about life as medicine and both have been invaluable. Charlie died with his mind and humor intact at age 93. I was honored to give his eulogy. At the end of my heme fellowship, I applied for a coveted group practice position in Washington DC. I was not offered the position, ostensibly because they needed a cardiologist, and was quite disheartened. Rath said "Why not stay here at Georgetown, you can always go into practice later" and indeed he doubled my salary from $6000 to $12,000 a year. It sounded good at the time. This was my introduction to academia, where I seemed to fit well and where I soon lost my interest in entering clinical practice. My career path had changed for the better as the result of failure to be selected. Another yin and yang of my trajectory.

Although, I liked my time in academia, I was becoming overwhelmed by it because I was heavily involved in teaching, had a high case load of hematology consults and private patients and was nominally head of hematology research. I found these three academic hats difficult to manage so that when I got a call from Paul Holland in 1969 to come back to NIH and resume my research on the Australia antigen and transfusion-associated hepatitis, I jumped on it. Hence, that serendipitous precipitin line, a Blumberg lecture that I heard only second-hand, and a failed job opportunity forged the rest of my working life and brought me to this Nobel Prize. Not even my mother could have predicted this and I'm so sad that she and my father were not alive to see me accept this award, albeit by Covid-dictated zoom.

I have been remiss not to write about my immediate family. I met my first wife, Barbara, at NIH. We had a seemingly happy marriage, but after 12 years it dissolved, in part because of the demands of my work and my compulsive response to those demands. After a period of deep remorse, I recovered my equilibrium and as a born-again bachelor I emerged into a phase of professional productivity and social awakening. During the early 80s, my career was expanding and I was being invited to speak at wonderful locations around the world that I never thought I would see. Those were heady times as HCV and Har-V were both coming into full blossom. From that point forward, I measured out my life in Power Point slides. In 1984, I met Diane Dowling and after an interminably long courtship, which will go unexplained, we married in 1995. It has been a wonderful relationship and seems to get better with time. As a psychotherapist,

Diane had a lot to work on within her marriage, namely me, and she has made me a better husband, a better parent and a better person. I am very grateful for her love, support and guidance. I had two children with Barbara, Mark and Stacey, of whom I am incredibly proud and who continue to give me joy into old age. Diane brought me two step-daughters, Lydia and Erinn, whom I consider daughters without the step. Wonderful and accomplished ladies. Each of these children married and have brought me 9 grandchildren that now age 11–21. My family cup runneth over.

My essay also runneth over, but I have one more topic to cover. Now, in old age, I measure out my life in losses. It is the curse of a long and healthy life that those closest to you disappear, one by one. First parents, aunts and uncles, then siblings and now first cousins. I have watched near two generations of the Alter clan depart; near 50 persons linked by blood, but no longer able to bleed, lying in scattered cemetery plots in Queens and remembered only sporadically by placing a small stone upon their grave; multiple lifetimes reduced to chemicals and molecules that no longer have cohesive function. And as I age, still in good health, my friends also begin to leave me one by one. I recite or listen to their eulogies, I learn new things about them that I wish I had known before, I console their survivors, but I find no consolation in their loss. I miss these friends deeply and wonder what friendship or what love I might lose next. The siphoning of other lives from my small residual well of being is hard to endure, but I do because this is the inexorable cycle of life. I am consoled by the fact that when my time comes, I can say these were my family and my friends and I was so lucky to have had them enrich my life. My last hope is that if there is a heaven, it will be filled with a limitless supply of pink Spalding balls.

I NEVER HAD NO NOBEL DREAMS
When I grew up in Ridgewood Queens
I never had no Nobel dreams
My goal was to hit the ball between the seams
To perhaps be a hero on one of my teams
To not have mom detect my schemes

High-school was unauspicious
I mixed with the mob, did nothing suspicious
I was small for my age and basically shy
Could have been the model for Catcher in the Rye
I was not voted most likely to succeed
But there was already a hint, my hair would recede
When I graduated high school, I gave no speeches

There was no sign that one day my deeds would exceed
 that of my teachers
Nothing about me was the fodder for memes
And certainly, I never had no Nobel dreams

In my 20's, it was medical school
Reading textbooks by the reams
Studying cadavers splayed at the seams
From massive texts, I was sucking the nectars
Studying all night and sleeping through lectures
At Rochester Med great teachers abounded
Everything was interesting, my career choices compounded
But of Nobel thoughts
I had naughts

Then came marriage and baby screams
My first house secure in its beams
I was ready to take on whatever life deems
Children growing
My career flowing
Lots of patients, lots of teaching
Deep research seemed beyond my reaching
Nobel aspirations, no one was preaching

Then an aboriginal antigen I had observed in my youth
Was linked to hepatitis B by Blumberg the sleuth
Of hepatitis viruses we were discerning the truth
For me a chance to return to my NIH roots
Non-A, non-B arose from long prospective studies
I tracked this agent along with bright NIH buddies
And with Chiron's cloning of C, there came the safest of bloodies
With this my career advanced
But Nobel dreams I never chanced

And now my non-Nobel dream has unfolded
With new credentials I'm suddenly embolded
A thousand congratulatory emails my computer uploaded
But for Nobel aspirations, I was not truly molded
I expect any day, Karolinska will have annulled it
And I will be the first Laureate to have his medal ungolded
To have the King of Sweden say, "from Alter we best hold it;
Get the medal back before he has sold it."

This award has given my life a massive shake-up
My heart and my mind are having a break-up
Strange forces are scrambling my genetic make-up
Though the leaves of my life, age will soon rake-up
I'm happy to be where I am because in life there's no make-up
From this Nobel dream, I'm afraid to wake up!

Hepatitis C Virus: From Hippocrates to Cure

Nobel Lecture, December 7, 2020 by
Harvey J. Alter
National Institutes of Health, Bethesda, MD, USA.

I had a vision, but not a dream. My vision was to better understand Transfusion-Associated Hepatitis (TAH) and to find ways to ameliorate this potentially devastating consequence of blood transfusion. I did not dream that this would lead to the discovery of a new viral agent, the near elimination of TAH and the potential to eradicate a virus chronically infecting 70 million persons world-wide. Certainly, I never dreamed of a Nobel Prize. Instead the Prize emerged, not from a eureka moment, but from a series of incremental observations that, together with the more basic discoveries of my co-recipients, Michael Houghton and Charles Rice, built a compelling story that in summation made my non-dream come true. It is the beginning of that story that I will relate in this lecture.

The hepatitis story actually began about 2000 years ago, when Hippocrates observed patients turning yellow, a condition he termed ikterus, and others that had hardening of the liver that he termed kirros. He considered these conditions the consequence of an imbalance in the four humors, in this case, an excess of yellow bile. While this hypothesis did not ultimately hold up to peer review, it emphasized Hippocrates great powers of observation and established him as the Father of Medicine. In truth, Hippocrates was making a profound statement, namely that disease was not the result of divine punishment, but rather the effect of external elements that perturbed the bodies delicate balance. He did not know that some of those external elements were infectious agents nor that icterus could be caused by a virus, but he made the critical distinction between divine intervention and earthly perturbations.

Nothing much else was learned in the realm of hepatitis for the next 2000 years except that there were recurrent wars and in every one, large numbers of combatants turned yellow and developed the classic symptoms of what we now call viral hepatitis. Yellow bile continued to be out of balance for centuries. Before any virus was observed and before any virus specific assays had been developed, epidemiologic investigations (1) discerned that there were two patterns of hepatitis, one highly infectious and transmitted by contaminated water and food, designated hepatitis A, and the other less contagious, primarily transmitted by sexual or percutaneous exposure and of longer incubation that was designated hepatitis B.

While clinical observations were numerous and liver enzyme tests and liver biopsies established hepatic inflammation, no specific viral marker emerged in those 2000 years. The breakthrough came serendipitously in 1962 when as an NIH fellow, I collaborated with Baruch Blumberg and, while looking for lipoprotein polymorphisms using Ouchterlony gel diffusion, we noted a precipitin line in agar that did not take up the usual lipid stain, but instead stained red for protein. (2) That immune-precipitate represented a reaction between the serum of a multiply-transfused patient with hemophilia and that of an Australian aborigine. This aboriginal serum was selected from a large repository of samples from indigenous populations that Blumberg had collected and stored and, by chance, was being tested that day. That antigen, originally called the "Red Antigen" for its staining characteristics was later renamed the Australia antigen in deference to the practice of naming new hemoglobin variants after the country or city of origin of the designated individual. The ultimate linking of that antigen to the hepatitis B virus is a fascinating story that resulted in a Nobel Prize for the late Baruch Blumberg and has been chronicled in the Nobel Lectures of 1976. The relevance to my treatise herein is that the Australia antigen, later called the hepatitis B surface antigen (HBsAg), provided the first serum marker for any hepatitis virus and laid the foundation for the events that ultimately led to the discovery of hepatitis C.

In the late 1960s, an NIH prospective study was initiated by Robert Purcell, Paul Holland and Paul Schmidt. I assumed responsibility for that study in 1970. The study aims were straightforward and in essence were to determine the incidence and causes of TAH and, if possible, to identify interventions that could reduce hepatitis incidence. To achieve this goal, transfused patients undergoing open heart surgery were enrolled and sampled weekly or bi-weekly for three months and then monthly for an additional three months. This population of cardiac surgery patients was chosen because they were heavily transfused, rarely had confounding liver diseases and were not immunosuppressed by disease or medication. Since there were no specific hepatitis viral markers when the study was initiated, hepatitis was diagnosed on the basis of liver enzyme abnormalities, namely an elevation of alanine amino transferase (ALT) to two and a half times the upper limit of normal (ULN) followed one or more weeks later by a second elevation at least two times ULN. Other measurements included aspartate amino transferase (AST), bilirubin and alkaline phosphatase. Non-viral causes of ALT, AST or bilirubin elevations were excluded to the extent possible. Importantly, all collected samples were stored frozen for later analyses. In addition, donor samples were similarly stored for donor-recipient linkage in patients who developed hepatitis in temporal relation to their blood transfusion.

Initial findings were both striking and surprising (Fig. 1). Approximately one-third of patients undergoing open heart surgery developed hepatitis based on the ALT elevations as defined above. This inordinately high incidence was predicated on three factors. First, the patients received large volumes of blood, averaging 17 units per case, because the heart lung machine used at the time required seven units of blood simply for priming. Second, the majority of cases were asymptomatic and occurred after hospital discharge. Such cases would not have been recognized if prospective follow-up post-discharge had not been in place. The third critical element for the inordinately high incidence proved to be the source of donor blood. Prior to 1970, a large proportion of transfused

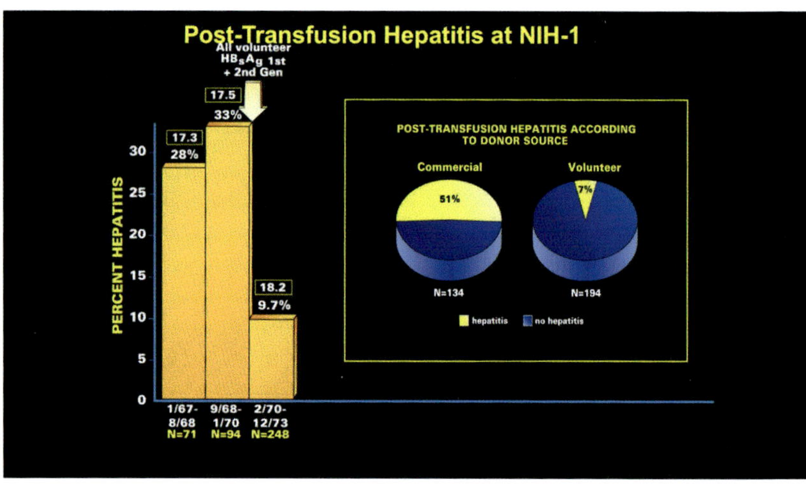

Figure 1. Bar graph showing early years of the prospective study. Rates of hepatitis exceeded 30% prior to 1970. Based on demonstrated risk of commercial donor blood (right figure), transition to an all-volunteer donor supply and introduction of hepatitis B virus (HBV) donor screening was instituted in 1970, resulting in a 70% reduction in hepatitis incidence.

blood at NIH was obtained from commercial sources. A study led by John Walsh (3) showed that if a patient received at least one unit of paid-donor blood, the risk of TAH was 51% whereas among those who received solely volunteer donor blood, only 7% developed hepatitis, even though the volume of blood received was similar. The use of paid donor blood at that time was predicated on limited supply.

In 1970, with these data in hand, the NIH blood bank was able to transition to an all-volunteer blood supply. Simultaneously, we introduced an unlicensed, home-brew assay for hepatitis B surface antigen. These combined measures (Fig. 1) resulted in a dramatic 70% reduction in TAH incidence to a new level of approximately 10%. (4) While we introduced two interventions simultaneously, extrapolation indicated that the major determinant of the marked decline in incidence was the conversion to an all-volunteer donor source. Indeed, nothing done since that time, including specific viral testing, has had such a dramatic impact.

In 1973, sensitive third generation assays for HBsAg became available and we tested stored samples from identified TAH cases. This revealed (Fig. 1) that only 25% of cases were related to hepatitis B virus (HBV) infection, leaving causation of the non-B cases yet to be explained. A breakthrough occurred in 1975 when Feinstone, Kapikian and Purcell at NIH discovered the hepatitis A virus using immune electron microscopy (5). We immediately sent our non-B case samples to their laboratory where testing revealed that not a single non-B hepatitis case was due to HAV infection. It was then, in a brilliant execution of deductive reasoning, that we decided that if these cases were not due to hepatitis A or hepatitis B infections, that we would call them non-A, non-B hepatitis (NANBH). (6) Actually, there was some rationale for this obtuse terminology in that we had not yet proven that we were dealing with a transmissible virus and, if so, how many there might be. Thus, we demurred on using "C", the next letter in the hepatitis alphabet.

The next critical step was to perform transmission studies using the chimpanzee model. We are no longer able to use chimpanzees in US government-sponsored research, but at the time they were an invaluable model and some suggest that a chimpanzee should have been the fourth recipient of this HCV-based Nobel Prize. Chimps were an ideal model for NANBH in that they were readily infected, developed elevated but low-level ALT levels, did not manifest any clinical illness or deleterious effects of the study and often became chronic asymptomatic carriers. Thus, their infection closely mimicked that of the majority of NANBH/HCV infected patients. In our initial chimpanzee studies (7), we were able to transmit hepatitis by inoculating only one milliliter of plasma, not only from patients with acute or chronic NANBH, but importantly, also from asymptomatic blood donors who had been implicated in human hepatitis

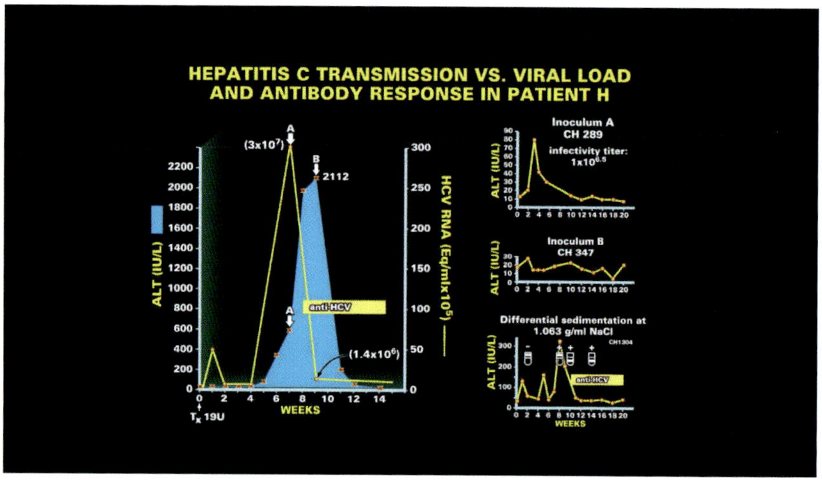

Figure 2. Graph depicting ALT elevations (blue) and HCV RNA levels (yellow line) in patient H. ALT levels began to rise 5 weeks post-transfusion and reached a peak of 2112 IU/ml. HCV RNA levels (yellow line) increased prior to ALT elevations and reached a peak of 3x10e7 copies/ml. A 500 ml plasmapheresis unit was obtained at point A and 1ml was shown to be infectious in a chimpanzee (upper panel right). Plasma obtained at the peak of ALT (point B) was not infectious in a second chimp (right middle panel) presumably because viral load was rapidly falling and residual virus was immuno-complexed (right lower panel).

transmission. Tabor and coworkers at the Food and Drug Administration (FDA) (8) performed parallel transmission studies and my primary collaborator, Robert Purcell, serially passaged NANBH from chimpanzee to chimpanzee. Hence, the presence of a blood transmissible NANB agent was firmly established in this animal model.

In 1977, the most important patient in my career came to NIH and was enrolled in our prospective study. The patient, Mr. H, had trailblazing as his avocation. One day, while marking a mountain trial he collapsed from a heart attack. His wife, who accompanied him, performed CPR and brought him back to life and ultimately to NIH where he was found to have triple vessel coronary disease and underwent one of the first coronary bypass surgeries. Six weeks after surgery, his ALT levels began to rise rapidly concomitant with classic symptoms of hepatitis and the onset of jaundice (Fig. 2). It had been my goal to obtain a large volume of blood from a patient in the early phase of acute NANBH, at which time the viral load was presumed to be at a high level. Thus, we obtained 500ml of plasma by apheresis on the ascending limb of his ALT curve. As seen in Figure 2, that plasma transmitted hepatis to a chimpanzee (upper panel on right). Subsequently, his ALT level continued to rise to a peak of 2112 IU/ml or 50 times the upper limit of normal, at which time we obtained a second apheresis unit. Surprisingly, that unit, obtained at point B, was not infectious in a second chimpanzee (middle right panel). In retrospect, we believe that unit was not infectious because the viral RNA level was falling precipitously and because by that time antibody had appeared that was complexing the virus and making it less infectious. Complexing was confirmed by ultracentrifugation as shown in the lower right panel. Purcell then took the plasma from point A, henceforth referred to as the H77 inoculum, and performed titration studies based on infectivity in chimpanzees. H77 was shown to have a titer of 10e6.5 chimp infectious doses per ml. Interestingly, this was almost identical to the HCV RNA titer of 10e7 copies/ml found in the patient sample when PCR subsequently became available. Hence, at this point we had two critical elements for further study, namely an infectious inoculum of known titer and a reliable animal model.

Armed with these critical elements, Steve Feinstone (9) performed chloroform extraction studies and showed that infectivity was abrogated by chloroform treatment whereas a mock chloroform extraction resulted in hepatitis transmission to a second chimp. This study implied that the NANBH agent had essential lipid in its envelope, narrowing the spectrum of potential viral etiologies. Le Fang He (10) in the Purcell lab then performed filtration studies wherein the H77 inoculum was passed through various size filters and the filtrates inoculated into chimps to assess residual infectivity. These experiments revealed that the NANBH virus was between 30 and 60 nanometers in diameter. Combining these studies, one could deduce that the NANB agent was small and lipid encapsulated, narrowing the taxonomic possibilities to that of a small RNA virus in the Togaviridae family, a hepatitis B-like agent or a new class of viral agents. We already had considerable evidence that NANBH was not related to the hepatitis B Hepadna family. My recollection is that Dan Bradley at the Centers for Disease Control (CDC) was the first to postulate that the NANBH virus was most likely a flavivirus, as it later proved to be.

While continuing to pursue the elusive direct assay for NANBH virus detection, we examined in greater detail the clinical consequences of chronic NANBH infection in collaboration with the NIH's National Institute of Diabetes and Digestive and Kidney Diseases (NIDDK) liver disease research branch, then under the leadership of Jay Hoofnagle. This collaboration continued for decades and included Adrian DiBesceglie, Marc Ghany, Jake Liang and a cadre of hepatology fellows, indeed too numerous to name individually, many of whom are now the current generation of leaders in the hepatology world. The key to this long collaboration was the accession of liver biopsies from chronically infected patients, most of whom were asymptomatic. As shown in Figure 3, initial biopsy of 39 individuals revealed that the majority had only mild histologic lesions, but that 10% already had cirrhosis and 13% had what was then termed severe

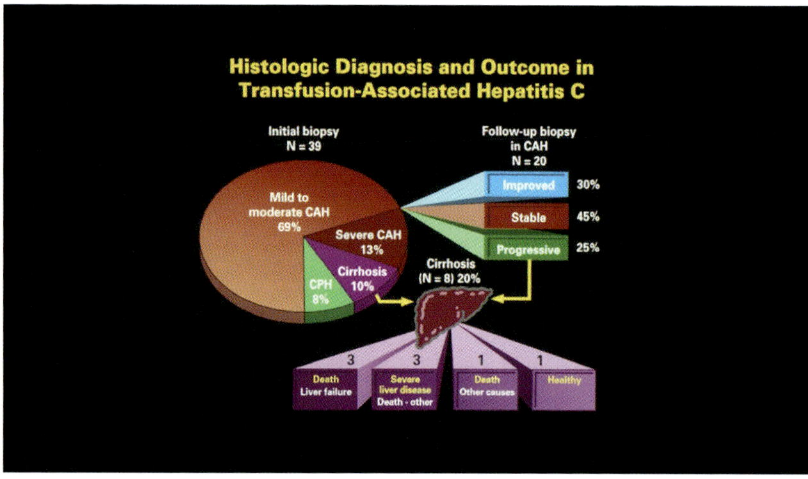

Figure 3. Liver biopsies were obtained from 39 patients. On initial biopsy, most had only mild to moderate inflammation, but 13% had chronic active hepatitis (CAH) and 10% had cirrhosis. Follow-up biopsy showed additional cases of cirrhosis such that overall, 8 of 39 (20%) had severe fibrosis progression. 3 died of liver failure and 3 others had severe liver disease when they died of their underlying heart disease.

chronic active hepatitis (CAH). Repeat liver biopsy, after varying intervals, in those with CAH showed that 25% progressed to cirrhosis while under observation resulting in a cumulative incidence of 20% (8 of 39) who developed severe fibrosis progression, terminating in cirrhosis. This 20% incidence of cirrhosis in chronic NANB/HCV hepatitis has withstood the test of time though the final number may be closer to 30%. Among the eight patients with cirrhosis, 3 died of liver failure and 3 others had severe liver disease when they died from complications of the heart disease that led to their initial surgery. This biopsy study led by Marvin Berman (11) dispelled the notion that NANBH was simply a benign transaminitis and proved that this primarily asymptomatic infection could lead to cirrhosis and death. From that point forward, the intensity of global study of this agent and its disease accelerated exponentially.

As we continued to prospectively follow the incidence of TAH, we found that the rate had declined from near 10% in 1970 to about 6% in 1985. The lowered incidence was probably related to the decreased volume of blood transfused, declining from an average of 18 units per patient in 1970–1973 to an average of 9 units per patient in the 1980s. During this long interval, we attempted several interventions to further decrease incidence, including ALT testing of blood donors, but this had no discernible impact. When HIV testing of donors was instituted in 1985, we thought this might serve as a surrogate marker for hepatitis carriers, but it did not do so because HIV at that time was transmitted primarily by sexual contact while NANBH was almost exclusively transmitted by percutaneous exposure. In the mid-80s we performed a retrospective analysis of this prospective cohort which suggested that donor screening for antibodies to hepatitis B core antigen (anti-HBc) might serve as a surrogate for the NANBH virus carrier state since HBV and NANB had overlapping routes of transmission. This retrospective study (12) and another independent study by the Transfusion Transmitted Virus Study Group (TTVS) (13) predicted that the introduction of anti-HBc donor screening might reduce the incidence of TAH by 25–30%. Hence, in 1987 we instituted routine donor screening for anti-HBc and indeed observed the predicted decline in TAH incidence to a level of 4% by 1989.

While the clinical severity of NANBH had become increasingly evident over the interval from 1975–1985, no serologic, enzymatic, radio-immunologic or early molecular method led to a specific NANBH assay or further elucidated the nature of the agent. It was then that I wrote a poem of frustration titled "I Can't See The Forest for the HBAgs":

I think that I shall never see
This virus called non-A, non-B
A virus I cannot deliver
And yet I know it's in the liver
A virus that we often blame,
But which exists alone by name
No antigen or DNA
No little test to mark its way.
A virus which in our confusion
Has forced us into mass collusion
To make exist just by exclusion
But is it real or an illusion?
O' Great Liver in the sky
Show us where and tell us why
Send us thoughts that will inspire us
Let us see this elusive virus
If we don't publish soon
They're going to fire us
Let us find this little beastie
Give us a sign ---- a star from the Easty
Thus, today, our quest we will begin
For this agent that plagues us like original sin
And perhaps someday, we all will agree
That indeed there exists a non-A, non-B

I do not know if this poem influenced Michael Houghton and his research group at the Chiron Corporation, but shortly after I recited it at a national meeting, Chiron announced that after a 6-year effort they had cloned the NANBH agent and called it hepatitis C virus (HCV). The description of the cloning experiments will be provided in a separate lecture by my Nobel co-recipient, Michael Houghton. This was a monumental breakthrough, utilizing a then novel cloning technique involving a gt-11 expression vector and immune-recognition of expressed antigens (14). Houghton assumed that antibody would be present in chronic carriers and/or recovered subjects even though such antibodies had not been detected in the previous decade. Although various accounts have been given, it is believed that 6 million clones were immune-screened before a single clone tested positive. By subcloning, the Chiron team, and principally Qui-Lim Choo, identified a genetic sequence that allowed them to express an antigen that then led to an antibody assay developed by George Kuo (15).

Subsequently, I received a call from Kuo requesting that I send him samples from what was called the Alter Non-A, Non-B panel. George implied that Chiron had developed an antibody test for NANBH virus carriers that he wished to validate with the notorious panel that had stymied 19 other laboratories claiming a specific NANBH assay. The panel was small, but not easy to decipher because every sample was present in duplicate and every duplicate was in a random position. The positive samples were all proven infectious in chimpanzee transmission studies and the negative controls were from blood donors who had donated at least 10 times and had never been implicated in hepatitis transmission. Kuo returned his test results promptly, but I was slow to analyze them because I expected another false claim. To my surprise and delight, Chiron broke the code perfectly, detecting 3 patients with chronic NANBH in 6 random positions and 2 implicated donors in 4 random positions (Fig. 4). They failed to detect 2 patients with acute hepatitis because they were testing for antibody rather than virus. Later samples from these patients tested positive in the Chiron antibody assay. Importantly, all 7 negative controls tested negative in 14 random positions.

With the Chiron Anti-HCV assay in hand, I then tested 15 of our most definitive NANBH cases and found that all 15 had been anti-HCV negative in their pre-transfusion sample and had seroconverted in post-transfusion samples (16). It was then evident that the NANBH virus and HCV were identical, as later confirmed by molecular sequencing. Further, in testing donors to 25 TAH cases, we detected an anti-HCV positive donor in 80% by a first-generation antibody assay and 88% by a more sensitive second-generation assay (Fig. 4). It was then evident that the introduction of anti-HCV testing to routine donor screening might prevent nearly 90%

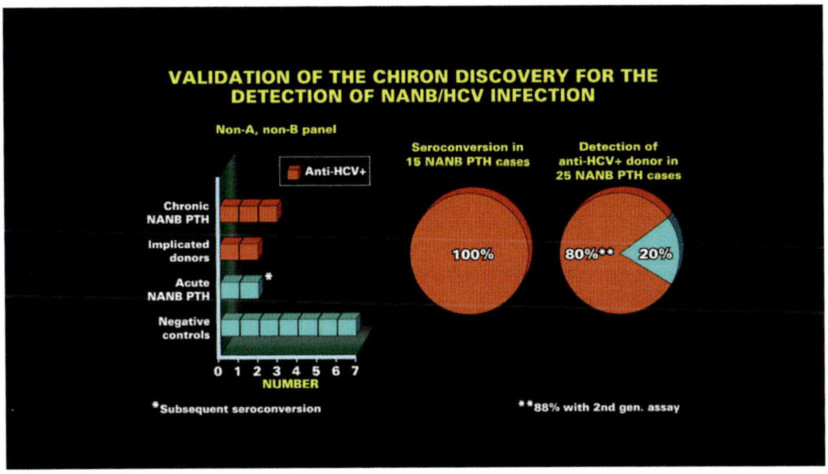

Figure 4. Testing of a coded NANBH panel by the Chiron anti-HCV antibody assay. All coded samples were in duplicate that were placed in random positions. The assay detected 3 chronic carriers and 2 implicated donors in all duplicate samples. Two patients with acute NANBH tested negative for antibody during the acute phase, but later seroconverted to antibody positive. 7 pedigreed negative controls were negative in all duplicate samples. 15 cases of NANBH in the NIH prospective study were all anti-HCV negative pre-transfusion and Anti-HCV positive post-transfusion. Donors to 25 NANBH cases were tested and an antibody positive donor was found in 80% of cases by initial testing and in 88% by a more sensitive 2nd generation anti-HCV assay. This predicted that anti-HCV donor screening might pr event nearly 90% of post-transfusion hepatitis (PTH).

of transfusion-associated hepatitis. We began such testing in 1990, and it was FDA-mandated nationally by 1992. Real life experience confirmed our prediction (Fig. 5). After the implementation of routine anti-HCV donor screening in 1990, hepatitis C incidence fell to 0.6% and then to zero when a more sensitive second-generation assay was introduced in 1992. Interestingly, a previously postulated, but never delineated, non-A, non-B, non-C form of hepatitis also disappeared after testing for HCV. I now believe that the relatively small number of presumed "Non-ABC" cases with generally mild post-transfusion ALT elevations were not representative of viral infection, but rather were the result of intercurrent events related to surgery, medications, or non-alcoholic fatty liver disease.

It can be estimated by extrapolating the incidence figures in our prospective study to the US population as a whole that blood transfusion may have transmitted 4.8 million cases of hepatitis in the two decades from 1970 to 1990 and, conversely, that an additional 2.4 million cases may have been prevented by HCV testing in the two decades from 1990 to 2010. Most of the transmitted cases prior to 1990 would have been asymptomatic and unrecognized, but 75% to 80% would have become chronic HCV carriers and 20% of those would have been at risk of ultimately developing cirrhosis if they did not first succumb to another illness.

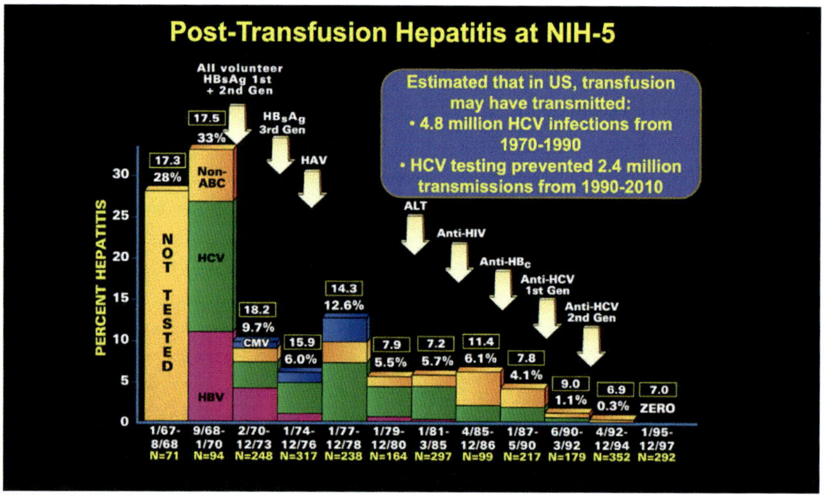

Figure 5. Bar graph showing the declining incidence of post-transfusion hepatitis over time and the interventions associated with that change. Anti-HCV donor testing was introduced in 1990, reducing incidence to 1.1% and a second generation assay introduced in 1992 resulted in the absence of HCV cases (green bar). No hepatitis cases were seen in the interval from 1995–1997. Estimates of cases caused by transfusion and cases prevented by testing are shown in the rectangle above the graph.

As previously noted, there was frustration in not being the first to unequivocally identify HCV, but I got over it fairly rapidly and wrote this poem of conciliation, titled, "There's No Sense Chiron Over Spilt Milk":

After more than a decade of severe frustration
and endless serologic constipation, the whole field in a state
of stagnation
the scientific equivalent of Faustian damnation.
At last, non-A, non-B has achieved emancipation
and enters an era of new creation
with enormous impact on blood donation.
But is this real or an aberration?
is it true love or just a flirtation?
Are we ready for delivery after this prolonged gestation?

The non-A, non-B story began to get brighter
When Bradley squirreled away chimp plasma of very high titer.
Seeking this agent with the passion of a zealot,
Houghton centrifuged particles to a concentrated pellet.
And the science moved forward from a state very placid
when the pellet released its ribo-nuclear acid;
A nucleic erection in a field formerly flaccid.

it was then Dr. Houghten's anti-viral prescription
to make clones by reverse transcription
While Chiron's finance department had another conniption.
But Michael said this was not a rehearsal
The RNA had to be changed by a transcription reversal.

Like Jack the Ripper at his most vindictive, he chopped up
the codon with an enzyme restrictive
Could these tiny pieces be of an antigen predictive?
Corporate executives prayed that it would on their knees
    benedictive
Showing that science and religion are not contradictive.

With the hope these pieces would become a viral detector
they inserted them with care into an expression vector,
Could these invisible codes recreate non-B architecture?

So that codon insertion might be most benefactory
they packaged the plasmid in an E. coli factory;
Man and bug in a collusive compactory
praying the result would not be anti-climactory.
Then out of the agar there spewed forth a clone
that for years of frustration would fully atone,
The non-A, non-B virus was suddenly known
and HCV became the name it would own
A viral peasant had ascended the throne.

Suddenly there was antigen expression,
Followed by detection of antibody and genome in rapid
    succession;
A major advance for our profession,
A triumphal ending to the non-B obsession.
Chiron opened a cloning concession;
9/10ths of the law is clearly possession.

As for Chiron I hold no resentment
in the tests support of my claims, I'll find my contentment.
For coming in second, I make no apologies;
I can always turn to the bench and do blood bank serologies.
For me there will be no Nobel prizen,
but there's always another virus on the horizon.

I like to think of the year 2000 as the end of the beginning. It marked the elucidation of a novel transfusion-transmitted agent (Non-A, Non-B hepatitis virus); the proof of its persistence in most infected individuals; the clinical description of its chronic sequelae, including cirrhosis and hepatocellular carcinoma; its cloning and characterization as a flavivirus and renaming as HCV; the development of assays for antibody detection and their implementation in blood donor screening, and; the virtual eradication of transfusion-associated hepatitis. There was still much more to come in the HCV story, but this was the end of the phase of discovery, the recognition of its global spread, its major role in chronic liver disease and liver-related mortality and its impact on blood transfusion safety. This was the end of 25 years of clinical observation and molecular characterization. It was a period of accelerating and often unexpected enlightenment, an end that was not anticipated at the beginning, but which played out in small increments, a cascade of observations that built one upon the other into what was ultimately a compelling story when curative therapies emerged from earlier genomic characterizations and the development of infectious clones. Charles Rice, the third recipient of this Nobel Prize, will describe the pathway to drug cure in a subsequent lecture.

While I have outlined the role that NIH played in the discovery phase and in the elimination of transfusion-associated hepatitis, I would now like to spend some time on ancillary studies that defined other aspects of HCV pathogenesis. We, and many others, have been interested in the persistent nature of HCV infection. What are the elements of the virus-host interaction that prevent viral clearance in the majority of HCV infected patients? One element is clearly the quasispecies nature of RNA viruses that lack efficient proof-reading capability for the elimination of evolving variants. Hence, HCV exists not as a single virus, but as a family of closely related, yet genomically and often antigenically distinct variants. Patrizia Farci, my closest collaborator over the past two decades, showed by sequencing 105 clones derived from a single sample of patient H's plasma, that 57% of clones were identical, but that 19 other variants were simultaneously present in smaller proportions (17). Hence, even if this patient had mounted a good immune response to the dominant strain, any of the other 19 variants could emerge to dominance and escape the antibody response. Farci also showed the rapid emergence of variants by cloning sequential samples over the course of the first 16 weeks of acute infection. Variants continued to appear in each successive sample and the rate of appearance accelerated once the anti-HCV response engaged and put additional immune pressure on the HCV envelope receptor sites, particularly in the envelope 2 region (18).

Jens Bukh, a major collaborator in Denmark, has worked extensively with the H77 inoculum and shown that antibodies that effectively neutralized the initial H strain inoculum failed to neutralize strains that evolved at a later time. (19). Shimizu, et al documented this neutralization escape in the chimpanzee model (20). Hence, rapidly evolving viral variation and escape from antibody neutralization combine to enhance viral persistence. In addition, Wedermeyer et al (21) have shown that CD8 T cell responses are blunted during HCV infection, presumably by antigen excess, and that T-cell activity recovers in those who spontaneously clear HCV. This blunting of the T-cell response is another characteristic element of HCV persistence. In another collaborative study, von Hahn in Jane McKeating's lab (22) showed that mutations that produced a change in only two amino acids in the E2 envelope protein resulted in escape from T cell immunity that was highly effective against the wild type strain. Hence, there is mutational escape from both antibody and T-cell immune responses that contribute to viral persistence.

In our prospective studies we observed that 60–70% of chronically infected patients had an indolent course with only minimal or moderate fibrosis progression over decades of observation whereas 20–30% progressed to cirrhosis in 15–40 years and 5% had rapid fibrosis progression resulting in cirrhosis and death in less than 10 years. Farci (23) examined the differences between rapid and slow progressors in my patient population and showed that slow progressors demonstrated initial containment of the virus with diminution in HCV RNA to very low levels followed by a rapid rise in HCV RNA as a presumed escape mutant became the dominant strain. In contrast, rapid progressors never showed immune containment as evidenced by persistently high-level HCV RNA from early on and throughout their chronic course. In addition, rapid progressors had high levels of monocyte chemotactic protein (MCP-1, CCL-2) a profibrogenic chemokine that both attracts and is produced by hepatic stellate cells, the most critical cells in hepatic collagen deposition. In a separate study (24), Kentaro Matsuura from the laboratory of Yasuhito Tanaka in Nagoya, Japan did an advanced fellowship in our laboratory and showed that the micro RNAs Let-7a and Let-7c were at low level in patients with advanced fibrosis and declined further in parallel with progression to cirrhosis. This made physiologic sense in that Let-7 is a suppressor of TGF-β, a major driver of hepatic stellate cell activation and concomitant fibrous tissue deposition. Thus, one can postulate that the production of MCP-1 and the release of TGF-β from Let-7 inhibition can combine to activate stellate cells and lead to rapid fibrosis progression. Clearly other factors may be involved but these were the key factors revealed in cytokine-chemokine arrays and in micro-RNA array screening.

Lastly, I want to discuss insights into hepatocellular carcinoma (HCC), a devastating consequence of HCV infection that occurs in 1–4% of patients who progress to cirrhosis. Yasuhito Tanaka and Masashi Mizokami in Japan performed a molecular clock analysis of samples from Mr. H, whose virus was HCV genotype 1a, and from a Japanese patient with genotype 1b infection (25). Sequencing serial samples from these patients allowed for mathematical analysis of when a virus diverged from a common ancestor and when it spread rapidly in a given population. The molecular clock estimated that genotype 1b emerged in Japan in 1880 and then persisted at low level until the 1930s and 40s when it spread rapidly within the population. This spread was coincident with the Pacific wars in which Japan engaged, at first with China and then with the US and it allies in World War II. We now learn that it was common for Japanese soldiers to "shoot-up" before battle, generally with amphetamines, and it was this needle sharing that was at the root of rapid HCV dissemination among Japanese combatants, ultimately involving much of the male population. In the US, genotype 1a emerged in 1910 and then lay dormant until the 1960s and 70s when it spread rapidly in the US population due to our illicit drug-use epidemic. HCV continues to spread by this mechanism to the current day and indeed would dissipate to large measure if the drug epidemic could be contained. Importantly, there is an approximate 30-year lag between the spread of HCV in the US and that in Japan. This is critically important because the incidence of HCC is 8 to 10-fold higher in Japan compared to the US and it has been postulated that the US incidence will ultimately match that of the Japan when the virus persists in our population for several more decades. Indeed, that appears to be the case. El-Serag (26) has shown that the incidence of HCC in the US tripled between the late 1970s and the late 1980s and that virtually all this increase was attributable to HCV infection. Hence, the difference between HCC incidence in Japan and the US appears to be primarily a function of time because the evolution to cirrhosis and then to HCC is a multi-decade phenomenon, generally exceeding 40 years. Given the trajectory of HCC in the US, it has been estimated that HCC incidence will increase 4-fold from a 2009 baseline to the year 2029, if there is no change in the standard of care. Fortunately, there has been a major change in the standard of care as direct-acting, HCV-specific antivirals have been developed that can now cure 95–100% of chronically infected individuals over an 8–12 week course of oral therapy with minimal side effects. Such high cure rates were unimaginable prior to 2010, when a combination of pegylated interferon and ribavirin were the standard of care. It is now conceivable that every identified HCV carrier can be treated to cure, with the caveat of the obstacles discussed below.

We have now reached the point where the eradication of HCV can be contemplated. With cure rates approaching 100%, from this time forward, no one with identified HCV chronic infection should develop cirrhosis or die from hepatitis C sequelae. However, there are caveats to this optimistic forecast. First, in the US, it is estimated that no more than half of HCV carriers have been identified and this proportion is significantly less in the developing world. Hence, intensive population screening for anti-HCV and/or HCV RNA is the first goal in global eradication. Second, the high cost of these life-saving drugs is a major impediment. Indeed, in the first years of DAA deployment, only a minor portion of identified patients were treated because insurance companies and national health services found the costs prohibitive. With time, there has been the realization that the cost of cirrhosis, end-stage liver disease, HCC and liver transplantation exceed the cost of 8–12 weeks of therapy and an increasing number of patients in the developed world have been treated, but there are huge gaps in developing nations even though the cost in these countries is markedly reduced. Cure is no longer constrained by science, but is a matter of dollars.

Given these caveats, is eradication possible? The usual path to viral eradication is vaccination. However, there is no current vaccine that has been licensed or even entered into clinical trials. Michael Houghton, in an accompanying lecture, will describe an HCV envelop protein vaccine that has been shown in the chimpanzee model to significantly decrease the frequency of chronic infection even though it does not prevent HCV infection per se. The success of messenger RNA-based vaccines for Covid-19 raises the hope that this approach might also be effective for HCV, but no such data are available at the time of this publication.

It is intriguing to speculate whether HCV could be eliminated globally in the absence of a vaccine. This would be a difficult task, but given that we have almost optimally sensitive screening assays and highly curative therapies, it is a goal that could be achieved over the course of the next 2–3 decades. The first essential step would be to develop a highly sensitive HCV screening assay that could be used in the field and provide results within minutes. A positive test would immediately trigger the provision of 1–3 month's supply of oral therapy at a price commensurate with the GNP of the nation. Starting therapy simultaneously with testing obviates the loss to follow-up that would occur with delayed treatment. To implement such a test and treat strategy on a global scale would be labor-intensive and costly, but it could be done in compliant nations if there was the political, corporate, philanthropic and moral will to make it happen. Chronic HCV is different from chronic HBV or HIV, because there is no need for life-long follow-up and treatment. It will be tragic if the cure of millions will be blocked, not by technologic capability, but by moral inertia and high cost. Vaccine development and this test and treat strategy are synergistic and can proceed in parallel. Of course, other preventive measures are also needed to reduce new infections, primarily by limiting shared needle and drug paraphernalia use among injection drug users and by standardizing global health care practices to insure single-needle use and proper sterilization of reusable instruments, to foster universal precautions, and where needed, to change practices of tattooing and ritual scarification.

Thus, the end of HCV is now in sight, even if one has to use a telescope to view it. Unless one can envision this goal, it will not happen. It is a long staircase to ascend, but will not happen unless we take that first step.

REFERENCES

1) Neeff JR, Stokes Jr J, Gellis SS. Homologous serum hepatitis and infectious (epidemic) hepatitis. *Am J Med Sci* 1945; **210**:561–75.
2) Blumberg BS, Alter HJ, Visnich S. A "new" antigen in leukemia sera. *JAMA* 1965; **191**:541–546.
3) Walsh JH, Purcell RH, Morrow AG, Chanock RM, Schmidt PJ. Post-transfusion hepatitis after open heart operations: incidence after the administration of blood from commercial and volunteer donor populations. *JAMA* 1970; **211**:261–265 .
4) Alter HJ, Holland PV, Purcell RH, Lander JJ, Feinstone SM, Morrow AG, Schmidt PJ. Postransfusion hepatitis after exclusion of the commercial and hepatitis B antigen positive donor. *Ann Int Med* 1972; **77**:691–699.
5) Feinstone SM, Kapikian AZ, Purcell RH. Hepatitis A: detection by immune electron microscopy of a virus like antigen associated with acute illness. *Science* 1973; **82**:1026–28.
6) Feinstone SM, Kapikian AZ, Purcell RH, Alter HJ, Holland PV. Transfusion-associated hepatitis not due to viral hepatitis type A or B. *N Engl J Med* 1975; **292**:767–770.
7) Alter HJ, Purcell RH, Holland PV, Popper H. Transmissible agent in "non-A, non-B" hepatitis. *Lancet* 1978; **1**:459–463.
8) Tabor E, Gerety RJ, Drucker JA, et al. Transmission of non-A, non-B hepatitis from man to chimpanzee. *Lancet* 1978; **1**:463–465.
9) Feinstone JM, Mihalik KB, Kamimura J, Alter HJ, London WT, Purcell RH. Inactivation of Hepatitis B Virus and Non-A, Non-B Virus by Chloroform. *Infect and Immun* 1983; **4**:816–821.
10) He L-F, Alling DW, Popkin TJ, Shapiro M, Alter HJ, Purcell RH. Determining the Size of Non-A, Non-B Hepatitis Virus by Filtration. *J Inf Dis* 1987: *156*:636–640.
11) Berman MD, Alter HJ, Ishak KG, Purcell RH, Jones EA. The chronic sequelae of non-A, non-B hepatitis. *Ann Int Med* 1979; **91**:1–6.
12) Koziol DE, Holland PV, Alling DW, Melpolder JC, Solomon RE, Purcell RH, Hudson LM, Shoup FJ, Krakauer H, Alter HJ. Antibody to Hepatitis B Core Antigen as A Paradoxical Marker for Non-A, Non-B Hepatitis Agents in Donated Blood. *Ann Intern Med* 1986; **104**:488–495,

13) Stevens CE, Aach RD, Hollinger FB. Hepatitis B virus antibody in blood donors and the occurrence of non-A, non-B hepatitis in transfusion recipients; an analysis of the transfusion-transmitted virus study. *Ann Intern Med* 1984; **101**:733–738.
14) Choo QL, Kuo G, Weiner AJ, Overby LR, Bradley DW, Houghton M. Isolation of a cDNA clone derived from a blood-borne non-A, non-B viral hepatitis genome. *Science* 1989; **244**:359–362.
15) Kuo G, Choo QL, Alter HJ, Gitnick GL, Redeker AG, Purcell RH, Miyamura T Dienstag JL, Alter MJ, Stevens CE, Tegtmeier GE, Bonino F, Colombo M, Lee W, Kou C, Berger K, Shuster JR, Overby LR, Bradley DW, Houghton M. An Assay for Circulating Antibodies to a Major Etiologic Virus of Human Non-A, Non-B Hepatitis. *Science* 1989; **244**:362–364.
16) Alter HJ, Purcell RH, Shih JW, Melpolder JC, Houghton M, Choo Q-L, Kou G. Detection of Antibody to Hepatitis C Virus in Prospectively Followed Transfusion Recipients with Acute and 00
17) Farci P, Alter HJ, Wong D, Miller RH, Shih JW, Jett B, Purcell RH. A Long Term Study of Hepatitis C Virus Replication in Non-A, Non-B Hepatitis. *New Engl J Med* 1991; **325**:98–104.
18) Farci P, Shimoda A, Coiana A, Diaz G, Peddis G, Melpolder JC, Strazzera A, Chien DY, Munoz SJ, Balestrieri A, Purcell RH, Alter HJ. The outcome of acute hepatitis C predicted by the evolution of the viral quasispecies. *Science* 2000; **288**:339–344.
19) Meuleman P, Bukh J, Farhoudi A, Desombere I, Vanwolleghem T, Verhoye L, Alter H, Purcell RH, Leroux-Roels G. In vivo evaluation of the cross-genotype neutralizing activity of polyclonal antibodies against hepatitis C virus. *Hepatology* 2011; **53**:755–762.
20) Shimizu YK, Yoshikuta H, Hijikata M, Iwamoto A, Alter HJ, Purcell RH. Neutralizing Antibodies Against Hepatitis C Virus and the Emergence of Neutralization Escape Mutants. *J Virol* 1994; **68**:1494–1500.
21) Wedermeyer H, He XS, Nascimbeni M, Davis AR, Greenberg HB, Hoofnagle JH, Liang TJ, Alter HJ, Rehermann B. Impaired Effector Function of Hepatitis C Virus-Specific CD8+ T Cells in Chronic Hepatitis C Virus infection. *J Immunology* **169**:3448–3454, 2002.
22) von Hahn T, Yoon JC, Alter H, Rice CM, Rehermann B, Balfe P, McKeating JA. Hepatitis C virus continuously escapes from neutralizing antibody and T-cell responses during chronic infection in vivo. *Gastroenterology* 2007;**132**: 667–678.
23) Farci P, Wollenberg K, Diaz G, Engle RE, Lai ME, Klenerman P, Purcell RH, Pybus OG, Alter HJ. Profibrogenic chemokines and viral evolution predict rapid progression of hepatitis C to cirrhosis. *PNAS* 2012;**109**:14562–14567.
24) Matsuura K, DeGiorgi V, Schechterly C, Wang R, Farci P, Tanaka Y, Alter HJ. Circulating let-7 levels in plasma and extracellular vesicles correlate with hepatic fibrosis progression in chronic hepatitis C. *Hepatology* 2016; **64**:732–45.
25) Tanaka Y, Hanada K, Mizokami M, Yeo AET, Shih JW, Gojobori T, Alter HJ. A Comparison of Molecular Clock of Hepatitis C Virus in the United States and Japan Predicts That Hepatocellular Carcinoma Incidence in the US Will Increase Over the Next Two Decades. *PNAS* **99**:15584–15589, 2002.
26) El-Serag HB, Kanwal F. Epidemiology of Hepatocellular Carcinoma In the United States: Where Are We? Where Do We Go? *Hepatology* 2014; **60**:1767–1775.

Michael Houghton. © University of Alberta

Michael Houghton

Biography

Born in London, England in 1951 to Leonard George and Elsie Cressy Houghton, he was raised in a working-class family along with elder brother Graham. Educated at an excellent government primary school (Lyndhurst Grove) until the age of 11, he then won a scholarship to Alleyn's School in Dulwich, London, founded by Edward Alleyn, a lead actor and producer of William Shakespeare's plays in the 16th and 17th centuries. Graduating with college entry examinations in physics, chemistry and math, he decided to study biology at college, having been heavily influenced by reading at age 17 about the life and works of Louis Pasteur and also watching BBC TV coverage of the discovery of the DNA double helix early on Sunday mornings. In 1969, he entered the school of Biological Sciences at the new University of East Anglia, where he was highly motivated by the teachings of Professors Balls, Wilden and Thane as well as by full extracurricular activities including playing for the college cricket and squash teams. Offered attractive PhD positions at a London cancer research center, Oxford University, and Glasgow University, he failed to secure a stipend so entered the research laboratories of GD Searle & Co. in High Wycombe, UK, which was engaged in basic molecular biology research. Registering for a PhD in Biochemistry at King's College, University of London in 1973 and co-supervised by Dr. Norman Carey (Searle) and Dr. James Chesterton (King's), he graduated with a PhD in 1977 in a ceremony at the Alberta Hall with his wife, Han Fong Ida, who received her B. Pharm degree at the same ceremony from the London School of Pharmacy.

He then continued as a Research Investigator at Searle, benefiting from the excellent mentorship of Dr. Norman Carey and Dr. Richard Palmiter, where he characterized the human fibroblast interferon gene using newly emerging recombinant DNA technology. Offered many positions at burgeoning genetic engineering companies in the UK and US, in 1982 he chose to join the Chiron Corporation in California, founded by University of California Professors Bill Rutter and Ed Penhoet. Intending originally to work on chimeric type 1 interferons and to identify what became known later as type 3 interferons, he was introduced to the problem of Non-A, Non-B hepatitis by Dr. Dino Dina, then Director of Virology at Chiron. As a result, he decided to devote his laboratory to the pursuit of the causative agent(s) of this infectious disease, using a molecular biological approach, which was known not to be caused by the already characterized hepatitis A & B viruses. After pursuing numerous avenues unsuccessfully for many years, along with Dr. Qui-Lim Choo in his own laboratory, Chiron collaborator Dr. George Kuo, and CDC collaborator Dr. Dan Bradley, he published on the discovery of the hepatitis C virus (HCV) genome in 1989 with his collaborators, having published on the novel structure of the HDV genome in 1986. Following the HCV discovery, he and collaborators developed a series of patient diagnostics and blood screening tests that prevented post-transfusion hepatitis globally, as well as contributing to the identification of key enzymes crucial to the virus life cycle, which then became drug development targets for the field.

HCV infection represents a serious pandemic and so again with Dr. Choo and Dr. Kuo along with Dr. Ralston and others, he published the first evidence for an efficacious HCV vaccine in 1994 and subsequently performed 4 clinical trials in 3 different countries. His work has indicated the feasibility of developing an HCV vaccine, which he hopes to roll out to high-risk groups by the mid-2020s using both adjuvanted recombinant protein approaches as well as by applying the new RNA vaccine technologies at the Li Ka Shing Applied Virology Institute at the University of Alberta in Edmonton, Canada. He is also working with and supporting other leaders at the University of Alberta to develop a vaccine for Group A Streptococcus, and better drugs to treat Alzheimer's disease and CMV infections. Now that HCV is a curable disease, he and others in the hepatology field are also turning their attention to the problem of non-alcoholic liver disease.

Education

1969–1972 BSc (Honors) Biological Sciences
University of East Anglia
Norwich, England

1973–1977 PhD Biochemistry
King's College, University of London
England

2019/7 Honorary Doctorate of Science
University of East Anglia
Norwich, England

Positions

1977–1982 Research Investigator, Human Interferon Genetics
Searle Research Laboratories, Buckinghamshire, England
1982–1988 Project Leader, Non-A, Non-B Hepatitis Discovery Research
Chiron Corporation, Emeryville, CA
1988–2000 Director, Hepatitis C Research
Chiron Corporation, Emeryville, CA
2000–2003 Vice-President, Hepatitis C Research
Chiron Corporation, Emeryville, CA
2003–2006 Vice-President, Hepatitis C & Virology Research
Chiron Corporation, Emeryville, CA
2006–2007 Vice-President, Hepatitis C & Virology Research
Novartis Vaccines & Diagnostics, Inc., Emeryville, CA
2007–2009 Chief Scientific Officer
Epiphany Biosciences Inc., San Francisco, CA
2010–2018 Canada Excellence Research Chair in Virology and Li Ka Shing Professor, Department of Medical Microbiology and Immunology, University of Alberta Edmonton, AB, Canada
2013 – present Director of the Li Ka Shing Applied Virology Institute, University of Alberta, Edmonton, AB, Canada

AWARDS/HONORS

The following honors were awarded for research on hepatitis C:
1992 Co-recipient of the Karl Landsteiner Award from the American Association of Blood Banks
1993 Co-recipient of the Robert Koch Medal from Germany
1993 Honoree of the Japanese Medical Congress
1993 Honoree of the Triennial International Hepatitis Meeting
1994 Co-recipient of the William Beaumont Prize from the American Gastroenterology Association
1994 Recipient of Beatrice Bitiello Award from Italian Association for Prevention of Viral Hepatitis
1994 Awardee of the Princess Takamatsu Cancer Research Fund from Japan
1998 Co-recipient of the International Hepatitis Foundation Award
1999 Co-recipient of the Hans Popper Award (Falk Foundation, Germany)
2000 Co-recipient of the Clinical Lasker Award (USA)
2005 Co-recipient of the Dale Smith Memorial Award of the American Association of Blood Banks
2009 Recipient of the Hepdart Lifetime Achievement Award (USA)
2011 Gold Medal from the Canadian Association for the Study of the Liver (CASL)
2011 The Australian Society for Microbiology Annual Scientific Meeting Bazeley Oration
2013 International Gairdner Award (Canada; declined)
2018 American Liver Foundation Distinguished Scientific Achievement Award (USA)
2020 The Nobel Prize in Medicine (The Nobel Foundation)
2021 Knighthood awarded by Queen Elizabeth II

Publications (by subject area and chronological order)

Transcriptional and Translational Control in Eukaryotes

Houghton M, Cox RF. (1974) The purification and properties of hen oviduct Form B DNA-dependent RNA polymerase. *Nucl. Acids Res.* **1**:299–308.

Doel MT, Houghton M, Cook EA, Carey NH. (1977) The presence of ovalbumin mRNA coding sequences in multiple restriction fragments of chicken DNA. *Nucl. Acids. Res.* **4**:3701–13.

Lilley DMJ, Houghton M. (1979) The interaction and RNA polymerase II from wheat with supercoiled and linear plasmid templates. *Nucl. Acids. Res.* **6**:507–23.

Lilley DMJ, Jacobs MF, Houghton M. (1979) The nature of the interaction of nucleosomes with eukaryotic RNA polymerase II. *Nucl. Acids Res.* **7**:377–99.

Lane CD, Colman A, Mohun T, Morser J, Champion J, Kourides I, Craig R, Higgins S, James TC, Applebaum SW, Ohlsson RI, Pauchas E, Houghton M, Matthews J, Miflin BJ. (1980) The Xenopus oocyte as a surrogate secretory system. *Eur. J. Biochem.* **111**:225–35.

Sumikawa K, Houghton M, Emtage JS, Richards BM, Barnard EA. (1981) Active multi-subunit ACh receptor assembled by translation of heterologous mRNA in Xenopus oocytes. *Nature.* **292**:862–4.

Sumikawa K, Houghton M, Smith JC, Bell L, Richards BM, Barnard EA. (1982) The molecular cloning and characterization of cDNA coding for the alpha subunit of the acetylcholine receptor. *Nucl. Acids Res.* **10**:5809–22.

Kenten JH, Molgaard HV, Houghton M, Derbyshire RB, Viney J, Bell LO, Gould HJ. (1982) Cloning and sequence determination of the gene for the human immunoglobulin epsilon chain expressed in a myeloma cell line. *Proc. Natl. Acad. Sci. USA.* **79**:6661–5.

Sumikawa K, Houghton M, Miledi R, Barnard EA. (1983) "A study of the mRNA and genes coding for the nicotinic acetylcholine receptor" *in* Cell Surface Receptors (Strange PG. ed.) pp. 249–69. Ellis Horwood Ltd., U.K.

Barnard EA, Houghton M, Miledi R, Richards BM, Sumikawa K. (1982) Molecular genetics of the acetyl choline receptor and its insertion and organization in the membrane. *Biol. Cell.* **45**:383.

Molecular Genetics of Human Fibroblast Interferon

Houghton M, Stewart AG, Doel SM, Emtage JS, Eaton MAW, Smith JC, Patel TP, Lewis HM, Porter AG, Birch JR, Cartwright T, Carey NH. (1980) The amino–terminal sequence of human fibroblast interferon as deduced primers. *Nucl. Acids Res.* **8**:1913–31.

Houghton M. (1980) Human interferon gene sequences. *Nature.* **285**:536.

Houghton M, Eaton MAW, Stewart AG, Smith JC, Doel SM, Catlin GH, Lewis HM, Patel TP, Emtage JS, Carey NH, Porter AG. (1980) The complete amino acid sequence of human fibroblast interferon as deduced using synthetic oligodeoxyribonucleotide primers of reverse transcriptase. *Nucl. Acids Res.* **8**:2885–94.

Houghton M, Jackson IJ, Porter AG, Doel SM, Catlin GH, Barber C, Carey NH. (1981) The absence of introns within a human fibroblast interferon gene. *Nucl. Acids Res.* **9**:247–66.

Houghton M, Doel SM, Catlin GH, Stewart AG, Porter AG, Tacon WCA, Eaton MAW, Emtage JS, Carey NH. (1981) "The cloning and expression of a human fibroblast interferon gene in bacteria" *in* Proceedings of the Battelle International Genetic Engineering Conference (Keenberg M. ed.). Battelle Seminars and Studies Program.

McCullagh KG, Davies JA, Sim IS, Dawson KM, O'Neill GJ, Doel SM, Catlin GH, Houghton M. (1983) Biological properties of human interferon beta 1 synthesized in recombinant bacteria. *J. Interf. Res.* **3**:97–111.

Porter AG, Bell LD, Adai JR, Catlin GH, Clarke JM, Davies JA, Dawson KM, Derbyshire RB, Doel SM, Dunthorne L, Finlay ME, Hall J, Houghton M, Hynes C, Lindley IJ, Nugent ME, O'Neill GJ, Smith JC, Stcwart AG, Tacon WC, Viney JH, Warburton N, Boseley PG, McCullagh KG. (1985) "Active hybrids formed between human beta and alpha interferons" *in* The Biology of the Interferon System (Schellekers and Stewart eds.).

Porter AG, Bell LD, Adai JR, Catlin GH, Clarke JM, Davies JA, Dawson KM, Derbyshire RB, Doel SM, Dunthorne L, Finlay ME, Hall J, Houghton M, Hynes C, Lindley IJ, Nugent ME, O'Neill GJ, Smith JC, Stewart AG, Tacon WC, Viney JH, Warburton N, Boseley PG, McCullagh KG (1986) Novel modified beta interferons: gene cloning, expression and biological activity in bacterial extracts. *DNA.* **5**:137–48.

Hepatitis Delta Virus

Wang KS, Choo QL, Weiner AJ, Ou JH, Najarian RC, Thayer RM, Mullenbach GT, Denniston KJ, Gerin JL, Houghton M. (1986) The structure, sequence and expression of the hepatitis delta (δ) viral genome. *Nature.* **323**:508–14. Erratum in Nature. (1987) 328: 456.

Wang K-S, Choo Q-L, Weiner AJ, Ou J-H, Denniston KJ, Gerin JL, Houghton M (1987) "The Viroid-like structure of the hepatitis delta genome: synthesis of a viral antigen in recombinant bacteria" *in* The Hepatitis Delta Virus and its Infection (Rizzetto M, Gerin JL, Purcell RH. eds.) pp. 71–82. Alan Liss Inc., New York.

Weiner AJ, Wang K-S, Choo Q-L, Gerin JL, Bradley DW, Houghton M. (1987) Hepatitis delta (δ) cDNA clones: Undetectable hybridization to nucleic acids from infectious Non-A, Non-B hepatitis materials and hepatitis B DNA. *J. Med. Virol.* **21**: 239–47.

Weiner AJ, Choo Q-L, Wang K-S, Govindarajan S, Redeker AG, Gerin JL, Houghton M. (1988) A single antigenomic open reading frame of the hepatitis delta virus encodes the epitope(s) of both hepatitis delta antigen polypeptides p24 delta and p27 delta. *J. Virol.* **62**:594–9.

Weiner AJ, Choo Q-L, Wang K-S, Govindarajan S, Redeker AG, Gerin JL, Houghton M. (1988) A single antigenomic open reading frame of the hepatitis delta virus encodes the epitope(s) of both hepatitis delta antigen polypeptides p24 delta and p27 delta. *J. Virol.* **62**:594–9.

Ponzetto A, Eckart M, D'Urso N, Negro F, Silvestro M, Bonino F, Wang KS, Chien D, Choo Q-L, Houghton M. (1993) Towards a vaccine for the prevention of hepatitis delta virus superinfection in HBV carriers. *Prog. Clin. Biol. Res.* **382**:207–10.

Eckart MR, Dong C, Houghton M, D'Urso N, Ponzetto A. (1993) The effects of using recombinant vaccinia viruses expressing either large or small HDAg to protect woodchuck hepadnavirus carriers from HDV superinfection. *Prog. Clin. Biol. Res.* **382**:201–05.

Nisini R, Paroli M, Accapezzato D, Bonino F, Rosina F, Santantonio T, Sallusto F, Amoroso A, Houghton M, Barnaba V. (1997) Human CD4+ T-cell response to hepatitis delta virus: identification of multiple epitopes and characterization of T-helper cytokine profiles. *J. Virol.* **71**:2241–51.

Non-A, Non-B Hepatitis / Hepatitis C Virus

Choo Q-L, Kuo G, Weiner AJ, Overby LR, Bradley DW, Houghton M. (1989) Isolation of a cDNA clone derived from a blood-borne non-A, non-B viral hepatitis genome. *Science.* **244**:359–62.

Kuo G, Choo Q-L, Alter HJ, Gitnick GI, Redeker AG, Purcell RH, Miyamura T, Dienstag JL, Alter MJ, Stevens CE, Tegtmeier GE, Bonino F, Colombo M, Lee W-S, Kuo C, Berger K, Shuster JR, Overby LR, Bradley DW, Houghton M. (1989) An assay for circulating antibodies to a major etiologic virus of human non-A, non-B hepatitis. *Science.* **244**:362–64.

Esteban JI, Viladomiu L, Bonzalez A, Roget M, Genesca J, Guardia J, Esteban R, Lopez-Talavera JC, Hernandez JM, Vargas V, Buti M, Kuo G, Choo Q-L, Houghton M. (1989) Hepatitis C virus antibodies among risk groups in Spain. *Lancet.* **334**:294–7.

Van Der Poel CL, Ressink HW, Lelie PN, Leentvaar-Kuypers A, Choo Q-L, Kuo G, Houghton M. (1989) Anti-hepatitis C antibodies and non-A, non-B post-transfusion hepatitis in the Netherlands. *Lancet.* **334**:297–8.

Parker T, de Medina M, Jeffers L, Reddy R, Bradley D, Schiff E, Houghton M, Choo Q-L, Kuo G. (1989) Hepatitis C virus HCV: A causative agent of cryptogenic cirrhosis CC among Cubans. *Hepatology.* **10**:685.

Saracco G, Houghton M, Kuo G, Choo Q-L, Rosina F, Lattore V, Torrani Cerenzia MR, Chiandussi L, Bonino F, Rizzetto M. (1989) Anti-hepatitis C virus in non-A non-B patients responding and non-responding to alpha 2A interferon. *J. Hepatol.* **9**:S219. Supplement 1.

Colombo M, Kuo G, Choo QL, Donato MF, Del Ninno E, Tommasini MA, Dioguardi N, Houghton M. (1989) Prevalence of antibodies to hepatitis C virus in Italian patients with hepatocellular carcinoma. *Lancet.* **334**:1006–8.

Alter HJ, Purcell RH, Shih JW, Melpolder JC, Houghton M, Choo Q-L, Kuo G. (1989) Detection of antibody to hepatitis C virus in prospectively followed transfusion recipients with acute and chronic non-A, non-B hepatitis. *N. Engl. J. Med.* **321**:1494–1500.

Takeuchi K, Boonmar S, Katayama T, Choo QL, Kuo G, Weiner AJ, Bradley DW, Houghton M, Saito I, Miyamura T. A cDNA fragment of hepatitis C virus isolated from an implicated donor of post-transfusion non-A, non-B hepatitis in Japan. *Nucl. Acids Res.* **17**:10367–72.

Miyamura T, Saito I, Kubo Y, Takeuchi K, Boonmar S, Katayama T, Kuo G, Choo Q-L, Houghton M. (1989) "Hepatitis C virus complementary DNA clones isolated from a single healthy carrier who was shown to be an implicated donor of post-transfusion non-A, non-B hepatitis" *in* Proceedings of the International Meeting on Non-A, Non-B Hepatitis, Tokyo (Shikata T, Purcell RH, Uchida T. eds.) Elsevier Science Publishers, Amsterdam.

McHutchison JG, Kuo G, Houghton M, Choo Q-L, Redeker AG. (1989) Autoimmune hepatitis is not associated with antibodies to hepatitis C virus (HCV). *Hepatology.* **10**:701.

Kamitsukasa H, Harada H, Yakura M, Fukuda A, Ohbayashi A, Saito I, Miyamura T, Choo Q-L, Houghton M, Kuo G. (1989) Intrafamilial transmission of hepatitis C virus. *Lancet.* **2**:987. Letter.

Colombo M, Kuo G, Choo Q-L, Houghton M, Donato MF, Tommasini MA, Bargiggia S, Piva A, Del Ninno E, Dioguardi N. (1989) High prevalence of antibody to hepatitis C virus in patients with primary liver carcinoma. *Hepatology.* **10**:700.

Colombo M, Kuo G, Choo Q-L, Houghton M, Tommasini MA, Rumi MG, Dioguardi ML, Donato MF, Del Ninno E. (1989) High prevalence of antibody to

hepatitis C virus in patients with hepatocellular carcinoma HCC. *J. Hepatol.* **9**:S20. Supplement 1.
Katkov WN, Cody H, Evans AA, Kuo G, Choo Q-L, Houghton M, Dienstag JL. (1989) The role of hepatitis C virus HCV in chronic liver disease. *Hepatology.* **10**:644.
Katkov WN, Fricdman LS, Cody H, Evans AA, Kuo G, Choo Q L, Houghton M, Huggins CE, Dienstag JL. (1989) Elevated serum alanine aminotransferase ALT in blood donors: The contribution of hepatitis C virus HCV. *Hepatology.* **10**:581.
McHutchison JG, Kuo G, Houghton M, Choo Q-L, Redeker AG. (1989) Circulating antibodies to hepatitis C virus HCV: A study of 160 cases of acute and chronic NANB hepatitis. *Hepatology.* **10**:645.
Jeffers L, de Medina M, Hasan F, Reddy R, Parker T, Silva M, Mendez L, Schiff E, Houghton M, Choo Q-L, Kuo G. (1989) Hepatitis C HCV associated idiopathic chronic hepatitis and cryptogenic cirrhosis. *Hepatology.* **10**:644.
Hasan F, Jeffers L, de Medina M, Reddy R, Parker T, Schiff E, Houghton M, Choo Q-L, Kuo G. (1989) Hepatitis C HCV associated hepatocellular carcinoma. *Hepatology.* **10**:608.
Evans AA, Cody H, Kuo G, Choo Q-L, Houghton M, Katkov WN, Dienstag JL. (1989) Seroepidemiology of hepatitis C virus HCV in selected population. *Hepatology.* **10**:644.
Krawczynski K, Kuo G, Dibisceglie A, Houghton M, Bradley DW. (1989) Blood-borne non-A, non-B hepatitis PT-NANB immunohistochemical identification of disease and hepatitis C virus-associated antigens. *Hepatology.* **10**:580.
De Bisceglie AM, Alter H, Kuo G, Houghton M, Hoofnagle JH (1989) Detection of antibody to hepatitis C virus in patients with various chronic liver diseases. *Hepatology.* **10**:581.
Prince AM, Brotman B, Huima T, Krauledat P, Houghton M, Kuo G, Kuo Q-L. (1989) "Distinction between chronic and self-limited forms of hepatitis C virus infection" *in* International Meeting on Non-A, Non-B Hepatitis, Tokyo, Japan (Shikata T, Purcell RH, Uchida T. eds.) pp.7–16. Elsevier Science Publishers, Amsterdam.
Houghton M. (1990) Discovery of hepatitis C virus and assay of non-A, non-B hepatitis virus. *Jikken Igaku.* **8**:203–6.
Mosley JW, Aach RD, Hollinger FB, Stevens CE, Barbosa LH, Nemo GJ, Holland PV, Bancroft WH, Zimmerman HJ, Kuo G, Choo Q-L, Houghton M. (1990) Non-A, non-B hepatitis and antibody to hepatitis C virus. *JAMA.* **263**:77–8.
Weiner AJ, Kuo G, Bradley DW, Bonino F, Saracco G, Lee C, Rosenblatt J, Choo Q-L, Houghton M. (1990) Detection of hepatitis C viral sequences in non-A, non-B hepatitis. *Lancet.* **335**:1–3.
Choo Q-L, Weiner AJ, Overby LR, Kuo G, Houghton M. (1990) Hepatitis C virus: The major causative agent of viral non-A, non-B hepatitis. *Br. Med. Bull.* **46**:423–41.
Miyamura T, Saito I, Katayama T, Kikuchi S, Tateda A, Houghton M, Choo Q-L, Kuo G. (1990) Detection of antibody against antigen expressed by molecularly cloned hepatitis C virus cDNA: Application to diagnosis and blood screening for posttransfusion hepatitis. *Proc. Natl. Acad. Sci.* USA. **87**:983–7.
Kew MC, Houghton M, Choo Q-L, Kuo G. (1990) Hepatitis C virus antibodies in southern African blacks with hepatocellular carcinoma. *Lancet.* **335**:873–4.
Chien D-S, Kuo GC, Sung J-L, Lai M-Y, Sheu J-C, Chen P-J, Yang P-M, Hsu H-M, Chang M-H, Chen C-J, Hahn L-C, Choo Q-L, Wang T-H, Houghton M. (1990) Hepatitis C virus infection in an area hyperendemic for hepatitis B and chronic

liver disease: The Taiwan experience. *J. Infect. Dis.* **162**:817–22.
Makris M, Preston FE, Triger DR, Underwood JCE, Choo Q-L, Kuo G, Houghton M. (1909) Hepatitis C antibody and chronic liver disease in haemophilia. *Lancet.* **335**:1117–9.
Stevens CE, Taylor PE, Pindyck J, Choo Q-L, Bradley DW, Kuo G, Houghton M. (1990) Epidemiology of hepatitis C virus. A preliminary study in volunteer blood donors. *JAMA.* **263**:49–53.
Jeffers L, Perez G, de Medina M, Schiff E, Ortiz-Interian C, Bourgoignie J, Vaamonde CA, Houghton M, Choo Q-L, Kuo G. (1990) Hepatitis C HCV infection in hemodialysis units. *Kidney Int.* **37**:303.
Oliveri F, Baldi M, Brunetto MR, Saracco G, Rosina F, Cerenzia MT, Rizzetto M, Soranzo ML, Colla L, Vallauri P, Verme G, Kuo G, Houghton M, Bonino F. (1990) Antibody to hepatitis C virus in the serum of patients with chronic hepatitis. Eur. J. Gastroenterol. *Hepatol.* **2**:347–50.
Houghton M, Richman K, Han J, Berger K, Lee C, Dong C, Overby L, Weiner A, Bradley D, Kuo G, Choo Q-L. (1990) "Hepatitis C virus (HCV): A relative of the pestiviruses and flaviviruses" *in* Viral Hepatitis and Liver Disease (Hollinger FB, Lemon SM, Margolis HS. eds.) pp. 328–33. Williams & Wilkins, Baltimore, MD.
Choo Q-L, Berger K, Kuo G, Houghton M. (1990) "Detection and mapping of immunologic epitopes expressed by bacterial cDNA clones of the hepatitis C virus" *in* Viral Hepatitis and Liver Disease (Hollinger FB, Lemon SM, Margolis HS. eds.) pp. 345–6. Williams & Wilkins, Baltimore, MD.
Kuo G, Choo Q-L, Shuster J, Kuo C, Berger K, Lee WS, Medina-Selby A, Houghton M. (1990) "Serodiagnosis of hepatitis C viral infection using recombinant-based assays for circulating antibodies to different viral proteins" *in* Viral Hepatitis and Liver Disease (Hollinger FB, Lemon SM, Margolis HS. eds.) pp. 347–9. Williams & Wilkins, Baltimore, MD.
Weiner AJ, Truett MA, Rosenblatt J, Han J, Quan S, Polito AJ, Kuo G, Choo Q-L, Houghton M. (1990) "HCV: Immunologic and hybridization-based diagnostics" *in* Viral Hepatitis and Liver Disease (Hollinger FB, Lemon SM, Margolis HS. eds.) pp. 360–3. Williams & Wilkins, Baltimore, MD.
Boonmar S., Takeuchi K, Kubo Y, Katayama T, Harada H, Ohbayashi A, Choo Q-L, Kuo G, Houghton M, Saito I, Miyamura T. (1990) "Molecular cloning of hepatitis C virus cDNA from plasma of an implicated donor or post-transfusion non-A, non-B hepatitis" *in* Viral Hepatitis and Liver Disease (Hollinger FB, Lemon SM, Margolis HS. eds.) pp. 371–4. Williams & Wilkins, Baltimore, MD.
Shimizu YK, Weiner AJ, Rosenblatt J, Wong DC, Shapiro M, Popkin T, Houghton M, Alter HJ, Purcell RH. (1990) Early events in hepatitis C virus infection of chimpanzees. *Proc. Natl. Acad. Sci. USA.* **87**:6441–4.
Van der Poel CL, Reesink HW, Lelie PN, Cuijpers MT, Leentvaar,-Kuypers A, Bakker E, Exel-Oehlers PJ, Polito A, Houghton M, Schaasberg W. (1990) "Impact of blood-donor screening for anti-HCV versus ALT, and cofactors for infectivity of anti-HCV-positive blood" *in* Viral Hepatitis and Liver Disease (Hollinger FB, Lemon SM, Margolis HS. eds.) pp. 427–30. Williams & Wilkins, Baltimore, MD.
Esteban JI, González A, Hernández JM, Madoz P, Muniz E, Torras J, Enriquez J, Buenestado J, Martin-Vega C, Sánchez C, Esteban R, Guardia J, Houghton M, Alter HJ. (1990) "Open prospective efficacy trial of anti-HCV screening of blood donors to prevent posttransfusion hepatitis: Interim report of the Barcelona PTH study" *in* Viral Hepatitis and Liver Disease (Hollinger FB, Lemon SM,

Margolis HS. eds.) pp. 431–3. Williams & Wilkins, Baltimore, MD.

Krawczynski K, Kuo G, Dibisceglie A, Bradley D, Houghton M, Alter M, Ebert J. (1990) "Blood-borne non-A, non-B hepatitis: Detection and identification of hepatitis C virus and disease-associated antigen HCV Ag in hepatocytes" *in* Viral Hepatitis and Liver Disease (Hollinger FB, Lemon SM, Margolis HS. eds.) pp. 434–5. Williams & Wilkins, Baltimore, MD.

Prince AM, Brotman B, Huima T, Krauledat P, Houghton M, Kuo G, Choo Q-L, Polito A, Di Nello R, Nelles MJ. (1990) Use of anti-HCV determinations for diagnosis of chronic HCV infection" *in* Viral Hepatitis and Liver Disease (Hollinger FB, Lemon SM, Margolis HS. eds.) pp. 450–5. Williams & Wilkins, Baltimore, MD.

Miyamura T, Saito I, Yoneyama T, Takeuchi K, Ohbayashi A, Watanabe Y, Choo Q-L, Houghton M, Kuo G. (1990) 'Role of hepatitis C virus in hepatocellular carcinoma" *in* Viral Hepatitis and Liver Disease (Hollinger FB, Lemon SM, Margolis HS. eds.) pp. 559–62. Williams & Wilkins, Baltimore, MD.

Takeuchi K, Boonmar S, Kubo Y, Katayama T, Harada H, Ohbayashi A, Choo Q-L, Kuo G, Houghton M, Saito I, Miyamura T. (1990) Hepatitis C viral cDNA clones isolated from a healthy carrier donor implicated in post-transfusion non-A, non-B hepatitis. *Gene.* **91**:287–91.

Watanabe J, Minegishi K, Mitsumori T, Ishifuji M, Oguchi T, Ueda M, Tokunaga E, Tanaka E, Kiyosawa K, Furuta S, Katayama T, Kuo G, Choo Q-L, Houghton M, Nishioka K. (1990) Prevalence of anti-HCV antibody in blood donors in the Tokyo area. *Vox Sang.* **59**:86–8.

Katayama T, Kikuchi S, Tanaka Y, Saito I, Miyamura T, Choo Q-L, Houghton M, Kuo G. (1990) Blood screening for non-A, non-B hepatitis by hepatitis C virus antibody assay. *Transfusion.* **30**:374–6.

Takeuchi K, Kubo Y, Boonmar S, Watanabe Y, Katayama T, Choo Q-L, Kuo G, Houghton M, Saito I, Miyamura T. (1990) The putative nucleocapsid and envelope protein genes of hepatitis C virus determined by comparison of the nucleotide sequences of two isolates derived from an experimentally infected chimpanzee and healthy human carriers. J. Gen. *Virol.* **71**:3027–33.

Kremsdorf D, Thiers V, Garreau F, Duclos H, Porchon C, Houghton M, Tiollais P, Brechot C. (1990) Cloning and sequence analysis of hepatitis B virus variants in non-A and non-B infections. *J. Hepatol.* **11**:S36. Supplement 2.

Saito I, Miyamura T, Ohbayashi A, Harada H, Katayama T, Kikuchi S, Watanabe Y, Koi S, Onji M, Ohta Y, Choo Q-L, Houghton M, Kuo G. (1990) Hepatitis C virus infection is associated with the development of hepatocellular carcinoma. *Proc. Natl. Acad. Sci. USA.* **87**:6547–9.

Hasan F, Jeffers LJ, De Medina M, Reddy KR, Parker T, Schiff ER, Houghton M, Choo Q-L, Kuo G. (1990) Hepatitis C-associated hepatocellular carcinoma. Hepatology. 12:589–91.

Takeuchi K, Kubo Y, Boonmar S, Watanabe Y, Katayama T, Choo Q-L, Kuo G, Houghton M, Saito I, Miyamura T. (1990) Nucleotide sequence of core and envelope genes of the hepatitis C virus genome derived directly from human healthy carriers. *Nuc. Acids Res.* **18**:4626.

Weiner AJ, Truett MA, Han J, Polito AJ, Choo, Q-L, Rosenblatt J, Quan S, Kuo G, Houghton M, Page E, Agius C, Nelles MJ. (1990) HCV testing in low-risk population. *Lancet.* **336**:695. Letter.

Makris M, Dewar MS, Preston FE, Choo Q-L, Kuo G, Houghton M. (1990) The relation of hepatitis C antibodies to acute non-A, non-B hepatitis NANBH in previously untreated hemophilic patients. *Br. J. Haematol.* **74**:44. Supplement 1.

Magrin S, Craxi A, Almasio P, Fabiano C, Fiorentino G, Palazzo U, Pinzello GB, Provenzano G, Pagliaro L, Houghton M, Han JH. (1990) HCV infection in autoimmune chronic active hepatitis. *J. Hepatol.* **11**:S99. Supplement 2.

Thaler MM, Landers DW, Wara DW, Houghton M, Veereman-Wauters G, Sweet RL, Brauer M, Han JH. (1990) Vertical transmission of hepatitis C virus detected by the polymerase chain reaction. *Hepatology.* **12**:849.

McHutchison JG, Kuo G, Houghton M, Choo Q-L, Redeker AG. (1990) Patients with acute and chronic NANB hepatitis. *Hepatology.* **12**:966.

Leal RJ, McHutchison JG, Houghton M, Choo Q-L, Kuo G, Redeker AG. (1990) Antibodies to hepatitis C virus anti-HCV in alcoholic liver disease: An analysis of risk factors. *Hepatology.* **12**:881.

Makris M, Preston FE, Triger DR, Underwood JCE, Kuo G, Choo Q-L, Houghton M. (1990) The role of hepatitis C virus in chronic liver disease in hemophilia. Br. J. *Haematol.* **74**:6. Supplement 1.

Nishioka K, Watanabe J, Furuta S, Tanaka E, Suzuki H, Iino S, Tsuji T, Yano M, Kuo G, Choo Q-L, Houghton M, Oda T. (1991) Antibody to the hepatitis C virus in acute hepatitis and chronic liver diseases in Japan. *Liver.* **11**:65–70.

Nishioka K, Watanabe J, Furuta S, Tanaka E, Iino S, Suzuki H, Tsuji T, Yano M, Kuo G, Choo Q-L, Houghton M, Oda T. (1991) A high prevalence of antibody to the hepatitis C virus in patients with hepatocellular carcinoma in Japan. *Cancer.* **67**:429–33.

Katkov WN, Dienstag JL, Cody H, Evans AA, Choo QL, Houghton M, Kuo G. (1991) Role of hepatitis C virus in non-B chronic liver disease. *Arch. Int. Med.* **151**:1548–52.

Choo Q-L, Han J, Weiner AJ, Overby LR, Bradley DW, Kuo G, Houghton M. (1991) "Hepatitis C virus is a distant relative of the flaviviruses and pestiviruses" *in* Proceedings of the International Meeting on Non-A, Non-B Hepatitis, Tokyo, Japan (Shikata T, Purcell RH, Uchida T. eds.) pp. 47–52. Elsevier Science Publishers, Amsterdam.

Di Bisceglie AM, Order SE, Klein JL, Wagoner JG, Sjogren MH, Kuo G, Houghton M, Choo Q-L, Hoofnagle JH. (1991) The role of chronic viral hepatitis in hepatocellular carcinoma in the United States. *Am. J. Gastroenterol.* **86**:335–8.

Choo Q-L, Richman KH, Han JH, Berger K, Lee C, Dong C, Gallegos C, Coit D, Medina-Selby A, Barr PJ, Weiner AJ, Bradley DW, Kuo G, Houghton M. (1991) Genetic organization and diversity of the hepatitis C virus. *Proc. Natl. Acad. Sci. USA.* **88**:2451–5.

Weiner AJ, Brauer MJ, Rosenblatt J, Richman KH, Tung J, Crawford K, Bonino F, Saracco G, Choo Q-L, Houghton M, Han JH. (1991) Variable and hypervariable domains are found in the regions of HCV corresponding to the flavivirus envelope and NS1 proteins and the pestivirus envelope glycoproteins. *Virology.* **180**:842–8.

van der Poel CL, Cuypers HTM, Reesink HW, Weiner AJ, Quan S, DiNello R, van Boven JJP, Winkel I, Mulder-Folkerts D, Exel-Oehlers PJ, Schaasberg W, Leentvaar-Kuypers A, Polito A, Houghton M, Lelie PN. (1991) Confirmation of hepatitis C virus infection by new four-antigen recombinant immunoblot assay. *Lancet.* **337**:317–9.

Reesink HW, Van der Poel CL, Plaisier ADD, Verstraten JW, Cuypers M. (1991) Comparison of first and second generation anti-HCV recombinant immunoblot assay with 5' UTR PCR. *Transfusion.* **31**:57S.

Han JH, Shyamala V, Richman KH, Brauer MJ, Irvine B, Urdea MS, Tekamp-Olson P, Kuo G, Choo Q-L, Houghton M. (1991) Characterization of the terminal

regions of hepatitis C viral RNA: Identification of conserved sequences in the 5' untranslated region and poly(A) tails at the 3'end. *Proc. Natl. Acad. Sci. USA.* **88**:1711–5.
Magrin S, Craxi A, Fabiano C, Fiorentino G, Almasio P, Palazzo U, Pinzello G, Provenzano G, Pagliaro L, Choo Q-L, Kuo G, Polito A, Han J, Houghton M. (1991) Hepatitis C virus replication in 'autoimmune' chronic hepatitis. J. *Hepatol.* **13**:364–7.
Cha T-A, Kolberg J, Irvine B, Stempien M, Beall E, Yano M, Choo Q-L, Houghton M, Kuo G, Han JH, Urdea MS. (1991) Use of a signature nucleotide sequence of Hepatitis C virus for detection of viral RNA in human serum and plasma. *J. Clin. Microbiol.* **29**:2528–34.
Houghton M. (1991) "Molecular virology of HCV" *in* Report on the Proceedings, Second International Symposium on HCV, Los Angeles (Bradley DW. ed.) pp.2–3. Advanced Therapeutics Communications, Secaucus, New Jersey.
Krawczynski K, Beach M, Mimms L, Meeks E, Vallari D, Taskar S, Kuo G, Houghton M, Bradley D. (1991) Replication and antigenic expression of HCV antiviral antibody response and liver pathology in acute and chronic HCV infection. *Hepatology.* **14**:78A.
Houghton M, Weiner A, Han J, Kuo G, Choo Q-L. (1991) Molecular biology of the hepatitis C viruses: Implications for diagnosis, development and control of viral disease. *Hepatology.* **14**:381–8.
Colombo M, Rumi MG, Donato MF, Tommasini MA, Del Ninno E, Ronchi G, Kuo G, Houghton M. (1991) Hepatitis C antibody in patients with chronic liver disease and hepatocellular carcinoma. *Dig. Dis. Sci.* **36**:1130–3.
Thudium K, Spaete R, Berger K, Choo Q-L, Houghton M, Ralston R. (1991) Expression and characterization of HCV structural proteins using in-vitro translation and recombinant vaccinia viruses. *J. Cell. Biochem.* **47**:92. Supplement 15E.
McHutchison JG, Kuo G, Houghton M, Choo Q-L, Redeker AG. (1991) Hepatitis C virus antibodies in acute icteric and chronic non-A, non-B hepatitis. *Gastroenterology.* **101**:1117–9.
Saracco G, Baldi M, Calvo PL, Manzini P, Abate M, Chiaberge E, Brunetto MR, Rizzetto M, Chien D, Kuo G, Houghton M, Bonino F. (1991) Hepatitis C virus markers for monitoring interferon therapy in chronic hepatitis C. *Hepatology.* **14**:75A.
Cuypers HT, Winkel IN, van der Poel CL, Reesink HW, Lelie PN, Houghton M, Weiner A. (1991) Analysis of genomic variability of hepatitis C virus. *J. Hepatol.* **13**:S15–9.
Rosina F, Fabiano A, Garripoli A, Smedile A, Mattalia A, Eckart MR, Houghton M, Bonino F. (1991) Rabbit-derived anti-HD antibodies for HDAg immunoblotting. *J. Hepatol.* **13**:130–3.
Weiner AJ, Christopherson C, Hall JE, Bonino F, Saracco G, Crawford K, Marion CD, Crawford KA, Venkatakrishna S, Miyamura T, McHutchinson J, Cuypers T, Houghton M. (1991) Sequence variation in hepatitis C viral isolates. *Hepatology.* **13**:S6–14.
Krawczynski K, Beach MJ, Bradley DW, Kuo G, di Bisceglie AM, Houghton M, Reyes GR, Kim JP, Choo Q-L, Alter MJ (1992) Hepatitis C virus antigen in hepatocytes: Immunomorphologic detection and identification. *Gastroenterology.* **103**:622–9.
Bresters D, Mauser-Bunschoten EP, Cuypers HT, Lelie PN, Han JH, Jansen PL, Houghton M, Reesink HW (1992) Disappearance of hepatitis C virus RNA in

plasma during interferon alpha-2B treatment in hemophilia patients. *Scand. J. Gastroenterol.* **27**:166–8.

Bresters D, Cuypers HT, Reesink HW, Schaasberg WP, van der Poel CL, Mauser-Bunschoten EP, Houghton M, Choo Q-L, Kuo G, Lesniewski R, Troonen H, Lelie PN. (1992) Enhanced sensitivity of a second generation ELISA for antibody to hepatitis C virus. *Vox Sang.* **62**:213–7.

Overby LR, Houghton M. (1992) "Hepatitis viruses" *in* Laboratory Diagnosis of Viral Infections (Lennette EH. ed.) pp. 403–44. Marcel Dekker, New York.

Spaete RR, Alexander DA, Rugroden ME, Choo Q-L, Berger K, Crawford K, Kuo C, Leng S, Lee C, Ralston R, Thudium K, Tung JW, Kuo G, Houghton M. (1992) Characterization of the hepatitis C virus E2/NS1 gene product expressed in mam-malian cells. *Virology.* **188**:819–30.

Weiner AJ, Geysen HM, Christopherson C, Hall JE, Mason TJ, Saracco G, Bonino F, Crawford K, Marion CD, Crawford KA, Brunetto M, Barr PJ, Miyamura T, McHutchinson J, Houghton M. (1992) Evidence for immune selection of hepatitis C virus (HCV) putative envelope glycoprotein variants: Potential role in chronic HCV infections. *Proc. Natl. Acad. Sci. USA.* **89**:3468–72.

Houghton M. (1992) "Heterogeneity of the HCV genome: Importance for control of the disease" *in* Hepatitis C Virus: Scientific and Clinical Status (Deinhardt F, Bradley DW, Houghton M. eds.) pp. 8–9. Advanced Therapeutics Communications, Secaucus, New Jersey.

Jeffers LJ, Hasan F, De Medina M, Reddy R, Parker T, Silva M, Mendez L, Schiff ER, Manns M, Houghton M. (1992) Prevalence of antibodies to hepatitis C virus among patients with cryptogenic chronic hepatitis and cirrhosis. *Hepatology.* **15**:187–90.

Weiner AJ, Christopherson C, Hall JE, Crawford K, Marion CD, Crawford KA, Barr PJ, Richman K, Kuo G, Houghton M. (1992) "The hypervariable amino terminus of the hepatitis C virus E2/NS1 protein appears to be under immune selection" *in* Vaccines 92 (Brown F, Chanock RM, Ginsberg HS, Lerner RA. eds.) pp. 303–8. Cold Spring Harbor Laboratory Press, New York.

Chien DY, Choo Q-L, Tabrizi A, Kuo C, McFarland J, Berger K, Lee C, Shuster R, Nguyen T, Moyer DL, Tong M, Furuta S, Omata M, Alter H, Schiff E, Jeffers L, Houghton M, Kuo G. (1992) Diagnosis of hepatitis C virus (HCV) infection using an immunodominant chimeric polyprotein to capture circulating antibodies: Re-evaluation of the role of HCV in liver disease. *Proc. Natl. Acad. Sci. USA.* **89**:10011–5.

Choo Q-L, Kuo G, Weiner A, Wang K-S, Overby L, Bradley D, Houghton M. (1992) Identification of the major, parenteral non-A, non-B hepatitis agent (hepatitis C virus) using a recombinant cDNA approach. *Sem. Liver Dis.* **12**:279–88.

Cuypers HT, Bresters D, Winkel IN, Reesink HW Weiner AJ, Houghton M, van der Poel CL, Lelie PN. (1992) Storage conditions of blood samples and primer selection affect the yield of cDNA polymerase chain reaction products of hepatitis C virus. *J. Clin. Microbiol.* **30**:3220–4.

Weiner AJ, Venkatakrishna S, Hall JE, Houghton M, Han J. (1992) "PCR: Application to hepatitis C virus (HCV) research and diagnostics" *in* Frontiers in Virology (Becker Y, Darai G. eds.) pp. 86–100. Springer Verlag, New York.

Van der Poel CL, Bresters D, Reesink HW, Plaisier AAD, Schaasberg W, Leentvaar-Kuypers A, Choo Q-L, Quan S, Polito A, Houghton M, Kuo G, Lelie PN, Cuypers HTM. (1992) Early antihepatitis-C virus response with second generation C200/C22 ELISA. *Vox Sang.* **62**:208–12.

Han JH, Houghton M. (1992) Group specific sequences and conserved secondary structures at the 3' end of HCV genome and its implication for viral replication. *Nucl. Acids Res.* **20**:3520.

Koziel MJ, Dudley D, Wong JT, Dienstag J, Houghton M, Ralston R, Walker B. (1992) Intrahepatic cytotoxic T lymphocytes specific for hepatitis C virus in persons with chronic hepatitis. *J. Immunol.* **149**:3339–44.

Yoo BJ, Spaete RR, Geballe AP, Selby M, Houghton M, Han JH. (1992) 5' end-dependent translation initiation of hepatitis C viral RNA and the presence of putative positive and negative translational control elements within the 5' untranslated region. *Virology.* **191**:889–99.

Botarelli P, Brunetto MR, Weiner AJ, Minutello MA, Unutmaz D, Calvo P, Bonino F, Houghton M, Abrignani S. (1993) T cell response to recombinant proteins of hepatitis C virus in blood and liver of patients with different clinical courses of infection. *Gastroenterology.* **104**:580–7.

Chien DY, Choo Q-L, Tabrizi A, Kuo C, McFarland J, Berger K, Lee C, Shuster JR, Nguyen T, Moyer DL, Tong M, Furuta S, Omata M, Fong CT, Tegtmeier G, Alter H, Schiff E, Jeffers L, Houghton M, Kuo G. (1993) Use of recombinant HCV antigen in the serodiagnosis of hepatitis C virus infection: Significant improvement in HCV antibody detection as compared with the first generation HCV C100-3 ELISA and the synthetic peptide EIA tests. *J. Gastroenterol. Hepatol.* **8**:S33–9.

Eckart MR, Selby M, Masiarz F, Lee C, Berger K, Crawford K, Kuo C, Kuo G, Houghton M. (1993) The hepatitis C virus encodes a serine protease involved in processing of the putative nonstructural proteins from the viral polyprotein precursor. *Biochem. Biophys. Res. Comm.* **192**:399–406.

Saracco G, Rosina F, Abate ML, Chiandussi L, Gallo V, Cerutti E, Di Napoli A, Solinas A, DePlano A, Tocco A, Cossu P, Chien D, Kuo G, Polito A, Weiner AJ, Houghton M, Verme G, Bonino F, Rizzetto M (1993) Long-term follow-up of patients with chronic hepatitis C treated with different doses of interferon-α2b. *Hepatology.* **18**:1300–5.

Koziel MJ, Dudley D, Afdhal N, Choo Q-L, Houghton M, Ralston R, Walker BD. (1993) Hepatitis C virus (HCV)-specific cytotoxic T lymphocytes recognized epitopes in the core and envelope proteins of HCV. *J. Virol.* **67**:7522–32.

Bresters D, Mauser-Bunschoten EP, Cuypers HT, Han JH, Jansen PL, Chamuleau RA, Houghton M, Reesink HW. (1993) Long term treatment of chronic hepatitis C with interferon alfa-2b: disappearance of HCV-RNA in a pilot study of eight haemophilia patients. *Gut (England).* **34**:S124–5.

Rosina F, Fabiano A, Maran E, Cozzolongo R, Smedile A, Mazzucco G, Garripoli A, Costa C, Eckart MR, Houghton M. (1993) Rabbit-derived anti-HD antibodies for HDAg immunoblotting. *Proc. Clin. Biol. Res.* **382**:189–91.

Selby MJ, Choo Q-L, Berger K, Kuo G, Glazer E, Eckart M, Lee C, Chien D, Kuo C, Houghton M. (1993) Expression, identification and subcellular localization of the proteinsencoded by the hepatitis C viral genome. *J. Gen. Virol.* **4**:1103–13.

Weiner AJ, Thaler MM, Crawford K, Ching K, Kansopon J, Chien DY, Hall JE, Hu F, Houghton M. (1993) A unique, predominant hepatitis C virus variant found in an infant born to a mother with multiple variants. *J. Virol.* **67**:4365–8.

Minutello MA, Pileri P, Unutmaz D, Censini S, Kuo G, Houghton M, Brunetto MR, Bonino F, Abrignani S. (1993) Compartmentalization of T lymphocytes to the site of disease: Intrahepatic CD4+ T cells specific for the protein NS4 of hepatitis C virus in patients with chronic hepatitis C. *J. Exp. Med.* **178**:17–25.

Lok ASF, Chien D, Choo Q-L, Chan T-M, Chiu EKW, Cheng IKP, Houghton M, Kuo G. (1993) Antibody response to core, envelope and nonstructural hepatitis C virus antigens: Comparison of Immunocompetent and immunosuppressed patients. *Hepatology*. **18**:497–502.

Chien DY, Choo Q-L, Ralston R, Spaete R, Tong M, Houghton M, Kuo G. (1993) Persistence of HCV despite antibodies to both putative envelope glycoprotein. *Lancet*. **342**:933. Letter.

Erickson AL, Houghton M, Choo Q-L, Weiner AJ, Ralston R, Muchmore E, Walker CM. (1993) Hepatitis C virus-specific CTL responses in the liver of chimpanzees with acute and chronic hepatitis C. J. *Immunol*. **151**:4189–99.

Ralston R, Thudium K, Berger K, Kuo C, Gervase B, Hall J, Selby M, Kuo G, Houghton M, Choo Q-L. (1993) Characterization of hepatitis C virus envelope glycoprotein complexes expressed by recombinant vaccinia viruses. *J. Virol.* **67**:6753–61.

Botarelli P, Brunetto MR, Minutello MA, Calvo P, Unutmaz D, Weiner AJ, Choo Q-L, Shuster JR, Kuo G, Bonino F, Houghton M, Abrignani S. (1993) T-lymphocyte response to hepatitis C virus in different clinical courses of infection. *Gastroenterology*. **104**:580–7.

Choo Q-L, Kuo G, Ralston R, Weiner A, Chien D, Van Nest G, Han J, Berger K, Thudium K, Kuo C, Kansopon J, McFarland J, Tabrizi A, Ching K, Moss B, Cummins LB, Houghton M, Muchmore E. (1994) Vaccination of chimpanzees against infection by the hepatitis C virus. *Proc. Natl. Acad. Sci. USA*. **91**:1294–8.

Simmonds P, Alfredo A, Alter HJ, Bonino F, Bradley DW, Brechot C, Brouwer JT, Chan S-W, Chayama K, Chen D-S, Choo Q-L, Colombo M, Cuypers HTM, Date T, Dusheiko GM, Esteban JI, Fay O, Hadziyannis SJ, Han J, Hatzakis A, Holmes EC, Hotta H, Houghton M, Irvine B, Kohara M, Kolberg JA, Kuo G, Lau JYN, Lelie PN, Maertens G, McOmish F, Miyamura T, Mizokami M, Nomoto A, Prince AM, Reesink HW, Rice C, Roggendorf M, Schalm SW, Shimotohno K, Stuyver L, Trépo C, Weiner A, Yap PL, Urdea MS. (1994) A proposed system for the nomenclature of hepatitis C viral genotypes. *Hepatology*. **19**:1321–4.

Houghton M, Selby M, Weiner A, Choo Q-L. (1994) Hepatitis C virus: Structure, protein products and processing of the polyprotein precursor. *Curr. Stud. Hematol. Blood Trans*. **61**:1–11.

Houghton M, Choo Q-L, Kuo G, Ralston R, Selby M, Weiner A, Chien D, Han J, Walker C, Abrignani S, Koziel M, Walker B, Cummins L, Muchmore E. (1994) "The hepatitis C virus: Genetic organization, persistence, and vaccine strategies" *in* Viral Hepatitis and Liver Disease (Nishioka K, Suzuki H, Mishiro S, Oda T. eds.) pp. 33–7. Springer-Verlag, Tokyo.

Weiner AJ, Thaler MM, Crawford K, Kansopon J, Ching K, Hall JE, Hu F, Chien D, Houghton M. (1994) "HCV-positive, HIV-1-negative mothers transmit HCV" *in* Viral Hepatitis and Liver Disease (Nishioka K, Suzuki H, Mishiro S, Oda T. eds.) pp. 463–7. Springer-Verlag, Tokyo.

Ray R, Khanna A, Lagging LM, Meyer K, Choo Q-L, Ralston R, Houghton M, Becherer PR. (1994) Peptide immunogen mimicry of putative E1 glycoprotein-specific epitopes in hepatitis C virus. *J. Virol.* **68**:4420–6.

Selby MJ, Glazer E, Masiarz F, Houghton M. (1994) Complex processing and protein: Protein interactions in the E2:NS2 region of HCV. Virology. 204:114–22.

Saracco G, Abate ML, Baldi M, Calvo PL, Manzini P, Brunetto MR, Oliveri F, Kuo G, Chien D, Houghton M, Verme G, Rizzetto M, Bonino F. (1994) Hepatitis C virus markers in patients with long-term biochemical and histological remission of chronic hepatitis. *Liver*. **14**:65–70.

Chien DY, McFarland J, Tabrizi A, Kuo C, Houghton M, Kuo G. (1994) "Distinct subtypes of hepatitis C virus defined by antibodies directed to the putative core, NS4, and NS5 region polypeptides" *in* Viral Hepatitis and Liver Disease (Nishioka K, Suzuki H, Mishiro S, Oda T. eds.) pp. 320–4. Springer-Verlag, Tokyo.

Yoo BJ, Selby MJ, Choe J, Suh BS, Choi SH, Joh JS, Nuovo GJ, Lee H-S, Houghton M, Han JH. (1995) Transfection of a differentiated human hepatoma cell line (Huh7) with in vitro-transcribed hepatitis C virus (IICV) RNA and establishment of a long-term culture persistently infected with HCV. *J. Virol.* **69**:32–8.

Cerny A, McHutchison JG, Pasquinelli C, Brown ME, Brothers MA, Grabscheid B, Fowler P, Houghton M, Chisari FV. (1995) Cytotoxic T lymphocyte response to hepatitis C virus – derived peptides containing the HLA A2.1 binding motif. *J. Clin. Invest.* **95**:521–30.

Weiner A, Erickson AL, Kansopon J, Crawford K, Muchmore E, Hughes AL, Houghton M, Walker CM (1995) Persistent hepatitis C virus infection in a chimpanzee is associated with emergence of a cytotoxic T lymphocyte escape variant. *Proc. Natl. Acad. Sci. USA.* **92**:2755–9.

Cerny A, Fowler P, Brothers MA, Houghton M, Schlicht HJ, Chisari FV. (1995) Induction in vitro of a primary human antiviral cytotoxic T cell response. *Eur. J. Immunol.* **25**:627–30.

Houghton M, Choo Q-L, Kuo G, Weiner A, Chien D, Ralston R, Urdea M, Moss B, Purcell R, Cummins L, Muchmore E. (1995) Prospects for prophylactic and therapeutic hepatitis C virus vaccines. *Princess Takamatsu Symp.* **25**:237–43.

Weiner AJ, Erickson AL, Kansopon J, Crawford K, Muchmore E, Houghton M, Walker CM. (1995) Association of cytotoxic T lymphocyte (CTL) escape mutations with persistent hepatitis C virus (HCV) infection. *Princess Takamatsu Symp.* **25**:227–35.

Lagging LM, Meyer K, Hoft D, Houghton M, Belshe RB, Ray R. (1995) Immune responses to plasmid DNA encoding the hepatitis C virus core protein. *J. Virol.* **69**:5859–63.

Koziel MJ, Dudley D, Afdhal N, Grakoui A, Rice C, Choo Q-L, Houghton M, Walker BD. (1995) HLA class I-restricted cytotoxic T lymphocytes specific for hepatitis C virus: Identification of multiple epitopes and characterization of patterns of cytokine release. *J. Clin. Invest.* **96**:2311–21.

Piazza M, Chien D, Quan S, Houghton M. (1996) Lack of antibodies to the envelope glycoproteins of hepatitis C virus in immunoglobulin preparations from screened donors. *Boll. Soc. Ital. Biol. Sper.* **72**:69–70.

Houghton M. (1996) "Hepatitis C viruses" *in* Fields Virology (Fields BN, Knipe DM, Howley PM et al. eds.) pp. 1035–58. Lippincott-Raven Publishers, Philadelphia.

Rosa D, Campagnoli S, Moretto C, Guenzi E, Cousens L, Chin M, Dong C, Weiner AJ, Lau JY, Choo Q-L, Chien D, Pileri P, Houghton M, Abrignani S. (1996) A quantitative test to estimate neutralizing antibodies to the hepatitis C virus: cytofluorimetric assessment of envelope glycoprotein 2 binding to target cells. *Proc. Natl. Acad. Sci.* USA. **93**:1759–63.

Missale G, Bertoni R, Lamonaca V, Valli A, Massari M, Mori C, Rumi MG, Houghton M, Fiaccadori F, Ferrari C. (1996) Different clinical behaviors of acute hepatitis virus infection are associated with different vigor of the anti-viral cell-mediated immune response. *J. Clin. Invest.* **98**:706–14.

Rehermann B, Chang KM, McHutchinson JG, Kokka R, Houghton M, Chisari FV. (1996) Quantitative analysis of the peripheral blood cytotoxic T lymphocyte response in patients with chronic hepatitis C virus infection. *J. Clin. Invest.* **98**:1432–40.

Rehermann B, Chang KM, McHutchinson J, Kokka R, Houghton M, Rice CM, Chisari FV. (1996) Differential cytotoxic T-lymphocyte responsiveness to the hepatitis B and C viruses in chronically infected patients. *J. Virol.* **70**:7092–102.

Houghton M, Choo Q-L, Chien D, Kuo G, Weiner A, Coates S, Cousens L, Wininger M, Selby M, Ralston R, Berger K, Dong C, Crawford K, Kansopon J, Chin M, Wong S, Tabrizi-Weight A, Purcell RH, Muchmore E, Morandi M, Rosa D, Abrignani S. (1996) "Development of an HCV vaccine" *in* Viral Hepatitis and Liver Disease (Rizzetto M, Purcell RH, Gerin JL, Verme G. eds.) pp. 656–9. Edizioni Minerva Medica, Torino, Italy.

Nelson DR, Marousis CG, Davis GL, Rice CM, Wong J, Houghton M, Lau YN (1997) The role of hepatitis C virus-specific cytotoxic T lymphocytes in chronic hepatitis C. *J. Immunol.* **158**:1473–81.

Pasquinelli C, Shoenberger JM, Chung J, Chang KM, Guidotti LG, Selby M, Berger K, Lesniewski R, Houghton M, Chisari FV. Hepatitis C virus core and E2 protein expression in transgenic mice. *Hepatology*. **25**:719–27.

Diepolder HM, Gerlach J-T, Zachoval R, Hoffmann RM, Jung M-C, Wierenga EA, Scholz S, Santantonio T, Houghton M, Southwood S, Sette A, Pape GR. (1997) Immunodominant CD4+ T-cell epitope within nonstructural protein 3 in acute hepatitis C virus infection. *J. Virol.* **71**:6011–9.

Calvo PL, Kansopon J, Sra K, Quan S, DiNello R, Guaschino R, Calabrese G, Danielle F, Brunetto, Bonino F, Massaro AL, Polito A, Houghton M, Weiner AJ. (1998) Hepatitis C virus heteroduplex tracking assay for genotype determination reveals diverging genotype 2 isolates in Italian hemodialysis patients. *J. Clin. Microbiol.* **36**:227–33.

Wong DKH, Dudley DD, Afdhal NH, Dienstag J, Rice CM. (1998) Liver-derived CTL in hepatitis C virus infection: Breadth and specificity of responses in a cohort of persons with chronic infection. *J. Immunol.* **160**:1479–88.

Ishii K, Rosa D, Watanabe Y, Katayama T, Harada H, Wyatt C, Kiyosawa K, Aizaki H, Matsuura Y, Houghton M, Abrignani S, Miyamura T. (1998) High titers of antibodies inhibiting the binding of envelope to human cells correlate with natural resolution of chronic hepatitis C. *Hepatology*. **28**:1117–20.

Pileri P, Uematsu Y, Campagnoli S, Galli G, Falugi F, Petracca R, Weiner AJ, Houghton M, Rosa D, Grandi G, Abrignani S. (1998) Binding of hepatitis C virus to CD81. *Science*. **282**:938–41.

Selby M, Erickson A, Dong C, Cooper S, Parham P, Houghton M, Walker CM. (1999) Hepatitis C virus envelope glycoprotein E1 originates in the endoplasmic reticulum and requires cytoplasmic processing for presentation by class I MHC molecules. *J. Immunol.* **162**:669–76.

Cooper S, Erickson AL, Adams EJ, Kansopon J, Weiner AJ, Chien DY, Houghton M, Parham P, Walker CM. (1999) Analysis of a successful immune response against hepatitis C virus. *Immunity*. **10**:439–49.

Colombatto P, Brunetto MR, Kansopon J, Oliveri F, Maina A, Aragon U, Bortoli ML, Scatena F, Baicchi U, Houghton M, Bonino F, Weiner AJ. (1999) High prevalence of G1 and G2 TT-virus infection in subjects with high and low blood exposure risk: identification of G4 isolates in Italy. *J. Hepatol.* **31**:990–6.

Abrignani S, Houghton M, Hsu HH. (1999) Perspectives for a vaccine against hepatitis C virus. *J. Hepatol.* **31** Suppl 1:259–63.

Hsu HH, Abrignani S, Houghton M. (1999) Prospects for a hepatitis C virus vaccine. *Clin. Liver Dis.* **3**:901–15.

Liberman E, Fong YL, Selby MJ, Choo QL, Cousens L, Houghton M, Yen TS. (1999) Activation of the grp78 and grp94 promoters by hepatitis C virus E2 envelope protein. *J. Virol.* **73**:3718–22.

Mustilli AC, Izzo E, Houghton M, Galeotti CL. (1999) Comparison of secretion of a hepatitis C virus glycoprotein in Saccharomyces cerevisiae and Kluyveromyces lactis. *Res. Microbiol.* **150**:179–87.

Rosen HR, Hinrichs DJ, Gretch DR, Koziel MJ, Chou S, Houghton M, Rabkin J, Corless CL, Bouwer HG. (1999) Association of multispecific CD4(+) response to hepatitis C and severity of recurrence after liver transplantation. *Gastroenterology.* **117**:926–32.

Allander T, Drakenberg K, Beyene A, Rosa D, Abrignani S, Houghton M, Widell A, Grillner L, Persson MA. (2000) Recombinant human monoclonal antibodies against different conformational epitopes of the E2 envelope glycoprotein of hepatitis C virus that inhibit its interaction with CD81. *J. Gen. Virol.* **81**:2451–9.

Alter HJ, Houghton M. (2000) Clinical Medical Research Award. Hepatitis C virus and eliminating post-transfusion hepatitis. *Nat. Med.* **6**:1082–6.

Fan XG, Tang FQ, Yi H, Liu WE, Houghton M, Hu GL. (2000) Effect of IL-12 on T-cell immune responses in patients with chronic HCV infection. *APMIS.* **108**:531–8.

Heile JM, Fong YL, Rosa D, Berger K, Saletti G, Campagnoli S, Bensi G, Capo S, Coates S, Crawford K, Dong C, Wininger M, Baker G, Cousens L, Chien D, Ng P, Archangel P, Grandi G, Houghton M, Abrignani S. (2000) Evaluation of hepatitis C virus glycoprotein E2 for vaccine design: an endoplasmic reticulum-retained recombinant protein is superior to secreted recombinant protein and DNA-based vaccine candidates. *J. Virol.* **74**:6885–92.

Houghton M. (2000) Strategies and prospects for vaccination against the hepatitis C viruses. *Curr. Top. Microbiol. Immunol.* **242**:327–39. Review.

Lee AY, Manning WC, Arian CL, Polakos NK, Barajas JL, Ulmer JB, Houghton M, Paliard X. (2000) Priming of hepatitis C virus-specific cytotoxic T lymphocytes in mice following portal vein injection of a liver-specific plasmid DNA. *Hepatology.* **31**:1327–33.

Lee AY, Polakos NK, Otten GR, Ulmer JB, Houghton M, Paliard X. (2000) Quantification of the number of cytotoxic T cells specific for an immunodominant HCV-specific CTL epitope primed by DNA immunization. *Vaccine.* **18**:1962–8.

Petracca R, Falugi F, Galli G, Norais N, Rosa D, Campagnoli S, Burgio V, Di Stasio E, Giardina B, Houghton M, Abrignani S, Grandi G. (2000) Structure-function analysis of hepatitis C virus envelope-CD81 binding. *J. Virol.* **74**:4824–30.

Weiner A, Chien D, Choo Q-L, Coates S, Kuo G, Houghton M. (2000) "Humoral response to HCV" *in* Hepatitis C. (Liang J, Hoofnagle J. eds.), pp. 125–145. Academic Press.

Schirren CA, Jung MC, Gerlach JT, Worzfeld T, Baretton G, Mamin M, Hubert Gruener N, Houghton M, Pape GR. (2000) Liver-derived hepatitis C virus (HCV)-specific CD4(+) T cells recognize multiple HCV epitopes and produce interferon gamma. *Hepatology.* **32**:597–603.

Bassett SE, Guerra B, Brasky K, Miskovsky E, Houghton M, Klimpel GR, Lanford RE. (2001) Protective immune response to hepatitis C virus in chimpanzees rechallenged following clearance of primary infection. *Hepatology.* **33**:1479–87.

Erickson AL, Kimura Y, Igarashi S, Eichelberger J, Houghton M, Sidney J, McKinney D, Sette A, Hughes AL, Walker CM. (2001) The outcome of hepatitis C virus infection is predicted by escape mutations in epitopes targeted by cytotoxic T lymphocytes. *Immunity.* **15**:883–95.

Merola M, Brazzoli M, Cocchiarella F, Heile JM, Helenius A, Weiner AJ, Houghton M, Abrignani S. (2001) Folding of hepatitis C virus E1 glycoprotein in a cell-free system. *J. Virol.* **75**:11205–17.

Polakos NK, Drane D, Cox J, Ng P, Selby MJ, Chien D, O'Hagan DT, Houghton M, Paliard X. (2001) Characterization of hepatitis C virus core-specific immune responses primed in rhesus macaques by a nonclassical ISCOM vaccine. *J. Immunol.* **166**:3589–98.

Schirren CA, Jung MC, Worzfeld T, Mamin M, Baretton G, Gerlach JT, Gruener NH, Zachoval R, Houghton M, Rau HG, Pape GR. (2001) Hepatitis C virus-specific CD4+ T cell response after liver transplantation occurs early, is multispecific, compartmentalizes to the liver, and does not correlate with recurrent disease. *J. Infect. Dis.* **183**:1187–94.

Tseng CT, Miskovsky E, Houghton M, Klimpel GR. (2001) Characterization of liver T-cell receptor gammadelta T cells obtained from individuals chronically infected with hepatitis C virus (HCV): evidence for these T cells playing a role in the liver pathology associated with HCV infections. *Hepatology.* **33**:1312–20.

Weiner AJ, Paliard X, Selby MJ, Medina-Selby A, Coit D, Nguyen S, Kansopon J, Arian CL, Ng P, Tucker J, Lee CT, Polakos NK, Han J, Wong S, Lu HH, Rosenberg S, Brasky KM, Chien D, Kuo G, Houghton M. (2001) Intrahepatic genetic inoculation of hepatitis C virus RNA confers cross-protective immunity. *J. Virol.* **75**:7142–8.

Legrand E, Neau D, Galperine T, Trimoulet P, Moreau JF, Pitard V, Lacut JY, Ragnaud JM, Dupon M, Le Bail B, Bernard N, Schvoerer E, Houghton M, Fleury H, Lafon ME. (2002) CD4 T lymphocyte proliferative responses to hepatitis C virus (HCV) antigens in patients coinfected with HCV and human immunodeficiency virus who responded to anti-HCV treatment. *J. Infect. Dis.* **186**:302–11.

Masciopinto F, Freer G, Burgio VL, Levy S, Galli-Stampino L, Bendinelli M, Houghton M, Abrignani S, Uematsu Y. (2002) Expression of human CD81 in transgenic mice does not confer susceptibility to hepatitis C virus infection. *Virology.* **304**:187–96.

Shoukry NH, Grakoui A, Houghton M, Chien DY, Ghrayeb J, Reimann KA, Walker CM. (2003) Memory CD8+ T cells are required for protection from persistent hepatitis C virus infection. *J. Exp. Med.* **197**:1645–55.

Seo MY, Abrignani S, Houghton M, Han JH. (2003) Small interfering RNA-mediated inhibition of hepatitis C virus replication in the human hepatoma cell line Huh-7. *J. Virol.* **77**:810–2.

Masciopinto F, Giovani C, Campagnoli S, Galli-Stampino L, Colombatto P, Brunetto M, Yen TS, Houghton M, Pileri P, Abrignani S. (2004) Association of hepatitis C virus envelope proteins with exosomes. Eur. *J. Immunol.* **34**:2834–42.

Houghton M, Abrignani S. (2004) "Vaccination against the hepatitis C viruses" *in* New Generation Vaccines 3rd Edition (Levine MM, Kaper JB, Rappuoli R, Liu M, Good MF. eds.) pp. 593–606. Marcel Dekker Inc., New York.

O'Hagan DT, Singh M, Dong C, Ugozzoli M, Berger K, Glazer E, Selby M, Wininger M, Ng P, Crawford K, Paliard X, Coates S, Houghton M. (2004) Cationic microparticles are a potent delivery system for a HCV DNA vaccine. *Vaccine.* **23**:672–80.

Coates S, Choo Q-L, Kuo G, Crawford K, Dong C, Wininger M, Weiner A, Abrignani S, Houghton M. (2004) "Hepatitis C" *in* Vaccines: Preventing Disease & Protecting Health (De Quadros CA. ed.) pp. 150–6. Pan American Health Organisation.

Song HC, Seo MY, Stadler K, Yoo BJ, Choo QL, Coates SR, Uematsu Y, Harada T,

Greer CE, Polo JM, Pileri P, Eickmann M, Rappuoli R, Abrignani S, Houghton M, Han JH. (2004) Synthesis and characterization of a native, oligomeric form of recombinant severe acute respiratory syndrome coronavirus spike glycoprotein. *J. Virol.* 78:10328–35.

Brazzoli M, Helenius A, Foung SK, Houghton M, Abrignani S, Merola M. (2005) Folding and dimerization of hepatitis C virus E1 and E2 glycoproteins in stably transfected CHO cells. *Virology.* **332**:438–53.

Coates S, Choo Q-L, Kuo G, Crawford K, Dong C, Wininger M, Weiner A, Berger K, Wong S, Ralston R, Morandi M, Pileri P, Rosa D, Muchmore E, Mahoney J, Brasky K, Purcell R, Abrignani S, Houghton, M. (2005) "Protection of chimpanzees against heterologous 1a viral challenge using a gpE1/gpE2 heterodimer vaccine" *in* Proceedings of the 11th International Symposium on Viral Hepatitis & Liver Disease (Jilbert AR, Grgacic EVL, Vickery K, Burrell CJ, Cossart YE. eds.) pp. 118–123. Australian Center for Hepatitis Virology.

Nattermann J, Schneiders AM, Leifeld L, Langhans B, Schulz M, Inchauspe G, Matz B, Brackmann HH, Houghton M, Sauerbruch T, Spengler U. (2005) Serum antibodies against the hepatitis C virus E2 protein mediate antibody-dependent cellular cytotoxicity (ADCC). *J. Hepatol.* **42**:499–504.

Houghton M, Abrignani S. (2005) Prospects for a vaccine against the hepatitis C virus. *Nature.* **436**:961–6.

Rosa D, Saletti G, De Gregorio E, Zorat F, Comar C, D'Oro U, Nuti S, Houghton M, Barnaba V, Pozzato G, Abrignani S. (2005) Activation of naive B lymphocytes via CD81, a pathogenic mechanism for hepatitis C virus-associated B lymphocyte disorders. *Proc. Natl. Acad. Sci. USA.* **102**:18544–9.

Vajdy M, Selby M, Medina-Selby A, Coit D, Hall J, Tandeske L, Chien D, Hu C, Rosa D, Singh M, Kazzaz J, Nguyen S, Coates S, Ng P, Abrignani S, Lin Y-L, Houghton M, O'Hagan D. (2006) Hepatitis C virus polyprotein vaccine formulations capable of inducing broad antibody and cellular immune responses. *J. Gen. Virol.* **87**:2253–62.

Stamataki Z, Coates S, Evans MJ, Wininger M, Crawford K, Dong C, Fong YL, Chien D, Abrignani S, Balfe P, Rice CM, McKeating JA, Houghton M. (2007) Hepatitis C virus envelope glycoprotein immunization of rodents elicits cross-reactive neutralizing antibodies. *Vaccine.* **25:**7773–84.

Manns MP, Foster GR, Rockstroh JK, Zeuzem S, Zoulim F, Houghton M. (2007) The way forward in HCV treatment – finding the right path. Nat. Rev. Drug Discov. 6:991–1000. Review. Erratum in: *Nat. Rev. Drug Discov.* (2008) **7**:102. *Nat. Rev. Drug Discov.* (2008) **7**:458.

Bowen DG, Shoukry NH, Grakoui A, Fuller MJ, Cawthon AG, Dong C, Hasselschwert DL, Brasky KM, Freeman GJ, Seth NP, Wucherpfennig KW, Houghton M, Walker CM. (2008) Variable patterns of programmed death-1 expression on fully functional memory T cells after spontaneous resolution of hepatitis C virus infection. *J. Virol.* **82**:5109–14.

Lin Y, Kwon T, Polo J, Zhu YF, Coates S, Crawford K, Dong C, Wininger M, Hall J, Selby M, Coit D, Medina-Selby A, McCoin C, Ng P, Drane D, Chien D, Han J, Vajdy M, Houghton M. (2008) Induction of broad CD4+ and CD8+ T-cell responses and cross-neutralizing antibodies against hepatitis C virus by vaccination with Th1-adjuvanted polypeptides followed by defective alphaviral particles expressing envelope glycoproteins gpE1 and gpE2 and nonstructural proteins 3, 4, and 5. *J. Virol.* **82**:7492–503.

Houghton M. (2009) Discovery of the hepatitis C virus. *Liver Int.* **29** Suppl 1:82–8. Review.

Houghton M. (2009) The long and winding road to the identification of the hepatitis C virus. *J. Hepatol.* **51**:939–48.
Drane D, Maraskovsky E, Gibson R, Mitchell S, Barnden M, Moskwa A, Shaw D, Gervase B, Coates S, Houghton M, Basser R. (2009) Priming of CD4+ and CD8+ T cell responses using a HCV core ISCOMATRIX vaccine: a phase I study in healthy volunteers. *Hum. Vaccin.* **5**:151–7.
Ray R, Meyer K, Banerjee A, Basu A, Coates S, Abrignani S, Houghton M, Frey SE, Belshe RB. (2010) Characterization of antibodies induced by vaccination with hepatitis C virus envelope glycoproteins. *J. Infect. Dis.* **202**:862–6.
Frey SE, Houghton M, Coates S, Abrignani S, Chien D, Rosa D, Pileri P, Ray R, Di Bisceglie AM, Rinella P, Hill H, Wolff MC, Schultze V, Han JH, Scharschmidt B, Belshe RB. (2010) Safety and immunogenicity of HCV E1E2 vaccine adjuvanted with MF59 administered to healthy adults. *Vaccine.* **28**:6367–73.
Houghton M. (2011) Prospects for prophylactic and therapeutic vaccines against the hepatitis C viruses. *Immunol. Rev.* **239**:99–108.
Stamataki Z, Coates S, Abrignani S, Houghton M, McKeating, J. (2011) Immunization of human volunteers with Hepatitis C virus envelope glycoproteins elicits antibodies that cross-neutralize heterologous virus strains. *J. Infect. Dis.* **204**:811–3.
Meunier JC, Gottwein JM, Houghton M, Russell RS, Emerson SU, Bukh J, Purcell RH. (2011) Vaccine-induced cross-genotype reactive neutralizing antibodies against hepatitis C virus. *J. Infect. Dis.* **204**:1186–90.
Houghton M. (2012) Chimp virus makes a savvy vaccine vector. *Sci. Transl. Med.* **4**:115fs1.
Law JLM, Chen C, Wong J, Hockman D, Santer DM, Frey SE, Belshe RB, Wakita T, Bukh J, Jones CT, Rice CM, Abrignani S, Tyrrell DL, Houghton M. (2013) A hepatitis C virus (HCV) vaccine comprising envelope glycoproteins gpE1/gpE2 derived from a single isolate elicits broad cross-genotype neutralizing antibodies in humans. PLoS ONE. 8:e59776. doi:10.1371/journal.pone.0059776.
Houghton M. (2013) Three isn't the magic number. Nat. Med. 19:807.
Santer DM, Ma MM, Hockman D, Landi A, Tyrrell DL, Houghton M. (2013) Enhanced activation of memory, but not naïve, B cells in chronic hepatitis C virus-infected patients with cryoglobulinemia and advanced liver fibrosis. *PLoS ONE.* **8**:e68308. doi: 10.1371/journal.pone.0068308.
Nourbakhsh M, Douglas DN, Pu CH, Lewis JT, Kawahara T, Lisboa LF, Wei E, Asthana S, Quiroga AD, Law LM, Chen C, Addison WR, Nelson R, Houghton M, Lehner R, Kneteman NM. (2013) Arylacetamide deacetylase: A novel host factor with important roles in the lipolysis of cellular triacylglycerol stores, VLDL assembly and HCV production. *J. Hepatol.* **59**:336–43.
Houghton M, Law J, Tyrrell DL. (2013) An inactivated hepatitis C virus vaccine on the horizon? *Gastroenterology.* **145**:285–8.
Grebely J, Bilodeau M, Feld JJ, Bruneau J, Fischer B, Raven JF, Roberts E, Choucha N, Myers RP, Sagan SM, Wilson JA, Bialystok F, Tyrrell DL, Houghton M, Krajden M. (2013) The Second Canadian Symposium on Hepatitis C Virus: A call to action. *Can. J. Gastroenterol.* **27**:627–32.
Steenbergen RH, Joyce MA, Thomas BS, Jones D, Law J, Russell R, Houghton M, Tyrrell DL. (2013) Human serum leads to differentiation of human hepatoma cells, restoration of very-low-density lipoprotein secretion, and a 1000-fold increase in HCV Japanese fulminant hepatitis type 1 titers. *Hepatology.* **58**:1907–17.
Law LMJ, Landi A, Magee WC, Tyrrell DL, Houghton M. (2013) Progress towards a

hepatitis C virus vaccine. *Emerg. Microbes Infect.* **2**:e79. doi:10.1038/emi.2013.79.

Colombatto P, Brunetto MR, Maina AM, Romagnoli V, Almasio P, Rumi MG, Ascione A, Pinzello G, Mondelli M, Muratori L, Rappuoli R, Rosa, D, Houghton M, Abrignani S, Bonino F. (2013) HCV E1E2-MF59 vaccine in chronic hepatitis C patients treated with PEG-IFNα2a and ribavirin: a randomized controlled trial. *J. Viral Hepatitis.* doi: **10**.1111/jvh.12163.

Houghton M. (2014) Hepatitis C: the next 25 years. *Antiviral Res.***110**:77–8. doi: 10.1016/j.antiviral.2014.06.018.

Wong JA, Bhat R, Hockman D, Logan M, Chen C, Levin A, Frey SE, Belshe RB, Tyrrell DL, Law JL, Houghton M. (2014) Recombinant hepatitis C virus envelope glycoprotein vaccine elicits antibodies targeting multiple epitopes on the envelope glycoproteins associated with broad cross-neutralization. *J Virol.* **88**:14278–88. doi: 10.1128/JVI.01911-14.

MacParland SA, Bilodeau M, Grebely J, Bruneau J, Cooper C, Klein M, Sagan S, Choucha N, Balfour L, Bialystok F, Krajden M, Raven J, Roberts E, Russell R, Houghton M, Tyrrell DL, Feld JJ. (2014) National CIHR Research Training Program in Hepatitis C. The 3rd Canadian Symposium on Hepatitis C Virus: expanding care in the interferon-free era. *Can J Gastroenterol Hepatol.* **28**:481–7.

Ahmed M, Wang F, Levin A, Le C, Eltayebi Y, Houghton M, Tyrrell L, Barakat K. (2015) Targeting the Achilles heel of the hepatitis B virus: a review of current treatments against covalently closed circular DNA. *Drug Discov Today.* **20**:548–61. doi: 10.1016/j.drudis.2015.01.008.

Logan M, Law J, Wong JA, Hockman D, Landi A, Chen C, Crawford K, Kundu J, Baldwin L, Johnson J, Dahiya A, LaChance G, Marcotrigiano J, Law M, Foung S, Tyrrell L, Houghton M.(2016) Native folding of a recombinant gpe1/gpe2 heterodimer vaccine antigen from a precursor protein fused with Fc IgG. *J Virol.* **91**:e01552-16 doi:10.1128/JVI.01552-16.

Huang W, Wang Y, Chen C, Law JL, Houghton M, Chen L. (2016) Fabrication of flexible self-standing all-cellulose nanofibrous composite membranes for virus removal. *Carbohydrate Polymers.* **143**:9–17. doi: 10.1016/j.carbpol.2016.02.011.

O'Shea D, Law J, Egli A, Douglas D, Lund G, Forester S, Lambert J, Law M, Burton DR, Tyrrell DL, Houghton M, Humar A, Kneteman N. (2016) Prevention of hepatitis C virus infection using a broad cross-neutralizing monoclonal antibody (AR4A) and epigallocatechin gallate. *Liver Transplantation.* **22**:324–32. doi: 10.1002/lt.24344.

Freedman H, Logan MR, Law JL, Houghton M. (2016) Structure and Function of the Hepatitis C Virus Envelope Glycoproteins E1 and E2: Antiviral and Vaccine Targets. ACS Infect Dis. 2:749–762. Epub 2016 Aug 16

Freedman H, Logan MR, Hockman D, Leman JK,, Law JL, Houghton M. (2017) Computational Prediction of the Heterodimeric and Higher Order Structure of gpE1/gpE2 Envelope Glycoproteins Encoded by the Hepatitis C Virus (HCV). *J Virol.* doi: **10**.1128/JVI.02309-16.

Sarhan MA, Abdel-Hakeem MS, Mason AL, Tyrrell DL, Houghton M. (2017) Glycogen Synthase Kinase 3β Inhibitors Prevent Hepatitis C Virus Release/Assembly through Perturbation of Lipid Metabolism. *Scientific Reports.* doi: **10**.1038/s41598-017-02648-6

Landi A, Law J, Hockman D, Logan M, Crawford K, Chen C, Kundu J, Ebensen T, Guzman CA, Deschatelets L, Krishnan L, Tyrrell DLJ, Houghton M. (2017) Superior immunogenicity of HCV envelope glycoproteins when adjuvanted with cyclic-di-AMP, a STING activator or archaeosomes. *Vaccine.* **35**:6949–6956 doi: 10.1016/j.vaccine.2017.10.072.

Bartenschlager R, Baumert TF, Bukh J, Houghton M, Lemon SM, Lindenbach BD, Lohmann V, Moradpour D, Pietschmann T, Rice CM, Thimme R, Wakita T. (2018). Critical challenges and emerging opportunities in Hepatitis C virus research in an era of potent antiviral therapy: Considerations for scientists and funding agencies. *Virus research.* **248**:53–62. doi: 10.1016/j.virusres.2018.02.016

Law JLM, Logan M, Wong J, Kundu J, Hockman D, Landi A, Chen C, Crawford K, Wininger M, Johnson J, Mesa Prince C, Dudek E, Mehta N, Tyrrell DL, Houghton M. (2018) Role of the E2 Hypervariable Region (HVR1) in the Immunogenicity of a Recombinant Hepatitis C Virus Vaccine. *Journal of virology.* **92**. pii: e02141-17. doi: 10.1128/JVI.02141-17.

Khera T, Behrendt P, Bankwitz D, Brown RJP, Todt D, Doepke M, Khan AG, Schulze K, Law J, Logan M, Hockman D, Wong JAJ, Dold L, Gonzalez-Motos V, Spengler U, Viejo-Borbolla A, Ströh LJ, Krey T, Tarr AW, Steinmann E, Manns MP, Klein F, Guzman CA, Marcotrigiano J, Houghton M, Pietschmann T. (2018) Functional and immunogenic characterization of diverse HCV glycoprotein E2 variants. *J Hepatol.* doi: **10**.1016/j.jhep.2018.11.003

Banda DH, Perin PM, Brown RJP, Todt D, Solodenko W, Hoffmeyer P, Kumar Sahu K, Houghton M, Meuleman P, Müller R, Kirschning A, Pietschmann T.(2019) A central hydrophobic E1 region controls the pH range of hepatitis C virus membrane fusion and susceptibility to fusion inhibitors. *J Hepatol.* **70**(6):1082–1092. doi: 10.1016/j.jhep.2019.01.033. Epub 2019 Feb 13.

Houghton M. (2019) Hepatitis C Virus: 30 Years after Its Discovery. *Cold Spring Harb Perspect Med.* 9(12). doi: 10.1101/cshperspect.a037069.

Akache B, Deschatelets L, Harrison BA, Dudani R, Stark FC, Jia Y, Landi A, Law JLM, Logan M, Hockman D, Kundu J, Tyrrell DL, Krishnan L, Houghton M, McCluskie MJ. (2019) Effect of Different Adjuvants on the Longevity and Strength of Humoral and Cellular Immune Responses to the HCV Envelope Glycoproteins. *Vaccines (Basel).* **7**(4). doi: 10.3390/vaccines7040204.

Immunoregulation

Egli A, Levin A, Santer DM, Joyce M, O'Shea D, Thomas BS, Lisboa LF, Barakat K, Bhat R, Fischer KP, Houghton M, Tyrrell DL, Kumar D, Humar A. (2014) Immunomodulatory Function of Interleukin 28B during primary infection with cytomegalovirus. *J Infect Dis.* **210**:717–27. doi: 10.1093/infdis/jiu144.

Egli A, Santer D, Barakat K, Zand M, Levin A, Vollmer M, Weisser M, Khanna N, Kumar D, Tyrrell L, Houghton M, Battegay M, O'Shea D. (2014) Vaccine adjuvants-understanding molecular mechanisms to improve vaccines. *Swiss Med Wkly.* **144**:w13940. doi: 10.4414/smw.2014.13940. eCollection 2014.

Egli A, Santer DM, O'Shea D, Barakat K, Syedbasha M, Vollmer M, Baluch A, Bhat R, Groenendyk J, Joyce MA, Lisboa LF, Thomas BS, Battegay M, Khanna N, Mueller T, Tyrrell DL, Houghton M, Humar A, Kumar D. (2014) IL-28B is a key regulator of B- and T-cell vaccine responses against influenza. *PLoS Pathog.* **10:**e1004556. doi: 10.1371/journal.ppat.1004556. eCollection 2014 Dec.

Egli A, Santer DM, O'Shea D, Tyrrell DL, Houghton M. (2014) The impact of the interferon-lambda family on the innate and adaptive immune response to viral infections. *Emerging Microbes Infections.* **3**:e51. doi: 10.1038/emi.2014.51.

Egli A, Humar A, Widmer LA, Lisboa LF, Santer DM, Mueller T, Stelling J, Baluch A, O'Shea D, Houghton M, Kumar D. (2015) Effect of Immunosuppression on T-Helper 2 and B-Cell Responses to Influenza Vaccination. *J Infect Dis.* **212**:137–46. doi: 10.1093/infdis/jiv015.

Syedbasha M, Linnik J, Santer D, O'Shea D, Barakat K, Joyce M, Khanna N, Tyrrell

DL, Houghton M, Egli A. (2016) An ELISA Based Binding and Competition Method to Rapidly Determine Ligand-receptor Interactions. *J Vis Exp*. doi: 10.3791/53575.

Santer DM, Minty GE, Mohamed A, Baldwin L, Bhat R, Joyce M, Egli A, Tyrrell DL, Houghton M. (2017) A novel method for detection of IFN-lambda 3 binding to cells for quantifying IFN-lambda receptor expression. *J Immunol Methods*. doi:10.1016/j.jim.2017.03.001.

Okoye IS, Houghton M, Tyrrell L, Barakat K, Elahi S. (2017) Coinhibitory Receptor Expression and Immune Checkpoint Blockade: Maintaining a Balance in CD8+ T Cell Responses to Chronic Viral Infections and Cancer. *Front Immunol*. doi: 10.3389/fimmu.2017.01215. eCollection 2017.

Santer DM, Minty GES, Golec DP, Lu J, May J, Namdar A, Shah J, Elahi S, Proud D, Joyce M, Tyrrell DL, Houghton M. (2020) Differential expression of interferon-lambda receptor 1 splice variants determines the magnitude of the antiviral response induced by interferon-lambda 3 in human immune cells. *PLoS Pathog*. **16**(4):e1008515. doi: 10.1371/journal.ppat.1008515

Freedman H, Kundu J, Tchesnokov EP, Law JLM, Nieman JA, Schinazi RF, Tyrrell DL, Gotte M, Houghton M. (2020) Application of Molecular Dynamics Simulations to the Design of Nucleotide Inhibitors Binding to Norovirus Polymerase. *J Chem Inf Model*. **60**(12):6566–6578. doi: 10.1021/acs.jcim.0c00742. Epub 2020 Dec 1. PMID: 33259199; PMCID: PMC7869559.

Autoimmune Liver Disease

Landi A, Weismuller TJ, Lankisch TO, Santer DM, Tyrrell DL, Manns MP, Houghton M. (2014) Differential serum levels of eosinophilic eotaxins in primary sclerosing cholangitis, primary biliary cirrhosis, and autoimmune hepatitis. *J Interferon Cytokine Res*. **34**(3):204–14. doi: 10.1089/jir.2013.0075.

Chronic Fatigue Syndrome

Steffen I, Tyrrell DL, Stein E, Montalvo L, Lee TH, Zhou Y, Lu K, Switzer WM, Tang S, Jia H, Hockman D, Santer DM, Logan M, Landi A, Law J, Houghton M, Simmons G. (2011) No evidence for XMRV nucleic acids, infectious virus or anti-XMRV antibodies in Canadian patients with chronic fatigue syndrome. *PLoS ONE*. **6**:e27870.

Jason LA, Unger ER, Dimitrakoff JD, Fagin AP, Houghton M, Cook DB, Marshall GD Jr, Klimas N, Snell C. (2012) Minimum data elements for research reports on CFS. *Brain Behav. Immun*. **26**:401–6.

Landi A, Broadhurst D, Vernon SD, Tyrrell DL, Houghton M. (2015) Reductions in circulating levels of IL-16, IL-7 and VEGF-A in myalgic encephalomyelitis/chronic fatigue syndrome. *Cytokine*. **78**:27–36. doi: 10.1016/j.cyto.2015.11.018.

Gindin Y, Chung C, Jiang Z, Zhou JZ, Xu J, Billin AN, Myers RP, Goodman Z, Landi A, Houghton M, Green RM, Levy C, Kowdley KV, Bowlus CL, Muir AJ, Trauner M. (2021) A Fibrosis-Independent Hepatic Transcriptomic Signature Identifies Drivers of Disease Progression in Primary Sclerosing Cholangitis. *Hepatology*. **73**(3):1105–1116. doi: 10.1002/hep.31488. Epub 2021 Feb 28. PMID: 32745270; PMCID: PMC8048608.

Computational Modeling

Barakat KH, Huzil JT, Jordan KE, Evangelinos C, Houghton M, Tuszynski J. (2013) A computational model for overcoming drug resistance using selective dual-inhibitors for aurora kinase A and its T217D variant. *Mol. Pharm*. **10**:4572–89.

Barakat KH, Law J, Prunotto A, Magee WC, Evans DH, Tyrrell DL, Tuszynski J, Houghton M. (2013) Detailed computational study of the active site of the hepatitis C viral RNA polymerase to aid novel drug design. *J. Chem. Inf. Model.* **53**:3031–43.

Anwar-Mohamed A, Barakat KH, Bhat R, Noskov SY, Tyrrell DL, Tuszynski JA,Houghton M. (2014) A human ether-á-go-go-related (hERG) ion channel atomistic model generated by long supercomputer molecular dynamics simulations and its use in predicting drug cardiotoxicity. *Toxicology Letters.* **230**:382–92. doi: 10.1016/j.toxlet.2014.08.007.

Barakat KH, Anwar-Mohamed A, Tuszynski JA, Robins MJ, Tyrrell DL, Houghton M. (2015) A Refined Model of the HCV NS5A protein bound to daclatasvir explains drug-resistant mutations and activity against divergent genotypes. *J Chem Inf Model.* **55**:362–73. doi: 10.1021/ci400631n.

Ahmed M, Pal A, Houghton M, Barakat K. (2016) A Comprehensive Computational Analysis for the Binding Modes of Hepatitis C Virus NS5A Inhibitors: The Question of Symmetry. *ACS Infect Dis.* **2**:872–881.

Ahmed M, Jalily Hasani H, Ganesan A, Houghton M, Barakat K. (2017) Modeling the human Nav1.5 sodium channel: structural and mechanistic insights of ion permeation and drug blockade. *Drug Des Devel Ther.* doi: 10.2147/DDDT.S133944. eCollection 2017.

Barakat KH, Houghton M, Tyrrell DL, Tuszynski JA. (2017) Rational Drug Design Rational Drug Design: One Target, Many Paths to It. *IGI Global.* doi: 10.4018/978-1-5225-1762-7.ch044

Freedman H, Winter P, Tuszynski J, Tyrrell DL, Houghton M. (2018) A computational approach for predicting off-target toxicity of antiviral ribonucleoside analogues to mitochondrial RNA polymerase. *The Journal of biological chemistry.* **293**:9696–9705. doi: 10.1074/jbc.RA118.002588

Ganesan A, Ahmed M, Okoye I, Arutyunova E, Babu D, Turnbull W, Kundu JK, Shields J, Agopsowicz K, Xu L, Tabana Y, Srivastava N, Zhang G, Moon T, Belovodskiy A, Hena M, Kandadai AS, Hosseini S, Hitt M, Walker J, Smylie M, West FG, Siraki AG, Lemieux MJ, Elahi S, Nieman JA, Tyrrell DL, Houghton M, Barakat K. (2019) Comprehensive in vitro characterization of PD-L1 small molecule inhibitors. *Scientific Reports.* **9**(1):12392. doi: 10.1038/s41598-019-48826-6.

Kalyaanamoorthy S, Lamothe SM, Hou X, Moon TC, Kurata HT, Houghton M, Barakat KH. (2020) A structure-based computational workflow to predict liability and binding modes of small molecules to hERG. Sci Rep. 10(1):16262. doi: 10.1038/s41598-020-72889-5.

Freedman H, Kundu J, Tchesnokov EP, Law JLM, Nieman JA, Schinazi RF, Tyrrell DL, Gotte M, Houghton M. (2020) Application of Molecular Dynamics Simulations to the Design of Nucleotide Inhibitors Binding to Norovirus Polymerase. *J Chem Inf Model.*; **60**(12):6566–6578. doi: 10.1021/acs.jcim.0c00742.

COVID-19

Law JLM, Logan M, Joyce MA, Landi A, Hockman D, Crawford K, Johnson J, LaChance G, Saffran HA, Shields J, Hobart E, Brassard R, Arutyunova E, Pabbaraju K, Croxen M, Tipples G, Lemieux MJ, Tyrrell DL, Houghton M. (2021) SARS-COV-2 recombinant Receptor-Binding-Domain (RBD) induces neutralizing antibodies against variant strains of SARS-CoV-2 and SARS-CoV-1. *Vaccine.* **39(**40):5769–5779. doi: 10.1016/j.vaccine.2021.08.081. Epub 2021 Aug 26. PMID: 34481699; PMCID: PMC8387217.

The Identification of the Hepatitis C Virus

Nobel Lecture, December 7, 2020 by
Michael Houghton
Li Ka Shing Applied Virology Institute, University of Alberta, Edmonton, Canada.

History of blood-borne viral hepatitis

After the discovery of the hepatitis B virus (HBV) by Dr. Baruch Blumberg and colleagues in the late 1960s (1–3), serological tests were developed to detect this virus in blood donors which led to the realization in 1974 that the majority of post-transfusion hepatitis cases were not due to blood transmission of this virus (4). With the discovery of the hepatitis A virus (HAV) in 1973 by Drs. Steve Feinstone, Bob Purcell and colleagues (5) and the introduction of tests to detect this virus, it was found in 1975 that the vast majority of non-HBV blood-borne hepatitis was not due to HAV (6). Originally perceived as of minor importance, the work of Dr. Harvey Alter and others in the field went on to show the important role of this disease in developing chronic liver disease (7). Termed blood-borne Non-A, Non-B hepatitis (NANBH), this triggered a long search spanning 15 years to identify the causative agent or agents.

The initial breakthrough in the discovery of HBV was the identification of an unusual antigen observed in the blood of Australian aborigines, later shown to be the major envelope glycoprotein of HBV (1–3). However, no such specific antigen was identified for the NANBH agent(s) prior to the discovery of HCV in 1989 (8). The visualization of

27nM diameter particles in the stools of infected animals and humans proved decisive in the discovery of HAV (5) but no such virus particle could be associated with NANBH (8). The critical discovery of HIV was enabled by its lytic propagation *in vitro* in peripheral blood mononuclear cells (PBMNCs) derived from AIDS patients by Drs. Françoise Barré-Sinoussi and Luc Montagnier (9) but the difficult *in vitro* propagation of HCV did not occur until 2005 by Dr. Takaji Wakita and colleagues (10) and to this day, it still remains difficult to grow efficiently in cell cultures. So, methods used to identify other hepatotrophic viruses and HIV proved unable to identify HCV.

The long molecular search for HCV (1982–1989)

My own quest to identify the causative agent(s) of NANBH began in 1982 with the belief that emerging molecular biological methodologies could be capable of success. Initially, Dr. Tatsuo Miyamura from the Japanese NIH visited and worked in my laboratory at the Chiron Corporation for one year bringing human liver samples derived from autopsied materials. However, the quality of the mRNA was poor resulting from degradation of the tissue after death. We then realized that to obtain intact mRNA, we would have to work with materials from the living chimpanzee model which had been shown to be a reliable model for NANBH by a few laboratories around the world (11–14). This animal model had proven to be of great value in HBV research. Consequently, we initiated a collaboration with Dr. Daniel Bradley of the US Centers for Disease Control (CDC) who was a leader in the NANBH chimpanzee model. Fig. 1 shows typical results from his laboratory in which acute NANBH could be passaged in the common chimpanzee (*pan troglodytes*) from a human factor 8 preparation contaminated with the NANBH agent(s).

From 1983 onwards, Dr. Bradley provided me with a continuous supply of blood and liver biopsy samples from many of his NANBH-infected and control, uninfected chimpanzees and for several years thereafter, one major theme of my laboratory was to identify rare NANBH-specific poly-A+ mRNAs derived from the etiological agent(s) of NANBH via interrogation of very large bacterial cDNA libraries derived from NANBH-infected chimpanzee liver samples. We used highly radioactive hybridization cDNA probes derived either from NANBH-infected liver poly-A+ mRNA (the + probe) or from control, uninfected livers (the – probe) to identify many genes apparently upregulated in NANBH-infected livers using duplicate blotted filters of the cDNA libraries (Fig. 2).

We now know that many genes within the liver are modulated (up- or down-regulated) by HCV infection, including those involved in the innate and adaptive host immune responses (15,16). Despite obtaining many upregulated NANBH-specific genes, none could be ascribed to being

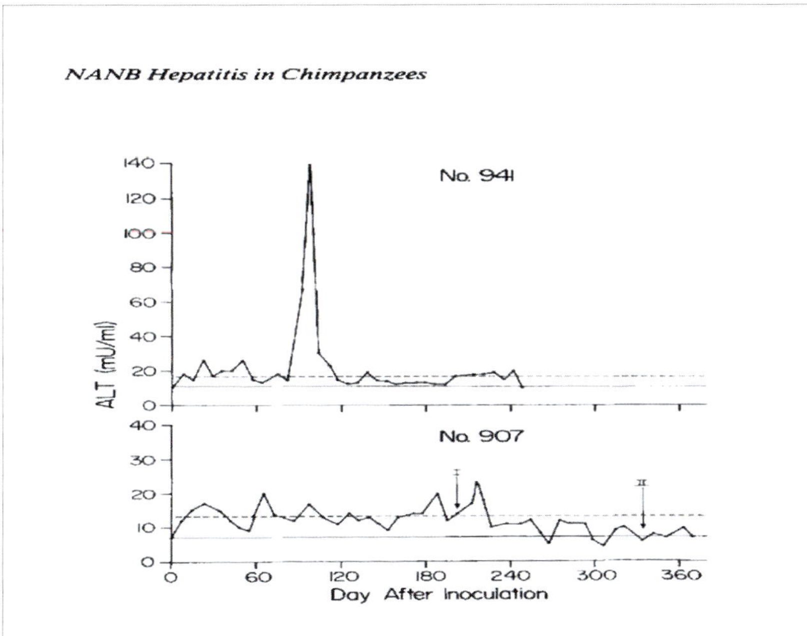

Figure 1. Alanine aminotransferase (ALT) activity (reflective of liver damage) in serial serum specimens from chimpanzees with non-A, non-B hepatitis. Chimpanzee no. 941 was inoculated with plasma II from chimpanzee no. 771 (122).

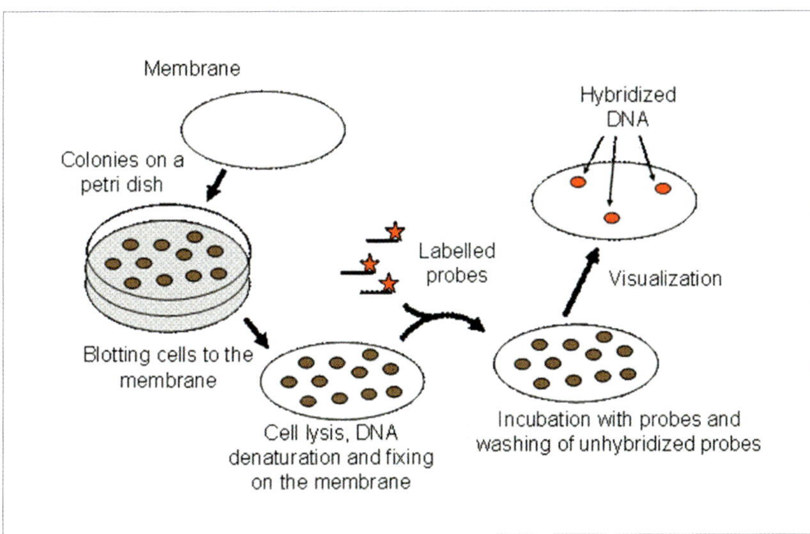

Figure 2. Method of identifying genes modulated by NANBH infection using radiolabeled cDNA hybridization probes derived from poly-A+ mRNA derived from NANBH-infected livers (the + probes) and control, uninfected livers (the – probes). In this way, genes up- and down-regulated by NANBH infection were identified.

derived from a NANBH etiological agent since they all hybridized to the chimpanzee and human genome and therefore were not extra-chromosomal, a criterion used for designating a possible pathogen derivation. In an attempt to increase our probability of success, we requested to Dr. Bradley that he try to produce chimpanzee liver and blood samples of similar infectivity titers to those reported by other NANBH research groups around the world (around 1 million chimpanzee infectious doses per ml of human plasma (17) or chimpanzee (18) plasma and similar infectious titers per gram of NANBH liver tissue from the same animals. We co-funded work in Dr. Bradley's laboratory to attain these goals in 1985 via his production of large plasma pools and liver samples from his chimpanzees #910 & #771 (19,20; Fig. 3). Unfortunately, interrogating cDNA libraries from these sources of tissue from 1985–1987 still failed to identify any clones derived from an etiological NANBH agent(s).

In 1979, Dr. Yohko Shimizu and colleagues at the US NIH demonstrated that one NANBH agent, termed the tubule-forming agent (tfa), elicited double layered membranes and membranous tubules within the endoplasmic reticulum of hepatocytes in infected chimpanzees (21). Dr. Bradley pointed out that such cytoplasmic membranous changes were reminiscent of changes within cell cultures infected by RNA viruses such as the flaviviruses and togaviruses, from which he suggested that the tfa may be related to such viruses (22). He also showed that the infectivity of the tfa was destroyed by treatment with organic solvents and that it could

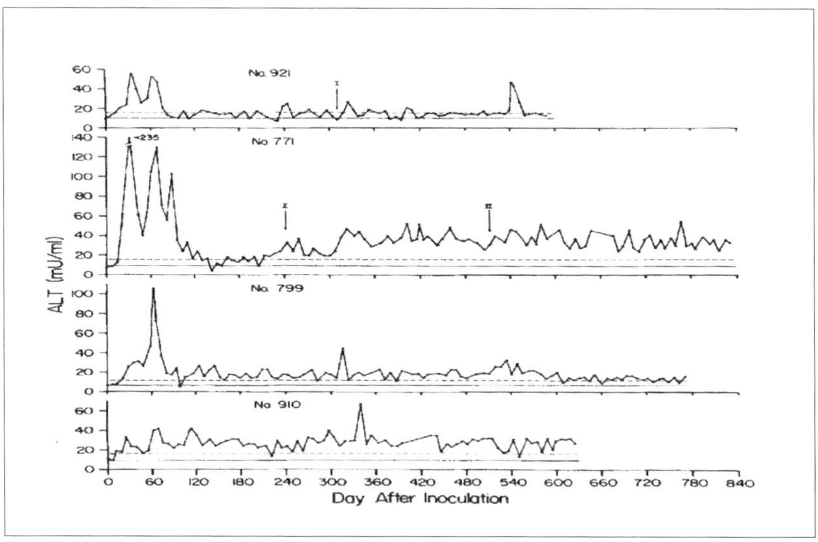

Figure 3. Alanine aminotransferase (ALT) activity in serial serum specimens from chimpanzees with non-A, non-B hepatitis. A plasma pool made from #910 was shown to have an infectivity titer of ~ 1 million chimp infectious doses per ml (19,20).

pass through a filter with pore sizes of 80nM, leading him to suggest that the tfa was either a small lipid-enveloped virus similar to flaviviruses and togaviruses or that it was HBV-like (a 40nM enveloped DNA virus), or that it was a novel type of virus (19,22,23).

This led my laboratory to interrogate his chimpanzee samples with hybridization probes derived from the HBV DNA genome as well as from various flaviviral and togaviral RNA genomes. There were papers published around this time suggesting that one NANBH agent could be HBV-like (24). However, we could not detect any hybridization signals between these known viral genomes and the NANBH chimpanzee samples. We also interrogated (again unsuccessfully) the chimpanzee NANBH samples using picornaviral and coxsackie viral genomes, since Dr. Bradley had observed 27nM viral particles resembling these non-enveloped viruses within some chimpanzee hepatocytes that received inocula previously-treated with organic solvents (19,22,23). None of this time-consuming work resulted in the identification of a specific molecular probe or handle for the NANBH agents.

In another attempt to identify a virus related to the NANBH agent(s) my laboratory, working with that of Dr. John Gerin's laboratory at the US NIH, identified the genome of the hepatitis D virus (HDV). Originally discovered in the form of an unusual antigen (the "delta antigen") by Dr. Mario Rizzetto within hepatocyte nuclei in some HBV patients experiencing severe hepatitis (25), its infectivity for non-human primates was shown to be associated with a RNA molecule that required HBV for infectivity and transmission (26). We showed that this RNA molecule was indeed the HDV genome being a covalently-closed-circular, single-stranded RNA whose complementary anti-genomic RNA encoded the delta antigen (27). Highly reminiscent of infectious plant pathogens like viroids and virusoids, the HDV genome exhibited strong intramolecular base-pairing under physiological conditions to form a double-stranded RNA rod-like structure that could only be visualized in the electron microscope under strongly denaturing conditions (27,28; Fig. 4).

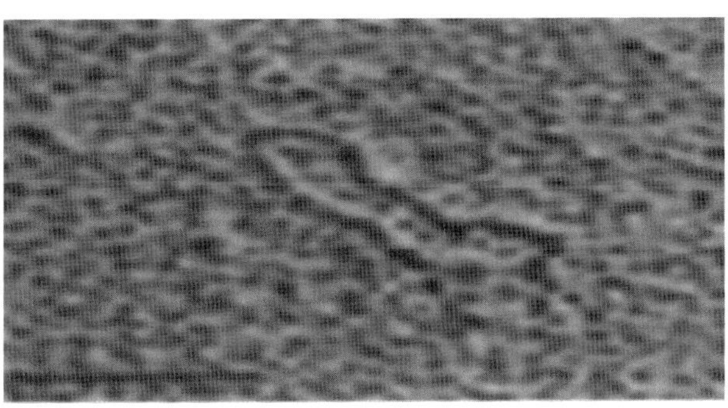

Figure 4. The covalently-closed-circular, single-stranded RNA genome of the HDV pathogen as observed in the EM under strongly denaturing conditions (27,28).

Replicated by host RNA polymerase II in a rolling circle mechanism, the HDV RNA genome, like the similar plant pathogens, was shown by Dr. Taylor's laboratory to possess an innate ribozymic activity, self-cleaving itself from a larger precursor RNA and then able to self-ligate (29,30). This HDV ribozyme has proven to be a useful tool in general molecular biological research and the availability of recombinant delta antigen and its specific antibodies led to improved diagnosis of HBV carriers superinfected with HDV. Therapeutic strategies have also been facilitated by further knowledge of its replication strategy (31).

Dr. Robert Purcell's laboratory at the US NIH was the first to propose that HDV may be related to HCV based on the similarity in tubular membranous structures observed within the hepatocytes of chimpanzees infected by either virus (32). However, Dr. Amy Weiner in my laboratory could not detect any cross-hybridization between the HDV genome and nucleic extracts derived from Dr. Bradley's NANBH-infected chimpanzee samples, thus ruling out the possibility of there being any close genetic relationship between the two agents (ref. 33; Fig. 5).

A possible relationship between the blood-borne NANBH agent and retroviruses was indicated by a report of reverse transcriptase activity in NANBH sera samples (34) and by the propagation of foamy-like viruses in cell cultures infected with NANBH samples (35), but despite intense efforts, we were unable to confirm these findings.

A new exciting era of NANBH research was heralded by Dr. Shimizu at the NIH who cloned B cells derived from NANBH individuals and showed that their secreted antibodies could bind specifically to thin liver sections derived from NANBH-infected livers but not to control, uninfected livers (36). Generally termed as the "Shimizu antibodies", this discovery spawned great activity in the field because hitherto, no NANBH-specific antibodies or antisera had been confirmed using similar histochemical binding assays.

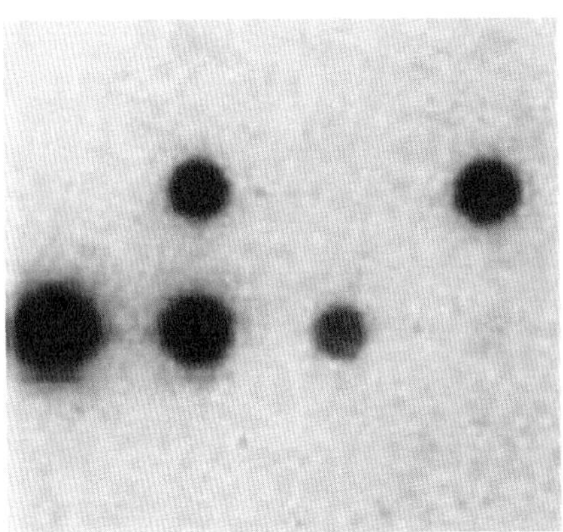

Figure 5. Dot blot hybridization of radio-labeled HDV cDNA to various standards and NANBH nucleic acids extracted from chimpanzee #910 plasma pool (top far right dot shows no detectable hybridization signal to NANBH-derived nucleic acids (33).

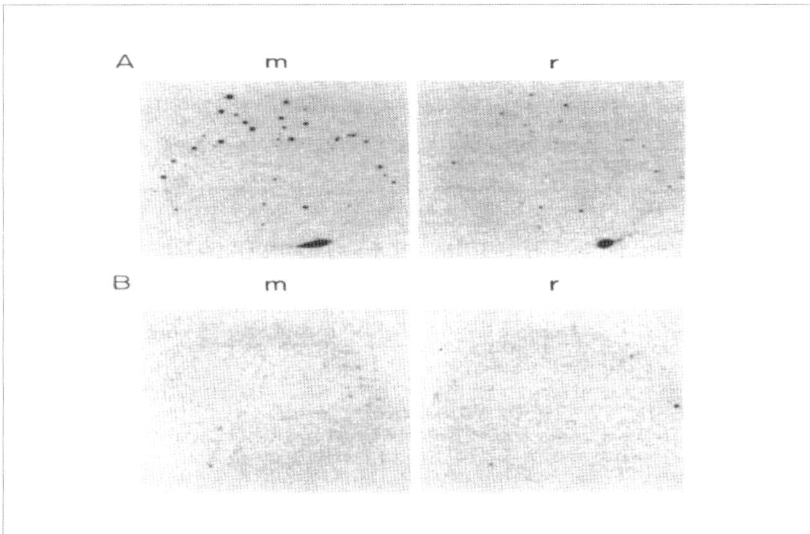

Figure 6. Detection of amylase genes using a recombinant immuno-screening method employing bacteriophage λgt11 lysogens and an amylase IgG probe (37).

In an effort to confirm the reactive antigen specific to the Shimizu antigens (hopefully encoded by the NANBH etiological agent), we employed the cDNA immuno-screening method devised by Drs. Young and Davis (37). Developing a highly efficient λgt11 bacteriophage cloning and expression vector, they demonstrated the ability to identify various genes by screening resulting proteomic cDNA bacterial libraries with specific monoclonal antibodies or polyclonal antisera of known high affinity and titer (Fig. 6).

Unfortunately, this approach was not successful when we screened cDNA libraries derived from NANBH-infected chimpanzee livers with the Shimizu antibodies. Indeed, it was well known at that time that immuno-screening cDNA libraries did not always meet with success. Earlier in our work, I had considered a similar immuno-screening approach using sera from NANBH patients and chimpanzees as a presumptive source of antibodies to the NANBH agent, but had deemed it too risky because no such antibodies had been demonstrated despite intensive efforts by the field. Also, the NANBH agent(s) was known to be highly persistent in most patients suggestive of a weak, inadequate humoral immune response, as known to be the case in cases of persistent HBV infection following exposure of immune-competent adults to the virus. However, in the context of our unsuccessful work on Shimizu antibodies (which Dr. Shimizu later showed, by purifying and sequencing the target protein, were targeting host proteins not encoded by HCV), my next-door laboratory neighbor Dr. George Kuo (then working on non-HCV projects) strongly encouraged me to gamble on screening cloned NANBH libraries with NANBH patient sera. I found his argument to be persuasive that the lack of demonstration of NANBH-specific antibodies to date could possibly be due more to limiting concentrations of NANBH antigen within the thin liver sections used in histochemical screening, rather than the true absence of circulating NANBH antibodies. Dr. Bradley was also of this opinion and was providing chimpanzee samples to another group pursuing this same approach (although this group subsequently terminated this approach due to lack of success and its perceived risk).

Initially, Dr. Qui-Lim Choo in my laboratory screened many of my NANBH-infected chimpanzee liver cDNA libraries derived from poly-A+ mRNA with sera derived from NANBH-infected individuals as a presumptive source of specific NANBH agent antibodies. Despite using many different chimpanzee and human sera samples derived from the minority of individuals that appeared to have spontaneously eradicated their acute NANBH infections based on normalization of serum ALT levels and who we therefore hoped had a strong protective immune response to the etiological agent, we could not identify any cDNA clones that were extra-chromosomal in origin and hence deemed unlikely to be derived from the causative NANBH genome. Based on the possibility that the NANBH agent could be togavirus-like which do not have a poly-A+ tail (or produced poly-A+ mRNAs during replication), I then turned to making λgt11 proteomic libraries derived from the NANBH-infected chimpanzee plasma pool from animal #910. In order to do this, this plasma pool was ultra-centrifuged such that the smallest virus known to man would be pelleted. Initially, RNA extracted from this pellet was then reverse-transcribed using random primers. The resulting cDNA was cloned into λgt11 and screened by Dr Choo using sera from apparently "convalescent" NANBH-infected chimpanzees and patients. Yet again, no molecular clones derived from the NANBH agent(s) genome could be identified. By now, after several years of trying unsuccessfully to track down the causative NANBH agent(s) using numerous different approaches, it felt like I was groping around in the dark looking for the proverbial "needle in a haystack".

As indicated earlier, it was well appreciated by molecular biologists at that time that cDNA immuno-screening often failed to identify genes of interest even when well-characterized antisera or monoclonal antibodies of high affinity were available. Therefore, wanting to reassure myself that this method really could be successful at identifying an infectious agent *de novo*, I used our previously characterized infectious HDV plasma extracts and cloned it into λgt11 prior to screening with HDV-infected patient sera that we already knew contained high titer antibodies to the delta antigen. Numerous HDV clones were successfully obtained in this way giving us confidence that this method could possibly work for identifying the NANBH genome(s). However, the infectivity titer of HDV plasma was known to be many orders of magnitude higher than the best NANBH chimpanzee plasma pool and so I knew that successful detection of NANBH clones was still very risky and uncertain (indeed, of at least 6 plasma-derived libraries that I made during this period without the assistance of PCR amplification technology , which had not yet emerged, only one was to prove successful). In addition, the worrying possibility of a weak humoral immune response to NANBH infection still existed.

I then made a second attempt to identify molecular HCV clones derived from the infectious chimpanzee #910 plasma pool, this time implementing two significant changes. The first involved cloning random-primed, reverse-transcribed cDNA from RNA extracted from the ultra-centrifuged plasma pellet, as well as cloning the DNA existing within this pellet. This was because we did not know if the NANBH agent(s) was a RNA or DNA virus. The second change was to screen this bacterial proteomic library with serum from patients diagnosed with chronic, persistent NANBH infection in the hope that perhaps these chronic patients might have higher titers and affinities of NANBH agent-specific antibodies as compared with apparent "convalescent" individuals experiencing acute, resolving NANBH infection (we later showed this indeed to be the case after we successfully identified HCV and developed anti-HCV antibody tests). Dr. Choo and myself selected a chronic NANBH patient with unusually high ALT levels that we hoped was indicative of a strong immune response to the NANBH agent(s). This resulted in the identification of several clones (Fig. 7).

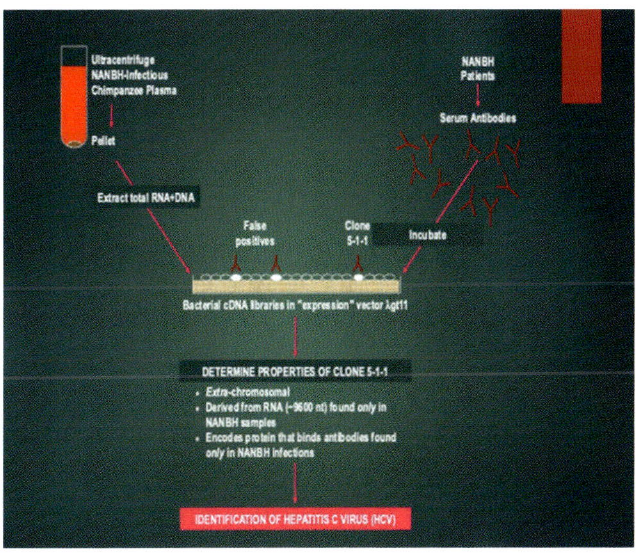

Figure 7. Identification of clone 5-1-1.

In addition to identifying positive clones encoding MS2 bacteriophage RNA (a carrier that was essential for me to use before the emergence of the powerful PCR nucleic acid amplification technique), Dr. Choo found clone 5-1-1, a small clone containing only ~ 100bps, but one that we could later prove was derived from an etiological agent of NANBH. We showed that clone 81 (overlapping with 5-1-1) was derived from a large RNA of around 10,000 ribonucleotides in length that was found only in NANBH-infected samples, not control uninfected samples, that was single-stranded and positive-stranded with respect to encoding the 5-1-1/81 antigen (Fig. 8).

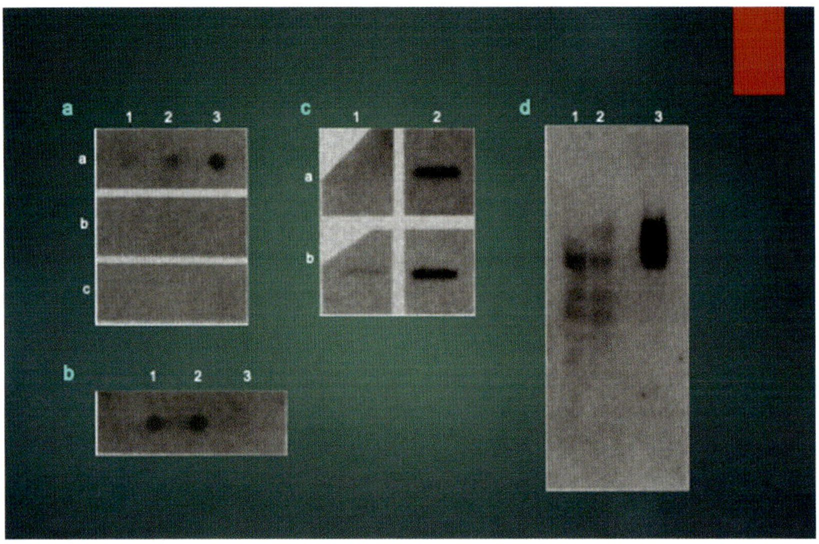

Figure 8. (38). Hybridization of clone 81 cDNA to RNA. (A) Spot hybridization of 2, 4, or 12 ug of total liver RNA extracted from either chronic NANBH-infected chimp 910 (al to a3) or from two control, uninfected animals (bl to b3 and cl to c3) with ^{32}P-labeled nick-translated clone 81 cDNA. (B) Spot hybridization of nucleic acid extracted from viral plasma pellets before (spot 1) or after treatment with either excess deoxyribonuclease 1 (spot 2) or ribonuclease A (spot 3). Hybridization probe as in (A). (C) Each strand of clone 81 cDNA was subcloned into phage M13mp18 and then labeled by incubating with Klenow Escherichia coli DNA polymerase 1 in the presence of hybridization probe primer and [a-^{32}P]dCTP. Each probe was then hybridized to slot blots containing either identical portions of viral RNA derived from infectious plasma (al and bl) or 2 pg of purified clone 81 double-stranded cDNA (a2 and b2). (D) Northern blot analysis of 30 pg of total RNA (track 1), 30 pg of unbound RNA (track 2), and 20 ug of bound RNA (track 3) after chromatography on oligo(dT)-cellulose. RNA was derived from the liver of infectious chimpanzee # 910. The top of the smear in track 3 corresponds to around 10,000 ribonucleotides in length. ^{32}P-labeled nick-translated clone 81 cDNA was used as the hybridization probe.

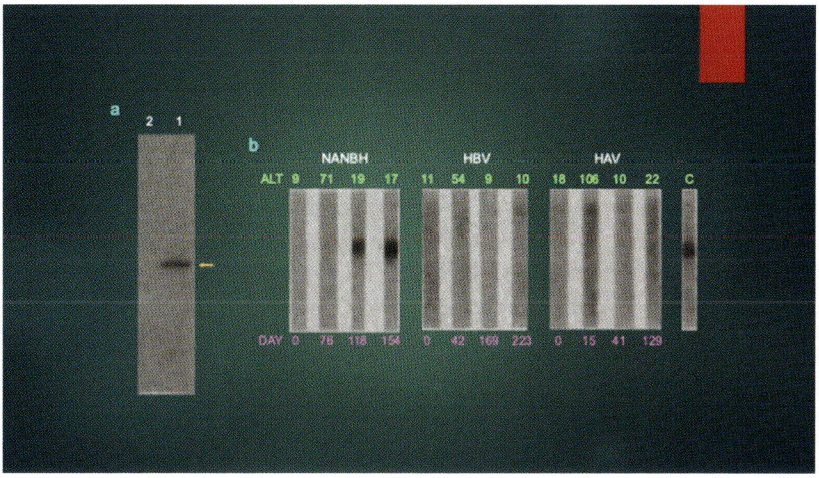

Figure 9. (38). Western blot assay for anti-5-1-1 antibodies. (A) Incubation of the chronic NANBH patient serum used to isolate done 5-1-1 with blots of total bacterial lysates expressing either a fusion of the 5-1-1 antigen with superoxide dismutase (SOD) (lane 1) or control SOD (lane 2). (B) Sequential serum samples from experimentally infected chimpanzees were reacted with identical strips cut from a preparative blot of total lysate containing the 5-1-1/SOD fusion protein. Day 0 represents the day of virus inoculation. Infections were monitored by serum ALT concentrations (international units per liter). Control strip C was incubated with the same patient serum used in (A).

Furthermore, using a Western blot analysis, only chimpanzees (4/4) infected with NANBH seroconverted to anti-5-1-1 antibodies unlike animals infected with HAV or HBV (0/7; representative examples shown in Fig. 9).

Furthermore, we showed that 7 of 11 USA NANBH patients had circulating antibodies to the 5-1-1 antigen but these were absent in control, uninfected individuals. Moreover, as we obtained more and more overlapping cDNA clones from the same cDNA library, we began to see distant but significant primary sequence identity with flaviviruses like Dengue virus. These data convinced us that we had indeed identified a major etiological agent of NANBH, which led to my first public disclosure of the discovery of HCV in the spring of 1988 at the University of California in San Francisco, work subsequently published in 1989 (38).

In the next stage of our work, Dr. Kuo developed a high throughput radioimmunoassay to detect antibodies targeting the antigen C100-3, which was encoded by a few adjacent clones overlapping with 5-1-1, which we used to assay a pedigreed panel of known infectious and known non-infectious NANBH samples from Dr. Harvey Alter. This blinded study revealed that our assay did indeed have specificity for NANBH and furthermore, demonstrated that such antibodies were present in infectious NANBH samples meaning that we could therefore use this assay to

Figure 10. The core HCV discovery team (40). Drs. Michael Houghton (upper left), Qui-Lim Choo (upper right), George Kuo (lower left), Daniel Bradley (lower right).

detect infectious blood donors and patients (39). In addition, using a highly pedigreed panel of 10 post-transfusion NANBH patients of Dr. Harvey Alter at the USA NIH, we showed that all 10 were positive for HCV antibodies and that there was at least one HCV-positive blood donor involved in 9 of these 10 cases (39). After assaying other NANBH patient cohorts from other collaborators from the USA, Italy and Japan, it became clear that the majority of NANBH samples were positive for HCV antibodies (39). This constituted confirmation that we (Fig. 10) had identified the major cause of blood-borne NANBH.

Ramifications of the molecular isolation of HCV

1. Diagnosis

Our first priority then was to protect the global blood supply by developing blood screening diagnostics to detect circulating HCV antibodies. Using the protein encoded by clone 5-1-1 and two adjacent clones, we were able to produce the C100-3 antigen in recombinant yeast which was used to detect circulating HCV antibodies in the first commercial EIA assays (initially produced by Ortho Diagnostics and shortly afterwards by Abbott Laboratories). This was shown to detect the majority of infectious blood donors and chronically-infected patients (39,41).

Meanwhile, we and many other laboratories around the world began to characterize the viral proteins encoded by the HCV RNA genome (42–47) from which it became clear that the virus encoded a large polyprotein precursor that was cleaved co- and post-translationally by a variety of cellular and viral proteases into structural virion proteins and a large number of proteins required for viral replication (Fig. 11).

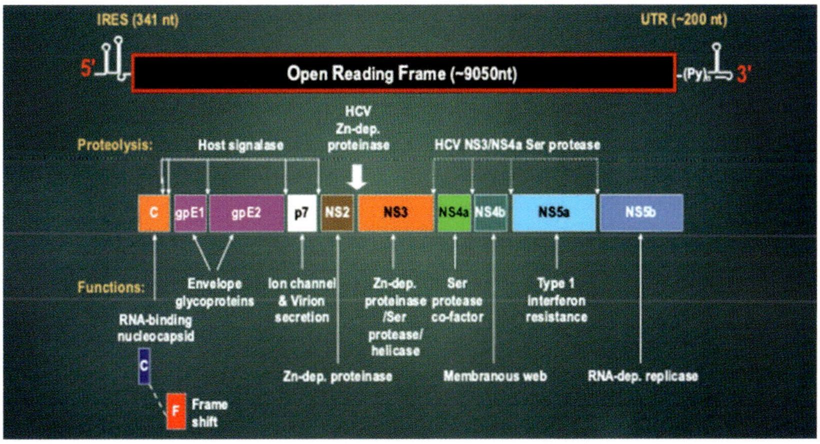

Figure 11. The HCV polyprotein precursor and its cleavage products (42–47).

By producing recombinant proteins from the various protein domains encoded by HCV, we were able to detect and map additional B cell epitopes, both linear and discontinuous epitopes conserved among the different strains of HCV, that we incorporated into a series of EIA tests of ever-increasing sensitivity (48,49). This led to commercial EIA tests detecting circulating HCV antibodies with >95% sensitivity as well as confirmatory tests (50). Dr. Weiner then developed the first PCR-based test to detect circulating HCV genomic RNA including in some individuals lacking HCV antibodies (51). Shortly afterwards, another laboratory colleague, Dr. Jang Han, showed that the 5'untranslated region (UTR) of the HCV genome was highly conserved among the many different genotypes and strains of HCV and was even highly conserved with the animal pestiviral 5'UTRs (52). It should be noted that although the genomic RNA of the HCV strain that we isolated (HCV1) does contain a short poly-A region, this is in the gene encoding the nucleocapsid C gene (43) and is not at the 3' terminus of the HCV genome (which can be a product of spurious PCR reactions; 52).

The highly conserved 5'UTR allowed very sensitive nucleic acid tests to be developed first by PCR methods (53) and later by transcription-mediated amplification (TMA) methods (54).

Over the course of a few years after the identification of the HCV RNA genome, the co-introduction of these HCV antibody and HCV genomic nucleic acid detection tests effectively arrested the global transmission of HCV from blood donors as well as allowing the diagnosis of all HCV patients. Such tools also became invaluable to monitor patient responses

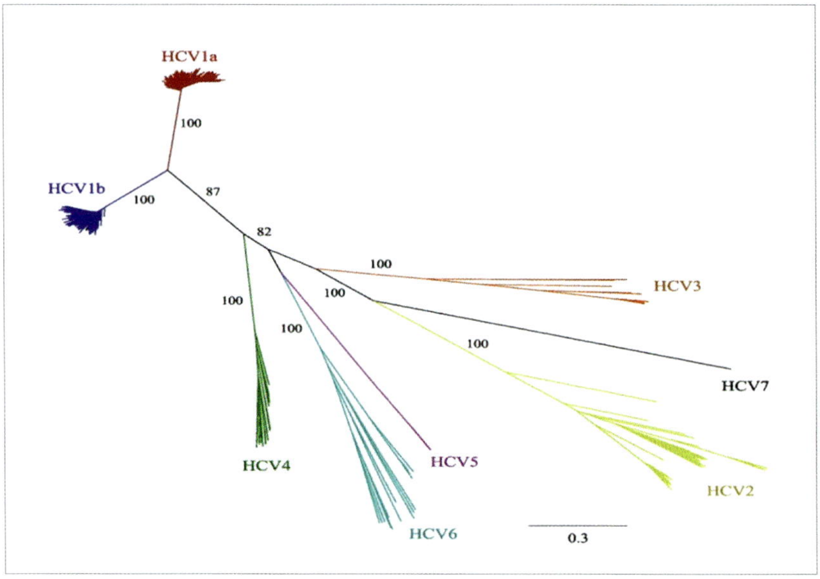

Figure 12. Phylogenetic tree of full-length HCV genomic RNA sequences (116).

to interferon and ribavirin therapy (55,56) and much later, to specific direct-acting-antivirals (DAAs; 57).

By this time, it became evident that HCV was a very heterogeneous virus that is now known to comprise at least 7 major genotypes around the world with each genotype comprising many related sub-types (Fig. 12; refs 58–60,114,116).

The first variant subtype identified was from the laboratory of Dr. Tatsuo Miyamura at the Japanese NIH who identified what is now known to be the most common global subtype (1b; 58). Subtype 1a is the most common in North America, meaning that the major genotype 1 (comprising subtypes 1a & 1b) is the most common around the world (61–63,114,116). Median inter-genotype diversity was calculated to be 32.39% at the nucleotide level and 25.02% at the amino acid level (116). The lowest nucleotide diversity is observed between HCV major genotypes 1 and 4 (29.03%) and the highest between HCV genotypes 2 and 3 (35.46%; 116).

2. Direct-acting-antiviral (DAA) development

Up until 2014, standard-of-care treatment of NANBH patients comprised type 1 interferon combined with the guanosine analogue, ribavirin (55,56). Requiring at least 6 months of treatment, it was associated with significant toxicity and resulted in virus eradication in only 40–50% of patients. The isolation and characterization of the HCV genome and its

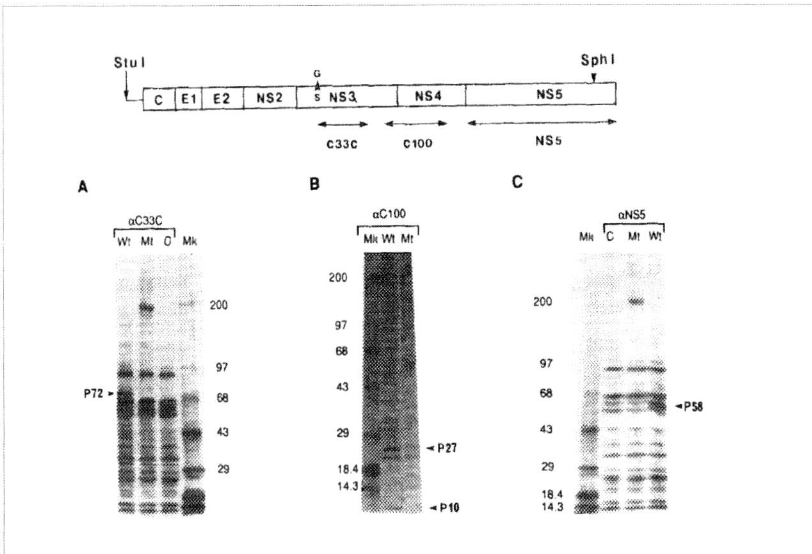

Figure 13. The HCV NS3-encoded serine protease activity. The polyprotein precursor fails to be properly cleaved in active site mutations (mt) versus wild-type enzyme (wt; 64).

encoded enzymes and proteins in 1989 enabled the HCV field, comprising numerous academic and corporate laboratories around the world, to pursue the development of direct-acting-antivirals (DAAs) to treat HCV patients. My laboratory (64) and those of Dr. Charles Rice (65) and Dr. Kunitada Shimotohno (66) first showed the presence of a serine chymotrypsin-like protease within the HCV NS3 protein (Fig. 13).

Later shown to complex with the NS4a co-factor (67,68), this activity was targeted by many groups and led to the approval of first generation antiviral DAAs, boceprevir and telaprevir, in 2011. While potent and of great value to patients, these drugs were associated with significant toxicity that led to the subsequent approval of more potent and better tolerated 2nd (Simeprevir and Asunaprevir) and 3rd generation drugs (Paritaprevir, Grazoprevir, Glecaprevir and Voxilaprevir; 69).

The HCV NS5b protein encodes the viral RNA-dependent RNA polymerase for which the nucleoside inhibitor prodrug, Sofosbuvir (Sovaldi), was approved for human use in 2013 (70). Non-nucleoside inhibitors of this enzyme have also been developed (Dasabuvir;71). All of this drug development work was greatly facilitated by the development of HCV cell culture systems, from the laboratories of Dr. Ralf Bartenschlager and Dr. Charles Rice, able to replicate the HCV genome (72,73). Although not capable of producing progeny virus, these systems served as valuable tools to screen for inhibitors of the HCV replication cycle and allowed the field to generate a huge amount of valuable information regarding HCV replication mechanisms. Furthermore, these replicon systems enabled the development of unconventional drugs targeting the NS5a protein. By screening for antiviral drugs and then selecting for drug-resistant mutations, the extremely potent Daclatasvir targeting the NS5a dimer was first developed by a group at Bristol-Myers-Squibb (74) followed by many active chemical derivatives from other pharmaceutical companies (Ledipasvir, Velpatasvir, Ombitasvir, Elbasvir). These drugs represent the most potent antivirals ever developed against any virus and they bind to a pocket formed by the NS5a dimer (Fig. 14; 75,76).

HCV drug development was also greatly facilitated by the ability to test efficacy in a mouse model developed by Drs. David Mercer, Lorne Tyrrell, Norman Kneteman and colleagues, able to support robust HCV replication in transplanted human hepatocytes (117).

The HCV DAAs are used together in various combinations to effectively shut down viral replication, resulting in eradication of HCV in the large majority of patients, usually within 2–3 months of oral administration. The first drug combination to accomplish this historic landmark was Harvoni (Ledipasvir + Sofosbuvir) from Dr. John McHutchison and colleagues at Gilead Sciences, which was approved for human use in 2014 followed later by Epclusa (Sofosbuvir + Velpatasvir) in 2016. The latter cocktail is very effective against all HCV genotypes. HCV represents the first chronically-infecting virus of man that can be eradicated using antivirals. This has led the World Health Organization (WHO) to issue the objective of eliminating HCV infection as a major public health threat by

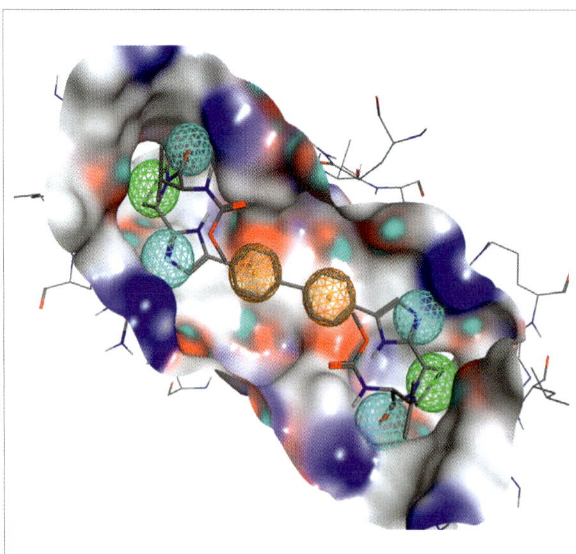

Figure 14. Essential pharmacophoric features for potent directly acting NS5A inhibitors (76). These features are a central biaryl core (orange spheres), two H-bond acceptor sites (cyan spheres), and two hydrophobic sites (green spheres). The pharmacophoric sites are mapped over Daclatasvir (green sticks).

2030 via a "test and treat" policy and by developing a much-needed vaccine. Notably, Egypt and other countries have already made substantial progress towards curing their HCV-infected peoples with DAAs (77).

3. Towards vaccines against HCV

Unfortunately, vaccine development for HCV has been slow due to many factors. Five years after we identified the HCV genome, Drs. Choo, Kuo, Ralston, Spaete and myself with other colleagues published strong evidence for prophylactic vaccine efficacy in the reliable chimpanzee model (78–80). Using adjuvanted recombinant envelope glycoproteins E1 & E2 (in the form of a native heterodimer, 79) derived from transfected mammalian cells, we could show that of 7 animals immunized 3 times over the course of 6–7 months, 5 were sterilized against homologous virus challenge while the other 2, although becoming acutely infected, eradicated the virus within a few months. In contrast, 4/4 control unimmunized animals became chronically infected after challenge (80).

Since HCV-associated disease manifests itself only in the chronic phase of infection, usually after many years, this data was most encouraging for the feasibility of developing an effective HCV vaccine. Unfortunately, the chimpanzee is an endangered species and very costly to use, such that private and public funding to extend this work took many years to become available. Perhaps vaccine over-pessimism from the field in general also contributed to this budget constraint despite the pandemic nature of HCV infection. Eventually, we showed that vaccinated chimpanzees challenged with a heterologous viral strain also significantly reduced the incidence of chronic infection (Table 1; 81,82). To date, our E1E2 vaccine remains the only HCV vaccine candidate shown to be efficacious at reducing chronic HCV infection in animal models.

After many years when the ability to produce infectious HCV pseudo-

Viral challenge	Group	Total	Acute infections	Chronic infection (%)	
Homolog	gpE1/gpE2	12	7	2(17)	P=0.03
HCV-1	Unimmunized	10	10	7(70)	
Heterolog	gpE1/gpE2	19	19	3(16)	P=0.02
H77	Unimmunized	14	14	8(57)	
Total	gpE1/gpE2	31	26	5(16)	P=<0.001
	Unimmunized	24	24	15(63)	

Table 1. Summary of outcome of chimpanzee challenge studies using animals vaccinated with adjuvanted recombinant gpE1/gpE2 vaccine or unimmunized control animals and then challenged with either homologous (HCV1) or heterologous (HCVH77) subtype 1a viruses that predominate in North America (ref. 82).

particles in the laboratories of Drs. Jane A, McKeating and François-Loïc Cosset (83,84), then infectious virus itself by Dr. Wakita and collaborators (85) finally became possible, we were able to show that adjuvanted recombinant E1E2 envelope glycoproteins derived from a single HCV 1a strain elicited broad cross-neutralizing antibodies in vaccinated mice (86), guinea pigs (86), chimpanzees (87) and then human volunteers (Fig. 15; 88,89).

The use of chimeric HCV pseudoparticles (pp) and cell-cultured (cc) viruses containing E1E2 derived from multiple genotypes has greatly facilitated our knowledge of cross-neutralizing antibody responses (90,115). However, antibody-mediated neutralization of infectivity is weaker against genotypes 2 and 3 relative to 1,4,5 and 6 (87–89) and we have demonstrated different neutralization potencies even within the same genotype and subtype (121), suggesting that an optimum vaccine may need to comprise E1E2 derived from multiple genotypes, as is common for vaccines against other viruses (and bacteria).

Both neutralizing antibodies and HCV-specific cellular immune responses correlate with protection against the development of chronic HCV infection (91–99) and we showed that in addition to cross-neutralizing antibodies, the HCV E1E2 vaccine also elicits strong T cell lymphoproliferative responses in vaccinated humans (89,123). Furthermore, in the last 20 years, various groups have isolated cross-neutralizing HCV monoclonal antibodies targeting multiple, discrete epitopes within E1 and E2 (100–112), some of which have been demonstrated to be protective in HCV animal models (102,107,109,110) and to synergize with each other (108). Using monoclonal antibody competition assays, we showed that human volunteers vaccinated with recombinant E1E2 elicited antibodies to many of these discrete cross-neutralizing epitopes (113).

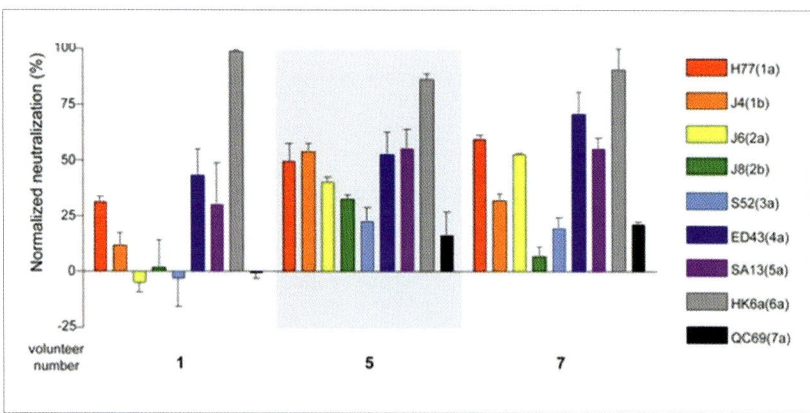

Figure 15. Human volunteers #s 1, 5 & 7 were vaccinated with adjuvanted gpE1/gpE2 and their antisera (diluted 1:50) assayed for the ability to neutralize different worldwide strains of HCVcc (89).

Along with the knowledge that acutely-infected chimpanzees (118,120) and people-who-inject-drugs (PWIDs; 119) who can eradicate HCV infection without therapy are then largely immune to developing chronic, persistent infection following re-exposure to homologous or heterologous HCV, these accumulated vaccine and immune correlates data do indicate the feasibility of developing a HCV vaccine. Further clinical studies of our HCV vaccine, designed to elicit broad cross-neutralizing HCV antibodies and broad, HCV-specific cross-reactive CD4+ & CD8+ T cell responses, are planned over the next few years with the objective of making it available first to high-risk PWIDs around the mid-2020s.

Together with the "test and treat" policy, successful vaccine development could mark the end of HCV being a major public health issue by 2030, if not before. The tremendously rapid development of SARS-2-Cov-2 vaccines using lipid/RNA and adenoviral vector technologies (124) are now also being applied to HCV vaccinology, giving the HCV field even more hope of delivering a global HCV vaccine within the 3rd decade of this millennium.

Acknowledgements

My sincere thanks to Professors Michael Balls, David Wilden and John Thane for their inspirational teaching at the University of East Anglia, UK, where I obtained my BSc in Biological Sciences, and to Professor Jim Chesterton, my PhD program supervisory Professor at King's College (Biochemistry), University of London, UK and to my co-supervisor Dr. Norman Carey at the G.D. Searle Research Laboratories, UK. I greatly benefited from the stimulatory mentorship of Dr. Richard Palmiter at the Searle Research laboratories (later a Professor at the University of Washington, Seattle, USA) and to senior technician, mentor and friend, Graham Catlin.

In addition to my key colleagues in the discovery of the HCV genome, Dr. Qui-Lim Choo, Dr. George Kuo and Dr. Dan Bradley, other valued contributors were Dr. Amy Weiner, Dr. Lacy Overby (original founder of the Abbott Infectious Disease Diagnostics division and a wonderful mentor) and Dr. Kang-Sheng Wang. I would also like to acknowledge Drs. Dino Dina, William Rutter, Ed Penhoet, Pablo Valenzuela, Maureen Powers, Sergio Abrignani, David Chien, Yiu-Lian Fong, Mark Selby, Xavier Paliard, Chris Walker, Michael Eckart, Konrad Stadler, Rick Najarian, Rich Thayer, Paul Luciw, Graham Bell, and my long-time friends and colleagues Guy Mullenbach and Rob Hallewell who all contributed either directly and/or indirectly to the work described above that was performed at the Chiron Corporation in California between 1982–2007.

I am indebted to Kim Berger, Cindy Lee, Christine Dong and Shirley Wong for their excellent research assistance and lasting friendships. Along with dedicated assistants in other Chiron Laboratories, Doris Coit,

Chica Medina-Selby, Carol Kuo & Carol Gallegos, they performed an immense amount of critical work over so many years. I have also greatly appreciated the life-long collaborations and support from Dr. Tatsuo Miyamura, Dr. Ferruccio Bonino, Dr. Maurizia Brunetto, and Dr. John McHutchison, and many others in the HCV field.

At the University of Alberta in Edmonton, Canada, I have greatly benefited from working with many excellent young scientists over the last decade including senior research assistants Drs. John Law, Mike Logan, and Amir Landi; research assistants Drs. Holly Freedman, Juthika Kundu and Joydeb Kundu; post-doctoral fellows Dr. Khaled Barakat (now Associate Professor, University of Alberta), Dr. Deanna Santer (now Assistant Professor, University of Manitoba), Dr. Max Rosario (now Assistant Professor, Johns Hopkins, Baltimore), and Dr. Anwar Mohammed ; excellent graduate students Jason Wong and Janelle Johnson; Darren Hockman (our poly-tasking lab manager); mammalian cell bioreactor and GMP specialists Mark Wininger, Kevin Crawford, Chao Chen, Catalina Mesa Prince, Drew Nahirney; Quality Control & Quality Assurance specialists Louise Tan, Nicole Brown, and other colleagues; Operations managers Drs. Wendy Magee and Ana Clementin; my superb administrator Iman Ibrahim; and to numerous other past and new talented members of our University team. My special thanks go to Professor Lorne Tyrrell for his never-ending, outstanding support, collaboration, enthusiasm and friendship over the last 10 years; and I am grateful to ex-University President Dr. Indira Samarasekera and ex-University Vice-President Research, Dr. Lorne Babiuk, who assisted Professor Tyrrell in bringing me to Canada from the USA on a Canada Excellence in Research Chair (CERC) grant from the Canadian federal government. Sincere thanks to Leeann Tyrrell, Meaghan and Anina Chiu and to their respective husbands Dr. John Chiu and Dr. Peter Chiu for their heart-felt support and friendships (and for so many enjoyable dinner evenings and great days on the golf courses of Alberta and California!).

I give my sincere thanks also to Mr. Li Ka Shing and his colleagues at the Li Ka Shing Foundation for their great kindness and generous financial and moral support from the Li Ka Shing Foundation, and also to the Government of Alberta and Alberta Innovates for their strong financial supports.

Finally, I have been so lucky to have had my dear wife Han Fong Ida alongside me for the last 47 years and counting. The quality of my life and career would not have been possible without her love and exceptional support, added to later by my much-loved sons, Thomas and Henry, daughter-in-law Sindy and grand-daughters Emmy and Gemma.

Thank you all for enriching my life so much!

REFERENCES

1. Particles associated with Australia antigen in the sera of patients with Leukaemia, Down's Syndrome and hepatitis. Bayer ME, Blumberg BS, Werner B. Nature. 1968 Jun 15; 218(5146):1057–9.
2. Australia antigen detected in the nuclei of liver cells of patients with viral hepatitis by the fluorescent antibody technic. Millman I, Zavatone V, Gerstley BJ, Blumberg BS. Nature. 1969 Apr 12;222(5189):181–4.
3. Australia antigen as a hepatitis virus. Variation in host response. Blumberg BS, Sutnick AI, London WT. Am J Med. 1970 Jan;48(1):1–8.
4. Long-incubation post-transfusion hepatitis without serological evidence of exposure to hepatitis-B virus. Prince AM, Brotman B, Grady GF, Kuhns WJ, Hazzi C, Levine RW, Millian SJ. Lancet. 1974 Aug 3;2(7875):241–6.
5. Hepatitis A: detection by immune electron microscopy of a viruslike antigen associated with acute illness. Feinstone SM, Kapikian AZ, Purcell RH. Science 1973 Dec 7;182(4116):1026–8.
6. Transfusion-associated hepatitis not due to viral hepatitis type A or B. Feinstone SM, Kapikian AZ, Purcell RH, Alter HJ, Holland PV. N Engl J Med. 1975 Apr 10;292(15):767-70.
7. The chronic sequelae of non-A, non-B hepatitis.Berman M, Alter HJ, Ishak KG, Purcell RH, Jones EA. Ann Intern Med. 1979 Jul;91(1):1–6.
8. Non-A, non-B hepatitis: advances and unfulfilled expectations of the first decade. Wai-Kuo Shih J, Esteban Mur JI, Alter HJ. Prog Liver Dis. 1986;8:433–52.
9. Isolation of a T-lymphotropic retrovirus from a patient at risk for acquired immune deficiency syndrome (AIDS). Barré-Sinoussi F, Chermann JC, Rey F, Nugeyre MT, Chamaret S, Gruest J, Dauguet C, Axler-Blin C, Vézinet-Brun F, Rouzioux C, Rozenbaum W. Science. 1983 May 20;220(4599):868–71.
10. Production of infectious hepatitis C virus in tissue culture from a cloned viral genome. Wakita T, Pietschmann T, Kato T, Date T, Miyamoto M, Zhao Z, Murthy K, Habermann A, Kräusslich HG, Mizokami M, Bartenschlager R, Liang TJ. Nat Med. 2005 Jul;11(7):791–6.
11. Transmissible agent in non-A, non-B hepatitis. Alter HJ, Purcell RH, Holland PV, Popper H. Lancet. 1978 Mar 4;1(8062):459–63.
12. Transmission of non-A, non-B hepatitis from man to chimpanzee. Tabor E, Gerety RJ, Drucker JA, Seeff LB, Hoofnagle JH, Jackson DR, April M, Barker LF, Pineda-Tamondong G. Lancet. 1978 Mar 4;1(8062):463–6.
13. Experimental infection of chimpanzees with antihemophilic (factor VIII) materials: recovery of virus-like particles associated with non-A, non-B hepatitis. Bradley DW, Cook EH, Maynard JE, McCaustland KA, Ebert JW, Dolana GH, Petzel RA, Kantor RJ, Heilbrunn A, Fields HA, Murphy BL. J Med Virol. 1979;3(4):253–69.
14. Transmission of non-A non-B hepatitis to chimpanzees by factor-IX concentrates after fatal complications in patients with chronic liver disease. Wyke RJ, Tsiquaye KN, Thornton A, White Y, Portmann B, Das PK, Zuckerman AJ, Williams R. Lancet. 1979 Mar 10;1(8115):520–4.
15. DNA microarray analysis of chimpanzee liver during acute resolving hepatitis C virus infection. Bigger CB, Brasky KM, Lanford RE. J Virol. 2001 Aug;75(15):7059–66.

16. Intrahepatic gene expression during chronic hepatitis C virus infection in chimpanzees. Bigger CB, Guerra B, Brasky KM, Hubbard G, Beard MR, Luxon BA, Lemon SM, Lanford RE. J Virol. 2004 Dec;78(24):13779–92.
17. Non-A, non-B hepatitis in chimpanzees and marmosets. Feinstone SM, Alter HJ, Dienes HP, Shimizu Y, Popper H, Blackmore D, Sly D, London WT, Purcell RH. J Infect Dis. 1981 Dec;144(6):588–98.
18. Non-A, non-B hepatitis viruses. Prince AM. Annu Rev Microbiol. 1983;37:217–32
19. The agents of non-A, non-B viral hepatitis. Bradley DW. J Virol Methods. 1985 Apr;10(4):307–19.
20. Etiology and natural history of post-transfusion and enterically-transmitted non-A, non-B hepatitis. Bradley DW, Maynard JE. Semin Liver Dis. 1986 Feb;6(1):56–66
21. Non-A, non-B hepatitis: ultrastructural evidence for two agents in experimentally infected chimpanzees. Shimizu YK, Feinstone SM, Purcell RH, Alter HJ, London WT. Science. 1979 Jul 13;205(4402):197–200
22. Posttransfusion non-A, non-B hepatitis: physicochemical properties of two distinct agents. Bradley DW, Maynard JE, Popper H, Cook EH, Ebert JW, McCaustland KA, Schable CA, Fields HA. J Infect Dis. 1983 Aug;148(2):254–65
23. Posttransfusion non-A, non-B hepatitis in chimpanzees. Physicochemical evidence that the tubule-forming agent is a small, enveloped virus. Bradley DW, McCaustland KA, Cook EH, Schable CA, Ebert JW, Maynard JE. Gastroenterology. 1985 Mar;88(3):773–9
24. Non-A, Non-B hepatitis virus: identification of a core antigen-antibody system that cross reacts with hepatitis B core antigen and antibody. Trépo C, Vitvitski L, Hantz O. J Med Virol. 1981;8(1):31–47
25. Immunofluorescence detection of new antigen-antibody system (delta/anti-delta) associated to hepatitis B virus in liver and in serum of HBsAg carriers. Rizzetto M, Canese MG, Aricò S, Crivelli O, Trepo C, Bonino F, Verme G. Gut. 1977 Dec;18(12):997–1003
26. Properties of delta-associated ribonucleic acid.Hoyer B, Bonino F, Ponzetto A, Denniston K, Nelson J, Purcell R, Gerin JL. Prog Clin Biol Res. 1983;143:91–7
27. Structure, sequence and expression of the hepatitis delta (delta) viral genome. Wang KS, Choo QL, Weiner AJ, Ou JH, Najarian RC, Thayer RM, Mullenbach GT, Denniston KJ, Gerin JL, Houghton M. Nature. 1986 Oct 9–15;323(6088):508–14 (erratum; Nature 1987 Jul 30–Aug 5;328(6129):456)
28. The hepatitis delta (delta) virus possesses a circular RNA. Kos A, Dijkema R, Arnberg AC, van der Meide PH, Schellekens H. Nature. 1986 Oct 9–15;323(6088):558–60
29. Antigenomic RNA of human hepatitis delta virus can undergo self-cleavage. Sharmeen L, Kuo MY, Dinter-Gottlieb G, Taylor J. J Virol. 1988 Aug;62(8):2674–9
30. The structure and replication of hepatitis delta virus. Taylor JM. Annu Rev Microbiol. 1992;46:253–76
31. Optimizing lonafarnib treatment for the management of chronic delta hepatitis: The LOWR HDV-1 study. Yurdaydin C, Keskin O, Kalkan Ç, Karakaya F, Çalişkan A, Karatayli E, Karatayli S, Bozdayi AM, Koh C, Heller T, Idilman R, Glenn JS. Hepatology. 2018 Apr;67(4):1224–1236
32. Cytoplasmic tubular structures in liver of HBsAg carrier chimpanzees infected with delta agent and comparison with cytoplasmic structures in

non-A, non-B hepatitis. Kamimura T, Ponzetto A, Bonino F, Feinstone SM, Gerin JL, Purcell RH. Hepatology. 1983 Sep–Oct;3(5):631–7

33. Hepatitis delta (delta) cDNA clones: undetectable hybridization to nucleic acids from infectious non-A, non-B hepatitis materials and hepatitis B DNA. Weiner AJ, Wang KS, Choo QL, Gerin JL, Bradley DW, Houghton M. J Med Virol. 1987 Mar;21(3):239–47
34. Detection of reverse transcriptase activity in association with the non-A, non-B hepatitis agent(s). Seto B, Coleman WG Jr, Iwarson S, Gerety RJ. Lancet. 1984 Oct 27;2(8409):941–3
35. Isolation of a virus from chimpanzee liver cell cultures inoculated with sera containing the agent of non-A, non-B hepatitis. Prince AM, Huima T, Williams BA, Bardina L, Brotman B. Lancet. 1984 Nov 10;2(8411):1071–5
36. Production of antibody associated with non-A, non-B hepatitis in a chimpanzee lymphoblastoid cell line established by in vitro transformation with Epstein-Barr virus. Shimizu YK, Oomura M, Abe K, Uno M, Yamada E, Ono Y, Shikata T. Proc Natl Acad Sci USA. 1985 Apr;82(7):2138–42
37. Efficient isolation of genes by using antibody probes. Young RA, Davis RW. Proc Natl Acad Sci USA. 1983 Mar;80(5):1194–8
38. Isolation of a cDNA clone derived from a blood-borne non-A, non-B viral hepatitis genome. Choo QL, Kuo G, Weiner AJ, Overby LR, Bradley DW, Houghton M. Science. 1989 Apr 21;244(4902):359–62
39. An assay for circulating antibodies to a major etiologic virus of human non-A, non-B hepatitis. Kuo GQ, Choo QL, Alter HJ, Gitnick GL, Redeker AG, Purcell RH, Miyamura TE, Dienstag JL, Alter MJ, Stevens CE. Tegtmeier GE, Bonino F, Colombo M, Lee W-S, Kuo C, Berger K, Shuster JR, Overby LR, Bradley DW, Houghton M. Science. 1989 Apr 21;244(4902):362–4
40. Clinical Lasker Medical Research Award. Hepatitis C virus and eliminating post-transfusion hepatitis. Alter HJ, Houghton M. Nat Med. 2000 Oct;6(10):1082–6
41. Detection of antibody to hepatitis C virus in prospectively followed transfusion recipients with acute and chronic non-A, non-B hepatitis. Alter HJ, Purcell RH, Shih JW, Melpolder JC, Houghton M, Choo QL, Kuo G. N Engl J Med. 1989 Nov 30;321(22):1494–500
42. Gene mapping of the putative structural region of the hepatitis C virus genome by in vitro processing analysis. Hijikata M, Kato N, Ootsuyama Y, Nakagawa M, Shimotohno K. Proc Natl Acad Sci USA. 1991 Jul 1;88(13):5547–51
43. Genetic organization and diversity of the hepatitis C virus. Choo QL, Richman KH, Han JH, Berger K, Lee C, Dong C, Gallegos C, Coit D, Medina-Selby R, Barr PJ, Weiner AJ, Bradley DW, Kuo G, Houghton M. Proc Natl Acad Sci USA. 1991 Mar 15;88(6):2451–5
44. Molecular structure of the Japanese hepatitis C viral genome. Kato N, Hijikata M, Nakagawa M, Ootsuyama Y, Muraiso K, Ohkoshi S, Shimotohno K. FEBS Lett. 1991 Mar 25;280(2):325–8
45. Expression, identification and subcellular localization of the proteins encoded by the hepatitis C viral genome. Selby MJ, Choo QL, Berger K, Kuo G, Glazer E, Eckart M, Lee C, Chien D, Kuo C, Houghton M. J Gen Virol. 1993 Jun;74 (Pt 6):1103–13
46. Expression and identification of hepatitis C virus polyprotein cleavage products. Grakoui A, Wychowski C, Lin C, Feinstone SM, Rice CM. J Virol. 1993 Mar;67(3):1385–95

47. Kinetic and structural analyses of hepatitis C virus polyprotein processing. Bartenschlager R, Ahlborn-Laake L, Mous J, Jacobsen H. J Virol. 1994 Aug;68(8):5045–55
48. Improved detection of hepatitis C virus antibodies in high-risk populations. McHutchison JG, Person JL, Govindarajan S, Valinluck B, Gore T, Lee SR, Nelles M, Polito A, Chien D, DiNello R, Quan S. Hepatology. 1992 Jan;15(1):19–25
49. Use of third-generation hepatitis C virus (HCV) enzyme immunoassay (EIA) to resolve second-generation HCV EIA-reactive and second-generation recombinant immunoblot assay-indeterminate blood samples: data to support current Food and Drug Administration guidance on HCV lookback. Busch MP, Tobler LH, Tegtmeier G, Polito A, Quan S, Hirschler NV, Dockter J, Giachetti C, Mimms L. Transfusion. 2000 Jan;40(1):10–4
50. New generation RIBA hepatitis C strip immunoblot assays. Polito AJ, DiNello RK, Quan S, Andrews W, Rose J, Lee F, Nelles M, Lee S. Ann Biol Clin (Paris). 1992;50(5):329–36.
51. Detection of hepatitis C viral sequences in non-A, non-B hepatitis.Weiner AJ, Kuo G, Bradley DW, Bonino F, Saracco G, Lee C, Rosenblatt J, Choo QL, Houghton M. Lancet. 1990 Jan 6;335(8680):1–3
52. Characterization of the terminal regions of hepatitis C viral RNA: identification of conserved sequences in the 5' untranslated region and poly(A) tails at the 3' end.Han JH, Shyamala V, Richman KH, Brauer MJ, Irvine B, Urdea MS, Tekamp-Olson P, Kuo G, Choo QL, Houghton M. Proc Natl Acad Sci USA. 1991 Mar 1;88(5):1711–5
53. Use of a signature nucleotide sequence of hepatitis C virus for detection of viral RNA in human serum and plasma.Cha TA, Kolberg J, Irvine B, Stempien M, Beall E, Yano M, Choo QL, Houghton M, Kuo G, Han JH. J Clin Microbiol. 1991 Nov;29(11):2528–34
54. Detection of residual hepatitis C virus RNA by transcription-mediated amplification in patients with complete virologic response according to polymerase chain reaction-based assays. Sarrazin C, Teuber G, Kokka R, Rabenau H, Zeuzem S. Hepatology. 2000 Oct;32(4 Pt 1):818–23
55. Treatment of chronic non-A, non-B hepatitis with recombinant human alpha interferon. A preliminary report. Hoofnagle JH, Mullen KD, Jones DB, Rustgi V, Di Bisceglie A, Peters M, Waggoner JG, Park Y, Jones EA. N Engl J Med. 1986 Dec 18;315(25):1575–8
56. Ribavirin and interferon for recurrent posttransplantation HCV infection: to treat or not to treat? McHutchison JG. Hepatology. 1997 Aug;26(2):505–6
57. Sofosbuvir and ledipasvir fixed-dose combination with and without ribavirin in treatment-naive and previously treated patients with genotype 1 hepatitis C virus infection (LONESTAR): an open-label, randomised, phase 2 trial. Lawitz E, Poordad FF, Pang PS, Hyland RH, Ding X, Mo H, Symonds WT, McHutchison JG, Membreno FE. Lancet. 2014 Feb 8; 383(9916):515–23
58. The putative nucleocapsid and envelope protein genes of hepatitis C virus determined by comparison of the nucleotide sequences of two isolates derived from an experimentally infected chimpanzee and healthy human carriers. Takeuchi K, Kubo Y, Boonmar S, Watanabe Y, Katayama T, Choo QL, Kuo G, Houghton M, Saito I, Miyamura T. J Gen Virol. 1990 Dec;71 (Pt 12):3027–33
59. Classification of hepatitis C virus into six major genotypes and a series of subtypes by phylogenetic analysis of the NS-5 region. Simmonds P, Holmes

EC, Cha TA, Chan SW, McOmish F, Irvine B, Beall E, Yap PL, Kolberg J, Urdea MS. J Gen Virol. 1993 Nov;74 (Pt 11):2391–9
60. Hepatitis C virus genotype 7, a new genotype originating from central Africa. Murphy DG, Sablon E, Chamberland J, Fournier E, Dandavino R, Tremblay CL. J Clin Microbiol. 2015 Mar;53(3):967–72
61. Global epidemiology and genotype distribution of the hepatitis C virus infection. Gower E, Estes C, Blach S, Razavi-Shearer K, Razavi H. J Hepatol. 2014 Nov;61(1 Suppl):S45–57
62. Global distribution and prevalence of hepatitis C virus genotypes.Messina JP, Humphreys I, Flaxman A, Brown A, Cooke GS, Pybus OG, Barnes E. Hepatology. 2015 Jan;61(1):77–87
63. Global epidemiology of hepatitis C virus infection: An up-date of the distribution and circulation of hepatitis C virus genotypes. Petruzziello A, Marigliano S, Loquercio G, Cozzolino A, Cacciapuoti C. World J Gastroenterol. 2016 Sep 14;22(34):7824–40
64. The hepatitis C virus encodes a serine protease involved in processing of the putative nonstructural proteins from the viral polyprotein precursor. Eckart MR, Selby M, Masiarz F, Lee C, Berger K, Crawford K, Kuo C, Kuo G, Houghton M, Choo QL. Biochem Biophys Res Commun. 1993 Apr 30;192(2):399–406
65. Characterization of the hepatitis C virus-encoded serine proteinase: determination of proteinase-dependent polyprotein cleavage sites. Grakoui A, McCourt DW, Wychowski C, Feinstone SM, Rice CM. J Virol. 1993 May;67(5):2832–43
66. Two proteinase activities in HCV polypeptide expressed in insect cells using baculovirus vector. Hirowatari Y, Hijikata M, Tanji Y, Nyunoya H, Mizushima H, Kimura K, Tanaka T, Kato N, Shimotohno K. Arch Virol. 1993;133(3–4):349–56
67. A central region in the hepatitis C virus NS4A protein allows formation of an active NS3-NS4A serine proteinase complex in vivo and in vitro. Lin C, Thomson JA, Rice CM. J Virol. 1995 Jul;69(7):4373–80
68. Hepatitis C virus-encoded nonstructural protein NS4A has versatile functions in viral protein processing. Tanji Y, Hijikata M, Satoh S, Kaneko T, Shimotohno K. J Virol. 1995 Mar;69(3):1575–81
69. Protease inhibitors for the treatment of hepatitis C virus infection. de Leuw P, Stephan C. GMS Infect Dis. 2017 Nov 28;5: Doc08
70. Enter Sofosbuvir: The Path to Curing HCV. Sofia MJ.Cell. 2016 Sep 22;167(1):25–29
71. Dasabuvir: A Non-Nucleoside Inhibitor of NS5B for the Treatment of Hepatitis C Virus Infection. Gentile I, Buonomo AR, Borgia G. Rev Recent Clin Trials. 2014;9(2):115–23
72. Replication of subgenomic hepatitis C virus RNAs in a hepatoma cell line. Lohmann V, Körner F, Koch J, Herian U, Theilmann L, Bartenschlager R. Science. 1999 Jul 2;285(5424):110–3
73. Efficient initiation of HCV RNA replication in cell culture. Blight KJ, Kolykhalov AA, Rice CM. Science. 2000 Dec 8;290(5498):1972–4
74. Chemical genetics strategy identifies an HCV NS5A inhibitor with a potent clinical effect. Gao M, Nettles RE, Belema M, Snyder LB, Nguyen VN, Fridell RA, Serrano-Wu MH, Langley DR, Sun JH, O'Boyle DR 2nd, Lemm JA, Wang C, Knipe JO, Chien C, Colonno RJ, Grasela DM, Meanwell NA, Hamann LG. Nature. 2010 May 6;465(7294):96–100

75. A Refined Model of the HCV NS5A protein bound to daclatasvir explains drug-resistant mutations and activity against divergent genotypes. Barakat KH, Anwar-Mohamed A, Tuszynski JA, Robins MJ, Tyrrell DL, Houghton M. J Chem Inf Model. 2015 Feb 23;55(2):362–73
76. A Comprehensive Computational Analysis for the Binding Modes of Hepatitis C Virus NS5A Inhibitors: The Question of Symmetry. Ahmed M, Pal A, Houghton M, Barakat K. ACS Infect Dis. 2016 Nov 11;2(11):872–881
77. Egypt's Ambitious Strategy to Eliminate Hepatitis C Virus: A Case Study. Hassanin A, Kamel S, Waked I, Fort M. Glob Health Sci Pract. 2021 Mar 31;9(1):187–200
78. Characterization of the hepatitis C virus E2/NS1 gene product expressed in mammalian cells. Spaete RR, Rugroden ME, Choo QL, Berger K, Crawford K, Kuo C, Leng S, Lee C, Ralston R, Thudium K, Tung JW, Kuo G, Houghton M. Virology. 1992 Jun;188(2):819–30
79. Characterization of hepatitis C virus envelope glycoprotein complexes expressed by recombinant vaccinia viruses. Ralston R, Thudium K, Berger K, Kuo C, Gervase B, Hall J, Selby M, Kuo G, Houghton M, Choo QL. J Virol. 1993 Nov;67(11):6753 –6761
80. Vaccination of chimpanzees against infection by the hepatitis C virus. Choo QL, Kuo G, Ralston RE, Weiner A, Chien D, Van Nest G, Han J, Berger K, Thudium K, Kuo C, Kansopon J, McFarland J, Tabrizi A, Ching K, Moss B, Cummins L, Houghton M, Muchmore E. PNAS February 15,1994 91 (4) 1294–1298
81. Prospects for a vaccine against the hepatitis C virus. Houghton M, Abrignani S. Nature. 2005 Aug 18;436(7053):961–6
82. Prospects for prophylactic and therapeutic vaccines against the hepatitis C viruses. Houghton M. Immunol Rev. 2011 Jan;239(1):99–108
83. Infectious hepatitis C virus pseudo-particles containing functional E1-E2 envelope protein complexes. Bartosch B, Dubuisson J, Cosset FL. J Exp Med. 2003 Mar 3;197(5):633–42
84. Characterization of infectious retroviral pseudotype particles bearing hepatitis C virus glycoproteins. Flint M, Logvinoff C, Rice CM, McKeating JA. .J Virol. 2004 Jul;78(13):6875–82
85. Production of infectious hepatitis C virus in tissue culture from a cloned viral genome. Wakita T, Pietschmann T, Kato T, Date T, Miyamoto M, Zhao Z, Murthy K, Habermann A, Kräusslich HG, Mizokami M, Bartenschlager R, Liang TJ. Nat Med. 2005 Jul;11(7):791–6
86. Hepatitis C virus envelope glycoprotein immunization of rodents elicits cross-reactive neutralizing antibodies. Stamataki Z, Coates S, Evans MJ, Wininger M, Crawford K, Dong C, Fong YL, Chien D, Abrignani S, Balfe P, Rice CM, McKeating JA, Houghton M. Vaccine. 2007 Nov 7;25(45):7773–84
87. Vaccine-induced cross-genotype reactive neutralizing antibodies against hepatitis C virus. Meunier JC, Gottwein JM, Houghton M, Russell RS, Emerson SU, Bukh J, Purcell RH. J Infect Dis. 2011 Oct 15;204(8):1186–90
88. Immunization of human volunteers with hepatitis C virus envelope glycoproteins elicits antibodies that cross-neutralize heterologous virus strains. Stamataki Z, Coates S, Abrignani S, Houghton M, McKeating JA. J Infect Dis. 2011 Sep 1;204(5):811–3
89. A hepatitis C virus (HCV) vaccine comprising envelope glycoproteins gpE1/gpE2 derived from a single isolate elicits broad cross-genotype neutralizing

antibodies in humans. Law JL, Chen C, Wong J, Hockman D, Santer DM, Frey SE, Belshe RB, Wakita T, Bukh J, Jones CT, Rice CM, Abrignani S, Tyrrell DL, Houghton M. PLoS One. 2013;8(3):e59776

90. Characterization of hepatitis C virus recombinants with chimeric E1/E2 envelope proteins and identification of single amino acids in the E2 stem region important for entry. Carlsen TH, Scheel TK, Ramirez S, Foung SK, Bukh J. J Virol. 2013 Feb;87(3):1385–99
91. Immunodominant CD4+ T-cell epitope within nonstructural protein 3 in acute hepatitis C virus infection. Diepolder HM, Gerlach JT, Zachoval R, Hoffmann RM, Jung MC, Wierenga EA, Scholz S, Santantonio T, Houghton M, Southwood S, Sette A, Pape GR. J Virol. 1997 Aug;71(8):6011–9
92. Analysis of a successful immune response against hepatitis C virus.Cooper S, Erickson AL, Adams EJ, Kansopon J, Weiner AJ, Chien DY, Houghton M, Parham P, Walker CM. Immunity. 1999 Apr;10(4):439–49
93. Memory CD8+ T cells are required for protection from persistent hepatitis C virus infection. Shoukry NH, Grakoui A, Houghton M, Chien DY, Ghrayeb J, Reimann KA, Walker CM. J Exp Med. 2003 Jun 16;197(12):1645–55
94. HCV persistence and immune evasion in the absence of memory T cell help. Grakoui A, Shoukry NH, Woollard DJ, Han JH, Hanson HL, Ghrayeb J, Murthy KK, Rice CM, Walker CM. Science. 2003 Oct 24;302(5645):659–62
95. Human serum facilitates hepatitis C virus infection, and neutralizing responses inversely correlate with viral replication kinetics at the acute phase of hepatitis C virus infection. Lavillette D, Morice Y, Germanidis G, Donot P, Soulier A, Pagkalos E, Sakellariou G, Intrator L, Bartosch B, Pawlotsky JM, Cosset FL. J Virol. 2005 May;79(10):6023–34
96. Rapid induction of virus-neutralizing antibodies and viral clearance in a single-source outbreak of hepatitis C. Pestka JM, Zeisel MB, Bläser E, Schürmann P, Bartosch B, Cosset FL, Patel AH, Meisel H, Baumert J, Viazov S, Rispeter K, Blum HE, Roggendorf M, Baumert TF. Proc Natl Acad Sci USA. 2007 Apr 3;104(14):6025–30
97. In vivo evaluation of the cross-genotype neutralizing activity of polyclonal antibodies against hepatitis C virus.Meuleman P, Bukh J, Verhoye L, Farhoudi A, Vanwolleghem T, Wang RY, Desombere I, Alter H, Purcell RH, Leroux-Roels G. Hepatology. 2011 Mar;53(3):755–62
98. Clearance of hepatitis C infection is associated with the early appearance of broad neutralizing antibody responses. Osburn WO, Snider AE, Wells BL, Latanich R, Bailey JR, Thomas DL, Cox AL, Ray SC. Hepatology. 2014 Jun;59(6):2140–51
99. Plasma deconvolution identifies broadly neutralizing antibodies associated with hepatitis C virus clearance. Kinchen VJ, Massaccesi G, Flyak AI, Mankowski MC, Colbert MD, Osburn WO, Ray SC, Cox AL, Crowe JE Jr, Bailey JR. J Clin Invest. 2019 Aug 13;129(11):4786–4796
100. Human monoclonal antibodies that inhibit binding of hepatitis C virus E2 protein to CD81 and recognize conserved conformational epitopes. Hadlock KG, Lanford RE, Perkins S, Rowe J, Yang Q, Levy S, Pileri P, Abrignani S, Foung SK. J Virol. 2000 Nov;74(22):10407–16
101. Human monoclonal antibody to hepatitis C virus E1 glycoprotein that blocks virus attachment and viral infectivity.Keck ZY, Sung VM, Perkins S, Rowe J, Paul S, Liang TJ, Lai MM, Foung SK. J Virol. 2004 Jul;78(13):7257–63
102. Broadly neutralizing antibodies protect against hepatitis C virus quasispecies challenge. Law M, Maruyama T, Lewis J, Giang E, Tarr AW, Stamataki Z,

Gastaminza P, Chisari FV, Jones IM, Fox RI, Ball JK, McKeating JA, Kneteman NM, Burton DR. Nat Med. 2008 Jan;14(1):25–7
103. Neutralizing monoclonal antibodies against hepatitis C virus E2 protein bind discontinuous epitopes and inhibit infection at a postattachment step. Sabo MC, Luca VC, Prentoe J, Hopcraft SE, Blight KJ, Yi M, Lemon SM, Ball JK, Bukh J, Evans MJ, Fremont DH, Diamond MS. J Virol. 2011 Jul;85(14):7005–19
104. Structural basis of hepatitis C virus neutralization by broadly neutralizing antibody HCV1.Kong L, Giang E, Robbins JB, Stanfield RL, Burton DR, Wilson IA, Law M. Proc Natl Acad Sci USA. 2012 Jun 12;109(24):9499–504
105. Structure of hepatitis C virus envelope glycoprotein E2 antigenic site 412 to 423 in complex with antibody AP33.Kong L, Giang E, Nieusma T, Robbins JB, Deller MC, Stanfield RL, Wilson IA, Law M. J Virol. 2012 Dec;86(23):13085–8
106. Human broadly neutralizing antibodies to the envelope glycoprotein complex of hepatitis C virus. Giang E, Dorner M, Prentoe JC, Dreux M, Evans MJ, Bukh J, Rice CM, Ploss A, Burton DR, Law M. Proc Natl Acad Sci USA. 2012 Apr 17;109(16):6205–10.
107. Human monoclonal antibody HCV1 effectively prevents and treats HCV infection in chimpanzees. Morin TJ, Broering TJ, Leav BA, Blair BM, Rowley KJ, Boucher EN, Wang Y, Cheslock PS, Knauber M, Olsen DB, Ludmerer SW, Szabo G, Finberg RW, Purcell RH, Lanford RE, Ambrosino DM, Molrine DC, Babcock GJ. PLoS Pathog. 2012;8(8):e1002895
108. Breadth of neutralization and synergy of clinically relevant human monoclonal antibodies against HCV genotypes 1a, 1b, 2a, 2b, 2c, and 3a. Carlsen TH, Pedersen J, Prentoe JC, Giang E, Keck ZY, Mikkelsen LS, Law M, Foung SK, Bukh J. Hepatology. 2014 Nov;60(5):1551–62
109. Broadly neutralizing antibodies abrogate established hepatitis C virus infection. de Jong YP, Dorner M, Mommersteeg MC, Xiao JW, Balazs AB, Robbins JB, Winer BY, Gerges S, Vega K, Labitt RN, Donovan BM, Giang E, Krishnan A, Chiriboga L, Charlton MR, Burton DR, Baltimore D, Law M, Rice CM, Ploss A. Sci Transl Med. 2014 Sep 17; 6(254):254ra129-.
110. Affinity maturation of a broadly neutralizing human monoclonal antibody that prevents acute hepatitis C virus infection in mice. Keck ZY, Wang Y, Lau P, Lund G, Rangarajan S, Fauvelle C, Liao GC, Holtsberg FW, Warfield KL, Aman MJ, Pierce BG, Fuerst TR, Bailey JR, Baumert TF, Mariuzza RA, Kneteman NM, Foung SK. Hepatology. 2016 Dec;64(6):1922–1933
111. Broadly neutralizing antibodies from an individual that naturally cleared multiple hepatitis C virus infections uncover molecular determinants for E2 targeting and vaccine design. Keck ZY, Pierce BG, Lau P, Lu J, Wang Y, Underwood A, Bull RA, Prentoe J, Velázquez-Moctezuma R, Walker MR, Luciani F, Guest JD, Fauvelle C, Baumert TF, Bukh J, Lloyd AR, Foung SKH. PLoS Pathog. 2019 May 17;15(5)
112. Antibody Responses in Hepatitis C Infection. Law M. Cold Spring Harb Perspect Med. 2021 Mar 1;11(3):a036962
113. Recombinant hepatitis C virus envelope glycoprotein vaccine elicits antibodies targeting multiple epitopes on the envelope glycoproteins associated with broad cross-neutralization. Wong JA, Bhat R, Hockman D, Logan M, Chen C, Levin A, Frey SE, Belshe RB, Tyrrell DL, Law JL, Houghton M. J Virol. 2014 Dec;88(24):14278–88.
114. Expanded classification of hepatitis C virus into 7 genotypes and 67 subtypes: updated criteria and genotype assignment web resource. Smith DB,

Bukh J, Kuiken C, Muerhoff AS, Rice CM, Stapleton JT, Simmonds P. Hepatology. 2014 Jan;59(1):318–27

115. Production and characterization of high-titer serum-free cell culture grown hepatitis C virus particles of genotype 1–6. Mathiesen CK, Jensen TB, Prentoe J, Krarup H, Nicosia A, Law M, Bukh J, Gottwein JM. Virology. 2014 Jun;458–459:190–208.
116. Genetic Diversity and Selective Pressure in Hepatitis C Virus Genotypes 1–6: Significance for Direct–Acting Antiviral Treatment and Drug Resistance. Cuypers L, Li G, Libin P, Piampongsant S, Vandamme AM, Theys K. Viruses. 2015 Sep 16;7(9):5018–39
117. Hepatitis C virus replication in mice with chimeric human livers. Mercer DF, Schiller DE, Elliott JF, Douglas DN, Hao C, Rinfret A, Addison WR, Fischer KP, Churchill TA, Lakey JR, Tyrrell DL. N M Kneteman Nat Med. 2001 Aug;7(8):927–33.
118. Intrahepatic genetic inoculation of hepatitis C virus RNA confers cross-protective immunity. Weiner AJ, Paliard X, Selby MJ, Medina-Selby A, Coit D, Nguyen S, Kansopon J, Arian CL, Ng P, Tucker J, Lee CT, Polakos NK, Han J, Wong S, Lu HH, Rosenberg S, Brasky KM, Chien D, Kuo G, Houghton M. J Virol. 2001 Aug;75(15):7142–8.
119. Protection against persistence of hepatitis C. Mehta SH, Cox A, Hoover DR, Wang XH, Mao Q, Ray S, Strathdee SA, Vlahov D, Thomas DL. Lancet. 2002 Apr 27;359(9316):1478–83.
120. Cross-genotype immunity to hepatitis C virus. Lanford RE, Guerra B, Chavez D, Bigger C, Brasky KM, Wang XH, Ray SC, Thomas DL. J Virol. 2004 Feb;78(3):1575–81.
121. A recombinant hepatitis C virus genotype 1a E1/E2 envelope glycoprotein vaccine elicits antibodies that differentially neutralize closely related 2a strains through interactions of the N-terminal hypervariable region 1 of E2 with scavenger receptor B1. Johnson J, Freedman H, Logan M, Wong JA, Hockman D, Chen C, He J, Beard MR, Eyre NS, Baumert TF, Tyrrell DL, Law JL, Houghton, M. J Virol. 2019 Nov 15;93(22):e00810–19.
122. Persistent non-A, non-B hepatitis in experimentally infected chimpanzees. Bradley DW, Maynard JE, Popper H, Ebert JW, Cook EH, Fields HA, Kemler BJ. J Infect Dis. 1981 Feb;143(2):210–8
123. Safety and immunogenicity of HCV E1E2 vaccine adjuvanted with MF59 administered to healthy adults. Frey SE, Houghton M, Coates S, Abrignani S, Chien D, Rosa D, Pileri P, Ray R, Di Bisceglie AM, Rinella P, Hill H, Wolff MC, Schultze V, Han JH, Scharschmidt B, Belshe RB. Vaccine. 2010 Aug 31;28(38):6367–73.
124. Progress of the COVID-19 vaccine effort: viruses, vaccines and variants versus efficacy, effectiveness and escape. Tregoning JS, Flight KE, Higham SL, Wang Z, Pierce BF. Nature Reviews Immunology. 2021 Aug 9; 21:626–636.

Charles M. Rice. © Nobel Prize Outreach AB. Photo: Stefan Bladh

Charles M. Rice

Biography

I was born on August 25, 1952 in Sacramento, California to Roberta Helen Rice and Charles Moen Rice Jr. (Fig. 1). My mother originated in Colorado Springs, Colorado, my father from Worcester, Massachusetts. I was the third Charles Moen Rice; we were all only children. I never met my paternal grandparents who were long gone by the time I was born. For most of my early years, we lived in Sacramento with a short interlude in San Jose (Fig. 2). My father worked as an insurance claims adjuster, my mother as a suburban housewife. My "brothers and sisters" growing up were a series of canine friends, largely dachshunds (Fig. 3). My parents and I loved dogs, a passion that endures to the present. A neighbor once said of my father, "if I believed in reincarnation, I would want to come back as one of your dad's dogs." No finer life could be imagined. I was an avid reader as a child, often staying up late into the night much to my parents' chagrin. Music was a large part of my early home life nourished in part from my father who was a fanatic stereophile and had to have the latest high-end equipment. I was exposed to classical music and musicals, but later degenerated to listening to the Rolling Stones and the Doors on a low-end record player in my room.

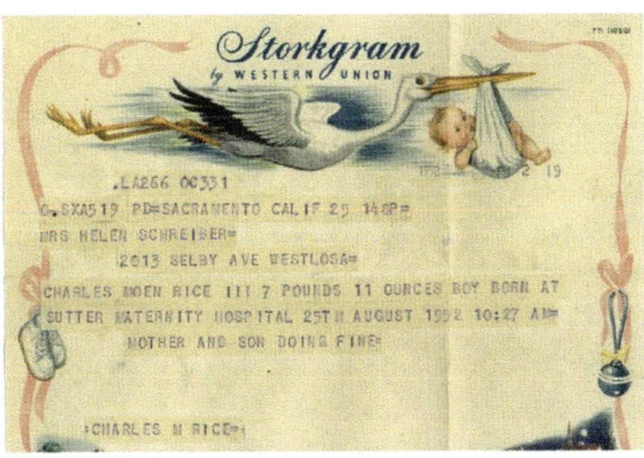
Storkgram
by WESTERN UNION
.LA266 0C331
SXA519 PD=SACRAMENTO CALIF 25
MRS HELEN SCHREIBER=
2013 SELBY AVE WESTLOSA=
CHARLES MOEN RICE III 7 POUNDS 11 OUNCES BOY BORN AT SUTTER MATERNITY HOSPITAL 25TH AUGUST 1952 10:27 AM=
MOTHER AND SON DOING FINE=
CHARLES M RICE=

Figure 1. Birth announcement telegram sent to my maternal grandmother.

Figure 2. My parents Roberta and Charles (middle and right) with Bill Easton and his mother, Tee (left).

Figure 3. My "brothers and sisters".

Amongst my most cherished boyhood items were the usual chemistry set and very basic microscope, but my real love of biology and nature came from frequent camping and hiking trips in the Sierra Nevada mountains. My education was exclusively in public schools. I did reasonably well, despite having a series of odd jobs including cleaning up after parties at a nearby clubhouse. My custodial skills are still quite good. All in all, I had a fairly normal trajectory for someone growing up in the 1960s, which was anything but a "normal" decade: Vietnam, the cold war, the moon landing, assassinations of JFK, MLK and RFK, sex, drugs and rock-n-roll, etc.

The last years of high school (Rio Americano), forced the choice of what to do next. The default trajectory for a middle-class kid was "go to college", so that's what I did. I applied to UC Davis and was accepted (with a small scholarship). For those of you unfamiliar with Northern California geography, Davis is about 15 miles west of Sacramento. So, for me, close to home, but far enough away. Given my love of animals, including even rats, I chose UC Davis given its veterinary medicine reputation, a vocation I thought I might pursue after finishing my undergraduate degree. In those days, going to a state university in your home state had an added bonus – it was cheap. I had no idea what major I was going to pursue but was leaning towards math until that was derailed by an advanced statistics course some years later. I blame this on my instructor, who had the appearance of wizard-like Gandalf, but the wisdom he was attempting to convey was lost on me.

An introductory biology course my freshman year turned out to be perhaps the most important chance encounter of my career. The course was taught by a Caltech-trained developmental biologist, Dennis Barrett. Dennis was/is a terrific biologist, and his lectures were captivating. Even more engaging were his exams, which were made up of multiple-choice questions. The exam was taken in class and the students could then take it home and revise their answers using any means available – textbooks, literature, or discussions with other students. This take-home portion of the exam, and the quest to get a perfect score, was an incredible learning experience because the correct answers to these multiple-choice questions were not the obvious ones and required a deeper dive and a real understanding of the biology. I don't recall the details now, but Dennis became a key advisor and advocate for me during my undergraduate years and beyond.

My first undergraduate research experience was in the chemistry department studying tunicates (sea squirts) who manage to concentrate vanadium from sea water over six orders of magnitude in specialized blood cells. While I didn't mind bleeding tunicates and doing elemental analysis, this wasn't my cup of tea. I switched to working with Dennis on his favorite organism, the purple sea urchin, *Strongylocentrotus purpuratas.* "Purps" as we would call them are commonly found off the US Western coast; we collected them on field trips in the area near the Bodega Bay Marine Station. These animals can be induced to shed their gametes with an electrical jolt or injection of potassium chloride. Fertilization ensues by mixing the eggs and sperm in sea water, and you can watch early development proceed with a simple microscope – amazing. I was hooked, but this was just a part-time, sporadic research experience interspersed with working at the UCD library, the Memorial Union food service, and spending my first few summers working at a cannery or as a tomato inspector. It was some years before I could stomach eating ketchup again.

The summer of 1973 provided the chance to really immerse myself in research. Dennis was an instructor in the Physiology Course at the Woods Hole Marine Biology Laboratory (MBL) in Massachusetts and suggested that I apply to their summer program. This worked out. I arrived and found myself immersed in science 24/7. The course covered developmental biology, immunology, biophysical biochemistry and was taught by world class scientists through lectures, reading material and intense hands-on lab work. I turned 21 at the MBL and had a memorable (that I don't actually remember) celebration (Fig. 4). After that, it was difficult to return to the undergraduate grind for my senior year at UCD.

Nonetheless, I survived and graduated in 1974 with a degree in Zoology. I've so far highlighted academic pursuits but in reality there was quite a bit going on outside of the classroom – soccer, softball, basketball, frequent backpacking trips and other activities that I will not detail here. Also in my academic wanderings, I had taken advantage of some "fun" classes, like the introduction to viticulture and enology anchored by the book "Wine: An Introduction" by UCD professors Maynard Amerine

Figure 4. My 21st birthday celebration at the Woods Hole Marine Biology Laboratory.

and Vernon Singleton. Growing up in California near the wine country, coupled with this course, seeded a real passion for wine. I found myself waffling between a career in wine and developmental biology research. Somewhat different trajectories...

Given this dilemma, I recruited a few college friends, we bought a used VW bus and departed for an extended trip to Central and South America. Before I left, I applied to two PhD programs strong in developmental biology: UC San Diego and Caltech. Uncertain as to my future, off we went, slowly making our way through Central America. After several months, a failing VW bus was abandoned in Costa Rica and we flew to Colombia to continue the South American leg of the adventure to Lima, Peru, where a telegram from my father awaited me: "you have been accepted at UCSD and Caltech and they want to know if you are coming". I was having a great time traveling and meeting an amazing cast of wanderers, but a decision had to be made. Continuing from Lima to Cuzco, to Machu Picchu, La Paz and Salta in Northern Argentina, and after an exchange of letters with Dennis, I decided on Caltech, despite its lower stipend. A transition back to reality was needed. Fortunately, Dennis needed a teaching assistant for the MBL physiology course and some help moving from Davis to Denver, where he was joining the University of Denver faculty. The cheapest flight back to the US was to Miami. From there, an attempt by a scruffy vagabond with a backpack to hitchhike back to California failed, requiring a multiday Greyhound bus ride to Sacramento. After a few days back home, we loaded up a rental truck, drove from Davis to Denver and after a short stop to unload, headed to Woods Hole for another memorable summer. Besides the science, there was healthy competition between the various MBL courses, including a long-standing rivalry between the Physiology Course, headed by immunologist John Cebra, and the Embryology Course headed by the intensely competitive Eric Davidson, a Caltech professor who studied sea urchin development. This rivalry was played out on a softball diamond and a flag football field. My recollection is that the Physiology Course usually triumphed at softball whereas Eric's team dominated the more physical football encounters. As you might have guessed, Eric's sea urchin lab was one of the main reasons I was going to Caltech.

The summer ended and I was off to Pasadena, driving a yellow-orange 1975 VW Rabbit that my parents helped me buy. Being raised in Northern California, I approached this transition with great trepidation. How could I defect to Southern California? It was going to be smoggy, crowded, and terrible. My goal was to get my PhD and move on to greener pastures ASAP. But...I ended up loving it and spent nearly a decade at Caltech. However, there was a surprise upon my arrival. Rather than being assigned to Eric's lab, as I'd hoped, I was placed in an animal virus lab, headed by James H. "Jim" Strauss and his wife Ellen Glowacki Strauss. In those days, laboratory rotations weren't necessarily required or even encouraged. As I learned more about the Davidson lab I began to have second thoughts – Eric was brilliant but rumored to be tough, demanding and he didn't suffer fools. I was pretty sure I belonged squarely in the "fools" category so I decided that this might not be the best fit. You could chalk it up to "no guts" and a lack of self-confidence, but at the same time the field of animal RNA viruses was still in its infancy and the Strauss lab was at the cutting edge of studying a model enveloped RNA virus, Sindbis virus, the prototype alphavirus. The lab was small with 3 PhD students, 2 techs, Jim and Ellen, and an excellent environment for someone entering a new field. Jim's office was inhabited by two free-flying parakeets who didn't necessarily take kindly to graduate students barging in for advice.

There had been extensive work in the lab on Sindbis virus RNA replication; less was known about the virus particle. Jim suggested that I work on a small, secreted glycoprotein called E3, which was processed from an envelope glycoprotein precursor during virion maturation and released into the cell culture supernatant. In the case of Sindbis, E3 was not virion associated whereas in the case of a related alphavirus, Semliki Forest virus, it stayed bound to the virus particle. I wasn't enamored with this project. Fortunately, we were given a great deal of freedom to come up with and test our own ideas. I did end up doing quite a bit of work on the Sindbis virus glycoproteins, and that comprised the bulk of my thesis. But the larger Caltech research environment was playing a key role in shaping my research interests. This included a remarkable group of fellow PhD students and postdocs, the Caltech faculty and, of course, the undergrads. Recombinant DNA was exploding, we could now "easily" sequence nucleic acids and our lab was conveniently positioned right down the hall from the Maniatis lab, with its freezer full of enzymes and cutting-edge molecular biology expertise. Rather than being left out of this revolution, it seemed obvious that we should determine the sequences of the Sindbis structural proteins. The only catch was that concerns about recombinant DNA cloning and safety prevented molecular cloning of these animal RNA viruses. I was left to figure out how to sequence the structural protein coding region without complementary DNA (cDNA) cloning. Without delving into the gory details, it turns out that some restriction enzymes can cleave single-stranded cDNA or RNA-cDNA hybrids, providing discrete fragments that can be end-labeled, gel purified and sequenced using the laborious Maxim-Gilbert chemical sequencing method. This eventually provided the sequence of Sindbis subgenomic mRNA revealing a single open reading frame (ORF) that encoded the capsid protein followed by E3 and the two virion glycoproteins, E2 and E1. This capped off my PhD work, which I defended in 1981.

At this point, if not before, I should have been charting my future and lining up a postdoc. But to be honest, I was having too much fun collaborating with other labs, finishing ongoing work, and starting new projects. Again, this was possible because of the freedom granted, as long as good science ensued. I made a few half-hearted attempts, sending out feelers to Ron Davis at Stanford and Tom Cech in Colorado, but they either didn't answer or weren't interested (which is perhaps not surprising for an applicant working on an obscure RNA virus). So, ignoring the "academic kiss of death" I just stayed in the Strauss lab.

There were two ongoing efforts that made this an easy decision. The first revolved around an undying belief in the power of genetics. For me, this had been nurtured at UC Davis as an undergraduate by the great evolutionary geneticists Theodosius Dobzhansky and G. Ledyard Stebbins, and at Caltech surrounded by Max Delbruck in his waning years and the "next generation" including Bill Wood, Seymour Benzer, and Ed Lewis. One of the many attractions of working on viruses is their small genome size yet a highly evolved ability to infect cells, replicate and survive in nature despite a complete dependence on cellular functions. Inert on their own, they provide a perturbation to probe cell and organismal biology. As highly efficient machines, we can learn much from viruses themselves and their replicative mechanisms. Classically, chemical mutagenesis, selection of conditional mutants and their characterization under permissive and non-permissive conditions were the mainstay of animal RNA virus genetics. For Sindbis, this had certainly been the case for an immensely useful panel of temperature-sensitive (ts) mutants. But now, given reverse transcriptase and the ability to copy RNA into cDNA and recombinant DNA methods, how could we exploit these new tools to allow site-directed genetic manipulation of positive-sense RNA viruses?

In concept, this should be simple: isolate viral genome RNA, make a cDNA copy, and clone it into a plasmid that allows transcription of a functional viral genome RNA using either cellular machinery or in vitro transcription. This was shown to work for the bacteriophage QB by Taniguchi and Weissman in 1978. Shortly thereafter we began working towards the same objective for Sindbis. This was a collaborative effort with Henry Huang, a brilliant postdoc in Leroy Hood's group. Cloning was now allowed and we were fortunate that Eugene Butler, a postdoc in the Maniatis lab, had purified and characterized the DNA-dependent RNA polymerase of Salmonella phage 6 (SP6) for his PhD and defined the 17-base promoter and initiating base. This turned out to be a highly efficient way to make unlimited quantities of synthetic Sindbis genome RNA with proper 5' and 3' terminal sequences. It took several more years, and additional work after I moved to St. Louis, before we published our first paper on the Sindbis infectious clone: "Production of infectious RNA transcripts from Sindbis virus cDNA clones: mapping of lethal mutations, rescue of a temperature-sensitive marker, and in vitro mutagenesis to generate defined mutants". In the meantime, Vincent Racaniello and David Baltimore had reported success with poliovirus and Paul Ahlquist for the plant RNA virus, bromegrass mosaic virus. Many other examples followed, including recent work with SARS-CoV-2. This opened up an almost limitless landscape of possibilities for virologists since in theory, for any positive-sense RNA virus, these so-called "infectious clones" could be manipulated to generate mutations and phenotypes could be tested.

The second area cooking in the Strauss lab was an effort to understand more about another group of enveloped RNA viruses, the flaviviruses. Given their similar virion morphology and physical properties, alphaviruses and flaviviruses had been lumped together in the family *Togaviridae*. However, there was no serological cross-reactivity between these groups, so the flaviviruses had recently become a separate family, the *Flaviviridae*. Not much was known about flavivirus genome structure and how this might be similar or different from the alphaviruses or other positive strand RNA virus families.

The prototype flavivirus was the famous yellow fever virus (YFV), shown to be a filterable agent, transmitted by the bite of infected mosquitoes, and the cause of the often fatal viscerotropic disease by Walter Reed and his colleagues at the turn of the last century. Decades later, by passaging a human isolate from a young Ghanaian named Asibi in mouse and chicken embryo tissues, Max Theiler derived an attenuated yellow fever strain, called 17D, that was no longer virulent in non-human primates (NHP) and protected them from lethal YFV infection. This vaccine, which has been administered to more than 500 million people, is still used today. A single shot confers broad and long-lasting protection. Max, who was then working at the Rockefeller Foundation, was awarded the Nobel Prize in Physiology or Medicine in 1951 for his work on the yellow fever vaccine.

Given the choice of working with virulent flaviviruses like dengue virus, yellow fever virus or the Japanese, St. Louis, Murray Valley, West Nile and tick-borne encephalitis viruses, the YF 17D vaccine strain seemed like a no brainer. With great help from Edith Lenches, a stoic Hungarian technician in the Strauss lab, we spent years optimizing conditions for virus growth and partial purification of the virus. With high quality genome RNA in hand, and the assistance of three Caltech undergrads, we were able to clone and determine the sequence of YF 17D and, a few years later, the Asibi strain. Our YF sequences, together with the West Nile virus sequences reported by Gerd and Gisa Wengler, revealed a genome organization more like the picornaviruses than the alphaviruses – a genome of ~11 kb consisting of a single long ORF flanked by two short non-coding regions. Work from the Wengler lab had shown that flavivirus genome RNAs were capped but not polyadenylated. Rather, the new sequences revealed a potential stable 3'-terminal RNA secondary structure. This set the stage for a new era in flavivirus molecular biology, but what was needed were "infectious clones".

Figure 5. Jim and Ellen Strauss with me and visiting Professor, Lynn Dalgarno.

Given our ongoing Sindbis work, which was beginning to show promise, I decided to make the YF infectious clone effort a priority. Lynn Dalgarno from Australia was on sabbatical in the Strauss lab (Fig. 5) and we had been working together on another flavivirus, Murray Valley encephalitis virus. We became quite good friends and he extended an offer for me to spend some time in his lab at Australia National University. This seemed reasonable since it would allow me to really focus on the YF project. Before I left, I interviewed at a few places for a faculty position and received an offer from Washington University. Earlier, Henry Huang had joined their Microbiology and Immunology department (later changed to Molecular Microbiology). This was an especially attractive option since Henry and I had active Sindbis projects with Sondra Schlesinger in the Micro department. We had gotten to know Sondra well during her visit to Caltech to do some sequencing of Sindbis defective-interfering particle RNAs. I decided that Wash U was the best option for hitting the ground running, with the added bonus of not going out on any more job interviews. This was despite the not-so-generous start-up package of 60K offered by then department chair, immunologist Joe Davie, and a chorus of warnings from coast-biased colleagues about taking a position in the Midwest. This was surely "career suicide".

Lynn warned me about the long lag time between ordering and receiving supplies at ANU so I packed up a complete collection of molecular biology enzymes and reagents before heading off to Canberra. My time in the Dalgarno lab was spent trying to assemble a full-length cDNA clone of YF 17D. I made literally hundreds of attempts but failed miserably. The reason: toxicity in *E. coli.* I did make some progress on learning which YF cDNA regions were tolerated versus toxic, which came in handy later. It was frustrating but the process was helped by frequent squash matches, an introduction to Aussie Rules football and a memorable stay on Heron Island, a beautiful coral gem on the Great Barrier Reef. There were some lasting upsides on the science front. With help from my future Wash U colleagues, we submitted a grant to the US Army to continue our YF efforts and I was selected to apply for the second class of the Pew Scholars Program. Fortunately, both were successful. Some years later, residual Pew funds helped jump start our work on hepatitis C.

Having survived left side drivers and a highly territorial Aussie magpie that swooped me whenever I wasn't looking, it was time to return to Pasadena, pack up and head to St. Louis. The VW Rabbit, now 10 years older and showing its age, was loaded up with precious lab samples and my wine collection. I drove straight to St. Louis, acquiring two speeding tickets on the way. Besides setting up the lab, my first task was to write an NIH grant on the Sindbis work for the February 1986 deadline. It was painful (I wanted to be doing experiments), but I was excited: we finally had a functional Sindbis clone and I was able to borrow an Apple Macintosh 512K computer to help streamline the writing. What fun! Temperature sensitive mutants could finally be assigned to specific viral proteins, the Sindbis machinery could be engineered to express heterologous proteins rapidly and at high levels, and variants could be selected that were non-cytopathic in vertebrate cells.

The early years of the lab were a bit slow but productive. We "worked" all the time but also took time off to have fun outside the lab. I also met my life partner, Peggy MacDonald, and her chocolate Labrador retriever, Bonnie. Most of the first crop of PhD students worked on Sindbis projects, I continued slogging away on YF 17D. I tried installing dual transcription terminators at the 5' and 3' YF-plasmid boundaries, low copy plasmids, different *E. coli* hosts, and finally I just abandoned plasmids altogether and tried lambda phage. I figured that if lambda was going to lyse the bacterial host anyway, maybe flavivirus cDNA toxicity wouldn't be an issue. Not so – it did work but the recombinant phage was unstable and the burst size was so small that DNA yields were too low to be useful. The polymerase chain reaction was seeing increased use but in those days the error rate was high, and it was difficult to amplify long templates. From my failed attempts in Australia and at Wash U, I knew that the full-length cDNA could be propagated in *E. coli* as two plasmids that could be designed to overlap by unique restriction enzyme sites. Why not try assembling the full-length template for SP6 transcription by in vitro ligation and completely avoid trying to amplify the full-length YF cDNA in bacteria? This took some optimization, but it worked. A modern version of this strategy is still used today to assemble flavivirus and coronavirus DNA templates for in vitro transcription of full-length genome RNAs.

Energized and hopeful, we sent the paper off to *Science*. As I recall, it went through 2 rounds of reviews that were largely favorable, but it was ultimately rejected because of a perceived "lack of general interest". A new journal, *The New Biologist*, offered to publish the work and it finally appeared in 1989. *The New Biologist* perished a few years later, making these reprints a real collector's item. I don't have any, so don't bother contacting me.

Nonetheless, we were finally poised to explore the molecular genetics of YFV. The top priority was to understand the molecular basis of the attenuated YF 17D vaccine strain. This would involve making an Asibi infectious clone, testing the resulting virus for virulence in rhesus macaques, and then making a series of 17D/Asibi recombinant viruses to map the changes responsible for attenuation. This was a collaboration with Joel Dalrymple at USAMRIID who had made the Asibi RNA that we used for cloning and sequencing. Joel would undertake the NHP work, my lab would make and characterize the chimeras in cell culture models and make stocks for animal inoculation. Tragically, Joel developed aggressive kidney cancer and passed away a few years later at the age of 53. Without our USAMRIID collaborator and internal advocate, the project stalled and remains unfinished to this day.

Other work on YF was progressing. We determined the order and boundaries of the mature flavivirus proteins and began to unravel the complex proteolytic processing scheme involved in their production. One key player was a viral serine protease embedded in the polyprotein that mediated cleavages required for assembly of infectious virus and the membrane associated RNA replication machinery. We could show that this protease activity was essential, as were each of its seven site-specific cleavages in the YF polyprotein.

At the time, we were not working on non-A, non-B post-transfusion hepatitis. From work in the chimpanzee model, the mystery agent appeared to be a small, enveloped virus, but its true identity was unknown; attempts to culture it in the laboratory had failed. Unexplained cases of hepatitis were on my mind, however, given the death of a close friend's mother from acute liver failure, cause unknown.

This changed shortly after the 1989 publication of landmark papers from Michael Houghton, Harvey Alter and their colleagues, identifying a new positive-strand RNA virus (called hepatitis C virus, HCV), with a genome of about 10 kb, as the cause of non-A, non-B post transfusion hepatitis. HCV infection was widespread, usually chronic and associated with mild to advanced liver disease, including cirrhosis and cancer. The sequence of the virus revealed a single long open reading frame encoding a polyprotein with features strikingly similar to members of the *Flaviviridae*, including the classical flaviviruses and the animal pestiviruses. A new virus member had joined our family.

As a relatively new Assistant Professor, it didn't make much sense to start working on a virus that couldn't be grown in the laboratory. This brief period of sanity was derailed by a phone call from Steve Feinstone at the FDA who had read our YF 17D infectious clone paper. Steve wondered if this couldn't be leveraged to produce a vaccine against HCV. Steve, a co-discoverer of HAV, had been characterizing the non-A, non-B agent in the chimpanzee model and, like many others, was attempting to identify the causal virus.

With Steve's nudge and encouragement, we decided to collaborate and started working on HCV, with the initial goal of learning more about HCV polyprotein processing and the functions of the resulting mature viral proteins – very similar to what we had done for YF. Steve had access to some of Harvey Alter's famous patient H plasma obtained in 1977 during the acute phase of infection and shown to be highly infectious in chimpanzees. The first step was to make HCV H-strain cDNA clones for

Figure 6. Arash Grakoui and I looking over data in the early HCV days.

expression studies in mammalian cells and in bacteria, to make region-specific antigens, and a panel of HCV-specific antisera. The problem? No one in the lab wanted to work on HCV. Generating antisera was boring and besides, there was no "virus" to work with in the lab.

Arash Grakoui (Fig. 6) stepped up to the plate in my lab, Steve and his team pitched in and together we managed to generate a near complete map of the HCV polyprotein cleavage products, identify two HCV-encoded proteases and their cleavage sites, and start some collaborative work on other predicted HCV enzymes. We weren't alone. Other groups in the US, Europe and Japan were reporting similar results and the HCV research community was growing. There was a sense of the field's importance, a mutual respect for our colleagues and competitors, and open sharing of ideas and materials. This nurturing atmosphere was seeded in large part by Michael Houghton, who initiated and anchored a yearly international HCV meeting where he would highlight and applaud every advance in the field.

But the central problem persisted. Advances were being made, but we still didn't have an HCV permissive cell culture system. I seemed to be on a downward spiral, starting with a virologist-friendly alphavirus, then a more challenging flavivirus and now, a virus that we couldn't even grow in the lab. We began to work on the animal pestiviruses, the closest relatives to HCV at the time, and which replicated in cell culture. This provided some needed experimental "relief", but we hadn't given up on HCV.

Early on, we had decided to try a "reverse" strategy. If we couldn't get infectious patient or chimpanzee serum to infect cells in culture, perhaps we could raise our chances by using infectious clone-derived HCV RNA. We would have an unlimited supply of infectious material to screen for permissive cells and conditions. The dilemma was how to assay infectivity – there was only the chimpanzee model and intra hepatic injection of a 9.6 kb RNA and hoping it would be taken up by hepatocytes, intact, seemed like a stretch. It wasn't completely far-fetched though, given that Susan Emerson and Robert Purcell had successfully taken this approach years earlier for hepatitis A virus RNA in marmosets.

We assembled what we believed was a full-length cDNA clone from Harvey Alter's H77 material, made RNA and Steve injected this intrahepatically into chimpanzees. Nothing happened – no evidence of HCV RNA in circulation but we did learn the input RNA was very rapidly degraded and quickly became undetectable. Disappointing but perhaps not too surprising given this crude and likely inefficient RNA transfection attempt.

At this time, we began to wonder about our full-length clone. Was it really full-length, or might there be something missing? HCV sequences were being reported around the world and while the original sequence reported by the Houghton group terminated with polyA, the subsequent sequences suggested polyU. A careful analysis of the cDNA cloning methods used suggested that they could be biased. Alexander "Sasha" Kolykhalov, a superbly talented molecular biologist who had immigrated from Russia, took on the challenging task of confirming or extending the 5' and 3' terminal sequences of the HCV genome. Had Sasha not joined my group, I doubt that I would be writing this piece. Immigrants have and continue to make some of the most impactful contributions to US and global success stories. Isolating ourselves from this flow of untapped talent is a sure formula for disaster.

I will keep the next specifics to a minimum since they are covered in my Nobel lecture. Sasha went on to discover a missing piece at the 3' end of the HCV genome RNA. This later turned out to be absolutely required for virus replication. However, modifying our existing clones with this new sequence still failed to initiate productive replication in chimpanzees. Why was this? We worried about everything. Missing sequences, transfection method, a requirement for RNA modifications or a viral protein for infectivity, lethal mistakes in our clone, and more. The list was daunting. Sasha cloned and assembled hundreds of full-length cDNA clones, tested them with restriction enzyme digestions and in culture for polyprotein translation. He sequenced a subset that passed these criteria and used this information to assemble a clone that reflected the dominant H77 consensus sequence. When Steve inoculated this RNA into two animals, we saw a rise in circulating HCV RNA, characteristic liver inflammation and delayed seroconversion; both animals went on to chronic infection. After eight years of working on HCV, we finally had an infectious clone; more examples followed soon thereafter. The journal *Science* did accept this one.

Despite being validated in vivo, transcripts from HCV consensus clones were unable to replicate in cell culture. Another breakthrough, another roadblock. Many laboratories tried different HCV isolates, diverse cell types and conditions, but nothing worked. The ability to launch infection in the chimp model opened up new possibilities to study HCV evolution, immunity and test the impact of mutations on virus viability but the cell culture roadblock remained. Overcoming this was important, not just for studying the virus lifecycle but also for developing antiviral drugs. Without a cell-based assay supporting virus replication, potent inhibitors could be developed using biochemical assays but there was no convenient pre-clinical assay to test and optimize their efficacy at blocking virus replication.

The next breakthrough, reported in *Science* in 1999, was from Ralf Bartenschlager's lab at the University of Mainz in Germany. Like many in the field, Ralf and his student, Volker Lohmann, were close to quitting their efforts on HCV and moving on but decided to try one more last-ditch effort. Starting with a consensus clone for a German isolate, they engineered an HCV RNA where the portion of the ORF encoding the virus structural proteins was replaced by the gene for neomycin resistance. This was followed by an internal ribosome entry site from encephalomyocarditis virus (EMCV) to drive translation of the HCV proteins needed for viral RNA amplification. If this engineered RNA could replicate and express the drug resistance gene, then cells would be resistant to the drug and able grow in its presence. It worked – a few drug-resistant colonies appeared and the HCV subgenomic replicon system was born. We had been trying a similar approach with our H77 infectious clone but failed to detect any drug-resistant colonies. In need of a positive control, we assembled the Bartenschlager subgenomic replicon using synthetic oligonucleotides and confirmed their results. Remarkably, when we and others sequenced the HCV RNA present in these rare colonies, it was not the sequence of the input RNA. When even single changes were engineered back into the original replicon sequence, some of them raised the efficiency of drug resistant colony formation by more that 10,000-fold. Ten years after HCV's discovery, we finally had a robust cell culture system for studying RNA replication and aiding drug development.

At this point, another phone call ushered in a major change in my life. Steve Goff, a virologist at Columbia, was on the line asking if I would visit Rockefeller University and advise a committee tasked with finding someone to head a new Tri-Institutional "Center for the Study of Hepatitis C". One thing led to another, and I was offered the position. But I was perfectly happy at Wash U. My stint as interim chair of Micro was finished so I could really focus on research, and, being an outdoor, non-city type, I had always done my best to avoid New York City. Not to mention that Rockefeller, with its stratospheric reputation, didn't seem like the right landing pad for a plodding molecular virologist. After some tough soul searching and a nudge from my scientific grandfather, Jim Darnell (you should do this!), Peggy and I decided to make the move. Peggy gave up her tenure track faculty position at Wash U, and we moved to NYC in mid-2000 to an apartment that was about 50 yards from the Rockefeller campus. Our two rescue dogs, Sadie and Wrangler, were not happy with the loss of their yard and the new digs. But we all managed to adjust aided by the vibrant scientific atmosphere at Rockefeller, wonderful colleagues who, after all, were not that scary, and a beautiful park-like campus.

Not many of the Wash U crew were able to relocate from St. Louis to NYC so the first years were spent rebuilding the group, closing the Wash U lab and waiting for our renovated space in the RU Hospital building to be finished. I was fortunate to have a few previous trainees return to the lab for postdocs as well as two senior scientists who joined the Center and formed groups working on HCV immunology and virus entry (using pseudovirus approaches since we still lacked cell culture infectious HCV). We also solidified interactions with clinical colleagues and the Clinical Director of the Center, Ira Jacobson, who is a renowned clinical hepatologist working at the time at Weill-Cornell and New York Presbyterian Hospital. Ira shared over the years a wealth of knowledge regarding important aspects of HCV diagnosis, treatment, and challenges that HCV infected individuals face, helping to keep the ultimate goal of the research clearly in sight.

From the HCV replicon work, the next step was obvious but when adaptive mutations were placed in full-length, otherwise unmodified HCV RNAs, they replicated and expressed the expected HCV proteins, but no virus was produced. A major advance, another roadblock. This became yet another annoyance for the field that persisted for another 5 years. This was finally solved by a serendipitous finding in Japan by Takaji Wakita. Takaji was studying an HCV isolate from a rare case of acute fulminant disease (Japanese fulminant hepatitis 1, or JFH1). Perhaps this virus, given its atypical pathogenesis in the patient might be able to replicate in culture. When he tested a JFH replicon, it replicated efficiently in human hepatoma cells with no need for adaptation. Later work by three groups, including ours, showed that JHF1 or chimeric HCV derivatives could produce virus that was infectious in cell culture, chimpanzees, and mice engrafted with human hepatocytes. More than 15 years after HCV's discovery, we finally had a complete cell culture replication system.

It was another six years before the first HCV protease inhibitors entered the clinic and several more years before IFN was abandoned in favor of effective drug cocktails that could eliminate the virus in virtually everyone treated, with minimal side effects. In the intervening years, much was learned about the intricacies of HCV RNA replication, virus entry, and the host cell factors important for virus infection. Efforts ramped up in the biotech and pharma sectors working to identify and optimize drugs for inhibition, to identify resistance mutations and their effect on viral fitness, and to test combinations of drugs for efficacy against the diversity of HCV genotypes present in the infected population and to determine the minimal duration to achieve a cure – elimination of the virus from an infected individual! The ultimate outcome, combination drug cocktails capable of achieving virological cure in two to three months across multiple HCV genotypes, was an amazing culmination of the hard work of so many individuals, across a diverse spectrum of fields. Looking back, I feel incredibly fortunate to have had, by a series of chance events in my life, the opportunity to be a part of this effort, and to see first-hand the role that basic science can play towards helping patients afflicted with a devastating disease. Not every basic scientist will be so lucky to be in such a "right place at the right time", but the future cannot be predicted, and random chance could strike at any time. I urge all young (and not so young) scientists to embrace their passion, follow their curiosity where it leads them, work hard, and to be fair and open in sharing ideas and reagents. No one discovery is likely to change the world, but the collective work of all of you will.

Hepatitis C: From Discovery to Cure[1]

Nobel Lecture, December 7, 2020 by
Charles M. Rice
Laboratory of Virology and Infectious Disease, The Rockefeller University, USA.

To begin, I want to thank the Nobel committee and organization for recognizing the hepatitis C story and allowing me and my esteemed colleagues and friends Harvey Alter and Michael Houghton to represent the many in this field who made this happen. I'm humbled and deeply honored to be awarded this prize but as you will learn shortly, this really came about through a series of chance events.

Now let's begin with some history and context on how we came to work on hepatitis C. Going back into the 17th and 19th centuries there were outbreaks of jaundice. This traces back even to Greek and Roman times of unexplained jaundice outbreaks. In the 1930s it was noticed that individuals vaccinated with an attenuated strain of yellow fever virus stabilized with human serum often came down with hepatitis or jaundice that was not characteristic of yellow fever. Then later, there was recognition of a link between hepatitis and people administered human serum. In the 1940s it was recognized that there were indeed two types of viral hepatitis – infectious, which was likely the form causing problems during

1. The references cited in this transcript are sparse and do not reflect the numerous seminal contributions from the many others who have contributed to the hepatitis C success story.

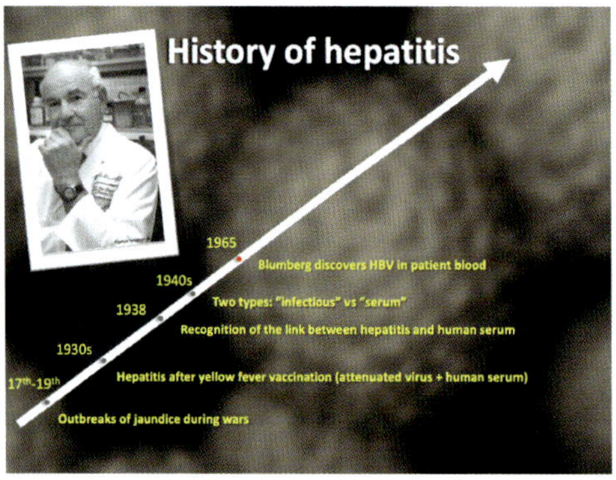

Figure 1. History of hepatitis. Events leading up to the breakthrough discovery by Baruch Bloomberg that the Australia antigen was the surface antigen of hepatitis B virus are depicted. These advances led to the development of a hepatitis B vaccine.

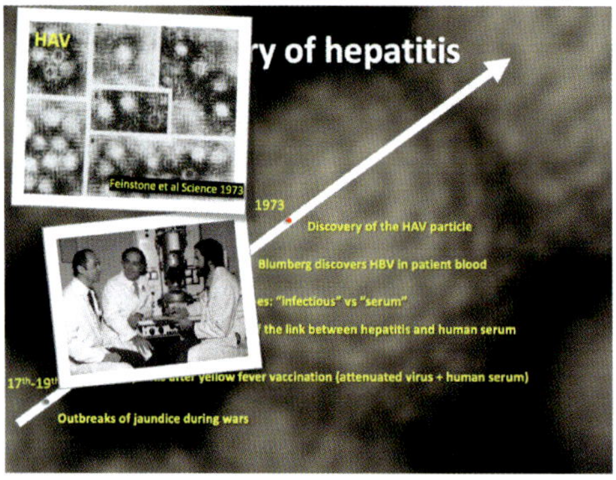

Figure 2. Discovery of the hepatitis A particle. Steve Feinstone, Al Kapikian, and Bob Purcell (lower photo) utilized immuno-electron microscopy to identify hepatitis A particles (upper photo) in 1973. These discoveries led to the development of assays to screen the blood supply for both hepatitis A and hepatitis B.

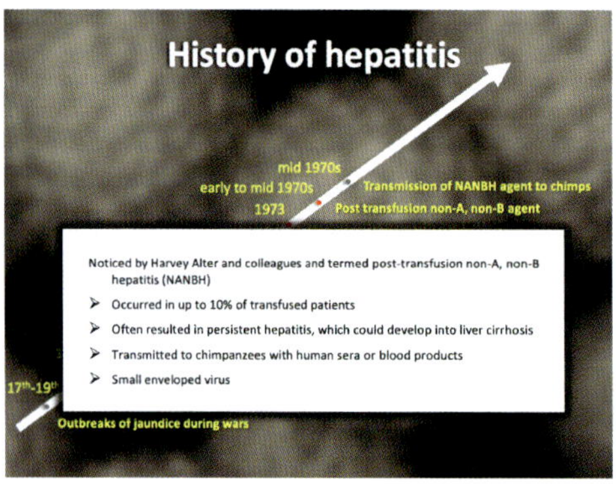

Figure 3. Post-transfusion hepatitis continues. Despite testing of the blood supply, Harvey Alter and colleagues noticed continued cases of hepatitis in up to 10% of transfused patients, which they termed non-A, non-B hepatitis (NANBH). The successful transmission of NANBH to chimpanzees led to the discovery that the agent was filterable, and efforts by many groups led to the conclusions that the agent was virus-like and probably enveloped given its sensitivity to organic solvents.

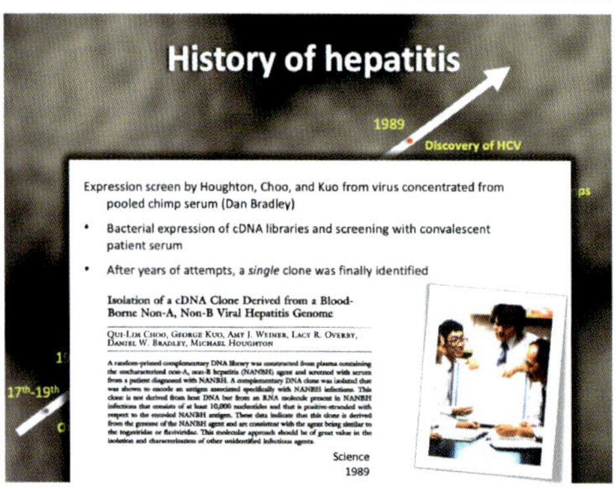

Figure 4. The discovery of hepatitis C virus. Using an expression screening approach, Michael Houghton and his team at Chiron Corporation and Dan Bradley at the Centers for Disease Control, used virus from infected chimpanzees and serum from a convalescent NANBH patient to isolate a cDNA clone.

wars, versus – serum, which I just mentioned. A major breakthrough came in 1965 when Baruch Bloomberg (Figure 1) discovered that the Australia antigen was actually the surface antigen of hepatitis B virus and this led to the development of a vaccine for hepatitis B and other advances. In 1973, Steve Feinstone, Al Kapikian, and Bob Purcell identified hepatitis A virus (1) using immuno-electron microscopy (Figure 2). With this breakthrough, it became possible to develop assays that could be used to screen blood products for both these viruses. However, that was not the whole story. Harvey Alter and others had noticed that there were still cases of post-transfusion hepatitis not associated with HAV and HBV, so they named this non-A, non-B hepatitis (2, 3). This occurred in about 10% of transfused patients and often resulted in persistent hepatitis that could lead to liver fibrosis, cirrhosis and was later associated with liver cancer (Figure 3). Several groups succeeded in transmitting this agent to our closest relative, the chimpanzee, but not to other non-human primates or more distant mammalian species. It was therefore pioneering chimpanzee studies that showed that the agent (or agents) was filterable, and virus-like, and probably enveloped given its inactivation by organic solvents. These findings spawned a great deal of effort among many groups to identify the responsible mystery virus. But that took almost 15 years from the time that post-transfusion non-A, non-B hepatitis was noticed until 1989. The transformative breakthrough came from Michael Houghton and his team at Chiron Corporation (Figure 4), collaborating with Harvey Alter at NIH and Daniel Bradley at the Centers for Disease Control. Using a cDNA expression screening approach, they identified a single short cDNA clone that was immunoreactive with serum from a convalescent non-A, non-B hepatitis patient. They extended this sequence to reveal a nearly 10 kb RNA viral sequence; serologic assays confirmed that this new virus, called HCV, was indeed the mysterious non-A, non-B agent (4). This was where we were in 1989. What caught my attention is highlighted in the abstract of the landmark Science paper (Figure 5). The HCV RNA genome was positive-sense with a size and organization reminiscent of the flaviviruses, which were also small, filterable, and inactivated by organic solvents.

At the time, we were intensively studying the prototype flavivirus, yellow fever virus. The lab phone rang and on the line was Stephen "Steve" Feinstone. As mentioned earlier, Steve co-discovered HAV. Still working with Robert "Bob" Purcell at the NIH, he had been a key player in several of the early chimpanzee studies with the non-A, non-B agent. He called me up out of the blue and said "Hey Charlie, how about making a yellow fever 17D – HCV chimera as a vaccine?" Steve was then at the FDA and interested in making a preventative HCV vaccine. We had been working with a tissue culture passaged, live-attenuated yellow fever virus strain,

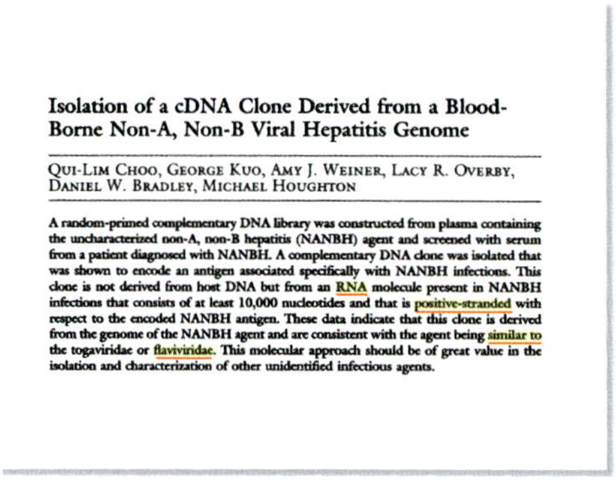

Isolation of a cDNA Clone Derived from a Blood-Borne Non-A, Non-B Viral Hepatitis Genome

QUI-LIM CHOO, GEORGE KUO, AMY J. WEINER, LACY R. OVERBY, DANIEL W. BRADLEY, MICHAEL HOUGHTON

A random-primed complementary DNA library was constructed from plasma containing the uncharacterized non-A, non-B hepatitis (NANBH) agent and screened with serum from a patient diagnosed with NANBH. A complementary DNA clone was isolated that was shown to encode an antigen associated specifically with NANBH infections. This clone is not derived from host DNA but from an RNA molecule present in NANBH infections that consists of at least 10,000 nucleotides and that is positive-stranded with respect to the encoded NANBH antigen. These data indicate that this clone is derived from the genome of the NANBH agent and are consistent with the agent being similar to the togaviridae or flaviviridae. This molecular approach should be of great value in the isolation and characterization of other unidentified infectious agents.

Figure 5. The NANBH agent is similar to the flaviviruses. The agent, named hepatitis C virus, was a positive-sense RNA virus with a size and genome organization similar to the flaviviruses, and also shared the characteristics of being small, filterable and inactivated by organic solvents.

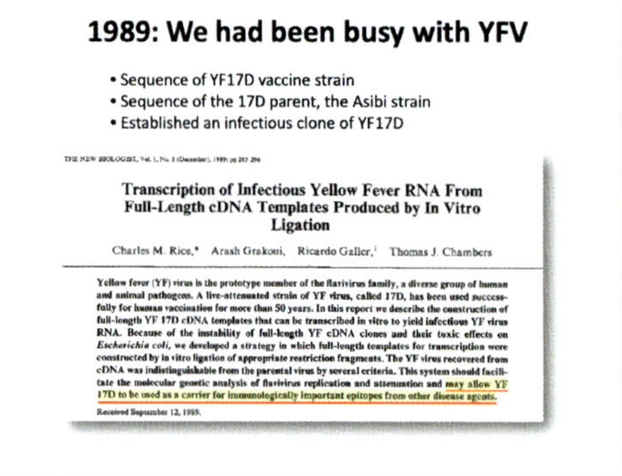

Transcription of Infectious Yellow Fever RNA From Full-Length cDNA Templates Produced by In Vitro Ligation

Charles M. Rice,* Arash Grakoui, Ricardo Galler,[1] Thomas J. Chambers

Yellow fever (YF) virus is the prototype member of the flavivirus family, a diverse group of human and animal pathogens. A live-attenuated strain of YF virus, called 17D, has been used successfully for human vaccination for more than 50 years. In this report we describe the construction of full-length YF 17D cDNA templates that can be transcribed in vitro to yield infectious YF virus RNA. Because of the instability of full-length YF cDNA clones and their toxic effects on *Escherichia coli*, we developed a strategy in which full-length templates for transcription were constructed by in vitro ligation of appropriate restriction fragments. The YF virus recovered from cDNA was indistinguishable from the parental virus by several criteria. This system should facilitate the molecular genetic analysis of flavivirus replication and attenuation and may allow YF 17D to be used as a carrier for immunologically important epitopes from other disease agents.

Received September 12, 1989.

Figure 6. Development of molecular tools for working with yellow fever virus. The discovery of HCV in 1989 coincided with work in my lab to sequence both the 17D vaccine strain and its virulent Asibi parent, and to develop an infectious clone system to generate wildtype and mutant derivatives for study. The underlined last sentence caught the attention of Steve Feinstone.

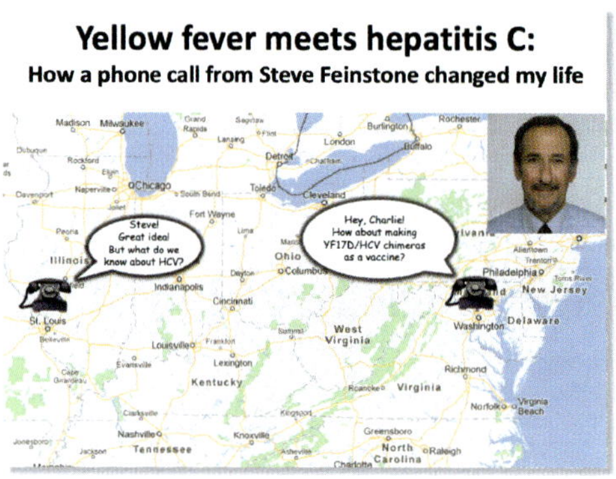

Figure 7. A phone call from Steve Feinstone propels me to work on HCV. During discussions with Steve about his idea to use YFV to make an HCV vaccine it became clear that we knew little about the HCV genome and its expression.

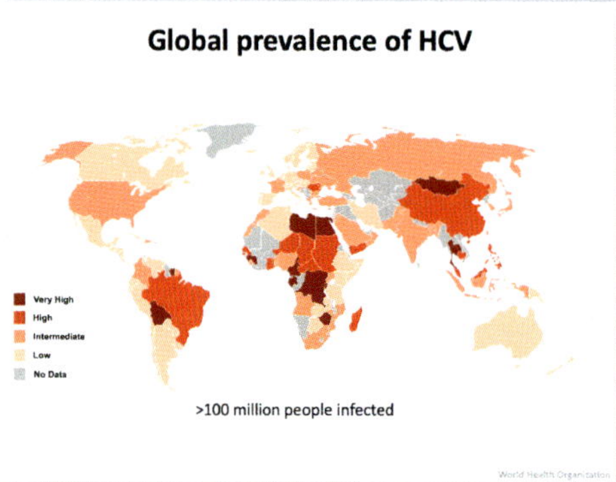

Figure 8. The global prevalence of HCV. At the time, HCV infection was present across the globe with estimates of more than 100 million chronically infected individuals.

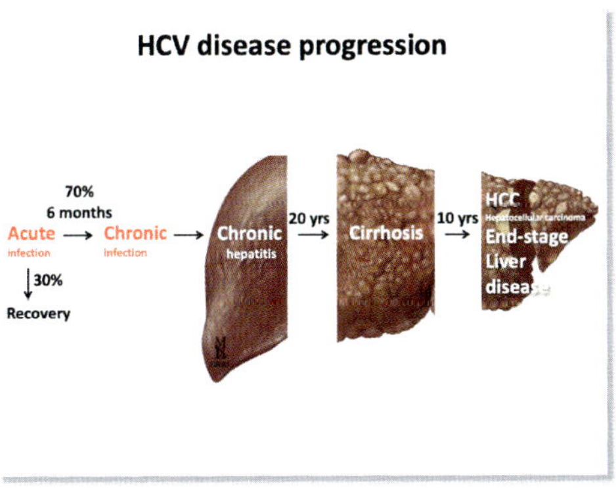

Figure 9. HCV disease progression. Acute HCV infection, usually asymptomatic, often progresses to chronic infection and chronic hepatitis. After many years, individuals can suffer from advanced liver fibrosis, cirrhosis, hepatocellular carcinoma, and end stage liver disease.

called 17D, that was developed by Max Theiler at the Rockefeller Foundation and reported in 1937. It has been used to vaccinate about half a billion people and is one of the most safe and effective human vaccines ever developed. Max was awarded the Nobel prize for this work in 1951. In 1985, we reported the 17D sequence and the sequence of the virulent parental virus, the Asibi strain, a few years later. After much work, we had established an infectious cDNA clone for 17D, which allowed manipulation of the viral genome as DNA to investigate the effects of specific mutations or other alterations on virus replication. This work was published (5) in *The New Biologist* – an aspiring new journal at the time, that went extinct shortly thereafter; the last sentence of the abstract (Figure 6) stated *"this may allow yellow fever 17D to be used as a carrier for immunologically important epitopes from other disease agents".* This is what caught Steve's attention (Figure 7). We got to talking about this but realized that we knew very little about the hepatitis C genome and how it was expressed. We did know by that time that HCV was a major problem, with more than 100 million people chronically infected (Figure 8). That became increasingly obvious when diagnostic tests were developed shortly after HCV's discovery. As Harvey and others had shown in the 1970s and 1980s, this was an infection that was usually asymptomatic in the acute phase, often went chronic, and chronic infection could lead to severe liver disease (Figure 9). Chronic HCV was a major public health problem. The immediate goals were clear: Clean up the blood supply, which was done quickly in the early 1990s; Try to educate high-risk groups – maybe develop a preventative vaccine; Identify those infected and improve treatment. The hope at that point, although this was a bit of a pipe dream, was to eradicate this exclusively human virus (Figure 10).

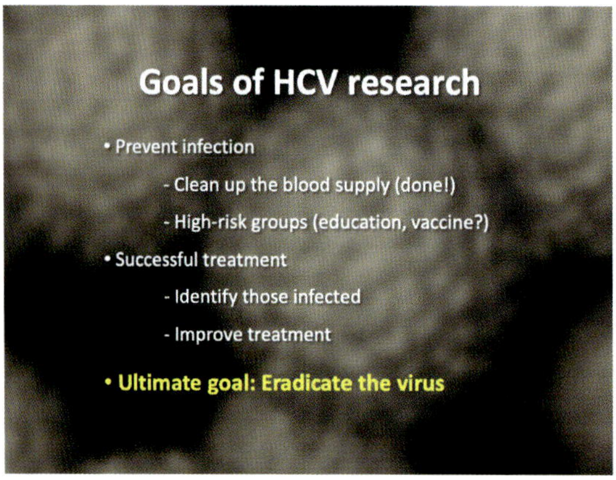

Figure 10. Goals of HCV research. At that time, the goals were clearcut. The blood supply was quickly cleaned up, but high-risk groups needed to be educated and potentially vaccinated, infected individuals needed to be identified, and treatment options needed improvement. The ultimate goal was to eradicate the virus.

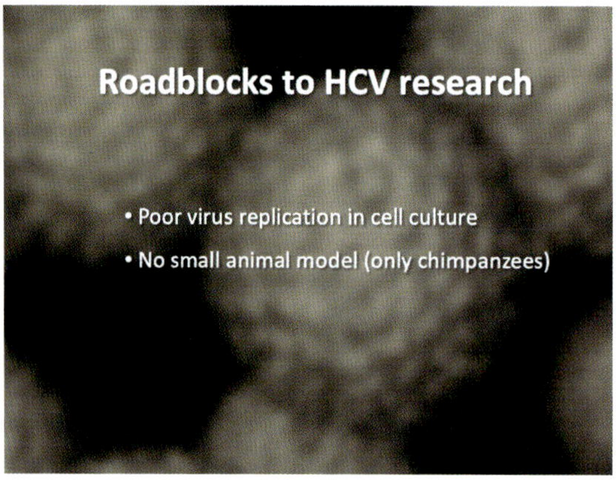

Figure 11. HCV research roadblocks. The lack of HCV replication in cell culture and in a small animal model limited researchers to working with clinical samples, chimpanzees, and information gleaned from the HCV sequence.

So, now what? There were some real challenges for hepatitis C (Figure 11). The virus had not been shown to replicate in any cell culture despite repeated attempts. There was no animal model other than chimpanzees, which were extremely important but available to only a few groups, and in very limited numbers. This left the field with clinical samples or chimpanzee passaged material, and what we could glean from the HCV genome sequence: A 9.6 kb RNA with a long open reading frame that could encode a polyprotein of over 3000 amino acids, flanked by short 5' and 3' non-coding sequences. An RNA structure near the 5' end of the genome was presumed (and later shown) to promote internal initiation of protein translation. But how was this polyprotein processed into the proteins the virus needed to replicate its RNA and produce progeny virus? Along with groups in Europe and Japan, Arash Grakoui who was a technician in the lab at that time took on the challenge. What he and others found were ten hepatitis C virus proteins including likely components of virus particles and the viral RNA replication machinery. Cleavages to produce the mature proteins were catalyzed by a combination of host and viral proteases. This defined the HCV protein boundaries and allowed investigators to begin to study viral enzymatic activities that could be attractive drug targets. Examples included an NS3-4A complex, which is a serine protease/helicase whose structure was later determined, and NS5B, the RNA-dependent RNA polymerase of the virus (Figure 12, reviewed in ref. 6).

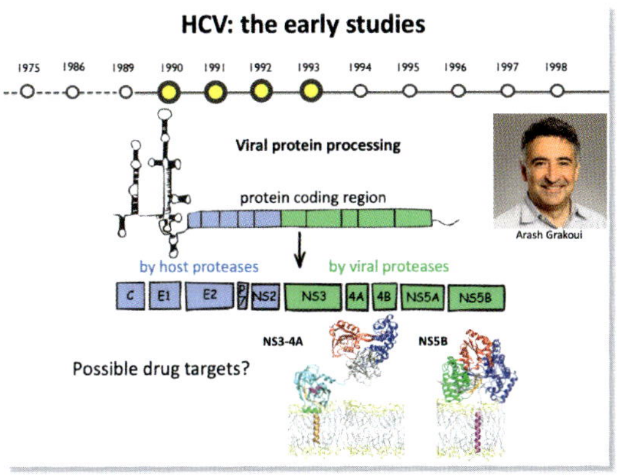

Figure 12. Early studies on HCV. Arash Grakoui in my lab, along with other investigators around the world, took on the task to study how the HCV polyprotein is processed. It was found that both host and viral proteases are required to yield 10 viral proteins, two of which had enzymatic activity that could be targeted for drug development. In subsequent years structures of the NS3-4A protease/helicase and the NS5B RNA-dependent RNA polymerase were determined by several groups to aid these efforts.

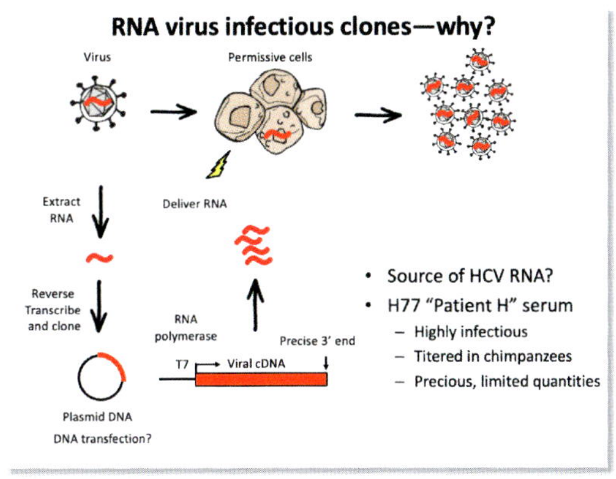

Figure 13. The utility of RNA virus infectious clones. In a normal virus infection (top) the virus particle efficiently delivers its genome RNA to the cell, and if permissive for infection, results in a burst of progeny viruses. For positive-strand RNA viruses like HCV, RNA isolated from the virus particle, if introduced into susceptible cells, can function as a messenger RNA and set off the entire virus life cycle (bottom). Generating a cDNA intermediate of the genome RNA allows virologists to use DNA manipulations to introduce desired changes. The resulting plasmid DNA construct can be transfected into cells, where if transcribed into a functional RNA, can launch the viral life cycle. Alternatively, the plasmid can be designed more precisely to use a phage-derived RNA polymerase to produce unlimited quantities of a synthetic mimic of the viral genome RNA. We utilized this latter approach using RNA from the highly infectious serum from "Patient H".

Advances were being made, but we still lacked the ability to test the importance of these findings for actual virus replication. The next obvious step was to make a cDNA clone of HCV that could be transcribed in vitro to produce synthetic HCV genome RNA that was infectious. If this succeeded, we would be able to produce unlimited quantities of infectious RNA and test different cell types and culture conditions in the hope of finding one that allowed HCV replication. Figure 13 displays the concept and possible routes to making functional cDNA clones. Success for this general approach was first reported in bacteria for the RNA phage QB (7) and a few years later for poliovirus (8). We had also succeeded in establishing such systems for the prototype alphavirus, Sindbis virus (9) and, as mentioned earlier, the vaccine strain of yellow fever virus (5).

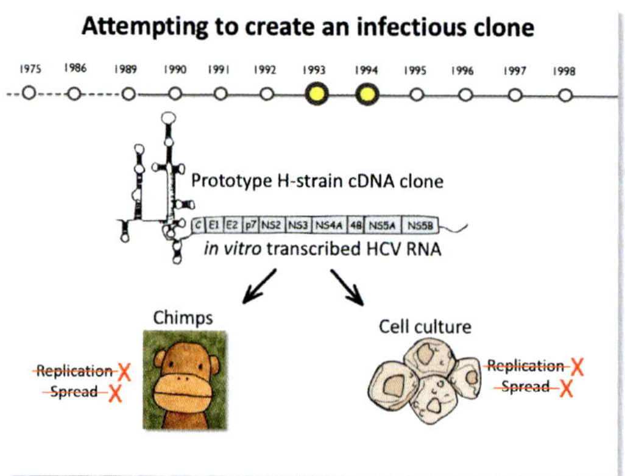

Figure 14. Launching HCV infection from an infectious clone. Using RNA from Harvey Alter's patient serum containing the genotype 1a "H77" strain of HCV, we generated a prototype HCV infectious clone from which we transcribed the RNA in vitro. However, no permissive cell lines had been established, and attempts to launch viral replication in cell culture failed. Working with Steve Feinstone, the RNA was injected into the liver of chimpanzees, but this also failed.

We began to try and accomplish this for hepatitis C (Figure 14). Our initial work had been carried out with the genotype 1a "H77" strain of HCV. This was obtained in 1977 by Harvey Alter at NIH from a post-transfusion patient, Mr. Hutchinson, in the acute, pre-seroconversion phase of infection (10). It had been titered in chimpanzees and was highly infectious, but it was precious and in limited quantities. The next dilemma was how to confirm that RNA transcripts from a cDNA clone were indeed infectious. The obvious and usual path was to simply transfect either a plasmid, or in vitro transcribed RNA, into cells shown to be permissive for virus replication. However, for HCV we didn't have a cell culture system that had been proven to work. Transfection of cells might not work, unless the block to infection was at virus entry, which would be bypassed. The only remaining possibility was the chimpanzee – by taking HCV genome RNA transcripts and injecting them directly into the liver and hoping that some intact RNA would successfully enter hepatocytes and initiate replication, which could then spread and amplify. This

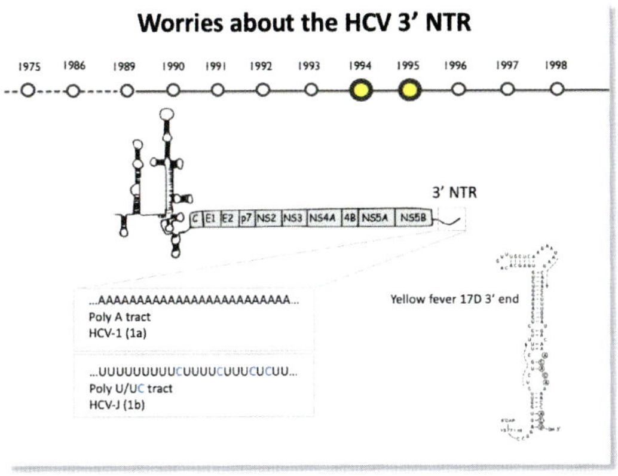

Figure 15. Conflicting reports on the sequence of the HCV 3'-non-translated region. The original 1989 report suggested a 3' terminal polyA tract, like normal cellular mRNAs, but sequences of other isolates reported later suggested a poly U/UC tract. However, it was known that distantly related yellow fever virus terminates with neither of these, but rather a stable stem loop RNA structure.

seemed like a crazy thing to try. However, Sue Emerson in Bob Purcell's lab, again at the NIH, had shown more than a decade earlier that you could use in vivo RNA transfection to launch hepatitis A virus infection in marmosets (11). It wasn't such a crazy idea – it had been shown to work, at least for one other virus. Steve Feinstone had access to chimpanzees and gave our prototype HCV-H full-length RNA a try. It failed. This was frustrating but we didn't give up.

We began to worry about the HCV genome sequence, might there be something missing? We focused on the 3' non-translated region given conflicting reports in the literature (Figure 15). The initial 1989 report suggested a 3' terminal polyA tract like normal cellular mRNAs, other, later isolate sequences, reported a polypyrimidine (poly U/UC) tract. We knew from our work on yellow fever that there was a different possibility. The yellow fever virus genome didn't terminate with either of these homopolymer tracts but rather a stable RNA stem loop structure. Alexander "Sasha" Kolykhalov, a remarkably talented molecular biologist from Novosibirsk, had immigrated to the US and joined my lab. Sasha took on this challenge and developed a fairly simple, at least on paper, 3' RACE protocol (Figure 16), which many of you reading this will think is just standard fare, but we were challenged by the fact that we had only tiny amounts of HCV RNA; every step had to be optimized in order to retrieve clones that might have arisen from the 3' end of HCV genome RNA. Sasha did all this, and what he found was quite a surprise (Figure 17). Following the stop codon of this long open reading frame there was a short region, quite variable among HCV genotypes, followed by an internal polyU/polyUC tract of variable length, and then a new sequence of about 98 bases (12), highly conserved among different HCV isolates (13). These kinds of terminal RNA elements in positive strand RNA virus genomes are absolutely required for replication. Simultaneously, the same sequence was discovered by Kunitada Shimotohno's lab in Japan (14).

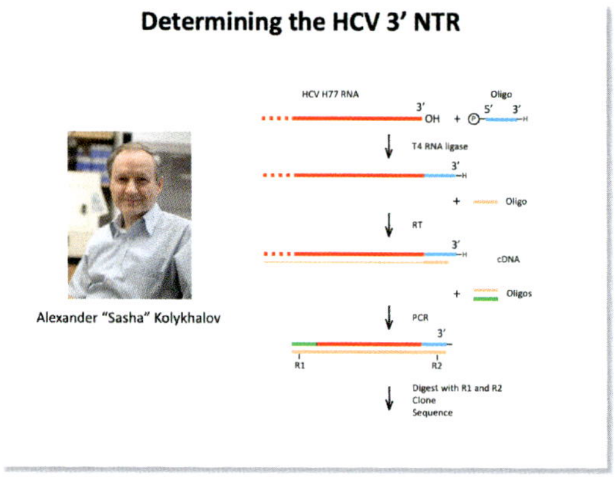

Figure 16. 3' RACE determination of the HCV 3'NTR sequence. Using optimized conditions for each of the steps, Sasha Kolykhalov performed 3' rapid amplification of cDNA ends (RACE) in order to determine the HCV 3'NTR sequence. Starting with HCV RNA from patient serum, T7 RNA ligase was used to attach an oligo at the 3' end. A complementary DNA oligo was then used to prime cDNA synthesis using reverse transcriptase (RT). The polymerase chain reaction (PCR) and the same oligo in combination with an oligo complementary to known upstream HCV sequences were used to amplify the cDNA. The DNA was then digested with restriction enzymes, cloned, and sequenced.

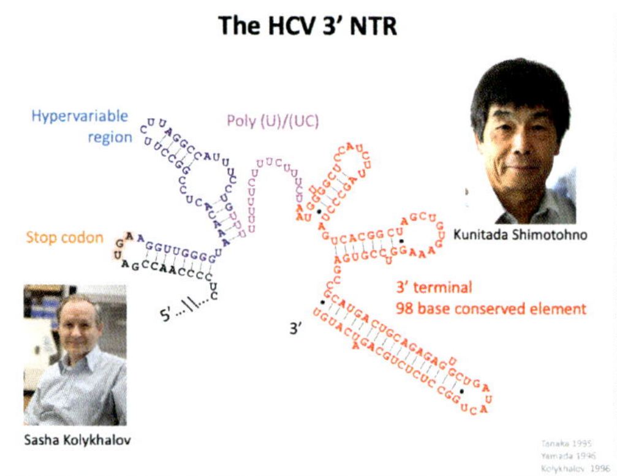

Figure 17. RNA structure of the HCV 3'NTR. Sequencing revealed that the HCV polyprotein stop codon is followed by a short sequence, highly variable between isolates, an internal polyU/polyUC tract of variable length, and a novel ~98 base pair sequence, highly conserved across different HCV isolates. This conserved sequence was simultaneously discovered in the laboratory of Kunitada Shimotohno.

The missing piece found, we modified our HCV-H clone to include the proper 3' non-coding region and repeated what we had tried before. No luck. We again failed to see any evidence for HCV replication in cell culture or in the chimpanzee model. After discovery of this new, highly conserved sequence element, likely to be essential, this was hugely disappointing.

We began to worry about everything. Perhaps this viral infection could not be initiated by injecting naked RNA into the liver of an animal. Could there be part of the genome still missing? Was HCV genome RNA not infectious? Could there be a virus-associated protein needed to initiate replication? What about RNA modifications deposited in an infected cell that would absent from synthetic HCV genome RNA made in a test tube? We knew that HCV was highly error prone, like all RNA viruses. There are

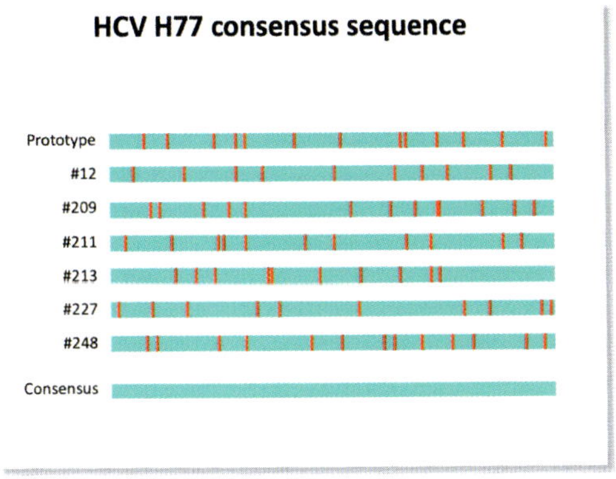

Figure 18. Determination of the HCV H77 consensus sequence. Sequences from 6 clones from a combinatorial library of more than 10^5 full length HCV cDNAs, and sequences from uncloned PCR products were all used to generate a consensus sequence. The prototype HCV clone and each of the 6 clones contained non consensus, potentially deleterious, sequence changes scattered across the genome, which likely explained our negative results.

an estimated trillion variants generated per day in a single chronically infected individual. That's a lot of variation; some of these mutations would be deleterious or lethal. In those days, the enzymes used for cDNA synthesis and cloning were also not as high fidelity as they are today. The one "worry" that we could address was to try and filter out possibly deleterious mutations.

Sasha began by constructing a combinatorial library with a complexity of $>10^5$ full-length HCV cDNAs using high-fidelity assembly reverse transcription–polymerase chain reaction (RT-PCR) followed by subcloning into a recipient plasmid vector containing the 5'-and 3'-terminal HCV consensus sequences. A flanking T7 promoter and an engineered specific restriction site allowed for production of runoff RNA transcripts; 233 of the clones were prescreened by restriction analysis, polyprotein processing, and production of the C-terminal NS5B protein (the HCV RNA-dependent RNA polymerase).

In yet another unsuccessful experiment, RNAs transcribed from 34 clones that passed these analyses were assayed for infectivity by direct intrahepatic injection in two HCV-naïve chimpanzees. As a negative control, a third animal received similar injections with transcripts from a clone containing a 20-amino acid in-frame deletion encompassing the NS5B polymerase active site. Serum samples were collected for 2 months after inoculation and analyzed for evidence of HCV replication. Neither experimental animal nor the negative control animal exhibited signs of productive infection. Of note was the absence of detectable circulating HCV RNA 2 days after inoculation. Missing terminal sequences, low RNA transfection efficiencies in vivo, and errors introduced during cDNA synthesis or PCR amplification might account for these negative results. Sasha sequenced six clones from the combinatorial library, as well as uncloned RT-PCR products, to determine a consensus sequence for the

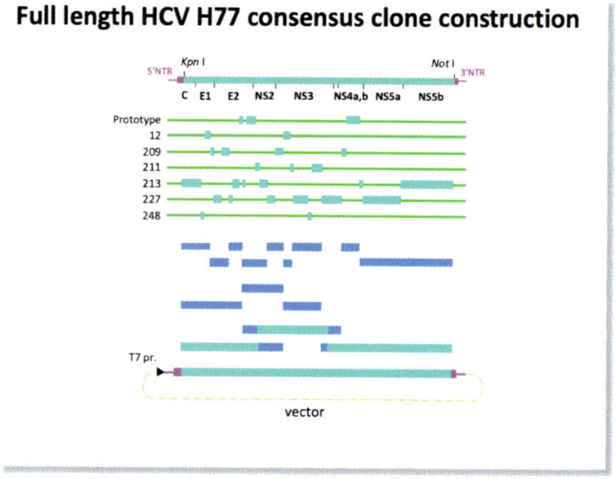

Figure 19. Assembly of an HCV H77 consensus clone. The prototype and 6 clones are shown schematically below the resultant consensus infectious clone. PCR and standard restriction enzymes were used to assemble the consensus clone. Regions from the clones used in the assembly are shown in teal rectangles. Blue rectangles below show intermediates generated for the assembly process. The teal and blue rectangles show the three final pieces that were assembled and cloned into the Kpn I and Not I sites of the prototype infectious clone to generate the consensus clone.

H77 isolate (Figure 18). From this analysis, it became clear that each of the six clones sequenced contained numerous non consensus sequence changes scattered throughout the genome, which could be deleterious and would explain the negative results. Sasha used this information to painstakingly direct the assembly of a full-length clone reflecting the H77 consensus sequence (Figure 19).

Aside from the sporadic changes corrected in this consensus clone, even greater sequence heterogeneity was noted in three regions of the HCV genome. During analysis of the extreme 5' terminus, we found a substantial number of clones that contained one or more bases in addition to the reported 5'-terminal sequence (5'-GCCAG . . . -3'). In the structural protein coding region, the N-terminus of the E2 virion envelope glycoprotein was highly variable, but a predominant sequence was identified. Near the 3' terminus, the poly(U/UC) tract was variable in length and composition. Because these differences might be functionally significant, variants of the consensus clone with additional bases at the 5' terminus and two different poly(U/UC) tracts were constructed (Figure 20). Silent nucleotide substitutions were engineered as markers so that virus recovered from transfected animals could be sequenced to identify which clones were infectious. From each of the 10 clones, full-length uncapped RNAs were transcribed from linearized template DNAs and again sent to Steve Feinstone at the FDA – Steve and his team did the chimpanzee work for the studies that I've mentioned and was involved in essentially all of our HCV work in the 1990s and beyond. This was probably going to be our final attempt.

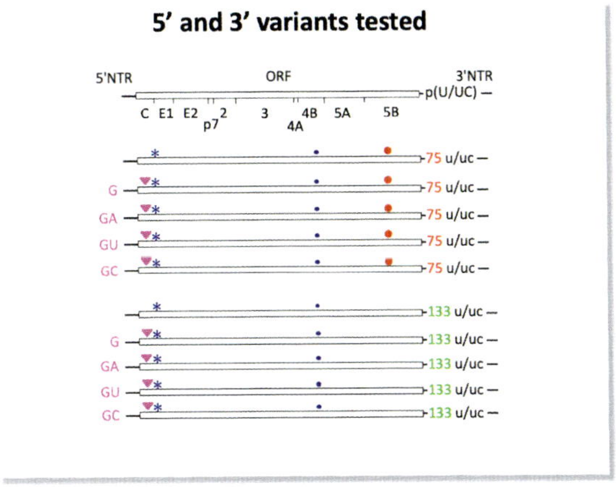

Figure 20. Variants constructed for testing. The HCV genome is shown schematically at the top. Variants constructed for testing are shown below. Pink indicates the 5′ terminal variants with extra bases, the asterisk represents the predominant hypervariable sequence of the E2 virion envelope glycoprotein N-terminus, and red and green show the two lengths of the poly (U/UC) tract. Red and blue circles indicate the location of silent mutations engineered to allow determination of which sequences were infectious.

Two chimpanzees were identified that were seronegative for all known hepatitis viruses, negative for HCV RNA by nested RT-PCR, and had normal baseline levels of liver enzymes. Animals were inoculated by direct intrahepatic injection at multiple sites (Figure 21). Serum samples and liver biopsies were taken before inoculation and at weekly intervals thereafter. For 3 months after inoculation, serum samples were assayed for liver enzymes, antibodies to HCV, and viremia, as assessed by quantifying circulating HCV RNA (15).

The results finally began trickling in and they were breathtaking (finally) (Figure 22).

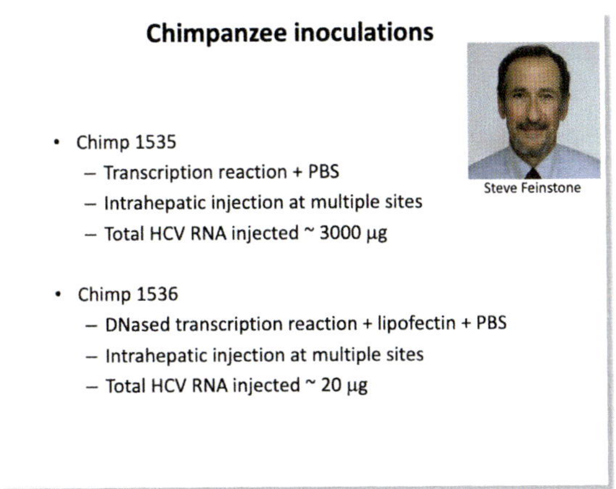

Figure 21. Inoculation of two chimpanzees with the 10 HCV clone variants. In collaboration with Steve Feinstone at the FDA, two chimps were inoculated with the pool of 10 HCV RNA variants. The conditions for each chimp are listed.

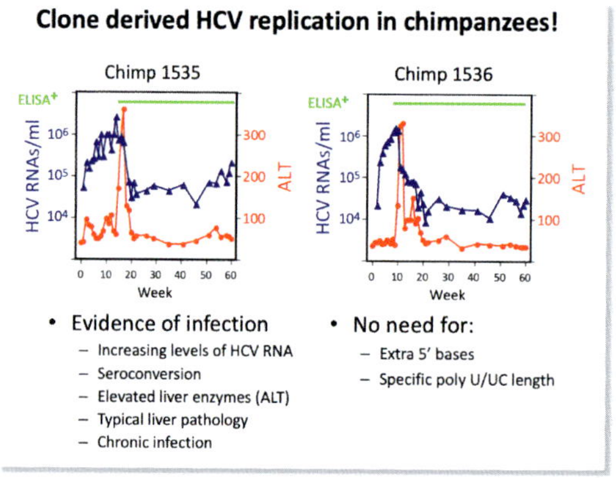

Figure 22. Clone-derived HCV replication in chimpanzees. Serum HCV RNA levels are shown in blue. ALT (red) and ELISA positivity for antibodies to HCV (green) are shown for each of the chimpanzees. Increasing HCV RNA, seroconversion, elevated liver enzymes, liver pathology and chronic infection all were evidence for HCV infection from the clones. Sequencing of the replicating variants determined that there was no need for the extra 5′ bases, nor a specific poly(U/UC) tract length.

Both chimpanzees had increased concentrations of serum alanine aminotransferase (ALT), which is a marker for liver damage; after a large spike, ALT gradually declined to the preinoculation baseline levels. HCV RNA titers were undetectable before inoculation, rapidly rose thereafter and peaked shortly before the ALT spike after which they rapidly declined. Histologic changes in liver biopsies were typical for hepatitis C in chimpanzees and were characterized predominantly by portal inflammation and focal necrosis. The severity of histologic lesions appeared to parallel the ALT elevations. These pathogenesis profiles were strikingly reminiscent of those obtained in chimpanzees inoculated with the infectious H77 material or other HCV-containing samples (16). Both animals eventually became chronically infected (17).

To prove that the HCV RNA signals detected were not due to residual template DNA from the transcription reactions, we assayed all the samples by PCR without reverse transcription. No products were obtained, which demonstrated that the signals detected were due to HCV RNA. If the HCV RNA detected in serum was from replicating virus, RNAs should be packaged in enveloped virions and resistant to degradation by ribonuclease (RNase), whereas residual transcript RNA should be RNase sensitive. Indeed, the HCV RNA in samples of chimpanzee serum was resistant to RNase digestion under conditions that completely degraded excess naked competitor RNA. Moreover, there was no correlation between the level of apparent viremia and the amount of inoculated RNA, which differed by ~150-fold between the two inoculated animals. Finally, in two previous animal experiments (a total of six animals), circulating HCV-specific nucleic acid was never detected, even as early as 2 days after inoculation. These experiments suggested that circulating RNA was a result of authentic virus replication rather than release of inoculated nucleic acid. Restriction enzyme digestion and sequence analysis of recovered viral RNA revealed the presence of engineered markers, proving that these infections stemmed from the inoculated transcript RNAs.

Silent markers were analyzed to identify the 5'-terminal sequence(s) and the length(s) of poly(U/UC) tract required or preferred for initiating infection. Although it was not possible to draw firm quantitative conclusions about differences in specific infectivity, the results clearly demonstrated that the RNA transcripts without any additional 5' nucleotides were infectious (15). Transcripts with additional 5' nucleotides could also initiate infection, although our analysis did not allow us to distinguish among the various derivatives tested. Transcripts containing either the 75-base or the 133-base poly(U/UC) tracts were infectious, but the 133-base poly(U/UC) tract was preferred.

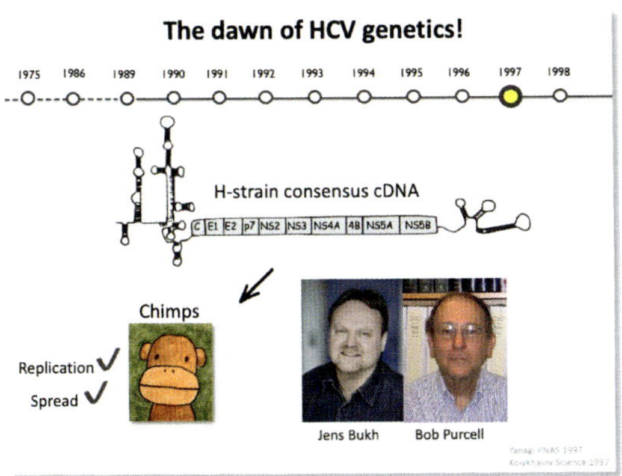

Figure 23. The dawn of HCV genetics. Nearly identical results were obtained by Jens Bukh and Bob Purcell. HCV genetics were now possible.

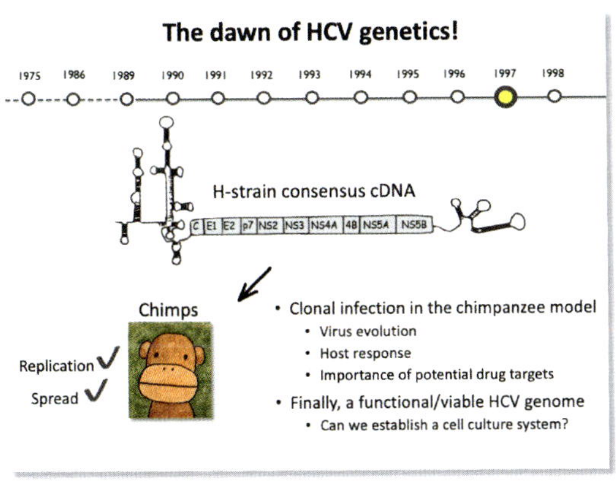

Figure 24. Potential uses of the HCV infectious clone. Given that the generated clones were infectious, clonal infection in the chimpanzee model would allow evaluation of virus evolution, host responses, and the importance of potential drug targets. Whether the functional viable HCV genome could be used to establish a cell culture system remained to be seen.

The demonstration that RNA transcribed from cloned HCV cDNA can initiate infection and cause hepatitis in transfected chimpanzees provided formal proof that HCV alone is the causative agent of this disease. Our experiments also defined the elements of a functional HCV genome RNA. Aside from a consensus genome sequence, no additional 5' sequences were required for infectivity and 3' poly(U/UC) tracts of variable length were tolerated. Very similar results were reported shortly thereafter by Jens Bukh and Bob Purcell (18), also using the HCV-H strain and the chimpanzee model (Figure 23). With this model, HCV infection could be launched from clonal derivatives to facilitate studies on virus evolution, pathogenesis, and host immune response relevant to understanding the factors that determine viral clearance versus chronic infection (Figure 24). Infectious HCV RNAs could be produced in unlimited quantities and used to identify and optimize cell-culture replication systems and to begin genetic analyses of HCV replication in vivo and in vitro. For example, the importance of potential drug targets could now be tested by mutagenesis of active site residues, which showed that each of the four HCV-encoded enzymes was absolutely required for infectivity (19).

With a functional, viable HCV genome sequence that we could make in unlimited quantities, could we use this to establish a cell culture replication system? One would hope so, but that didn't turn out to be the case. These RNAs, that were infectious in vivo in the most stringent model that you could imagine, – the chimpanzee model – were unable to replicate when transfected into a variety of different cell types. Similar results were being seen by other groups around the world, working with different HCV isolates including chimpanzee-validated infectious clones.

The next breakthrough came from Ralf Bartenschlager's lab (Figure 25). Ralf and Volker Lohmann engineered what is called a subgenomic replicon using a consensus sequence from a German genotype 1b isolate. This modified HCV RNA was engineered to express an antibiotic resistance gene product under the control of the HCV RNA replication machinery. The structural protein coding region was substituted with the neomycin resistance gene, followed by an internal ribosome entry site to drive translation of the downstream non-structural proteins believed to comprise the HCV RNA replication machinery. If such an RNA found itself in a cellular environment where it could replicate, not kill the cell, and express the neomycin resistance gene product, then those cells should become neomycin resistant and propagate in the presence of the antibiotic. Such an approach had the advantage of employing positive selection for even very rare events. After electroporation of this subgenomic replicon RNA into a human hepatoma cell line, Huh-7 cells, and selection with the neomycin analog G418, they were able to isolate drug resistant colonies that harbored continuously replicating HCV RNAs (20). This was a

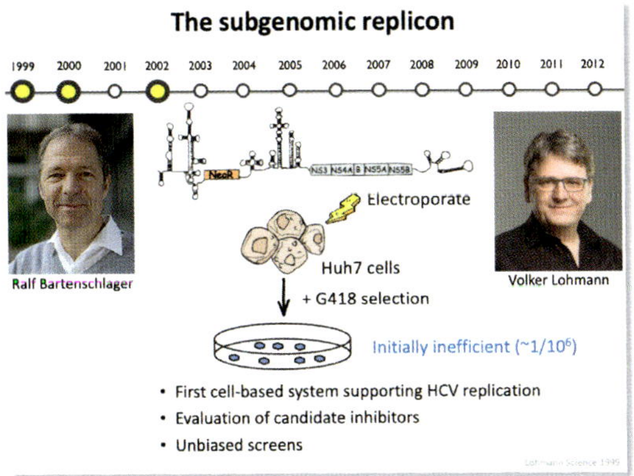

Figure 25. The HCV subgenomic replicon. Using a consensus sequence from an HCV genotype 1b isolate, Ralf Bartenschlager, Volker Lohmann and colleagues engineered a replicon that expressed the neomycin resistance gene in place of the structural protein coding region and under the control of the HCV replication machinery. Translation of the HCV proteins believed to be responsible for RNA replication was driven by an internal ribosome entry site from encephalomyocarditis virus. After delivery of the RNA into Huh-7 cells by electroporation, cells that survived and allowed replication of the RNA were selected by growth in the presence of a neomycin analog (G418), giving rise to colonies harboring the replicating HCV RNA. Although initially inefficient, this was the first cell-based system supporting HCV replication and opened the door for viral target-based and unbiased inhibitor screens.

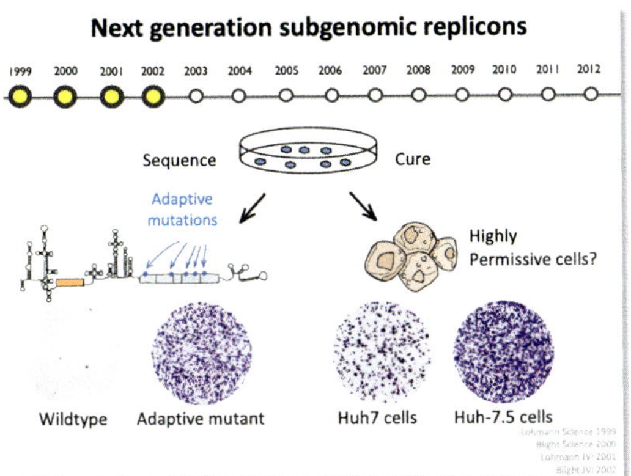

Figure 26. Next generation subgenomic HCV replicons. Sequencing of the replicon RNA within the surviving colonies revealed mutations, which when engineered back into the replicon clone resulted in markedly increased colony formation efficiency, as shown by the purple stain. On the other hand, curing of cells that were permissive for HCV replication by treatment with the antiviral cytokine, interferon, resulted in a cell line (Huh-7.5) more permissive for initiation of HCV replication.

major breakthrough – the first cell-based system supporting HCV RNA replication. This enabled the evaluation of inhibitors that were being developing against HCV enzymes using biochemical assays, and unbiased, cell-based screens to uncover new antiviral targets.

However, a remaining problem was the transduction efficiency of these replicons – only about one in a million transfected cells successfully initiated replication. Next generation replicons emerged from the efforts of several groups (21–23), including the Bartenschlager lab and ours (Figure 26). When we sequenced the HCV RNA present in these drug-resistant colonies, the recovered sequences were not the same as the input replicon

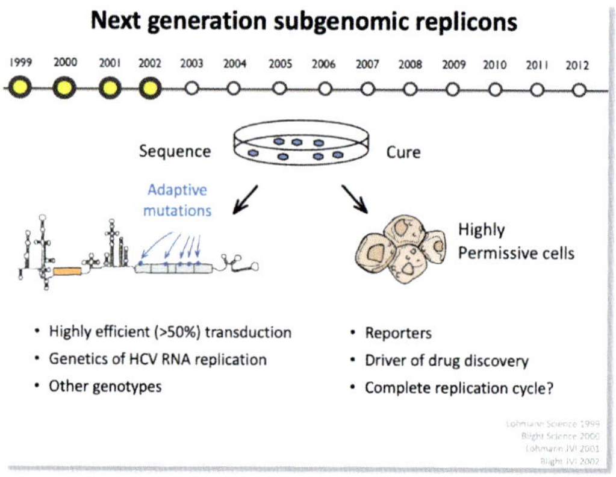

Figure 27. Applications of next generation subgenomic HCV replicons. With improved replicons and highly permissive cells, highly efficient transduction now allowed study of the genetics of HCV RNA replication, the evaluation of other genotypes, and generation of reporter replicons to allow facile drug discovery. But would adaptive mutations and permissive cells allow the complete HCV replication cycle?

RNA. Rather, they harbored mutations, which when re-engineered back into the parental replicon, enhanced the ability of these RNAs to initiate replication. Some single amino acid substitutions could increase the efficiency by more than 10,000-fold. You could cure replicon-containing cells with the antiviral cytokine interferon and identify hepatoma cells that were inherently more permissive for HCV replication. Replicons for other HCV genotypes and subtypes followed, as did replicons that expressed convenient reporter genes. Ten years after the initial discovery of HCV, we now had efficient HCV cell culture systems. It was finally possible to study HCV RNA replication in the laboratory. This was a major driver of drug discovery – compounds could now be tested and improved in a physiologically relevant context (Figure 27).

But were these cell culture adaptive mutations really the secret to getting a complete replication cycle? The subgenomic replicons lacked the HCV proteins important for assembling infectious virus particles, a basic capsid protein and two envelope glycoproteins. The obvious next step was to engineer replicon adaptive mutations into full length genomes that were otherwise unmodified. Upon delivery of these full-length RNAs to hepatoma cells, they replicated efficiently and expressed the expected structural and non-structural viral proteins. But no virus was produced, another roadblock was reached (Figure 28).

Enter Takaji Wakita, a hepatitis virus investigator in Japan. Takaji had been studying HCV isolates from different cases of post-transfusion hepatitis. One patient exhibited a rare case of acute fulminant liver disease. This HCV genotype 2a isolate was called Japanese fulminant hepatitis 1, or JFH-1. When the Wakita lab constructed and tested a subgenomic replicon of JFH-1, it was able to replicate with high efficiency without adaptive mutations. Puzzled by the lack of virus produced by full-length HCV genomes with cell culture adaptive mutations, many in the field were getting suspicious that adaptive mutations, selected for persistent HCV RNA

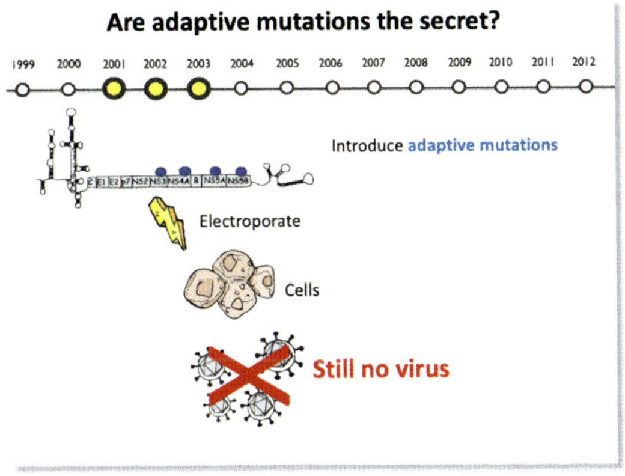

Figure 28. Testing adaptive mutations in the context of the complete HCV genome. Adaptive mutations (blue circles) were cloned into the infectious clone. After electroporation of HCV RNA into cells, the RNA was able to replicate, viral proteins were expressed, but no virus was produced.

replication in hepatoma cells, were somehow deleterious for downstream steps leading to the assembly of infectious virus. Three papers published in 2005 demonstrated that JFH-1 full length RNAs, or HCV chimeras substituting a structural region from other genotype 2a isolates, could produce HCV in cell culture (24-26). The released virus was infectious in naive hepatoma cells, transmissible to chimpanzees and liver-humanized mice, and virus recovered from these animals was infectious in cell culture (27, 28) (Figure 29).

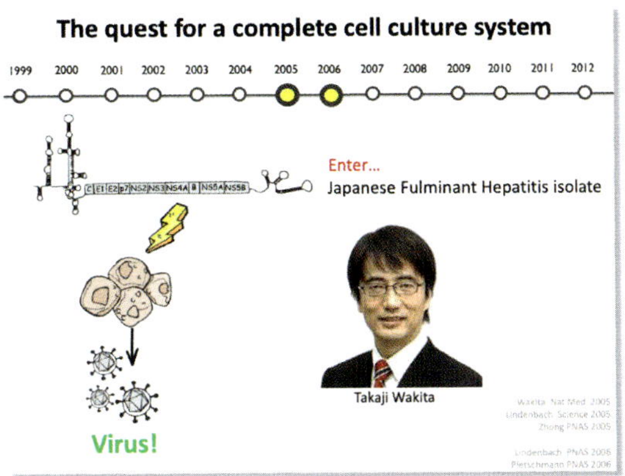

Figure 29. A unique Japanese HCV isolate. Takaji Wakita used HCV sequences from an unusual case of fulminate acute hepatitis to generate a genotype 2a based replicon that was able to replicate efficiently without adaptive mutations. This isolate, designated JFH-1, or chimeric genomes containing the structural region sequences from this or other genotype 2a isolates, resulted in full-length HCV RNA genomes that were able to replicate and produce infectious virus. The released virus was infectious in naïve cells, transmissible to chimpanzees and liver-humanized mice, and virus recovered from the animals was infectious in cell culture.

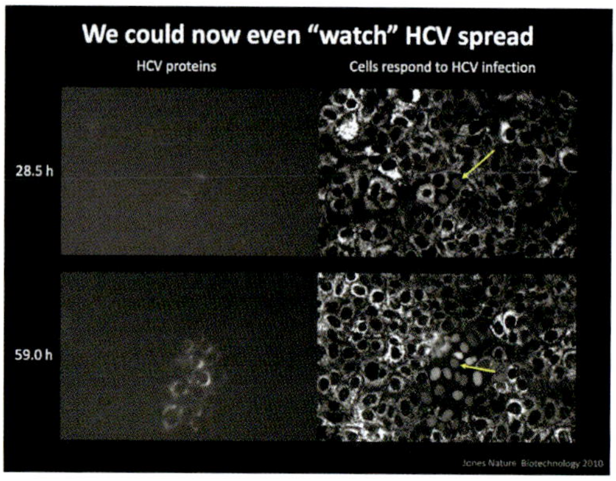

Figure 30. Watching HCV spread in cell culture. Two stills from a movie taken at 28.5 and 59 hours after infection show the expression of the viral protein (left) and the movement of a reporter protein domain that is released upon cleavage by the HCV protease and traverses to the cell nucleus. Arrows point to the same cell at the two different times.

We finally had a complete replication system for HCV – more than 15 years after its discovery! This ushered in a golden age in HCV research. We could study the complete virus life cycle, begin to look at virus – host interactions, the association of viral replication and assembly with membranes and lipid droplets in cells, study innate and adaptive immune responses, and develop neutralization assays useful for vaccine development. We found that HCV and related viruses were not that rare and commonly found in other species in nature. HCV became one of the best studied positive strand RNA viruses – surpassed only recently by SARS-CoV-2. Figure 30 shows you an example of how far we had come – we could devise systems where we could monitor hepatitis C virus spread in culture in real time (29). After this long struggle, not having a virus to work with in the laboratory, this was very gratifying.

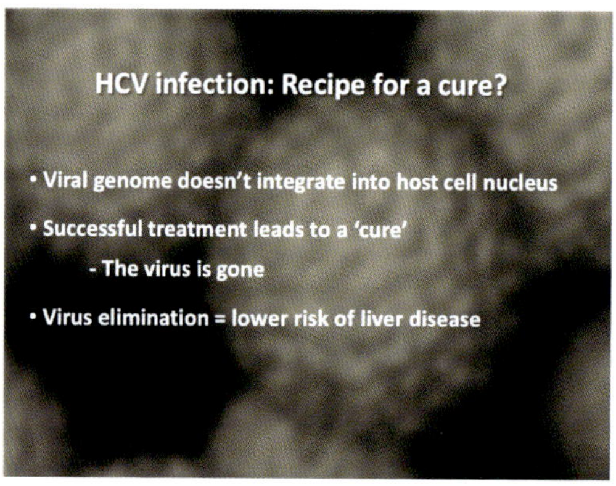

Figure 31. Can HCV be cured? The HCV features suggesting that a cure can be achieved are listed.

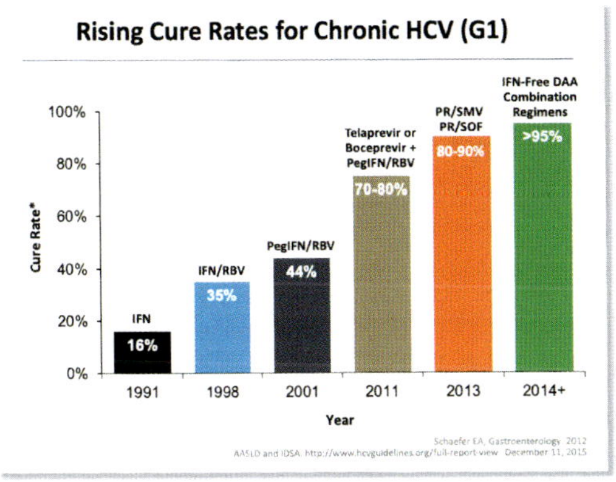

Figure 32. HCV treatment history. Treatments introduced in the indicated year and their cure rates are shown.

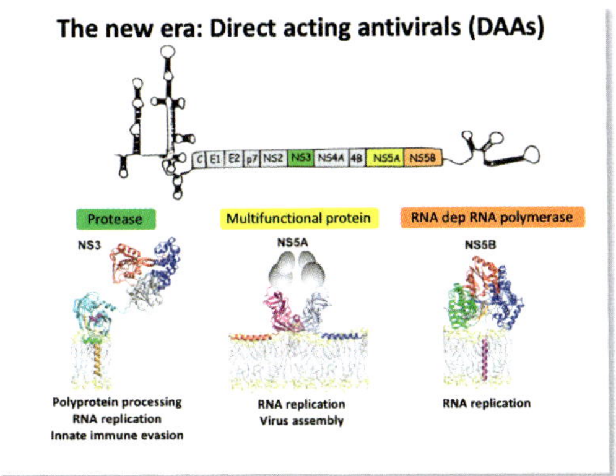

Figure 33. HCV direct acting antiviral targets. The location of the drug targets in the HCV genome is shown schematically at the top. Below, the viral protein functions and their experimentally determined structures are shown.

A few concluding comments about treatment. HCV is a virus that was ripe for coming up with a cure (Figure 31). We already knew this from the development and progression of interferon-based therapies for treating hepatitis C: The viral genome doesn't integrate, it's a cytoplasmic RNA virus, and successful treatment had been already shown to eliminate the virus. When the virus is gone there was accumulating evidence for lower risk of liver disease. To provide a brief history of HCV treatment (Figure 32), back in the non-A, non-B agent era, physicians began treating with systemic type I interferon with poor single digit sustained response rates. By 1991, with higher and longer interferon regimens, the efficacy of virus elimination was about 16%. That improved in subsequent years by combining interferon with ribavirin and making a pegylated, stabilized, version of type I interferon. In 2011, the first HCV serine protease inhibitors, telaprevir and boceprevir, were approved in combination with the current interferon-based regimens. Addition of these direct-acting antivirals (drugs that directly target viral functions) further improved efficacy, but also led to severe side effects. Finally, 2014 marked the end of interferon-based therapies. We had interferon-free, direct-acting, antiviral com-

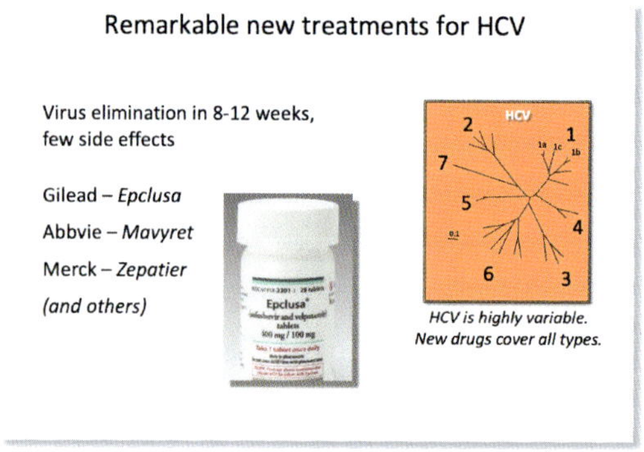

Figure 34. Effective treatments for genetically diverse HCV. Examples of effective drugs are shown on the left, while the genetic diversity of HCV strains is highlighted on the right.

binations with more than 95% cure rates. These potent inhibitors targeted several different essential viral functions (Figure 33); the NS3-4A serine protease, involved in polyprotein processing, RNA replication, as well as innate immune evasion, the NS5B RNA-dependent RNA polymerase, the engine of the HCV RNA replication machine, and NS5A, another non-structural protein involved in RNA replication and virus assembly. Different combinations of these inhibitors are now in widespread use and HCV can now be eliminated in two or three months. Remarkably, these combination therapies are broadly effective despite the HCV's immense global diversity (Figure 34).

It has been a remarkable journey (Figure 35) from a mystery virus in the 1970s and 1980s with a low cure rate, to the landmark discovery in 1989 – the virus identified, diagnostics in place, 2011 with the introduction of the first direct-acting antivirals, with a 75% cure rate, to 2014 and beyond – with highly effective regimens. But there are still challenges and the story is not finished (Figure 36). To get to 100% cure and eradication, which theoretically should be possible for HCV because it is an exclusively human virus, we need to identify those infected. Since this virus is often asymptomatic, there is a large segment of the world's population that is unaware of their infection. Once identified, getting patients into care and successfully treated can be a challenge. Affordable treatment is quite varied across the globe but lowering the price and making treatment universally available are important goals. With interferon-based therapies we were challenged by certain patient groups who were more difficult to treat, like HIV/HCV co-infected individuals. Fortunately, this seems to have disappeared in this new era of direct-acting antivirals. Virus resistance, always a potential problem, has not been a major issue with these next generation drug combinations. Finally, we do not have a preventative HCV vaccine, which many of us think is going to be needed to achieve global eradication. Hopefully the current efforts being put into developing COVID-19 vaccines will provide useful lessons for making a vaccine for hepatitis C.

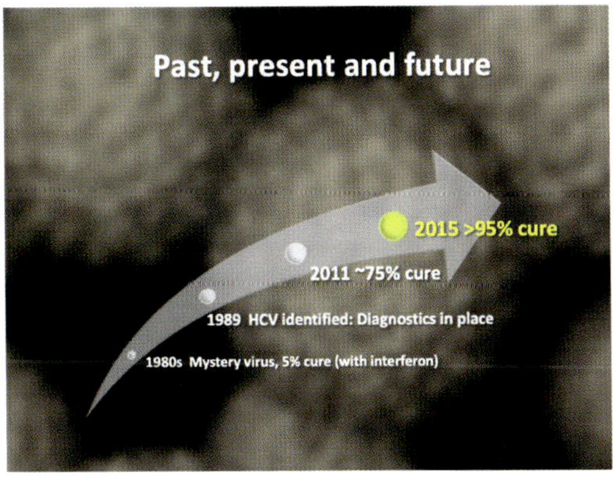

Figure 35. HCV cure trajectory. Key events and times in the development of an HCV cure are indicated.

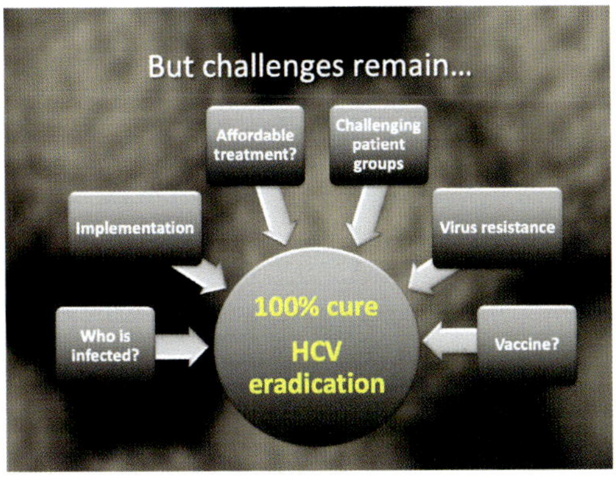

Figure 36. Remaining challenges on the path to HCV eradication.

Figure 37. The HCV success story. Success required a long road, traveled by many.

I want to sum up with just a few thoughts. The HCV success story, at least from a molecular virology standpoint, has been a long road (Figure 37), but there have been many, many people involved – basic scientists, clinicians, people in biotech and pharma, people running these clinical trials, and the patients themselves. It has been a great team effort and wonderful to have been involved. I do want to thank my mentors and special friends (Figure 38): Dennis Barrett who got me interested in a

Figure 38. Mentors and special friends. A few of the many mentors and special friends who were instrumental in my career choices and the ultimate path it took are listed. Of course, I am indebted to my parents who were always supportive, and to Dennis Barrett who got me hooked on biology and gently guided me on the right path. I appreciate the incredible mentorship of Jim and Ellen Strauss, who allowed me the freedom to explore my scientific interests, and other friends and colleagues at Caltech, whose work inspired me. At Washington University, I was surrounded by amazing virologists, Sondra and Milt Schlesinger, Henry Huang, and John Majors, and developed an enduring friendship with Stan and Claire Birge, who generously shared their Wyoming cabin for much needed rest and relaxation. Jim Darnell was instrumental in convincing me to accept the offer to come to Rockefeller, and I have immensely enjoyed working with many colleagues in the city, including Ira Jacobson and Andrea Branch.

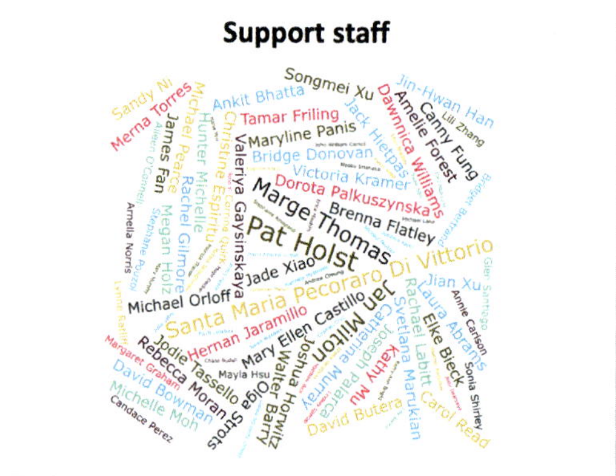

Figure 39. The many important support staff. Support staff members critical for our lab's work are shown in a Word Cloud.

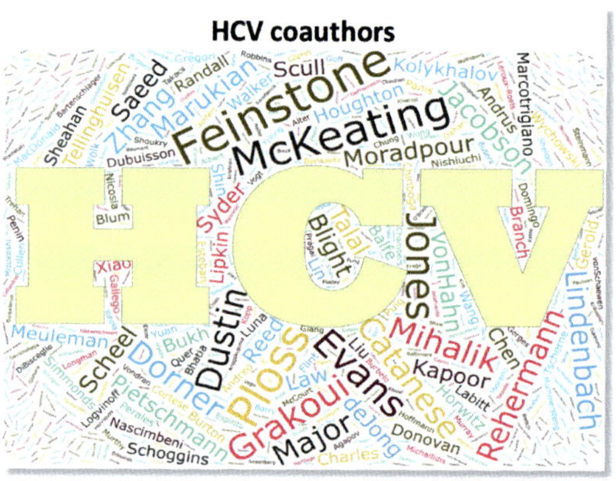

Figure 40. The many important collaborators. Names of co-authors who contributed to our HCV work are shown in a Word Cloud.

research career, and Jim and Ellen Strauss at Caltech, who were the ones that introduced me to the wonders of viruses and virology. In order to accomplish anything, we all depended on an incredible support staff (Figure 39) and a wonderful cast of collaborators (Figure 40). More broadly, I am profoundly grateful to all of my colleagues in this field for the many years of helpful input, encouragement, shared advances and failures, and friendship.

I'd also like to thank Peggy MacDonald, my partner in life and in science, and two of our canine support staff (Figure 41).

Finally, I would like to leave you with a quote that I love, written by Lewis Thomas from a book of essays that he published more than 30 years ago (30). He writes:

> In the course of scientific progress things tend to pop up. Most of the celebrated achievements had their origins in moments of surprise in the laboratory, unplanned for, unanticipated, unpredicted. No committee convened ... to survey the future prospects for biomedical research could possibly have guessed at the things that lay ahead. No branch of government could have laid out detailed plans for scientific needs of the future, beyond asserting ... that what the country needed was more fundamental knowledge about the human form and function, about the agents and influences responsible for disease, and, in a certain sense, about nature itself.
>
> Scientific research works, it is the only way to get at the underlying mechanisms of disease, and the only way to learn what to do about them.

So, with that I will end this story, and thank those of you who have managed to make it this far without dozing off. It's been a real pleasure for a basic scientist to be involved in this medical success story.

Figure 41. Other very important players. Peggy MacDonald and two of our former dogs, Wrangler and Sadie, are shown outside of the Rockefeller University Faculty-Student Club.

REFERENCES

1. Feinstone SM, Kapikian AZ, Purceli RH. 1973. Hepatitis A: detection by immune electron microscopy of a viruslike antigen associated with acute illness. *Science* **182**:1026–8.
2. Alter HJ, Holland PV, Purcell RH. 1975. The emerging pattern of post-transfusion hepatitis. *Am J Med Sci* **270**:329–34.
3. Dienstag JL, Purcell HR, Alter HJ, Feinstone SM, Wong DC, Holland PV. 1977. Non-A, non-B post-transfusion hepatitis. *Lancet* **1**:560–2.
4. Choo QL, Kuo G, Weiner AJ, Overby LR, Bradley DW, Houghton M. 1989. Isolation of a cDNA clone derived from a blood-borne non-A, non-B viral hepatitis genome. *Science* **244**:359–62.
5. Rice CM, Grakoui A, Galler R, Chambers TJ. 1989. Transcription of infectious yellow fever RNA from full-length cDNA templates produced by in vitro ligation. New Biol 1:285–96.
6. Moradpour D, Penin F. 2013. Hepatitis C virus proteins: from structure to function. *Curr Top Microbiol Immunol* **369**:113–42.
7. Taniguchi T, Palmieri M, Weissmann C. 1978. QB DNA-containing hybrid plasmids giving rise to QB phage formation in the bacterial host. *Nature* **274**:223–8.
8. Racaniello VR, Baltimore D. 1981. Cloned poliovirus complementary DNA is infectious in mammalian cells. *Science* **214**:916–9.
9. Rice CM, Levis R, Strauss JH, Huang HV. 1987. Production of infectious RNA transcripts from Sindbis virus cDNA clones: Mapping of lethal mutations, rescue of a temperature-sensitive marker, and in vitro mutagenesis to generate defined mutants. *J Virol* **61**:3809–19.
10. Alter HJ, Purcell RH, Holland PV, Popper H. 1978. Transmissible agent in non-A, non-B hepatitis. *Lancet* **1**:459–63.
11. Emerson SU, Lewis M, Govindarajan S, Shapiro M, Moskal T, Purcell RH. 1994. In vivo transfection by hepatitis A virus synthetic RNA. *Arch Virol Suppl* **9**:205–9.
12. Kolykhalov AA, Feinstone SM, Rice CM. 1996. Identification of a highly conserved sequence element at the 3' terminus of hepatitis C virus genome RNA. *J Virol* **70**:3363–71.
13. Yamada N, Tanihara K, Takada A, Yorihuzi T, Tsutsumi M, Shimomura H, Tsuji T, Date T. 1996. Genetic organization and diversity of the 3' noncoding region of the hepatitis C virus genome. *Virology* **223**:255–61.
14. Tanaka T, Kato N, Cho MJ, Shimotohno K. 1995. A novel sequence found at the 3' terminus of hepatitis C virus genome. *Biochem Biophys Res Commun* **215**:744–9.
15. Kolykhalov AA, Agapov EV, Blight KJ, Mihalik K, Feinstone SM, Rice CM. 1997. Transmission of hepatitis C by intrahepatic inoculation with transcribed RNA. *Science* **277**:570–4.
16. Shindo M, Di Bisceglie AM, Biswas R, Mihalik K, Feinstone SM. 1992. Hepatitis C virus replication during acute infection in the chimpanzee. *J Infect Dis* **166**:424–7.
17. Major ME, Mihalik K, Fernandez J, Seidman J, Kleiner D, Kolykhalov AA, Rice CM, Feinstone SM. 1999. Long-term follow-up of chimpanzees inoculated with the first infectious clone for hepatitis C virus. *J Virol* **73**:3317–25.
18. Yanagi M, Purcell RH, Emerson SU, Bukh J. 1997. Transcripts from a single full-length cDNA clone of hepatitis C virus are infectious when directly transfected into the liver of a chimpanzee. *Proc Natl Acad Sci U S A* **94**:8738–43.

19. Kolykhalov AA, Mihalik K, Feinstone SM, Rice CM. 2000. Hepatitis C virus-encoded enzymatic activities and conserved RNA elements in the 3' nontranslated region are essential for virus replication in vivo. *J Virol* **74**:2046–51.
20. Lohmann V, Korner F, Koch J, Herian U, Theilmann L, Bartenschlager R. 1999. Replication of subgenomic hepatitis C virus RNAs in a hepatoma cell line. *Science* **285**:110–3.
21. Blight KJ, Kolykhalov AA, Rice CM. 2000. Efficient initiation of HCV RNA replication in cell culture. *Science* **290**:1972–4.
22. Blight KJ, McKeating JA, Rice CM. 2002. Highly permissive cell lines for subgenomic and genomic hepatitis C virus RNA replication. *J Virol* **76**:13001–14.
23. Lohmann V, Korner F, Dobierzewska A, Bartenschlager R. 2001. Mutations in hepatitis C virus RNAs conferring cell culture adaptation. *J Virol* **75**:1437–49.
24. Lindenbach BD, Evans MJ, Syder AJ, Wolk B, Tellinghuisen TL, Liu CC, Maruyama T, Hynes RO, Burton DR, McKeating JA, Rice CM. 2005. Complete replication of hepatitis C virus in cell culture. *Science* **309**:623–6.
25. Wakita T, Pietschmann T, Kato T, Date T, Miyamoto M, Zhao Z, Murthy K, Habermann A, Krausslich HG, Mizokami M, Bartenschlager R, Liang TJ. 2005. Production of infectious hepatitis C virus in tissue culture from a cloned viral genome. *Nat Med* **11**:791–6.
26. Zhong J, Gastaminza P, Cheng G, Kapadia S, Kato T, Burton DR, Wieland SF, Uprichard SL, Wakita T, Chisari FV. 2005. Robust hepatitis C virus infection in vitro. *Proc Natl Acad Sci U S A* **102**:9294–9.
27. Lindenbach BD, Meuleman P, Ploss A, Vanwolleghem T, Syder AJ, McKeating JA, Lanford RE, Feinstone SM, Major ME, Leroux-Roels G, Rice CM. 2006. Cell culture-grown hepatitis C virus is infectious in vivo and can be recultured in vitro. *Proc Natl Acad Sci U S A* **103**:3805–9.
28. Pietschmann T, Kaul A, Koutsoudakis G, Shavinskaya A, Kallis S, Steinmann E, Abid K, Negro F, Dreux M, Cosset FL, Bartenschlager R. 2006. Construction and characterization of infectious intragenotypic and intergenotypic hepatitis C virus chimeras. *Proc Natl Acad Sci U S A* **103**:7408–13.
29. Jones CT, Catanese MT, Law LM, Khetani SR, Syder AJ, Ploss A, Oh TS, Schoggins JW, MacDonald MR, Bhatia SN, Rice CM. 2010. Real-time imaging of hepatitis C virus infection using a fluorescent cell-based reporter system. *Nat Biotechnol* **28**:167–71.
30. Thomas L. 1986. *The Lasker awards: Four decades of scientific medical progress.* Raven Press, New York.